NASA SP-4008

ASTRONAUTICS AND AERONAUTICS, 1967

Chronology on Science, Technology, and Policy

Text by
Science and Technology Division
Library of Congress

Sponsored by
NASA Historical Staff
Office of Policy

Scientific and Technical Information Division   1968
NATIONAL AERONAUTICS AND SPACE ADMINISTRATION
Washington, D.C.

Foreword

The first decade of the Space Age is behind us and now in the hands of the historians. It was a period of dynamic advance, bringing new appreciations of our solar environment and new capabilities to develop and exploit advancing knowledge and techniques. Since the space challenge of the next decade is likely to be just as demanding and rewarding as that of the last one, perhaps we may gain useful perspective by reflecting upon our experiences even as we contemplate tomorrow.

Ten years ago our planning for the future required the piecing together and development of the organization and management as well as disciplinary and technical capability for space exploration. Today our planning for the future can be based upon well-established and versatile capabilities in science, engineering, and administration. Reliable space vehicles are available ranging from small sounding rockets to the Titan III and Saturn V class vehicles. The use of automated techniques in space has been demonstrated by the Mariner, Ranger, Surveyor, Lunar Orbiter, Explorer, Geophysical and Solar Observatories, Tiros, Nimbus, Syncom, the Applications Technology Satellite, and many others. The capability of man himself to function usefully in space has been demonstrated in the successes of Mercury and Gemini and expands rapidly as Apollo proceeds.

Ten years ago we could only assert from intuition and prophecy the value of space experimentation to science. Now we can point to the profound influence it has had on the geosciences, on astronomy, and is beginning to have in life sciences. Thanks to our planetary probes, we have known since 1965 that the planet Mars has craters like the Moon. Thanks to our lunar probes, a detailed lunar atlas can be prepared.

Ten years ago we could only point forward hopefully to space applications in meteorology, communications, navigation, and geodesy. Today these are all operational areas. Global weather photographs are received daily even in some of the most undeveloped areas of the world. Reliable, live TV broadcasts bring worldwide historic events and Olympic games into our living rooms routinely and in real time. Ships in all oceans use signals from satellites to refine their knowledge of their position; soon aircraft will be doing the same. Specially instrumented satellites and probes are daily filling in details on the momentous geophysical discoveries of Van Allen and others.

Ten years ago there was serious question as to whether there would be enough skilled manpower or public interest to carry out a sustained program of space exploration. Today great teams of competent scientists, engineers, and managers are engaged in developing our national space and aeronautical capability and applying its hard-earned technology to the most rewarding applications for commercial, industrial, social, and military purposes. Unparalleled scientific and technological success in our space venture during the past decade not only vastly increased our skills and resources but also developed high confidence in our ability to solve other problems in today's society. Problems of national defense, oceanography, earth resources, the cities, transportation, population, air pollution, and food—these are challenges that cannot be sidestepped and for which the innovative enthusiasm and know-how of the space program offers guidance, inspiration, and technique.

Ten years ago the American space program began by almost doubling its effort each year for several years; today the program is shrinking and has been since 1966. Ten years ago mastering space technology was indeed a major problem. Success was a rare and precious commodity. Today success is routine and failures are rare. A decade ago the American people were caught up in a wave of interest in the space venture and in a determined drive to catch up with the Soviet Union's bold initiative in space. Today Americans can take satisfaction in what our Nation has so far accomplished in space; they should also know that what has been done is only the merest beginning of what can be done. Of course the values of space research, and indeed of all research, must be weighed against their contribution to national needs. In making this judgment it is very important that we fully appreciate the investment we have made and the nature of the dividends we have earned.

This useful chronology on Astronautics and Aeronautics, following the pattern of its predecessor volumes, offers first documentation upon the multitudinous events, people, and circumstances of the last year of the first decade of the Space Age. All of us note with sadness the tragic loss of Astronauts Grissom, White, and Chaffee on January 27. They joined the ranks of other brave explorers like Magellan and Amundsen who also pioneered man's first steps into a newly accessible environment. The world's shock and grief at their passing was eloquent testimony to the sense of identification that all men feel with those few who explore new realms in their behalf. Other events, though significant, were not so memorable. The first colored pictures of Earth from the first Applications Technology Satellite or a new world's speed record of 4,534 miles per hour by an X-15 rocket aircraft may not be recalled when one begins to assess the events of 1967. The chronology logs them for you. The United States launched a total of 87 spacecraft into Earth orbit or on escape trajectories during 1967. Each one is cataloged here in the chronology, one measure of the vast sweep of scientific and technical accomplishment during the year just past. It is at our own peril that we forget or ignore what we have already experienced; we will simply have to relearn it if we do.

Today, as we confer a solid "well done" on the gains of the first decade of space exploration and take the first steps into the second decade, we find these steps hampered by uncertainty, reservations, and in some quarters even dismay. Here again words out of history can offer renewed perspective. Speaking a decade ago, before NASA was created and the national space commitment was undertaken, a wise statesman of science and technology, the late Hugh L. Dryden, observed:

> We must understand that the kind and magnitude of space program that our national interest requires will cost hundreds of millions of dollars each year for many years to come. I know that some knowledgeable people fear that although we might be willing to spend a couple of billions for space technology in 1958, because we still remember the humiliation of Sputnik last October, next year we will be so preoccupied by color television, or new style cars, or the beginning of another national election campaign that we'll be unwilling to pay another year's installment on our space conquest bill. For that to happen—well, I'd just as soon we didn't start.

Fortunately, for the sake of our children's future if not for the protection of our own skin, I do not think we are that grasshopper-minded. . . . We can and we must succeed in finding our destiny in space.

HOMER E. NEWELL,
Associate Administrator,
National Aeronautics and Space Administration.

Contents

	PAGE
FOREWORD	III
Associate Administrator, Homer E. Newell	
PREFACE	IX
JANUARY	1
FEBRUARY	33
MARCH	59
APRIL	95
MAY	133
JUNE	173
JULY	199
AUGUST	229
SEPTEMBER	260
OCTOBER	287
NOVEMBER	326
DECEMBER	363
APPENDIX A: SATELLITES, SPACE PROBES, AND MANNED SPACE FLIGHTS, 1967	395
APPENDIX B: MAJOR NASA LAUNCHES, 1967	431
APPENDIX C: ABBREVIATIONS OF REFERENCES	435
INDEX AND LIST OF ABBREVIATIONS AND ACRONYMS	439

Preface

In 1961 the first publication by the one-man NASA Historical Staff was a well-received chronology—*Aeronautics and Astronautics: An American Chronology of Science and Technology in the Exploration of Space, 1915–1960.* The following year a volume covering the year 1961 was prepared by the two-man Historical Staff and published by the House Committee on Science and Astronautics. Also that year began a practice still existent. Monthly draft chronologies were distributed for use and comment throughout NASA and the Government, with the object of providing a timely reference and revising them for an annual printed edition. The 1962 chronology volume, considerably enlarged over 1961, was again printed by the House Committee. From 1963 to the present, NASA has published each annual *Astronautics and Aeronautics* chronology volume. As far as is known it is the only substantial, continuing chronology effort in the fields of aeronautics and astronautics. The reference value of each volume has increased with the passage of time.

In 1964 the NASA Historical Staff joined in partnership with the Science and Technology Division of the Library of Congress to produce the Astronautics and Aeronautics chronology. Funded by NASA, employees of the Library drafted the monthly chronology segments, validated critical comments, added new items as later documentation became available, prepared the revised manuscript for publication, and compiled the intensive index. This partnership continues.

Thus it was that by the end of the first decade of the Space Age the greatest share of NASA historical publications had been chronologies. Only a modest beginning was made on publishing full-fledged volumes in the NASA Historical Series. The second decade of NASA begins with more narrative histories in process than chronologies. The reader will find on the inside back cover of this volume a complete list of NASA historical publications to date.

This annual volume remains the product of a number of hands. The entire NASA Historical Staff participated in source collection, review, and publication. At the Library Mrs. Anne Horton (through August 1967), Charles Thibault (from September on), Mrs. Carmen Brock-Smith, and Mrs. Gay Arnelle drafted the monthly texts. Mr. Thibault and Mrs. Brock-Smith revised the monthly drafts for annual publication. Mr. Arthur Renstrom prepared the index. In the NASA Historical Staff, the general editor was Dr. Frank W. Anderson, Jr., Deputy NASA Historian. Technical editors were Mrs. Helen T. Wells (through July 1967) and Mrs. Carrie Karegeannes. Appendix A, "Satellites, Space Probes, and Manned Space Flights, 1967," and Appendix B,

"Major NASA Launches, 1967," were prepared by Dr. Anderson. Appendix C, "Abbreviations of References," was prepared by Mrs. Karegeannes. Creston Whiting, of NASA's Information Services Branch, Scientific and Technical Information Division, kept the process abreast of Russian releases. At the NASA centers the historians and historical monitors submitted local material for the chronology. Validation was the work of many busy people throughout NASA and in other relevant branches of the Federal structure.

A chronology is but the first step toward history as an intellectual discipline and even it is never really completed. Comments, additions, and criticisms are always welcomed.

EUGENE M. EMME,
NASA Historian.

January 1967

January 1: NAS had received over 400 applications for NASA astronaut training with one week remaining before Jan. 8 deadline. NAS had been asked by NASA to recruit and nominate a group of scientists and engineers for final selection and training as astronauts "to conduct scientific experiments in manned orbiting satellites and to observe and investigate the lunar surface and circumterrestrial space." Applicants were required to be U.S. citizens no taller than six feet, born after Aug. 1, 1930; have a doctorate degree in the natural sciences, medicine, or engineering; and meet physical requirements for pilot crew members. (NASA Release 66-333)

- Twentieth anniversary of Atomic Energy Commission (AEC).

January 2: Spanish scientists had submitted to the Spanish government a five-year, $10,000,000 space research program, the *Chicago Tribune* reported. According to Comisión Nacional de Investigación del Espacio (CONIE) President Gen. Rafael Calvo Rodes, primary program objectives would be to gather scientific data and gain practical experience in designing and launching rockets. During separate, $1,500,000 preparatory program which was completed in December 1966, Spain launched her first rocket, collected data, conducted preliminary experiments, trained technical personnel, and prepared permanent launching sites; during next phase, Spain would build her own Skua rockets. (Darrah, *C Trib*, 1/3/67)

- "The moon is not a dead body, but a living, breathing organism," Tass reported following successful completion of lunar surface studies by U.S.S.R.'s *Luna XIII*. Prof. Nikolay Kozyrev's 1964 claim that Aristarchus crater had smoke, gas, or dust rising from it had been confirmed by observations from earth of possible lunar volcanoes or eruptions of lunar dust; results of lunar-surface studies; a "special glow" discovered by Pulkovo Observatory near Kepler crater; and "manifestation of moon's geological life," recorded in 1966. (UPI, *NYT*, 1/3/67, 10)

- Aerospace industry must develop new type of management to cope with its record $26-billion workload during 1967, Robert Hotz wrote in *Aviation Week*. "What makes the current management problem so acute is the fact that military demands are reaching a peak at the same time that commercial aviation is soaring on the greatest boom in its history. During 1965, the aerospace industry actually manufactured slightly more commercial airframe pounds than military—a historic milestone. The industry also has under development six new types of airline jet transports for which there were no prior military programs and which must be financed from privately raised funds. This is another unprecedented task for aerospace managements to tackle...." (Hotz, *Av Wk*, 1/2/67, 11)

- U.S. experts believed establishment of permanent lunar colonies and exploitation of lunar resources were among long-range goals of Soviet space program, Heather M. David reported in *Technology Week*. Analysis of research reports had indicated that future Soviet missions directed toward these objectives would include: (1) orbiting of three or four cosmonauts for three weeks; (2) extensive practice of extravehicular activity and assembly in orbit; (3) orbiting of animals to continue studying effects of long-term weightlessness and radiation; and (4) lunar landing of animals for observation. David predicted preparations for interplanetary travel would be initiated shortly after establishment of permanent lunar colonies. (David, *Tech Wk*, 1/2/67, 18)

- The *New York Times* called for fuller and more open debate on cost and consequences of SST program: "The SST's priority is debatable and so is its financing. The industry may be right in asserting that the costs of development are so huge that no company can afford to take the risk on its own. It points out that the backing of both Britain and France has been required to launch the Concorde. But if the combined efforts of two countries and their aircraft industries are needed for that much less ambitious project, the American SST might be more safely and soundly developed if it commanded the combined resources of the entire American industry and the Government.

    "What is at stake is not the commitment to proceed with the SST. That decision has been made and cannot—and should not—be reversed. But if the SST is to be sound economically as well as technologically, the Administration and the industry must be less secretive and indefinite about the costs and how they are to be financed. An open debate is needed on how to get the most for the taxpayers' money." (*NYT*, 1/2/67, 18)

January 3: Group of NASA officials, headed by MSFC Director Dr. Wernher von Braun, arrived at McMurdo Station in the Antarctic beginning 10-day expedition to observe environmental conditions—including temperatures, isolation factors, and survival techniques—for comparison and possible application to problems of space flight. (MSC *Roundup*; *NYT*, 1/4/67, 10; AP, *W Post*, 12/22/67, A21)

- Joseph J. Tymczyszyn, a senior FAA test pilot and chief of FAA's West Coast Supersonic Transport Development Field Office, had emerged as the test pilot most likely to certify the SST, Mitchell Gordon disclosed in the *Wall Street Journal*. Tymczyszyn, certifier of both the Boeing 707 and the Douglas DC–8 jetliners for U.S. service, had already spent four years familiarizing himself with SST operation, and had logged more than 1,000 hrs in SST simulators. (Gordon, *WSJ*, 1/3/67, 1)

- *Philadelphia Inquirer* praised success of U.S.S.R.'s *Luna XIII* and urged that U.S. continue its intensive lunar-landing program: "Space treaty or no space treaty, the competitive moon race goes on and even though nobody really likes the idea, the U.S. has to try to do better than the Soviet Union's Luna 13. . . .

    "No one can guess for certain what strategic role it [the moon] might play if . . . the Soviet Union were to make technological advances that would enable it to deny the use of space to the rest of the world.

    "Luna 13 has chalked up a meaningful achievement for the Russian space program. Americans should not be tempted to underrate it." (*P Inq*, 1/3/67)

- Underground nuclear detonation conducted Dec. 3, 1966, by AEC for ARPA had confirmed decoupling theory—that an underground explosion could

be muffled significantly by firing it in a cavity—DOD announced. Further experiments would be conducted under Project Sterling to "develop a better understanding of the problems of seismic detection." (DOD Release 10–67)
- USAF announced plans to train 3,247 pilots a year—an increase of 15% over present level—and provide a ninth Undergraduate Pilot Training (UPT) facility at Randolph AFB. Increase was required because of Vietnam war. (DOD Release 2–67; AP, *NYT*, 1/4/67, 30)

January 4: XB–70 No. 1 research aircraft piloted by NASA test pilot Fitzhugh Fulton and North American Aviation, Inc., pilot Van Shepard reached mach 2.53 and 60,400-ft altitude during flight at Edwards AFB for national sonic boom program. (NASA Proj Off)
- A record number of visitors—246,838—toured MSFC in 1966. (MSFC Release 67–1)
- British government would launch an all-British satellite within 2½ yrs, Reuters reported. According to a *Daily Mail* article, a 200-lb satellite would be launched from Woomera Rocket Range with new Black Arrow rocket. (Reuters, *NYT*, 1/4/67, 40)
- Unarmed Mace A missile, launched by USAF from Eglin AFB, Fla., during target-interception practice, went out of control and accidentally flew over the western tip of Cuba before crashing into the sea. (Wilson, *W Post*, 1/5/67, A1)
- Replacement of obsolete missiles by more advanced types had resulted in the largest surplus-disposal program since the end of World War II, Frederick Taylor asserted in the *Wall Street Journal*. DOD was reportedly trying to dispose of more than 200 phased-out Atlas and Titan I missiles and their sites, and it seemed probable that within a few years 800 Minuteman I, 600 Polaris, and 54 Titan II missiles also would become obsolete, adding to the surplus. (Taylor, *WSJ*, 1/4/67, 1)

January 5: NASA Administrator James E. Webb addressed Armed Forces Reserves Joint Assembly at NASA Hq.

"It is a mistake to think of the space program only in terms of the rockets and spacecraft we launch. These are the focal points of our attention, but we are also developing the ability to organize and manage vast scientific and technical enterprises that involve large segments of our economic, social, and political decision-making process. This complex interdevelopment of technological power and of the social organization necessary to control, direct, and exploit it . . . is a central element of the space age." (Text)
- Press briefing on future OSSA programs was held at NASA Hq. Edgar M. Cortright, Deputy Administrator for Space Science and Applications, served as moderator.

Associate Administrator for Space Science and Applications Dr. Homer E. Newell stressed the importance of basic research to all national scientific programs. ". . . the results of basic research are vital to the realization of practical applications. When we began the NASA program, we had the sounding rocket, we had the capability of putting together small satellite launchers and to apply those techniques, and we built the best program on those techniques that we could. Now we have manned flight capabilities; we have large boosters, the ability to get out further into the solar system than ever before. . . . Now is the time to profit by those capabilities, to continue the basic research."

Sunblazer probe that could be launched into orbit around the sun and transmit radio signals back to earth was discussed by Jesse L. Mitchell, Director of Physics and Astronomy Programs. Spacecraft would determine electron density in solar atmosphere by measuring time it took signals to travel from spacecraft back to earth. NASA was also considering a plan to launch a satellite toward Jupiter, then use Jupiter's magnetic field to push the spacecraft toward other planets and eventually hurl it into space 1-10 trillion miles away.

Oran W. Nicks, Director of Lunar and Planetary Programs, outlined plans for unmanned missions to Mars and Venus, beginning with a small instrumented probe scheduled for launch in June 1967. In the 1971 Mars and 1972 Venus missions, teams of two spacecraft would be launched by a modified Saturn V booster. One spacecraft would penetrate planet's atmosphere; take measurements as it sped toward planet's surface; and flash data to the second spacecraft—programed to fly past the planet—before crashlanding. Data would be transmitted to earth by the flyby spacecraft. Missions would pave the way for later spacecraft with improved equipment—possibly including TV cameras—to softland on the planets.

Thumb-sized desert mice fitted with tiny radio transmitters and telemetry equipment which could broadcast mouse's temperature up to six months were being conditioned for future space flight, Dr. Orr E. Reynolds, Director of Bioscience Programs, disclosed. "Astromouse"—which ate only sunflower seeds and never drank water—might be orbited with three astronauts in Apollo spacecraft. Others might be launched into interplanetary space in a Pioneer study of biological specimens outside earth's gravitational field.

Plans for a multi-purpose satellite that would combine the functions of many unmanned spacecraft currently in orbit were revealed by Dr. Morris Tepper, Deputy Director of Space Applications and Director of Meteorology. Developed under unified space applications program, satellite would carry equipment to observe weather conditions; relay communications to distant earth stations; provide navigation traffic control for ships and aircraft; and compile data on earth's natural resources.

No mention of unmanned scientific lunar probes was made during the briefing. When asked later in an interview whether he thought interest in unmanned exploration of the moon would diminish after astronauts landed, Dr. Newell replied: "Not at all." Although astronauts would dominate lunar exploration, unmanned Surveyor spacecraft would probably land in unsafe areas, and Lunar Orbiter photography missions would continue for mapping. Discussion of unmanned lunar missions had been omitted because "we wanted to tell about the newer of new things in our thinking." (Transcript; Hines, W *Star*, 1/8/67, A16)

- U.S. must "answer the post-Apollo question . . . very soon," Douglas Aircraft Co. Vice President and General Manager J. P. Rogan told AIAA/AAS Space Forum in Washington, D.C. He suggested that U.S. develop five "building blocks" which would maintain options for any future course of action chosen: (1) long-duration orbital experience; (2) reusable spacecraft; (3) reusable launch vehicles; (4) nuclear-powered stages; and (5) improved secondary power systems. (Text)

- U.S.S.R. had developed a device that enabled jet aircraft to take off almost straight up, reaching the stratosphere "in a matter of minutes," *Komsomolskaya Pravda* reported. Article did not include detailed de-

scription of device but indicated it consisted of rocket engines that provided sudden thrust of tremendous force. (AP, *NYT*, 1/7/67, 10)
- At least 200 Anglo-French Concorde supersonic aircraft would have already been sold by 1974 when the U.S. supersonic transport was scheduled to go into operation, Concorde sales manager E. H. Burgess speculated at a London news conference. Construction of first prototype, which would make its maiden flight Feb. 28, 1968, was two or three weeks ahead of schedule, he said. (AP, *NYT*, 1/6/67, 62M)
- Refunds totaling $35 million "probably will be paid" by FAA to Lockheed Aircraft Corp. and Pratt & Whitney Div., United Aircraft Corp. for their losses in SST design competition, the *Wall Street Journal* speculated. Boeing Co. and General Electric Co. had been selected Dec. 31, 1966, to continue development and refinement of their SST designs. (*WSJ*, 1/5/67)
- Object, believed to be "probable" 10th moon of Saturn discovered by French astronomer Dr. Audouin Dollfus of Meudon Observatory in December 1966, had been sighted by Richard L. Walker of the U.S. Naval Observatory, Walter Sullivan reported in the *New York Times*. New moon, believed to be a 150-mi-dia "snowball" of very light, frozen material, was in orbit 52,000 mi from Saturn with a period of 18 hrs. Confirmation was uncertain, however; Walker, in a telephone interview, said the object was so close to the rings of Saturn that it could be a lump in the rings, rather than a moon. (Sullivan, *NYT*, 1/5/67, C17)

January 6: Harold B. Finger, Director of Nuclear Systems and Space Power in NASA Hq. OART and Manager of AEC–NASA Space Nuclear Propulsion Office, temporarily assigned additional duty of heading group of NASA officials who would analyze needs for procedural revisions and functional alignments within NASA Hq. (NASA Release 67-2)
- NASA's *Surveyor I* spacecraft, which softlanded on the moon June 2, 1966, was reactivated on the lunar surface for 12 hrs to transmit data on the motion of the moon. JPL said the new data would be used to help "refine the Moon's distance from the Earth, its position in orbit at a given time, and other elements of its velocity and motion." (*SBD*, 1/11/67, 42)
- USAF reported that an SR–71 strategic reconnaissance aircraft was missing and presumed down after test flight from Edwards AFB. (UPI, *W Post*, 1/7/67, 3)

January 7: Patent for a method of determining whether life exists on other planets had been granted to aeronautical engineer Albin M. Nowitzky, Stacy V. Jones reported in the *New York Times*. Unlike previous proposals involving distant observations, Nowitzky's concept called for an automated spacecraft landing on a planet's surface to make tests free from earthly contamination and influence. A soil sample would be taken, mixed with distilled water, and sterilized. Part of the resulting "nutrient" would be transferred to a test chamber, exposed to the planet's atmosphere, and sealed, while the remaining part would be used as a "control." Sensors in the test chamber would register any metabolic reaction, and these data would be telemetered to earth or to an orbiting satellite. (Jones, *NYT*, 1/7/67, 34)
- French scientists were teaching monkeys that pressure on one out of five buttons produced a gum drop, in preparation for series of space flights with Vesta sounding rocket from Hammaguir Range, AP reported. Researchers hoped to determine effects of weightlessness on animals' behavior patterns. (AP, *W Star*, 1/7/67, A2)

January 8: Explosive growth of world's population in the 20th century was an instinctive preparation of mankind for challenges of space travel and resettlement on other planets, Soviet scientist Igor M. Zabelin contended in *Pravda*. "... two very striking facts—the explosive growth of mankind and the exploration of space—do not coincide by accident. Some inner motivations are leading mankind to new and unknown shores and, therefore, it is gathering its strength and reorganizing [its social structure]." This explosion was imperative to supply the manpower necessary for the space age, he asserted, because the solar system could not be conquered with "only three billion human beings." (Anderson, *NYT*, 1/9/67, 1)

- *New York Times* questioned FCC's ability to investigate adequately the organization of a domestic comsat system. "The questions faced by the Federal Communications Commission and Congress ... go far beyond the obvious ones of whether Comsat's present monopoly in the international satellite field should be extended to the new area or whether private corporations should be empowered to launch separate systems for television or press transmission. . . .

  "Perhaps the most worrisome aspect of the tangle ... is the incapacity of the F.C.C. to grapple with such problems in total terms. . . . [It] has neither the research staff nor the funds to discharge its planning responsibilities with anything approaching adequacy. The country must now look to it for decisions it is ill-equipped to give on the organization of a new satellite system and on the future of noncommercial TV." (*NYT*, 1/8/67, 12E)

January 9: William M. Magruder, Lockheed Aircraft Corp., had been selected to receive Flight Safety Foundation's 1967 Richard Hansford Burroughs Test Pilot Award, *Aviation Week* reported. (*Av Wk*, 1/9/67, 13)

January 10: U.S. space program could benefit from experience and scientific research in the Antarctic, group of NASA officials headed by MSFC Director Dr. Wernher von Braun concluded after studying U.S. installations and activities at McMurdo Station since Jan. 3. Interest in the Antarctic was based on research being conducted there which was applicable to space flight and on way Americans worked under conditions of stress, isolation, and extreme cold—conditions similar to those on the moon and in space flight. The group also observed minute fungi and algae growing in Antarctic and studied methods being used to detect living organisms, as preparation for detecting and studying such growths possibly existing in harsh environments of other planets. (Durdin, *NYT*, 1/15/67, 87)

- MSFC had exercised one-year $10,451,092 renewal option of cost-plus-award-fee contract with Mason-Rust Co. for continued support services at Michoud Assembly Facility. (MSFC Release 67-6)

- Designs of both American SST and Anglo-French Concorde would have to be modified to meet standards proposed by their potential customers, International Air Transport Assn. (IATA) Director General Knut Hammarskjold asserted. He advised that aircraft be modified to meet all of the organization's requirements, even if it meant postponing delivery date.

  The *Wall Street Journal* later commented: "Since IATA's members include most of the world's major airlines ... Mr. Hammarskjold's words plainly deserve attention. . . . [He] appears to have shot a large hole in the SST supporters' main argument for haste. It is that the British and French are already well ahead of the U.S. . . . Calm consideration would appear even more mandatory for the Government, saddled as it is with a war and a budget whose income side long ago since has lost touch with outgo. . . . A late blooming success is far better than a dismal early failure." (UPI, *NYT*, 1/11/67, 73; *WSJ*, 1/17/67, 18)

- President Johnson, in his State of the Union address before joint session of Congress, deferred decision on deploying a Nike-X antimissile missile system despite evidence that U.S.S.R. was building a defensive system near Moscow. "We have a solemn duty to slow down the arms race . . . if that is at all possible, in both conventional and nuclear weapons and defenses. . . . I realize any additional race would impose on our peoples and on all mankind for that matter, an additional waste of resources with no gain in security to either side." The President said he intended to seek international agreements "bearing directly on this problem." He did not mention the U.S. space program in his address. (*PD*, 1/16/67, 26–39)

January 10–11: Dr. George E. Mueller, NASA Associate Administrator for Manned Space Flight, discussed achievements and objectives of U.S. space program in a series of lectures at the Univ. of Sydney Summer School, Australia. After emphasizing intimate interrelationship between NASA's unmanned and manned programs, Dr. Mueller assessed Mercury and Gemini flights: "In the Mercury program, we established man's capabilities in short space flights and we laid the foundation for manned space flight technology. In the Gemini program, we gained operational proficiency, learned about man's capabilities in flights lasting up to two weeks, and developed new techniques." The Apollo program, he said, was proceeding on schedule, and several alternative post-Apollo programs were under consideration: "(1) Direct economic benefits, with emphasis on extensive earth-orbital activities; (2) Lunar exploration and science; (3) Planetary exploration and science; (4) Maximum effort aimed at pre-eminence in earth-orbital, lunar, and planetary activities; (5) Balanced combination of economic benefits, lunar and planetary exploration, and science." (Text)

January 11–14: NASA launched ComSatCorp's *Intelsat II–B* comsat from ETR by three-stage Thrust Augmented Improved Delta booster into elliptical transfer orbit in preparation for geostationary orbit. Transfer orbit had 22,904-mi (36,875-km) apogee; 185-mi (298-km) perigee; 10-hr 54-min period; and 26° inclination. On Jan. 14 ComSatCorp fired apogee motor transferring satellite into geostationary orbit over the Pacific.

Scheduled to enter commercial service Jan. 27, *Intelsat II–B* would provide 24-hr communications service between U.S. and earth stations in Hawaii, Japan, and Australia; assist in fulfilling Project Apollo communications requirements; and provide military communications between Hawaii and Japan, and Thailand and the Philippines for National Communications System (NCS). It was second satellite in ComSatCorp's INTELSAT II program to place two comsats in synchronous orbit—one over the Pacific, one over the Atlantic; first satellite—*Intelsat II–A*—was launched by NASA Oct. 27, 1966, but failed to achieve synchronous orbit because apogee motor malfunctioned. Subsequent "systems tests" of the apogee motor at Arnold Engineering Development Center indicated malfunction had been thermal in nature. INTELSAT II comsats, larger and improved versions of Early Bird (INTELSAT I) satellite, were capable of handling television and data transmissions of up to 240 voice channels. (NASA Proj Off; ComSatCorp Release; UPI, *NYT*, 1/8/67, 18; 1/12/67, 8; 1/15/67, 57; AP, W *Star*, 1/8/67, A18; Reddig, W *Star*, 1/11/67, A5; W *Post*, 1/12/67, C1)

January 11: Initial flight test of USAF's Scramjet (supersonic combustion ramjet)—forerunner of hypersonic aircraft—was partially successful. Launched by Scout booster from Vandenberg AFB, vehicle accomplished primary goal by demonstrating it could separate properly from booster, but failed to perform some secondary missions. (AP, *NYT*, 1/14/67, 4; *Tech Wk*, 1/23/67, 3)

- French research balloon, released from southwest France Jan. 5 and last heard from over Mongolia, was found in North Carolina. Purpose of flight was to test onboard equipment: radio transmitter, barometer, and device for measuring angle of sun above horizon. (AP, *NYT*, 1/13/67, 43)
- Rep. Frank T. Bow (R-Ohio) introduced bill (H.R. 112) to establish a Government-owned Supersonic Transport Authority to supervise and privately finance with Government guarantee the development and construction of the SST prototype. (*CR*, 1/11/67, H104)
- Senate Commerce Committee unanimously approved President Johnson's nomination of Alan S. Boyd as Secretary of Transportation. (NASA LAR VI/2)
- Sonic booms from military aircraft had caused damage to prehistoric cliff dwellings in Canyon de Chelly National Monument in Arizona and to geological formations in Bryce Canyon National Park in Utah, National Park Service said in report to Interior Secretary Stewart L. Udall. Report also cited potential damage to cliff dwellings in Mesa Verde National Park in Colorado. Secretary Udall announced he had received promise of cooperation from USAF, and would seek aid of FAA and special NAS sonic panel. (*NYT*, 1/12/67, 1)
- AFSC had awarded Hughes Aircraft Co. $9,000,000 initial increment to contract for R&D of an experimental comsat. (DOD Release 21-67)
- Latest Trendex poll indicated 69% of Americans believed U.S. space program should continue at its present pace even if U.S.S.R. were to achieve first manned lunar landing. (Jackson, LA *Herald-Examiner*, 1/11/67)
- Communist China had been placing considerable emphasis on large-scale production of weapons, particularly medium-range missiles capable of delivering atomic warheads, CIA Director Richard Helms disclosed in a closed briefing of the Joint Congressional Committee on Atomic Energy. (*NYT*, 1/12/67, 4)

January 12: JPL had selected TRW Systems, Inc., for negotiations on a $1-million contract for design, modification, fabrication, and testing of the propulsion subsystem for the 1969 Mariner Mars mission. (NASA Release 67-4)

January 13: Lunar Orbiter news briefing was held at NASA Hq.

Dr. Leonard Reiffel, Apollo Program, OMSF, said that 635 *Lunar Orbiter I* and *II* photos of the 1,000-mi Apollo strip near moon's equator had "removed any doubt that there are numbers of suitable places for Apollo to land satisfactorily." Of the regions photographed, 10 sites appeared to have smooth areas; detailed analysis of three of these had revealed two areas that were only about 6% crater-marked. Dr. Reiffel said Apollo landing site should be 3- to 5-mi-dia oval area, relatively free of craters and boulders, with 30 mi of smooth terrain approaching it and no more than 7° slope. In addition, since moon's optimum light ranged from 7° to 20° and sun moved across moon 13° per day, landing sites ideally should be spaced 26° apart at this longitude to allow for launch delays. "We . . . need not one site or a couple of sites ideally, but we would like, in order to make life as simple as possible, to have a number of sites. Then at the specific time of the launch . . . we will pick a set of three. . . ."

Dr. Lawrence Rowan, U.S. Geological Survey, said Lunar Orbiter photos had shown that moon's dark mare was smoothest terrain; bright rays had variable roughness and upland plains appeared very rough. He said photos indicated that mass movement on the moon modified both mare and upland.

LaRC Lunar Orbiter Project Manager Dr. Clifford H. Nelson, discussing future Lunar Orbiter missions, said Lunar Orbiter 3 would be "site confirmation" flight to photograph areas already determined by *Lunar Orbiter I* and *II* as promising for Apollo landings. Spacecraft 4 and 5 might be used to photograph 400 lunar sites of primary scientific interest. Capt. Lee R. Scherer, Lunar Orbiter Project Manager, OSSA, noted that one Lunar Orbiter in polar orbit around the moon could survey 80% of moon's front face with 10 times resolution of earth-based photos. "These types of missions would provide a framework upon which to base major decisions concerning our total lunar exploration program after these initial manned landings," Scherer said.

- Display of one *Lunar Orbiter I* photo—first ever taken of earth from interplanetary distances—reopened discussion about possibility of life on Venus. Dr. Homer E. Newell, NASA Associate Administrator for Space Science and Applications, said that photo, which showed earth as an indistinctly detailed, cloud-shrouded crescent, bore striking resemblance to Venus as depicted in earth-based telescopic photos. If Venus could be photographed as clearly as earth, "We might see holes in the clouds," Newell suggested. Current prevailing scientific theory was that solid clouds surrounded Venus trapping infrared radiation and making planet too hot to support life. (Transcript; Clark, *NYT*, 1/14/67, 12)
- Pilot Joseph Cotton (Col., USAF) and co-pilot Van Shepard, North American Aviation, Inc., flew XB-70 No. 1 research aircraft to mach 2.57 and 61,000-ft altitude in flight at Edwards AFB for national sonic boom program. (NASA Proj Off)

January 14: USAF launched an unidentified satellite from WTR with Thor-Agena D booster; satellite reentered Feb. 2. (UPI, *NYT*, 1/16/67, 18; *Pres Rep 1967*; GSFC *SSR*, 2/15/67)

- U.S.S.R. tested first missiles in November 1947 and first multistage ballistic missile 10 yrs later, *Pravda* disclosed in tribute to Sergey Korolev, "chief constructor of space systems for U.S.S.R.," on first anniversary of his death. (Reuters, *NYT*, 1/16/67, 20)
- New U.S. Ambassador to the U.S.S.R. Llewellyn E. Thompson carried special message from President Johnson to Soviet leaders in Moscow, which reportedly appealed for mutual agreement not to deploy antimissile systems. Ambassador Thompson's mission was first diplomatic followup to President Johnson's Jan. 10 State of the Union address in which the President indicated that U.S. had decided to defer deployment of such a system and hoped U.S.S.R. would do the same. (*NYT*, 1/15/67, 23)

January 15: Senate Minority Leader Everett M. Dirksen (R-Ill.) suggested on ABC's "Issues and Answers" TV program that Government expenditures on the space program be cut to avoid a tax increase. "I've been looking at that space program, I don't mind telling you. Does it really make a difference whether we get to the moon this year or next year, or the following year?

"I doubt very much whether it makes a great deal of difference. And if we reconcile ourselves to that thesis, then of course there is a place where you can make a very substantial budget cut." (W Star, 1/16/67, A5)

- Prolonged pause in Soviet manned space flights had been causing considerable speculation in U.S. political and scientific circles, Stephen S. Rosenfeld wrote in the *Washington Post*. Speculation ranged from predictions that the Soviets were moving on to a new "family" of larger spacecraft to suggestions they had paused to solve serious problems encountered in prolonged weightlessness. (Rosenfeld, W Post, 1/15/67, E3)
- Laser applications in industry and space were rapidly increasing, William Smith noted in the *New York Times*. A solar-powered laser was being developed for communication across the 50 million mi between earth and Mars, and Dr. Henry Lewis, director of RCA's David Sarnoff Research Center, predicted the laser would also play a very important role in tracking space vehicles. Said Dr. Lewis: "The laser fits the bill with greater accuracy than radar is capable of obtaining." (Smith, NYT, 1/15/67, F1)

January 16: GSFC engineers had successfully used an electric "screwdriver" to restore power to *Explorer XXXIII* satellite, in orbit 252,900 mi from earth, saving the spacecraft from an almost certain power blackout. Emergency repair—believed to be the most distant satellite "fix" ever accomplished—was conducted via Rosman, N.C., tracking station. Engineers turned off the spacecraft transmitter so that an increased power surge through the other electronic systems eliminated the suspected short circuit. (NASA Release 67–5)

- Undulating, banana-shaped barium cloud, released from a Nike-Iroquois sounding rocket launched from Eglin AFB in USAF experiment on air currents, touched off dozens of reports of UFO's throughout southeastern U.S. (UPI, NYT, 1/18/67, 48)
- U.S. could have a permanent manned space station in orbit by 1970, predicted Charles W. Mathews, Director of Saturn/Apollo Applications, NASA Hq. OMSF, in *Space Business Daily* interview. Mathews said that by clustering various modules, instrument packages, and spacecraft in orbit for reuse and utilizing resupply of expendables and crew replacement, it would be possible to extend capabilities of hardware currently under development to the point where such a space station could be permanently manned in about three years. (SBD, 1/16/67, 63)
- *This New Ocean: a History of Project Mercury* (SP–4201) by Loyd S. Swenson, Jr., James M. Grimwood, and Charles C. Alexander was published by NASA. It was NASA's first major program history published and was available to public from Superintendent of Documents, GPO. (NASA Release 67–1)
- MSFC awarded Sanders Associates, Inc., $2,149,548 contract for logistic support of Saturn V operational display systems used to present information on vehicle's status during simulated and actual launch preparations. (MSFC Release 67–8)
- V/Adm. Hyman G. Rickover (USN) received Printing Industries of Metro-

politan New York's Franklin Award for "advancement of scientific knowledge and furtherance of democracy." In his acceptance speech, Admiral Rickover said citizens in a democracy had a duty to become educated so they could support social action against harmful technologies. "There is a knowledge gap of vast dimension between the public and that small elite of highly intelligent, highly educated experts who understand science and have the use of technology. . . . A free society cannot, in the long run, bear the burden of having a mass of voters who lack the education they need to make them responsible citizens. Society ceases to be free if a pattern of life develops where technology, not man, becomes central to its purpose. We must not permit this to happen lest the human liberties for which mankind has fought, at so great a cost of effort and sacrifice, be extinguished." (Gilroy, *NYT*, 1/17/67, 27)

- Photos taken Jan. 9 by two teenage brothers, purportedly showing a helicopter-sized UFO hovering over Lake St. Clair near Detroit, did not "indicate an obvious hoax" and were being studied further for authenticity, Dr. J. Allen Hynek, USAF consultant on UFO's and Chairman of Northwestern Univ. Astronomy Dept., disclosed in a telephone interview. (Hofmann, *NYT*, 1/17/67, 10)
- Dr. Robert J. Van de Graaff, inventor of the Van de Graaff particle accelerator used in nuclear physics research and cancer therapy, died at age 65. (AP, *NYT*, 1/17/67, 35)

January 17: XB-70 No. 1 research aircraft was piloted by Col. Joseph Cotton (USAF) and NASA test pilot Fitzhugh Fulton to mach 2.54 and 60,200-ft altitude in flight at Edwards AFB for national sonic boom program. (NASA Proj Off)

- Feasibility of using nuclear engines on USAF's C-5A transport so it could stay aloft for several months was being studied by LeRC, Charles Harper, Director of Aeronautics Div., NASA OART, reported in *Washington Post* interview. NASA was concentrating on the hardware necessary to transfer nuclear reactor heat to the jet engines; if this problem could be solved, Harper said the next question would be "what good does it do the Air Force" to have engines of unlimited endurance. Plans specified that nuclear reactors operate only after takeoff to avoid danger of radiation on the ground. (Wilson, *W Post*, 1/18/67, 1)
- *Explorer XXXII* (Atmosphere Explorer B) aeronomy satellite, launched May 25, 1966, suffered complete depressurization, causing immediate and continuing degradation of unsealed batteries and satellite's demise. NASA surmised that internal pressure loss could have been caused by meteoroid impact or weld rupture of spacecraft shell. Satellite mission objectives had been achieved and the mission declared a success in December 1966. All instruments had been operational until time of depressurization with exception of two neutral particle mass spectrometers; they had provided quality data for the first 100 orbits. (NASA SP-4007, 192; NASA Proj Off)

January 18: USAF Titan III-C booster launched from ETR successfully inserted eight Initial Defense Communications Satellite Program (IDCSP) comsats into random equatorial orbits.

Powered flight of Titan III-C was close to planned parameters. Transtage and payload were inserted into parking orbit with 113-mi (182-km) apogee and 106-mi (171-km) perigee where first transtage burn made necessary course corrections. Second transtage burn 66 min later moved stage and load into transfer orbit with 21,000-mi (33,810-km) apogee and 120-mi (193-km) perigee. Third burn put satellite dispenser frame and eight satellites into near-synchronous, 21,000-mi-altitude orbit. At 6 hours 11 min GET, the 100-lb satellites were ejected separately into slightly different orbital paths. USAF officials reported all satellites were functioning properly.

Eight new satellites, combined with seven others launched by USAF June 16, 1966, would reinforce DOD's worldwide communications system and provide reliable, full-time radio link between Washington, D.C., and U.S. troops in Vietnam and other distant outposts. (UPI, *NYT*, 1/19/67, 7; AP, B *Sun*, 1/19/67, 1; *Av Wk*, 1/23/67, 32)

- NASA appointed Francis W. Kemmett to succeed Dr. James T. Hootman as Executive Secretary of the Inventions and Contributions Board, Office of Industry Affairs. (NASA Ann)
- NASA Astronauts M. Scott Carpenter and Edwin E. Aldrin, Jr., tested space simulator at Martin Co.'s plant near Littleton, Colo. Unit, which provided 6° of freedom, was linked to computer that determined exertion forces on astronaut as he performed simple tasks in simulated weightless environment. (*Denver Post*, 1/19/67)
- NASA–USAF XB–70 flight research program should be exploited because it could contribute greatly to understanding operational problems of the SST, Alvin S. White, formerly an NAA XB–70 test pilot and currently TWA's Manager of Flight Research and Development, told *Aviation Daily*. "I think that airplane [XB–70] has tremendous potential . . . it can provide the least expensive way to do a lot of things. We could feed the B–70 into the air traffic control system to learn how it can cope with SSTs. We could test fly prototype engines at very high speeds, and it could contribute a lot to understanding sonic boom problems over air routes." (*Av Daily*, 1/18/67)
- Najeeb Halaby, Pan American World Airways, Inc., Vice President and former FAA Administrator, praised the Johnson Administration's "carefully calculated risk" in the development of the SST, in a speech before the Wings Club in New York. He suggested that as a result both the Government and industry involved in producing the aircraft "are in a position to make a better decision and a better plane than . . . France, Britain, and the Soviet Union." Halaby said aviation industry had a duty "to help the President and the Congress make the most prudent and the most temporary investment of public funds in a national interest venture." (*NYT*, 1/19/67, 57)
- Sun's illumination of a missile's shock wave had caused strange aerial phenomenon seen throughout southwestern U.S. Jan. 17, USAF officials at Vandenberg AFB reported. (AP, *NYT*, 1/19/67)

January 19: U.S.S.R. launched *Cosmos CXXXVIII* into earth orbit with 293-km (182-mi) apogee; 193-km (119-mi) perigee; 89.2-min period; and 65° inclination. Equipment performed satisfactorily. Satellite reentered Jan. 27. (*SBD*, 1/20/67, 98; GSFC *SSR*, 1/31/67)

- USAF ramjet-powered missile, called the low-altitude supersonic vehicle (LASV), was air-launched from F–100 Super Sabre aircraft in successful flight test at Holloman AFB. Marquardt Corp., designer of the ramjet, was conducting the flight test program for the vehicle under an $11-million USAF contract. (UPI, *W Post*, 1/20/67, A15)
- Charges that many specifications for construction of $25-million tank chamber at Plum Brook Station were being compromised were denied by

Dr. Abe Silverstein, Director of NASA Lewis Research Center which operated station. Charges, made by Plum Brook employee and referred to Rep. Charles A. Mosher (R-Ohio), were being investigated by GAO. Silverstein said almost all the allegations were "not true" or represented a difference in judgment between himself and the complaining employee: "It's a matter of his judgment against ours, and I don't have to accept it." (Thomson, Cleveland *Plain Dealer*, 1/20/67)

- Project Themis, a university-based program to establish new academic centers of excellence in research areas important to DOD's long-range scientific and technological goals, was announced by Secretary of Defense Robert S. McNamara. McNamara said more than 400 universities had been asked to submit proposed programs by Feb. 15. During 1967, $20 million had been authorized for the project, and DOD expected to initiate up to 50 new departmental centers before the end of the year. (DOD Release 44-67)
- United Air Lines President George E. Keck, in a speech before International Aviation Club in Washington, D.C., echoed similar pleas by executives of Pan Am and TWA in calling for "expeditious but orderly" advancement of U.S. SST program. Although SST "is a pretty real one on the drawing boards," Keck said there were "certainly some important decisions and actions that have to be taken here to make it a reality." Among these, determination of a "sound, reasonable" financing plan was of paramount importance. (*NYT*, 1/22/67, 50)

January 20: Saturn V 3rd stage (S-IVB) was completely destroyed in explosion 10 min before it was scheduled to be ignited in test at Douglas Aircraft Corp.'s Sacramento plant. NASA later determined accident had been caused by faulty welding in high-pressure helium sphere. (UPI, *W Post*, 1/27/67, A2; UPI, *NYT*, 1/22/67, 29; UPI, *W Post*, 1/28/67, A7)

- NASA should prepare a list of priorities for its future programs to guide Congress in making any budget cuts required by the Vietnam war, Chairman of the Senate Committee on Aeronautical and Space Sciences Clinton Anderson (D-N. Mex.) told Baltimore *Sun* reporter Albert Sehlstedt, Jr. Senator Anderson insisted he was not necessarily predicting cuts in the FY 1968 budget, but observed that "something might have to give a little" while so much was being spent on the war. (Sehlstedt, Jr., B *Sun*, 1/21/67)
- F-111A aircraft, lent by USAF to NASA to obtain basic flight research data for the design and development of advanced variable sweep wings, was delivered to FRC. (NASA Release 67-7)
- MSFC awarded $100,000 six-month contract to Bell Aerosystems Co. for flight testing of Bell's "pogo stick" one-man flying vehicle at LaRC's 1/6 gravity test facility. Pogo stick was being evaluated for Apollo lunar surface mission use. (MSFC Release 67-12; *Marshall Star*, 1/25/67, 3)
- ComSatCorp advised FCC of plans to award four-month study contracts— one to Lockheed Missiles and Space Co. for $272,000; one to Hughes Aircraft Co. for $299,000—for research on multi-purpose synchronous comsats. Lockheed had proposed an inertia-wheel stabilization concept, and Hughes a spin-stabilized approach for stabilizing a spacecraft in synchronous orbit. (ComSatCorp Release)
- Theory that chlorophyll—a key compound of terrestrial life—exists in outer space was proposed by Dr. Fred M. Johnson of Electro-Optical Systems, Inc., at Univ. of California science symposium. Dr. Johnson

said that light absorption patterns of interstellar dust obtained by spectroscopic examination had clearly demonstrated the presence of chlorophyll. Confirmation of his theory could indicate that life in forms familiar to man exists on other planets both in the earth's galaxy and throughout the universe. (*NYT*, 1/21/67, 9)

January 21: Luna XII, first Soviet satellite to successfully transmit photos to earth from lunar orbit, had completed its mission and broken radio contact with ground stations, Tass announced. Satellite had been launched Oct. 22, 1966. (*W Post*, 1/22/67, A23; UPI, *NYT*, 1/23/67, 7)

- "Serious difficulties" had developed in Soviet program to land a man on the moon and it would take "no little time" to solve them, Cosmonaut Gherman Titov said in *Aviatsiya i Kosmonavtika*. (UPI, *NYT*, 1/23/67, 21)

- Report that Communist China would launch a "space vessel" in 1967 appeared in *Red Flag*, newspaper of the Red Guard of the Peking Aeronautical Institute. U.S. experts had been crediting Communist China with capability to launch a satellite since her successful guided missile nuclear weapon test Oct. 27, 1966. (*NYT*, 1/22/67, A4)

January 22: William E. Zisch, Vice Chairman of the Board of Aerojet General Corp., was sworn in as a part-time consultant to NASA Administrator James E. Webb on technology development and utilization of NASA-developed technology by nonaerospace industries. (NASA Release, 67–8)

- Lifting-body vehicles, with their ability to maneuver and land on conventional runways, might become the spacecraft of the future, J. V. Reistrup suggested in the *Washington Post*. Because a naval recovery fleet would not be required, space missions could become a day-to-day operation; crews might use the vehicles to inspect foreign spacecraft, repair U.S. satellites, make reconnaissance flights, fly in search and rescue operations, or take replacement crews and supplies up to manned space stations. NASA and USAF had been testing various types of lifting-body vehicles: M2–F2 and HL–10 in piloted glide tests, and a small automated model of SV–5 as part of Prime—Precision Recovery Including Maneuvering Entry—program. (Reistrup, *W Post*, 1/22/67)

- Growing confidence that U.S. would accomplish a manned lunar landing by 1970 had plunged space planners into an urgent debate over post-Apollo plans, John Wilford wrote in the *New York Times*. Should the U.S. concentrate on unmanned and manned exploration of Mars or Venus? An expanded lunar-oriented program? Giant manned laboratories in earth orbit for research and surveillance? Networks of communications and meteorological satellites? Or a series of instrument probes to comets, asteroids, and outer planets? Such questions had opened a searching reexamination into the value of space exploration, and some critics were urging a less ambitious and costly program, with more emphasis "on the problems that confront us here on earth." While the debate continued, NASA was said to favor concentrating on manned orbiting laboratories that would exploit present technology. Although a broad range of space goals was considered attainable in the next two decades, NASA officials said failure to define post-Apollo objectives soon could cause U.S. to find itself unprepared to achieve any major space goal before the mid-1970s. (Wilford, *NYT*, 1/22/67, 1)

January 23: New extraterrestrial mineral had been analyzed with electron microprobe and named "niningerite" by ARC geochemists Drs. Klaus Keil and Kenneth Snetsinger. Named for Dr. H. H. Nininger, whose research on meteorites paralleled and confirmed early ARC research on blunt-body reentry shapes, the iron magnesium sulfide mineral was found in six stony meteorites estimated to be about 4.5 billion years old. (ARC Release; *Science*, 1/27/67, 451–3)

- *Technology Week* marked beginning of its second decade of publishing by asking some of the Nation's leading educators, scientists, and engineers to project their goals for technology. AEC Chairman Dr. Glenn T. Seaborg emphasized the need for man to first choose goals for technology: ". . . we must give more thought to guiding the direction of our technological future.

    ". . . if there are any ultimate goals for technology we will know what they are only when we all agree on the goals of man."

    MSFC Director Dr. Wernher von Braun proposed an earth resources management system employing space technology—principally, resource-sensing satellites—to meet the needs of the world's rapidly growing population. Said Dr. von Braun: ". . . I am firmly convinced that one of the great future contributions of our space program will be in helping to manage more effectively the utilization of our world resources."

    In discussing long-delayed peaceful application of nuclear explosions—Project Plowshare—Dr. Edward Teller, Associate Director of Lawrence Radiation Lab., suggested that questions of safety, fear of the unknown, limited nuclear test ban, and international politics were chief obstacles to program. Until these obstacles were removed, Plowshare would continue to lie dormant.

    Prof. Isaac Asimov, Boston Univ. School of Medicine, foresaw that mankind, now faced with the practical need for working under low-gravity conditions, would have to develop "low-gravity engineering." Low-gravity engineering might also prove useful in future exploration beneath the sea.

    Dr. Joshua Lederberg, Professor of Genetics at Stanford Univ. School of Medicine, proposed that man's central technological goal was the harmonization of his technical goals, a process Dr. Lederberg called "eutechnics." He went on to declare: "In the eutechnical society, the penultimate crime may become to introduce any technological innovation as a subsystem benefit without analyzing its impact on the whole future of man. The ultimate one would be to deny man his humanity by denying him the chance to think, to know himself."

    Dr. Charles S. Draper, head of MIT's Aeronautics and Astronautics Dept., emphasized the need to establish an adequate balance between "directly beneficial" and "remedial" technologies. As air and water pollution, land damage from mining, soil erosion, and other problems become acute, man must give more thought to remedial actions for disturbances inflicted on the environment. Dr. Draper suggested space vehicle technology as ". . . an excellent model for the technology of overall environmental control for the earth."

    Gen. Bernard A. Schriever (USAF, Ret.), former AFSC Commander, believed that the big challenge of the next 25 yrs would be the effective management of technology for the public good. Schriever suggested that: "the systems management concept [developed in Nation's military and space programs] will be required to bring these technologies to bear on these problems. . . ."

    Clarence L. Johnson, Lockheed Aircraft Corp. Vice President, listed 10 aerospace projects which, he felt, should receive major attention

January 23

during next 20 yrs. Among these projects were use of nuclear power plants in aerial transports to reduce fuel-to-cargo ratios, solution of sonic boom problems, and development of high-density fuels.

Dr. Athelstan F. Spilhaus, dean of Univ. of Minnesota's Institute of Technology, explored the objectives and benefits of geotechnology—use of geophysical and geochemical knowledge "to intervene in the natural processes of our environment on a massive—and even global—scale." Dr. Spilhaus said he believed the public would ". . . welcome . . . an imaginative experiment in city living designed to remove the chores, dirt, delays, and human wastage, and emphasize the opportunities for social, cultural, economic and recreational advantages that high-density living provides."

In final article, TRW Vice Chairman Simon Ramo wrote: "We are in the science era of mankind; we are not yet in the social era of mankind. . . . A quick look at our society shows that science's greatest challenge now and for the next few decades is for the application of science to human needs." (*Tech Wk*, 1/23/67, 32–79)

- U.K. and France had agreed to develop and build a variable-sweep-wing fighter aircraft and three types of helicopters, *Aviation Week* reported. Development costs were expected to total $560 million. Total world market for the fighter was estimated at 1,000 aircraft; and for the helicopters, 750. (*Av Wk*, 1/23/67, 35)
- *Gemini XII* Astronaut Edwin E. Aldrin, Jr., speaking to the 15th annual Women's Forum on National Security in Washington, D.C., urged that critics of the U.S. space program be cognizant of its byproducts, notably its contributions to science, medicine, and education. "As a parent, I'd think any effort that was beamed at raising the sights of youth would be worth the money," Aldrin said. (Dean, *W Star*, 1/24/67, B10)
- France announced plans to launch her fourth satellite from Hammaguir Range Feb. 3 to test tracking device based on the use of laser beams. (UPI, *WJT*, 1/23/67, 10; UPI, *P EB*, 1/23/67)
- Soviet newspapers revealed that two cities devoted entirely to scientific research were being constructed near Moscow, Theodore Shabad reported in the *New York Times*. The first, Pushchino, would be devoted entirely to research in biology and related fields; the second, Krasnaya Pakhra, would specialize in earth sciences. (Shabad, *NYT*, 1/23/67, 1, 8)

January 23–27: Texas Symposium on Relativistic Astrophysics, sponsored by Yeshiva Univ., Univ. of Texas, Southwest Center for Advanced Studies, and NASA's Goddard Institute for Space Studies, was held in New York City. In opening session, Dr. Allan R. Sandage of Mt. Palomar Observatory announced observation of what was apparently the greatest explosion ever recorded by man. Explosion occurred in quasar 3C–446 many million or billions of years ago, and caused it to increase in brilliance 20-fold in a matter of weeks or months.

Cal Tech's Dr. Maarten Schmidt told conference of a new quasar, PKS 0237–23, that appeared to be farther from earth than any other reported to date. PKS 0237–23, receding at 80% the speed of light, was most remarkable, however, because it displayed two "red shifts"—displacement toward red end of the light spectrum as objects speed away. If both shifts were indicators of velocity, it would mean that part of the quasar was receding from earth 17,000 mps more slowly than the rest.

Dr. Riccardo Giacconi of American Science and Engineering, Inc., and Dr. Sandage reported that second starlike source of intense x rays had been discovered by Oct. 11 sounding rocket and later telescopic observation. Of 30 known x-ray sources, only one had previously been linked to a starlike object. Finding of second source, located in Constellation Cygnus, persuaded a number of astronomers that such visible objects represented a new type of celestial phenomenon, perhaps related to the manner in which stars are formed.

Princeton Univ. physicist Dr. Robert H. Dicke revealed an observation that might invalidate Einstein's general theory of relativity. Dr. Dicke told the conference that his experiments showed the sun to be flattened at the poles and that the oblateness was sufficient to explain a significant portion of Mercury's orbital behavior, without recourse to relativity. It was the precise conformity of Mercury's orbit to Dr. Einstein's predictions that was the chief pillar of his theory. "It wouldn't surprise me if general relativity is just plain wrong," Dr. Dicke commented. (*NYT*, 1/24/67, 22; 1/26/67, 18; 1/28/67, 1, 12; 1/29/67, 62)

January 24: President Johnson's message to Congress on the FY 1968 budget said in part: "In 1961, this Nation resolved to send a manned expedition to the moon in this decade. Much hard work remains and many obstacles must still be overcome before that goal is met. Yet, in the last few years we have progressed far enough that we must now look beyond our original objective and set our course for the more distant future . . . we have no alternative unless we wish to abandon the manned space capability we have created.

"This budget provides for the initiation of an effective follow-on to the manned lunar landing. We will explore the moon. We will learn to live in space for months at a time. Our astronauts will conduct scientific and engineering experiments in space to enhance man's mastery of the environment. . . ." The President recommended that NASA proceed with the Voyager program for an unmanned landing on Mars in 1973 and continue "other unmanned investigations nearer the earth. . . .

"These new ventures are the result of careful planning and selectivity," Mr. Johnson said. "We are not doing everything in space that we are technologically capable of doing . . . we are choosing those projects that give us the greatest return on our investment. . . ." (Text, *CR*, 1/24/67, S755)

- President Johnson submitted his FY 1968 Budget Request to Congress. Total space budget recommended was $7.242 billion. Of this sum, NASA would receive $5.050 billion; DOD, $1.998 billion; AEC, $151.6 million; ESSA, $34.8 million; U.S. Geological Survey, $4.8 million; Agricultural Research Service, $.4 million; and NSF, $2.8 million.

NASA FY 1968 budget would keep Project Apollo on schedule and provide major hardware funding for several new programs: (1) $454.7 million for Apollo Applications (AA); (2) $71.5 million for Voyager; (3) $68.9 million for Mariner to provide for Mariner/Mars 1971 program; (4) $2 million for five Sunblazer missions to study solar corona; (5) $35.5 million for Applications Technology Satellite (ATS) to provide new configuration with large directable antenna; and (6) $2.3 million for Voice Broadcast Satellite. Funding for aeronautics research almost doubled from $35.9 million in FY 1967 to $66.8 million with most of increase attributable to increased study of X-15 and XB-70 aircraft, hypersonic ramjet, aircraft noise reduction, delta X-15 aircraft, and SST. Programs terminated or deferred by FY 1968 budget included Jupiter probe, Surveyor and Lunar Orbiter (no follow-on funds), large solid motor project, Atlas-Agena combination, and new launch vehicle development. $60–$100 million were placed in President's Contingency Fund pending decision on development of Nerva nuclear propulsion system.

Major portions of DOD's space budget would be spent on: (1) Manned Orbiting Laboratory (MOL)—$430 million; (2) Titan III–C launch vehicle; and (3) military comsats. Most of ESSA's funds would support increased activity on satellite sensors and systems in support of World Weather Watch and improved techniques for weather forecasts, river and flood prediction, and other warning services. (Text, *NYT*, 1/25/67, 20–23; *Tech Wk*, 1/30/67, 14–33; *Av Wk*, 1/30/67, 23–38)

- Briefing on NASA FY 1968 budget was held at NASA Hq.

    NASA Administrator James Webb said: ". . . I believe that on this budget we can build an excellent foundation for the work that we as a nation will need to do in 1970 and beyond.

    "I believe it is a balanced program that makes the best use of the resources that we have created since 1958. I believe it includes those items that enable the Congress and the country to clearly see and judge the issues which are basic to our future in space.

    "I believe it assumes success. That is, that it is a minimum budget for the work that is required and does not provide for losses of stages or for unexpected catastrophes such as are always a possibility in this program.

    ". . . finally, I believe that this budget indicates that we have proven in the National Aeronautics and Space Administration that the space capability of this nation can be developed, that this job he is putting in this budget can be done, and I believe it indicates clearly that he believes it is important that we continue to develop our national space capability. . . ." (Transcript)

- Six Nike-Tomahawk sounding rockets launched from KSC carried 134-lb Marshall/Univ. of Michigan Probe (Mump) instrumented payloads to altitudes up to 206 mi (331 km) into the thermosphere. Rockets and equipment functioned satisfactorily and data—including atmosphere density and temperature measurements and ion and electron density measurements—were telemetered to ground; no recovery was attempted. Objective of program was to obtain additional data needed to develop a more accurate model of earth's upper atmosphere for orbital-lifetime and space-vehicle-dynamics analyses. (*Marshall Star*, 2/1/67, 4; *W Post*, 1/26/67, A5)

- Dr. James I. Vette, staff scientist at Aerospace Corp.'s Space Science Lab., was named Director of the National Space Science Data Center (NSSDC)— a $1,491,600 central facility for collection, organization, storage, retrieval, and dissemination of space science data obtained from satellites, sounding rockets, balloons, and high-altitude aircraft—nearing completion at GSFC. (NASA Release 67–6)

- New members of House Committee on Science and Astronautics were named: Reps. Jack Brinkley (D–Ga.); Bob Eckhardt (D–Tex.); Larry Winn, Jr. (R–Kan.); Guy Vander Jagt (R–Mich.); Jerry L. Pettis (R–Calif.); Donald E. Lukens (R–Ohio); and John E. Hunt (R–N.J.). (NASA LAR VI/8, 11; Committee Ofc.)

- USAF Avionics Lab., Wright-Patterson AFB, was conducting experiments with 6-mi-long laser beam to learn how atmospheric conditions affect laser communications, AFSC announced. Experimental results to date indicated that communications were severely limited by fog and heavy rain, although transmissions could penetrate such conditions for a few thousand feet at reduced volume. (AFSC Release 254.66)

- Secretary of State Dean Rusk, in keynote address to annual meeting of House Committee on Science and Astronautics' Panel on Science and Technology, suggested that many of the potential perils of advancing technology might be avoided if a group of experts were impanelled every five years "to explore our technological future. . . . The values of this type of forecasting to policy judgments is obvious," Rusk said. "In most cases a true technological innovation does not reach full bloom for some years—the first basic patent on the transistor was, after all, issued in 1930. . . ." He recommended that panel members be selected from the natural sciences, the social sciences, and industry. (Text)

- President of Aerospace Industries Assn. Dr. Karl G. Harr, Jr., testifying before Senate Special Subcommittee on Scientific Manpower Utilization, urged that the systems approach—"an extremely high degree of capability to evaluate, plan and do certain complex and difficult things"—be applied beyond the national defense and space effort to "problems connected with improving the quality of American life. . . . Many of these problems, such as air pollution and water pollution control, relief of traffic congestion, provision of adequate food supply, school systems, housing, and crime control are already fully identified and very much on the front burner of public concern." Harr pointed out the need for experimentation in the application of the systems approach to gain breadth of experience and emphasized that efforts should be aimed at regional, state, and municipal levels. "It remains for us as a nation to find the various political formulae for its application wherever needed. Whatever the cost it will be far cheaper to solve these problems than not to. . . ." (Testimony)

- At the AIAA Aerospace Sciences Meeting in New York, the 1967 Goddard Award—AIAA's highest—was presented to LeRC engineers Irving A. Johnsen and Seymour Lieblein and former LeRC engineer Robert O. Bullock for their work 10 yrs ago which led to development of the transonic compressor for jet-powered aircraft. (LeRC Release 67-4)

January 25: U.S.S.R. launched *Cosmos CXXXIX* into earth orbit with 210-km (130-mi) apogee; 144-km (89-mi) perigee; 87.5-min period; and 50° inclination. Equipment functioned satisfactorily. Satellite reentered same day. (*W Post*, 1/26/67, E1; GSFC *SSR*, 1/31/67)

- XB-70 No. 1 research aircraft was flown by NASA test pilot Fitzhugh Fulton and North American Aviation, Inc., test pilot Van Shepard to mach 1.41 and 35,000-ft altitude to obtain data on stability and control maneuvers. (NASA Proj Off)

- NASA Administrator James E. Webb, testifying at Senate Committee on Aeronautical and Space Sciences' hearings on aeronautical R&D, recommended establishing a "civil aircraft technology" to counteract growing divergence between requirements for civil and military aircraft systems. Webb observed that most modern military aircraft could not be readily redesigned into economical commercial systems: "When unit costs were low and interaction between the elements of the design were small, a manufacturer could afford a substantial element of uncertainty to developing a new airplane, since it was not too difficult or costly to correct deficiencies exposed during flight operations. . . ." With the complexities of modern aircraft, however, deficiencies would lead to "major escalations in cost and could mean financial disaster for a commercial venture. . . . Economics thus require the designer to take an increasingly conservative approach in applying new and advanced concepts. . . ." This conservatism was causing an increasing gap between "the advanced technology that research indicates is possible and the technology actually being used in commerce. To assure pre-eminence in aeronautics, advanced technology must continually be incorporated into new designs. . . ."

Webb said NASA was working closely with Dept. of Transportation and FAA to identify and solve critical and specialized civil aviation technology requirements. He cited major efforts in supersonic transport field, including work on second-generation propulsion, sonic boom phenomena, and aircraft handling qualities; and study of jet engine noise-suppression problems. (Testimony)

- Orbiting satellites could survive Van Allen radiation belts 10 to 100 times longer than originally estimated if their outer shells were shaped like spheres, Dr. Charles Mack, MIT Lincoln Lab, suggested at AIAA's Aerospace Sciences Meeting in New York. Basing his prediction on results from new MIT-developed technique to simulate Van Allen belts, Dr. Mack said that radiation would always strike the surface of a sphere at wide angles, improving the chances of scattering and thereby decreasing the amount of absorption. Ideal spacecraft would be 10-ft-dia sphere with instrumentation in the core, Dr. Mack said. Most current satellites were smaller and carried instruments near surface. *New York Times* writer John Wilford reported that GSFC planned to have MIT-type simulator in operation by 1969. (Wilford, *NYT*, 1/26/67, 10)

- *The Washington Post* commented on NASA's FY 1968 budget: "Exploration of the moon. Giant orbiting space stations, with as many as a dozen men living and working in them for as long as a year. Unmanned landings on Mars.

"These are the goals the United States has set for itself in outer space in the 1970s. With a budget request for fiscal 1968 of $5.05 billion, more than $500 million of it for dramatic new programs to carry us beyond a manned landing on the moon, President Johnson has made it clear that the United States is in space to stay, no matter what the Soviet Union does or does not do. . . ." (*W Post*, 1/25/67, A18)

January 26: NASA successfully launched *Essa IV*, fourth meteorological satellite in ESSA's Tiros Operational Satellite (TOS) system, from WTR using three-stage Thor-Delta booster. Satellite achieved nearly-polar, sun-synchronous orbit with 888-mi (1,429-km) apogee; 822-mi (1,323-km) perigee; 113-min period; and 102° inclination. Wheel orientation maneuver was scheduled for completion during 18th orbit, at which time first photos would be programed and two-week spacecraft checkout and evaluation program would begin.

An advanced version of the cartwheel configuration, 290-lb *Essa IV* carried two Automatic Picture Transmission (APT) camera systems which would photograph earth's cloudcover and immediately transmit pictures to local APT ground stations. *Essa IV* was replacing *Essa II* in the TOS system because orbital drift limited *Essa II*'s usefulness.

ESSA financed, managed, and operated TOS system; GSFC was responsible for procurement, launch, and initial checkout of spacecraft in orbit. *Essa I* was launched Feb. 3, 1966; *Essa II*, Feb. 28, 1966; and *Essa III*, Oct. 2, 1966. (NASA Proj Off; ESSA Release 67–17)

- Messages from President Johnson and Japan's Prime Minister Eisaku Sato were exchanged between Washington, D.C., and Tokyo in ceremony inaugurating commercial service via ComSatCorp's *Intelsat II–B* comsat. In President Johnson's message, delivered by Chairman of the Senate Commerce Committee Warren G. Magnuson (D-Wash.), he said: "The beginning of Pacific satellite service tonight is more than a great technical feat: It is a promise of deeper understanding between the peoples of East and West. . . ." Ceremony also included live color television exchanges and telephone, teletype, facsimile, and photograph transmissions between the two capital cities. *Intelsat II–B* was launched by NASA from ETR Jan. 11. (ComSatCorp Release)
- Apollo Applications (AA) briefing was held at NASA Hq.

   Dr. George E. Mueller, Associate Administrator for Manned Space Flight, presented most detailed statement to date on the program. He said plans were to form an "embryonic space station" in 1968–69 by clustering four AA payloads launched with Uprated Saturn I boosters. First mission would be launch of manned spacecraft, followed several days later by launch of spent S–IVB stage converted into a workshop. After two spacecraft had docked, crew would enter workshop through an airlock. They would prepare workshop for storage and return to earth in their spacecraft 28 days later. In three to six months, second manned capsule would be launched on 56-day mission to deliver resupply module to workshop and rendezvous with unmanned Apollo Telescope Mount (ATM), fourth and last launch in series. Cluster would be joined with multiple docking launched on S–IVB workshop. Emphasizing the importance of manning the ATM, Dr. Mueller said: ". . . if there is one thing the scientific community is agreed on it is that when you want to have a major telescope instrument in space it needs to be manned.

   "First of all you need him to point it. Second, you need him to be able to change the films and so on. Thirdly, you need him to maintain it so when something goes wrong he can fix it instead of having to sit here on the ground and be frustrated by the fact that some little gadget didn't quite trip when it should have. . . ." Dr. Mueller said principal areas toward which $454.7 million FY 1968 post-Apollo effort would be directed were "development of extended flight capability, the conduct of manned astronomical and Earth observations from space, and continued exploration of the Moon." (Transcript; *Marshall Star*, 2/1/67, 7–8; Clark, *NYT*, 1/30/67, 2; Reistrup, *W Post*, 1/27/67, A7)
- In his annual "defense posture" statement to the Senate Armed Services and Appropriations Committees, U.S. Secretary of Defense McNamara made a determined attempt both to persuade Congress not to insist on a U.S. antimissile defense and to dissuade the U.S.S.R. from continuing her efforts to deploy such a system. McNamara said U.S. and Soviet nuclear missiles were so effective that neither country could hope to protect itself against the other in time of war, and an increase in defensive capability by one power could easily be offset by an increase in the offensive capability of the other.

   A major policy statement by Soviet Defense Minister Malinovsky published in *Kommunist* had stressed strategic offensive missiles, without referring to U.S.-U.S.S.R.'s burgeoning antimissile defense system. (Frykland, Gwertzman, *W Star*, 1/26/67, D3)

January 27: Three-man crew for NASA's first manned Apollo spaceflight (AS–

January 27: Apollo Command Module at Launch Complex 34, Kennedy Space Center, is a silent, grim reminder of the flash fire that took the lives of the three-man crew when it swept through the spacecraft during launch rehearsal.

204) died, apparently instantly, when flash fire swept through their Apollo I spacecraft mated to an Uprated Saturn I booster 218 ft above the ground. Crew was Virgil I. Grissom, one of seven original Mercury

astronauts; Edward H. White II, first American to walk in space; and Roger B. Chaffee, preparing for his first space flight. Accident, worst in the history of the U.S. space program, occurred at 6:31 p.m. EST at KSC's Launch Complex 34 during first major rehearsal for scheduled Feb. 21 mission.

Immediately after tragedy, MSC Director Dr. Robert Gilruth, Chairman of the Board of North American Aviation, Inc., Lee Atwood, and NASA's Apollo Program Director M/G Samuel C. Phillips flew to KSC from Washington, D.C. General Phillips described the accident at a press conference Jan. 28: "The facts briefly are: at 6:31 p.m. (EST) the observers heard a report which originated from one of the crewmen that there was a fire aboard the spacecraft . . ." At T-10 and holding in the countdown, observers saw a "flash fire" break out inside the spacecraft, penetrate its shell, and surround its exterior with smoke. Rescue workers rushed to the pad area but could not open the hatch for five minutes. Phillips said astronauts' only hope of escape would have to open hatch manually with a crank tool. He refused to speculate on how long accident would delay the Apollo program or on source of the fire but admitted that "if there is an ignition source . . . in pure oxygen this, of course, has flash-fire potential." His statement that spacecraft was operating on internal power when fire broke out was corrected Feb. 3 by NASA Deputy Administrator Dr. Robert C. Seamans: "The fuel cells in the service module were not in use, and the so-called internal power was being supplied by batteries having the same characteristics as the fuel cells but located external to the spacecraft."

NASA immediately appointed a board of inquiry (see Jan. 28) and announced that all data had been impounded pending its formal investigation. Unmanned Apollo flights would proceed on schedule, but manned flights were postponed until board's inquiry was completed. Apollo I backup crew—Walter M. Schirra, Jr., Donn F. Eisele, and Walter Cunningham—were expected to be named prime crew for rescheduled mission. (NASA Proj Off; NASA Release 67-16; *NYT*, 1/28/67, 1, 18; 1/29/67, 1, 48, E1; *W Post*, 1/28/67, A1, A7; 1/29/67, A1, A18; *W Star*, 1/28/67, A1, A4; 1/29/67, A1, A6)

- Representatives of 62 nations signed space law treaty [see Dec. 8, 1966] at separate ceremonies in Washington, D.C., London, and Moscow. Notable absentees were Communist China, Albania, and Cuba. Representatives of France said they would sign after studying treaty further. At White House ceremony, attended by Soviet Ambassador Anatoly Dobrynin, British Ambassador Sir Patrick Dean, and U.S. Ambassador to the U.N. Arthur Goldberg, President Johnson described treaty as the "first firm step toward keeping outer space free forever from the implements of war" and said it would assure that American and Soviet astronauts "will meet someday on the moon as brothers and not as warriors." Treaty, which limited military activities in space, had been agreed upon by U.S. and U.S.S.R. Dec. 8, 1966, and unanimously approved by U.N. General Assembly Dec. 19. It would become effective when ratified by U.S., U.S.S.R., U.K., and two other countries. (Kilpatrick, *W Post*, 1/28/67, 1; Frankel, *NYT*, 1/28/67, 1)

- NASA awarded Boeing Co. net bonus of $1,895,312 in accordance with incentive contract provisions for superior performance of *Lunar Orbiter I* spacecraft. Boeing Co. was prime contractor for *Lunar Orbiter I*, launched by NASA Aug. 10, 1966, on mission to orbit the moon and

January 27: Secretary of State Dean Rusk signs the space law treaty at White House ceremonies while (seated, left to right) Soviet Ambassador Anatoly Dobrynin, British Ambassador Sir Patrick Dean, U.S. Ambassador to the U.N. Arthur J. Goldberg, and President Lyndon B. Johnson look on.

photograph possible landing sites for Apollo astronauts. (NASA Release 67-14)
- JPL issued RFP's to industry on contracts for preliminary design and definition studies of an unmanned Voyager landing capsule. Two identical Voyager spacecraft would be launched by a single Saturn V booster in 1973 and 1975 to conduct scientific studies of Mars and search for extraterrestrial life. From the industrial proposals, due in March, two to four contractors would be chosen. (NASA Release 67-15)
- U.S.S.R. was definitely trying to achieve the first manned lunar landing, concluded a 920-page report on the Soviet space program prepared by Library of Congress Legislative Reference Service for Senate Committee on Aeronautical and Space Sciences: ". . . materials gathered from Soviet sources on future lunar missions reveal this pattern of complete unanimity: the lunar mission is desirable and necessary; it must be a manned mission; it is certain to succeed. . . ." Report characterized

Soviet space program as "well-planned, orderly, and vigorously pursued with concentration on specific, limited objectives, each achieving a marked advance beyond the one preceding." There was said to be no evidence "of either Soviet disenchantment with the program or a desire . . . to cut it back. On the contrary, space specialists have underscored the durability of the Soviet commitment to space exploration. . . ." Report was entitled: *Soviet Space Programs, 1962–65; Goals and Purposes, Achievements, Plans, and International Implications.* (Text)

January 28: NASA Nike-Tomahawk sounding rocket launched from Churchill Research Range carried GSFC-instrumented payload to 128-mi (209-km) altitude to gather data on charged particle fluxes and to investigate distribution of electric fields in the ionosphere during auroral displays. Rocket and instrumentation performance was satisfactory, but vehicle spin rate was excessive. Most scientific data were lost because nose cone failed to eject properly and antennas failed to deploy. (NASA Rpt SRL)

- NASA Deputy Administrator Dr. Robert C. Seamans, Jr., appointed Apollo 204 Review Board to investigate Jan. 27 accident at KSC which killed Astronauts Virgil I. Grissom, Edward H. White II, and Roger B. Chaffee: Dr. Floyd L. Thompson, Director of LaRC, Chairman; Astronaut Frank Borman, MSC; Maxime Faget, MSC; E. Barton Geer, LaRC; George Jeffs, North American Aviation, Inc.; Dr. Frank A. Long, Cornell Univ.; Col. Charles F. Strang, Norton AFB, Calif.; George C. White, Jr., NASA Hq.; and John Williams, KSC. Three advisory members were added later: Charles W. Mathews, Director of Saturn/Apollo Applications, NASA Hq. OMSF; John Yardley, McDonnell Co. executive; and L/Col. William D. Baxter (USAF). LaRC Chief Counsel George Mallay would serve as counsel to the board. (NASA Release 67–16; Wilford, *NYT*, 1/30/67, 2)

- Senate Aeronautical and Space Sciences Committee would conduct a "full review" of the Jan. 27 accident at KSC which killed three Apollo astronauts, Chairman Clinton P. Anderson (D-N. Mex) announced. (*NYT*, 1/29/67, 47; *W Post*, 1/29/67, A18)

- The Nation mourned the deaths of NASA Apollo Astronauts Grissom, White and Chaffee.

    President Johnson: "Three valiant young men have given their lives in the nation's service. We mourn this great loss and our hearts go out to their families."

    Vice President Humphrey: "The deaths of these three brilliant young men . . . is a profound and personal loss to me. . . . The United States will push ever forward in space and the memory of the contributions of these men will be an inspiration to all future space-farers."

    NASA Administrator James Webb: "We've always known that something like this would happen sooner or later, but it's not going to be permitted to stop the program. . . . Although everyone realized that some day space pilots would die, who would have thought the first tragedy would be on the ground?"

    Former President Eisenhower: "The accident that took the lives of three of our highly trained, skilled and courageous American astronauts is a tragic loss to our entire nation. . . ." (UPI, W *Star*, 1/28/67, 1; AP, *NYT*, 1/29/67)

- *New York Times* praised signing of space law treaty and suggested further cooperation in space.

    "Surely the present is an appropriate time for another effort . . . the President could suggest to the Soviet Union that a precedent from Antarctic practice be followed in the future. Both the United States and the U.S.S.R. plan in the next few years to orbit capsules containing more than three astronauts and to put exploratory parties of men on the moon for stays of several weeks' duration.

"It would be in the spirit of the Space Treaty to have such groups contain at least one representative of the other nation, just as in the Antarctic today Soviet and American scientists work at each other's bases. Agreement on such a first step would do much to build the mutual trust and confidence needed to make possible the far greater degree of international cooperation that the inexhaustible challenges of space require of all nations." (*NYT*, 1/28/67, 26C)

- Continuing shortage of physicists was threatening the Nation's scientific progress, American Institute of Physics warned in its biennial report on physics manpower. Study, prepared under an NSF grant, said U.S. colleges and universities were not producing enough graduates to meet growing demand of education, research, and industry. In recent years—during a time of rapid growth in total college enrollment—there had been a continuing decline in the number of physics majors and a leveling off of physics baccalaureate degrees awarded. (*NYT*, 1/29/67, 1)
- Communist China charged that U.S.S.R. had betrayed Vietnamese people by signing space law treaty with U.S. and other nations Jan. 27. (Reuters, *NYT*, 1/29/67, 9)

January 28–29: Press commentary on deaths of Astronauts Grissom, White, and Chaffee.

Washington *Evening Star:* "There is bitter irony in the fact that the first disaster in our space program came during a simulated launch, a checkout in preparation for the real thing. From the beginning, it has been feared that some fatal mishap was inevitable. But most people had expected it during an actual launch, in space, during re-entry or in landing—hardly during a routine test with the rocket still on its pad...." (*W Star*, 1/28/67, A6)

New York Times: "By chance, Grissom, White and Chaffee died on the day the space treaty was formally signed in Washington, London, and Moscow. Behind the unprecedented international cooperation represented by that treaty's conclusion was the knowledge that space holds more than enough risks for men without adding the horrors arising from strife among nations. The tragedy at Cape Kennedy underlines that somber reality." (*NYT*, 1/29/67, 10E)

Baltimore *Sun:* "They were brave men. When a man finally does put a foot on the surface of the moon, Grissom and White and Chaffee will be notable among those who assisted him in getting there." (B *Sun*, 1/29/67, 30)

- Most Soviet reporting of Jan. 27 flash fire in which three Apollo astronauts died was factual. A few members of Soviet and Italian press, however, charged that "haste" had contributed to the accident.

Radio Moscow: "We in the Soviet Union are deeply grieved at the news of the tragedy at Cape Kennedy.

"The courage of Virgil Grissom, Edward White, and Roger Chaffee had won our esteem and we join in paying homage to their memories. . . ."

Trud (Moscow): "The astronauts became the victims of the space race created by the leaders of the United States space program. Recently, the hurry, the haste in space flights has continued to grow. There were a number of flaws in the Apollo system."

La Stampa (Turin, Italy): "Technical revisions must be brought about so that the tragedy will not repeat itself . . . perhaps the rigorous time schedule . . . ought to give way to slower and more secure rhythm."

Il Popolo (Rome): The accident offered "questions for reflection, above all an invitation to weigh in proper measure the margin of risk that accompanies all the conquest of man. . . . It is ironic that this happened, just when any danger seemed to be less proximate, during a normal exercise on earth." (AP, *NYT*, 1/30/67, 2, 3; AP, B *Sun*, 1/30/67; *Trud*, 1/29/67, 3, USS-T Trans.)

January 29: Memorial service for Astronaut Roger B. Chaffee was held in Houston. Among those attending were Dr. Wernher von Braun, Director of MSFC; Dr. Robert Gilruth, Director of MSC; and Dr. George E. Mueller, NASA Associate Administrator for Manned Space Flight. (Bloom, *W Post*, 1/30/67)

- Greatest damage to Apollo I spacecraft in which three astronauts died in Jan. 27 flash fire seemed to have occurred near point where electrical cables from launch pad entered cabin, reported *Aviation Week* writer George Alexander, who represented news media when NASA authorized one person to visit KSC's Launch Complex 34. Alexander said the spacecraft "looked like the inside of a furnace . . . the interior . . . is a darkened, dingy compartment. Its walls are covered with a slate-gray deposit of smoke and soot; its floor and couch frame are covered with ashes and debris—most of it indeterminate. . . ." (UPI, *NYT*, 1/30/67; *W Post*, 1/30/67, A1, A2; W *Star*, 1/30/67, A1, A3)

January 30: Separate memorial services were held in Houston for Astronauts Virgil I. Grissom and Edward H. White II. Among those attending were Dr. Wernher von Braun, Director of MSFC; Dr. Robert Gilruth, Director of MSC; Dr. George E. Mueller, NASA Associate Administrator for Manned Space Flight; Astronaut Alan Shepard; and former Astronaut John Glenn. (Hines, W *Star*, 1/30/67, 1)

- NASA had awarded Electro Optical Systems, Inc., a $2-million contract for design, fabrication, and testing of the power subsystem for the 1969 Mariner Mars mission. (NASA Release 67-17)

- Princeton Univ. physicist Dr. Roman Smoluchowski told annual meeting of American Physical Society in New York that huge amounts of energy emitted by Jupiter—three times the amount it received from the sun—could be accounted for by a gradual shrinking of the planet. Dr. Smoluchowski said that Jupiter was losing gravitational energy because it was shrinking about 1/50th of an inch a year, and that it was this lost energy that it was emitting. Shrinkage was occurring, he reported, because liquid and solid molecular forms of hydrogen, which comprise planet's outermost layer, were gradually being compressed into metallic hydrogen, which comprises the next innermost layer. Rate of compression, according to Dr. Smoluchowski's calculations, would yield just the right amount of shrinkage to account for the observed amount of energy Jupiter emits. (*NYT*, 1/31/67, C52)

- U.S.S.R. had perfected and tested an antimissile device that used x-rays generated by a nuclear blast to paralyze or disintegrate incoming missiles hundreds of miles from their targets, *U.S. News & World Report* said. There was no comment from DOD. (*US News*, 1/67)

January 30–31: NASA launched 18 sounding rockets in sets of two at two-hour intervals from Churchill Research Range and Pt. Barrow, Alaska, between 6:15 p.m. and 8:20 a.m. EST to gather meteorological data between 12- and 124-mi altitudes. Twelve Nike-Apaches launched from Churchill ejected vapor cloud experiments between 43- and 124-mi altitudes; drift of vapor trails was recorded photographically to obtain data on wind directions and speeds. Six Nike-Cajuns launched from Pt. Barrow carried acoustic grenades which were ejected and detonated at programed altitudes to provide correlative data on atmospheric conditions. Experiment series was being conducted by GSFC under OSSA direction. (NASA Release 67–13)

January 31: Nike-Apache sounding rocket carrying 55-lb payload to obtain electron density profile was launched from NASA Wallops Station to 121-mi (193-km) altitude. Minutes later, Nike-Cajun grenade experiment was launched over the Atlantic for atmosphere temperature support data at 50- to 56-mi (80- to 90-km) altitudes. Launchings, first two in series of three to study relation of high absorption radio waves in ionosphere to seasonal temperature variations in upper atmosphere, were conducted by NASA for Univ. of Illinois, GCA Corp., and GSFC. Rockets and equipment functioned satisfactorily. (Wallops Release 67–4)

- In related experiments, Nike-Apache sounding rocket was launched from Ft. Churchill, Canada, to obtain wind profiles by release of chemical and photograph of chemical cloud at 96-mi (154-km) altitude; another Nike-Apache sounding rocket was launched from Ft. Churchill to study shortterm variation of density, pressure, and temperature during 12-hr period to 80-mi (127-km) altitude; Nike-Cajun sounding rocket was launched from Point Barrow, Alaska, to determine shortterm variations in upper atmosphere temperatures with payload of standard grenades. All rockets and payloads performed well. (NASA Proj Off)

- XB–70 No. 1 research aircraft was piloted by NASA test pilot Fitzhugh Fulton and Col. Joseph Cotton (USAF) to mach 1.35 and 37,000-ft altitude to conduct stability and control maneuvers. (NASA Proj Off)

- Ninth anniversary of first U.S. satellite, *Explorer I.* In defiance of the original predicted lifespan that should have ended six years ago, the satellite was circling the earth 14 times daily and had completed 43,000 orbits. (MSFC Release 67–17)

- President Johnson transmitted his annual *Report to the Congress on United States Aeronautics and Space Activities.*

    He wrote in covering letter: "America's space and aeronautics programs made brilliant progress in 1966. We developed our equipment and refined our knowledge to bring travel and exploration beyond earth's atmosphere measurably closer. And we played a major part in preparing for the peaceful use of outer space. . . ." The President cited accomplishments of NASA's Gemini, Lunar Orbiter, and Surveyor programs and noted that during 1966 U.S. had put into earth orbit or escape missions 100 spacecraft—"a record number of successful launches for the period. We launched weather satellites, communications satellites and orbiting observatories. . . .

    "These accomplishments—and the promise of more to come—are the fruits of the greatest concerted effort ever undertaken by any nation to advance human knowledge and activity. Space, so recently a mystery, now affects and benefits the lives of all Americans. . . ."

In separate chapter, National Aeronautics and Space Council (NASC) said that although U.S. continued to lead space technology competition in number and variety of successful missions, "USSR maintains its lead in weight of payloads orbited. . . . Certainly the hiatus in Soviet manned space activity during 1966 is no basis for complacency on our part as their preparations for future flights appear to be ambitious and the resources being devoted to space competence and performance are absolutely and relatively impressive."

Dept. of State said it was pleased to be able to forward *Surveyor I* and *Lunar Orbiter I* photos to missions abroad for appraisal by foreign scientists. It also said foreign reaction to receiving cloud-cover photos from U.S. satellites in Automatic Picture Transmission (APT) system had been "encouraging." (Text)

- Flash fire swept through oxygen-filled pressure chamber at Brooks AFB, Tex., killing Airman 2/C William F. Bartley, Jr., and Airman 3/C Richard G. Harmon in accident similar to the one Jan. 27 which killed three Apollo astronauts at KSC. Fire struck at 9:45 a.m. EST, about 12 min. after airmen had entered the chamber to take blood samples from rabbits under observation to determine effects of pure oxygen on the blood. Cabin pressure was brought down to ground level within 13 sec after fire had ignited, but both men died from burns within several hours. Col. James B. Nuttall, commander of USAF School of Aerospace Medicine, said there were no plans to change research methods because of accident. "As you have seen, it is not safe," but he said one-gas systems were less complicated and more reliable than two-gas systems. (AP, *W Star*, 2/1/67, A3; O'Toole, *W Post*, 2/1/67, 1)

- Eric Sevareid offered a eulogy on CBS–TV News for three Apollo astronauts who died at KSC Jan. 27: "Grissom and White and Chaffee—mortals who aspired to the moon and eternal space—were returned to the earth today from which they came and to which we all belong.

    "They had lived life more intensely in a very few years than most of us do in our lifetimes and they shall be remembered far longer.

    "They were among the men who wield the cutting edge of history and by this sword they died. . . .

    "We are told they will be replaced. This only means that other such men will take their places. The three cannot be replaced. There never was a replaceable human being." (Text)

- Funeral services were held for Apollo astronauts who died in Jan. 27 flash fire at KSC. All three men were buried with full military honors: Virgil I. Grissom (L/Col., USAF) and Roger Chaffee (LCdr., USN) at Arlington National Cemetery, and Edward H. White II (L/Col., USAF) at West Point. President Johnson and hundreds of dignitaries attended services at Arlington; Mrs. Johnson and Vice President Humphrey were among the attendees at West Point. (Hines, *W Star*, 1/31/67, A1, A6; Secrest, *W Post*, 2/1/67, B1; Casey, *W Post*, 2/1/67, A1, A4)

- Operation of Wilhelm Forster Observatory's tracking station in West Berlin was described by its director, Harro Zimmer, in an AP interview. The station, built on a 250-ft mound of rubble, was in a geographic position enabling it to detect Soviet signals from launching pads in the U.S.S.R. Zimmer said he wired tracking data to Smithsonian Astrophysical Observatory, Cambridge, Mass., on various assignments involving both U.S. and U.S.S.R. satellite launches, but received no pay from either the U.S. or the Observatory. (AP, *NYT*, 1/31/67)

During January: British government awarded Manchester Univ. a $126,000 grant to design a successor to radiotelescope at Jodrell Bank Experimental Station. New radiotelescope would have an aerial dish 400 ft in dia, and would cost about $11.2 million to build. (AP, W *Star*, 2/1/67, A3; AP, *NYT*, 2/2/67, 2)

- Chimpanzees could survive sudden exposure to near vacuum according to tests performed by 6751 Aeromedical Research Laboratory at Holloman AFB. Visual inspection of the EEG as well as power spectral density computer analysis indicated the expected greater subcortical resistance to anoxia when compared to cortical responses. Results showed that all subjects not only survived the exposures in excellent health but recovered within 4 hr after rapid decompression and were able to execute a complex behavioral schedule at a preexposure level of performance. Central nervous system damage was absent or negligible. (Text)

- Edgar M. Cortright, NASA Deputy Associate Administrator for Space Science and Applications, reviewed in *Astronautics & Aeronautics* the achievements of U.S. unmanned spacecraft: "For exploring the farthest reaches of space, for probing its most unknown and hazardous regions, for continuous monitoring of many diverse phenomena, for day-in, day-out practical utility, and as precursors of all space activity, automated spacecraft remain without peer." Among unmanned spacecraft worthy of future development, Cortright named Voyager, Mariner, a Jupiter probe, a new Applications Technology Satellite (ATS), a Synchronous Meteorological Satellite, a Voice/TV Broadcast Satellite, and a Navigation Satellite. (*A&A*, 1/67, 22–8)

- Earth resource evaluation satellites could give ". . . greatest immediate return of all space expenditures," *Space/Aeronautics* reported. Applications were listed: spotting indications of petroleum and mineral deposits; providing land use maps for urban renewal, agricultural and industrial development; monitoring underground streams; checking crop and forest vigor; and monitoring pollution dynamics. (*S/A*, 1/67, 85)

- In spite of unleased capacity in two commercial communications satellites now in orbit, optimism prevailed, and ComSatCorp's entry into "global marketplace has led carriers to slash transoceanic rates," *Space/Aeronautics* reported. As technical manager of INTELSAT, ComSatCorp had been establishing operational position in "aeronautical and domestic communications market." (*S/A*, 1/67, 80)

- Dr. Karl G. Harr, Jr., President of the Aerospace Industries Assn., emphasized the importance of U.S. space effort to our total national interest in *Aerospace:* "As a nation . . . we have probably never learned as much in so short a length of time as from our space effort. . . . Leaving aside the advantages in scientific knowledge, national prestige, defense insurance, domestic economics, and everything else, the task of conquering space requires far and away the greatest technological reach man has ever attempted. . . . As it has in the past, this will set the pace for our total national technological advance. Every aspect of our national life which is influenced in any degree by such advances will be the beneficiary." Harr cited a range of future alternatives that would extend the Nation's technological reach: (1) more ambitious, wholly automated program, including extraterrestrial probes and new applications for earth-orbiting spacecraft; (2) expanded manned spacecraft program which might involve construction of a "permanent" space station or a manned lunar base; and (3) manned expedition to Mars. (*Aerospace*, 1/67, 2-7)

- *Wall Street Journal* staff reporters interviewed experts in many fields for series of articles on probable developments between 1967 and 2000. Jonathan Spivak, in lead-off article, reported that MSFC Director Dr. Wernher von Braun and more than a score of prominent U.S. scientists and engineers were making optimistic forecasts of future U.S. space achievements. Insufficient money and lack of nuclear rockets could present serious problems, yet most space experts foresaw the following major possibilities for the next few decades: (1) Early 1970s: extended lunar exploration by astronauts; earth orbital flights of three months' duration to determine effects of prolonged weightlessness. (2) Mid-1970s: landing of unmanned spacecraft on Mars to search for signs of life; launching of first civilian space station; operation of satellite systems to aid civilian ship and aircraft navigation and to survey earth's natural resources; development of improved comsats; use of meteorological satellites to produce accurate long-range global weather forecasts. (3) Late 1970s: first flight of six-man spacecraft, circling Mars and returning to earth in 600 days; orbiting of giant astronomical telescope above earth's atmosphere; establishment of semipermanent base on moon. (4) Mid-1980s: landing of men on Mars for 10-20 days of exploration. (5) The 1990s: launching of unmanned probes to Jupiter, Saturn, and perhaps other planets; longer-duration astronaut exploration of Mars; maintenance of small colony on moon; manned flights around Jupiter and Saturn.

    Jerry Bishop, focusing specifically on electronic communications in second article, said that researchers were envisioning creation of a vast network of facilities over the next three decades that would use present technology—notably, comsats—to make instant audio, visual, and facsimile transmissions available worldwide. Same network would also enable computers to exchange information with each other and communicate with users located elsewhere. Although such services were not considered economically feasible at present, communications experts expressed confidence that demand for picture phones and facsimile would be high enough and costs low enough to bring about widespread introduction of the services before the end of the century. (Spivak, *WSJ*, 1/6/67, 1; Bishop, *WSJ*, 1/16/67, 1)

- Press commented on President Johnson's failure to authorize SST prototype construction.

    New York Times: "Now the basic size and shape of the plane have been settled. . . . Yet neither [company] has been told when to quit refining the designs that resulted from a three-year, Government-sponsored competition and begin 'cutting metal' on flight test models—a step that had been expected on Jan. 1. The companies can take only intermediate steps until the President and the Congress decide how fast the plane should be built, how industry and Government are to divide the financial risk, and whether another, smaller plane should be built for domestic routes. . . ." (Clark, *NYT*, 1/8/67, E2)

    Wall Street Journal: "President Johnson evidently has decided to postpone for some time the $4.5 billion [SST] program. . . . This decision . . . hasn't been publicly disclosed, but is nonetheless obvious to Congressional and Administration officials. . . . How long the delay will last depends on whether it has resulted solely from current budget pressures generated by the Vietnam war or is based on a belief that the growing criticism of the SST—as an allegedly costly and technically unsound project that will 'benefit the jet set' and be an 'ear shattering' nuisance to everyone else—is justified." (Zimmerman, *WSJ*, 1/3/67, 2)

Washington *Evening Star*: "Competitors in the SST sweepstakes had been hoping . . . that the winner would receive the green light to build a flyable prototype airplane right away. What Boeing and GE got instead, was a month-to-month commitment to do more of what they have been doing for 2½ years—study, design, analyze. . . . One thing is certain about the SST program at this point: there will be no big new government funding until fiscal 1969 at the earliest." (Hines, *W Star*, 1/2/67, A10)

- An advanced concept in solid rocketry was demonstrated when Lockheed Propulsion Co. static-fired large-scale motor with two interconnected combustion chambers for 7.5 min. Firing—longest ever achieved by high-performance solid-propellant gas generator—was 80% successful. Lockheed Vice President Kenneth H. Jacobs commented: "The firing demonstrates the feasibility of a large-scale, all-solid rocket propulsion system for spacecraft attitude control and trajectory adjustment." (*Marshall Star*, 2/1/67, 5)
- Sen. Edward W. Brooke (R-Mass.) and Sen. Charles H. Percy (R-Ill.) were assigned to vacancies on the Senate Committee on Aeronautical and Space Sciences created by the resignations of Harry F. Byrd, Jr. (D-Va.), and Sen. George D. Aiken (R-Vt.). (Senate Comm.)
- Dr. Herbert Friedman, Superintendent of NRL's Atmosphere and Astrophysics Div., was appointed chairman of NRC's Committee on Solar-Terrestrial Research. (*A&A*, 1/67, 18)
- Responsibilities of the Defense Communications Agency were described by its director, L/G Alfred Starbird (USA). General Starbird said that DOD's Communications Satellite Program had been divided into three phases: (1) use of existing SYNCOM satellites, (2) installation of Initial Defense Communications Satellite System, and (3) investigation of Advanced Defense Communications Satellite Project for later implementation. General Starbird outlined progress in implementing these steps and in meeting the "special operational requirements" for communication support of NASA's Apollo program and for increased circuits to Southeast Asia. (*Data*, 1/67, 5-9)
- McDonnell Co. and Douglas Aircraft Co. announced plans to "proceed immediately" with merger negotiations. (*WSJ*, 1/16/67, 2)

February 1967

February 1: NASA Administrator James E. Webb told Washington, D.C., women's club that Deputy Administrator Dr. Robert C. Seamans, Jr., would report to KSC Feb. 2 to meet with Apollo 204 Review Board investigating Jan. 27 accident which killed three Apollo astronauts. Dr. Seamans would report Board's preliminary findings to Sen. Clinton P. Anderson (D–N.Mex.), Chairman of Senate Aeronautical and Space Sciences Committee, and Rep. George P. Miller (D–Calif.), Chairman of House Committee on Science and Astronautics. Webb said NASA's decision to use 100% oxygen atmosphere in spacecraft had been made after "a long series of tests and evaluations. Any change would be made only after a most careful examination of all alternatives." (Sehlstedt, Jr., B *Sun*, 2/2/67; *W Post*, 2/2/67)

- Three sets of experiments were continued at Ft. Churchill, Canada, and Point Barrow, Alaska [see January 31]. Last five of a series of six Nike-Apache sounding rockets were successfully launched from Ft. Churchill to obtain wind profiles of upper atmosphere to 96-mi (154-km) altitude. Last five of a series of six Nike-Apache sounding rockets were launched from Ft. Churchill to study short-term variation of density, pressure, and temperature during 12-hr period where altitudes ranged from 94 mi (152 km) to 102 mi (164 km). In first launch, Nike booster malfunction resulted in payload malfunction; for other four launches, rocket and payload performance were satisfactory. Last five of a series of six Nike-Cajun sounding rockets were successfully launched from Point Barrow at two-hour intervals to determine short period variations in upper atmosphere temperatures. (NASA Rpt SRL)

- $5.050-billion NASA authorization bill (H.R. 4450) was introduced to House by Rep. George P. Miller (D–Calif.) and referred to the Committee on Science and Astronautics. Bill was superseded March 2, by $5.100-billion authorization bill (H.R. 6470). (NASA LAR VI/15; Committee Off.)

- Announcement of major breakthrough in Soviet space program—probably an orbiting manned space platform—was imminent according to unidentified U.S. scientist recently returned from two weeks in U.S.S.R., Marquis Childs reported in *Washington Post*: "The likelihood of a dramatic new Soviet achievement takes on particular weight in view of the tragedy to the three American astronauts at Cape Kennedy. It will dramatize the degree to which the Soviets are ahead in heavy boosters essential to carry out a landing on the moon. Repeated denials from both sides that this is not a race to the moon are all very well. The fact remains that the first power to land men on that silver orb will gain enormous prestige." (Childs, *W Post*, 2/1/67)

- U.S.S.R. planned to fly a supersonic transport ". . . probably into Kennedy Airport, now that the Moscow-London-New York run has been opened—probably early in 1968," stated Sen. Warren G. Magnuson (D-Wash.), Chairman, Senate Committee on Commerce, speaking on floor of the Senate. He inserted into *Congressional Record*, *Look* interview of Najeeb E. Halaby, PAA's senior VP and former FAA administrator, which stated that Soviets planned to flight test Tu-144 "SST" next winter and begin commercial service three years later. (*CR*, 2/1/67, A383-5)

- Apollo accident at KSC, January 27, 1967, was discussed in press circular released by Soviet Embassy in Washington, D.C.: "The sorrow of [the] American people is shared by peoples of all countries. In reality, cosmonauts are somehow representatives of the whole Earth, of the entire mankind in boundless Cosmos, no matter what . . . country has dispatched them." (U.S.S.R. Press Cir/GSFC Trans.)
- MSC Director Dr. Robert R. Gilruth, in welcoming address at MSC's two-day Gemini Summary Conference, urged space scientists to avoid "second-guessing" Apollo 204 Review Board because undue speculation on the Jan. 27 accident was a disservice both to the Nation and the Board. (MSC *Roundup*, 2/3/67, 8; *W Post*, 2/2/67, A2)
- Report entitled "Status of the Federal Aircraft Noise Abatement Program," in response to President Johnson's direction in Mar. 2, 1966, Transportation Message, covered initiation of ten-point action program aimed at alleviating jet aircraft noise in vicinity of airports. Report showed participating agencies to include OST, FAA, NASA, HUD, and Dept. of Commerce. Meeting program objectives would require continuing program of research performed in-house and by contractors, several projects being reported as underway.

NASA and FAA jointly pursued investigations on: noise measuring methods and techniques; sociological noise assessment; noise level and duration criteria; LeRC Quiet Engine project effects of blade characteristics on compressor noise levels; compressor noise suppression and absorption; scale model studies in support of compressor acoustics investigations; landing approach and climb-out procedures; and implication of introduction of Boeing 747 and SST (and others) on aggravation of aircraft noise problem.

HUD and FAA jointly studied: delineation of aircraft noise exposure and land use values in vicinity of major international airports; relative economic and social costs and benefits through application of noise reduction and land-use modification; stimulation of airport community development in directions which would anticipate or ameliorate community aircraft noise problems; noise alleviation through insulation of houses; and compatible land use.

In relating U.S. and international efforts in the broad field of aircraft noise abatement, the U.S. would have to proceed with a sense of urgency to develop its national standards to the point where they could be proposed for international adoption. The forum in which such standards must ultimately be agreed upon was the International Civil Aviation Organization (ICAO).

In the field of legislating proper control of aircraft noise, both 89th and 90th Congresses had received "aircraft noise" bills from the Administration, empowering Secretary of Transportation ". . . to promulgate noise standards and to exercise all of the regulatory and certification authority . . . on the basis of such noise standards." (Text; NASA SP-4007, 80; Hudson, *NYT*, 4/9/67, 1)

February 1

- FAA announced that contracts for development of SST would be extended on month-to-month basis with funds already available, "pending negotiations . . . with the manufacturers and the airlines for airline financial participation in the Phase III (Prototype Construction) program." Evert Clark commented in the *New York Times:* "The move to have the airlines share the cost is intended to lessen budgetary pressures on the Administration, make the program more appealing to Congress and have the airlines, the ultimate beneficiaries of the project, show good faith by joining in the risk to a greater degree. . . ." (Text; Clark, *NYT,* 2/2/67, 1)

- Transportation men were still the great "risk-takers of our society," Secretary of Transportation Alan S. Boyd told Transportation Assn. of America meeting in Chicago. "Most of the entrepreneurs associated with America's great moments in transportation seem to have lost their shirt.

    "The builder of the greatest clipper ship went broke.

    "The builder of the greatest steamboat of the nineteenth century went broke.

    "The builder of the first monorail went broke.

    "Most of the early railroads went broke.

    "Most of the early canals went broke.

    "For that matter, the builder of the first Ford automobile went broke, at least on the first try.

    ". . . our society must try to hold open financial rewards which are commensurate with such risks. Especially when you realize how brief . . . the life of many important transportation innovations has been. . . ." (Text)

- USAF's $3.3-million solar vacuum telescope, designed to predict more accurately solar proton showers which endanger astronauts and spacecraft, was meeting its timetable for construction at Sacramento Peak Observatory near Sunspot, N. Mex., Office of Aerospace Research announced. Telescope—of which 193 ft would be underground and 135 ft above ground—would be largest solar vacuum telescope in the world when completed in 1968. In addition to predicting "safe periods" for space flight, it would study solar phenomena and solar-terrestrial relationships. (OAR Release)

- Organizational changes became effective at AFSC Hq.: (1) creation of new Deputy Chief of Staff (DCS) for Operations; (2) reassignment of functions and responsibilities of the Office of the Deputy Commander for Space to other appropriate staff agencies; (3) redesignation of DCS for Foreign Technology to DCS for Intelligence and DCS/Plans to DCS Development Plans; (4) establishment of Office of the Headquarters Commandant as a special staff office; and (5) reassignment of GAO activities function to Deputy Chief of Staff, Procurement and Production. (AFSC Release 28.67)

- Entire universe might be immersed in a sea of tiny particles known as neutrinos whose presence had thus far eluded observation, Dr. William A. Fowler of Cal Tech told American Physical Society meeting in New York. His theory sought to explain absence of helium in the spectra of certain older stars within the framework of the "big bang" theory of the universe. If the exploding fireball from which the universe was formed had been packed with neutrinos, these particles would have interacted with neutrons to produce a cloud of protons and electrons, Dr. Fowler said. Latter would then have combined to form hydrogen atoms which would have stringently limited amount of helium formed. "Neutrino astronomy is just getting started," he said, "but it might ultimately verify or deny presence of the hypothetical neutrino sea." (Sullivan, *NYT*, 2/2/67, 25)

February 2: USAF launched an unidentified satellite from WTR using an Atlas-Agena D booster; satellite reentered Feb. 12. (*Tech Wk*, 2/13/67, 10; *Pres Rep 1967*; GSFC *SSR*, 2/15/67)

- Aerospace industry had the knowledge to build quieter aircraft and "it is time to apply it," M. Carl Haddon, Lockheed Aircraft Corp. executive, said at AIAA meeting in Washington, D.C. ". . . we have reached a turning point beyond which we as designers and managers must see to it that no plane is designed and built that is not significantly quieter than existing aircraft." Haddon predicted that failure by aerospace industry to produce quieter aircraft could result in legislation that would curfew air traffic. (*W Star*, 2/3/67, C6)

February 3: Astronauts Virgil I. Grissom, Edward H. White II, and Roger B. Chaffee had died of "asphyxiation due to smoke inhalation" but cause of Jan. 27 flash fire was still unknown, NASA Deputy Administrator Dr. Robert C. Seamans, Jr., reported in memorandum to NASA Administrator James E. Webb. Dr. Seamans had met Feb. 2 with Apollo 204 Review Board at KSC. ". . . clear identification of the source of ignition or of its possible source will depend upon detailed step-by-step examination of the entire spacecraft and its relative test support equipment." Dr. Seamans said duplicate Apollo spacecraft had been flown to KSC from North American Aviation, Inc.'s Downey, Calif., plant to permit a "parallel step-by-step disassembly process." In addition, Board was "defining a series of investigative tasks . . . and assigning these to teams for execution. . . ." Report confirmed that spacecraft had been operating on external power when fire occurred, but there was no evidence "up to this time that the source of power whether simulated internal or external was related to the accident." Dr. Seamans emphasized that his statement was preliminary.

NASA Associate Administrator for Manned Space Flight Dr. George E. Mueller announced that NASA would proceed with launching of three unmanned Apollo flights scheduled for 1967: AS–206, AS–501, and AS–502. Manned Apollo missions were postponed indefinitely pending the outcome of Apollo 204 Review Board's investigation. (NASA Releases 67–21, 67–22)

- MSFC awarded Boeing Co. a $120-million contract modification for five Saturn V 1st stages. Modification increased total contract value to $977 million for fabrication and assembly of 15 stages. (NASA Release 67–20)

- MSC Apollo Program Manager Dr. Joseph F. Shea briefed major Apollo contractors and subcontractors at MSC about Jan. 27 flash fire, John Wilford reported in *New York Times*. MSC made no announcement of the meeting. (Wilford, *NYT*, 2/4/67, 15)

- Two-to-six percent weight loss in astronauts returning from space flights might reflect natural readjustment of body water content, not simple dehydration, Paul Webb of Webb Associates reported in *Science*. He based his theory on similar losses which had occurred during experiments in simulated weightlessness; blood normally pooled in the extremities returned to circulation, increasing central blood volume and causing excretion of water not replaced during space flights. Webb said it was not yet clear whether reduced water content during weightlessness has harmful effects. (Webb, *Science*, 2/3/67, 558–9)

- NASA's decision to launch *Ats I* satellite Dec. 6, 1966, with one of its important experiments purposely rendered inoperative, was praised by G. W. Swenson, Jr., National Radio Astronomy Observatory, and R. N. Bracewell, Stanford Univ. Radio Astronomy Institute, in *Science*. "The NASA people made a courageous and farsighted decision . . . [when] they disconnected the beacon transmitter" in *Ats I* which would have operated on the same frequency band as many of the world's most important telescopes. (*Science*, 2/3/67, 518–21)

- Lunar Receiving Laboratory, under construction at MSC, would provide scientists with an unparalleled opportunity to examine extraterrestrial materials under controlled conditions, MSC staff members James C. McLane, Jr., Elbert A. King, Jr., Donald A. Flory, Keith A. Richardson, James P. Dawson, Walter W. Kemmerer, and Bennie C. Wooley declared in *Science*. The laboratory would have four major functions: "(i) distribution of lunar samples to the scientific community for detailed investigations after a period of biologic quarantine; (ii) performance of scientific investigations of samples that are time-critical and must be accomplished within the quarantine period; (iii) permanent storage under vacuum of a portion of each sample; and (iv) quarantining and testing of the lunar samples, spacecraft, and astronauts for unlikely, but potentially harmful, back-contamination (contamination of extraterrestrial origin)." (*Science*, 2/3/67, 525–9)

- Lewis Research Center issued RFP's on system employing 2,000- to 6,000-lb comsats which could broadcast directly to home TV sets from 22,300-mi-altitude synchronous orbits. Successful proposal would receive study contract. (LeRC Release 67–6)

- Sixty-day Sealab III experiment would be conducted 430 ft below the Pacific off San Clemente Island in winter 1967 as part of USN's "Man-in-the-Sea" program. Underwater laboratory would be staffed by five eight-man diving teams rotating on 12-day shifts. Aquanauts would conduct experimental salvage techniques and oceanographic and marine biology research; and undergo series of physiological and human performance tests. Sealab III's long-range goal was to develop capability for rescue and salvage operations under ocean's surface and to determine how submerged continental shelf could be used militarily. In Sealab I experiment, conducted July 20–30, 1964, four divers were submerged 192 ft beneath the Atlantic; in Sealab II, Aug. 28–Oct. 10, 1965, three teams of 10 divers alternated in spending two-week periods 205 ft beneath the Pacific off La Jolla, Calif. (DOD Release 72–67)

February 3–4: Mars symposium sponsored by NASA Institute for Space Studies, Yeshiva Univ., and New York Univ. was held in New York. Discoveries reported at meeting were later summarized by Walter Sullivan in *New York Times*: (1) Martian surface, although free of water-carved valleys, is so heavily eroded that wind alone cannot account for it. It is possible that ground is soaked with liquid water even though air of Mars is too thin and dry to produce rain; (2) new observations seem to confirm that air of Mars contains methane or methane-based gases, presence of which is difficult to explain unless they are generated by living organisms; and (3) evidence that carbon dioxide is predominant component of Martian atmosphere implies that oxygen is being continuously injected into air. Most likely source of oxygen is plant life performing photosynthesis. Sullivan concluded: "Despite the perceptible swing of the pendulum toward the possibility of life on Mars, a wide disparity of views on the nature of the Martian environment was presented at the meeting. It was evident that the information is still insufficient to assess the likelihood of life on Mars or even to design experiments for landings there that would most effectively test for the existence of life. . . ." (Sullivan, *NYT*, 2/12/67, 12E)

February 4: NASA's *Lunar Orbiter III* (Lunar Orbiter C) unmanned spacecraft was successfully launched by Atlas-Agena D booster from ETR in mission to photograph possible lunar landing sites for Apollo astronauts (see Feb. 8–28).

Agena 2nd stage fired to boost 850-lb spacecraft into 100-mi (161-km) altitude parking orbit, reignited after 15-min coast period, injecting spacecraft on 92-hr translunar trajectory, and separated. On schedule *Lunar Orbiter III* deployed its four solar panels and two antennas, locked its five solar sensors on the sun, and fixed its star-tracker on Canopus.

Primary objectives of NASA's *Lunar Orbiter III* mission, third in series of five, were (1) to place three-axis stabilized spacecraft into lunar orbit; (2) to obtain high-resolution pictures of previously photographed lunar surface areas to confirm their suitability as landing sites for Apollo and Surveyor spacecraft; and (3) to improve knowledge of the moon. Photos would cover 12 primary target sites located within the Apollo zone of interest on the moon's front face. Spacecraft would also monitor micrometeoroids and radiation intensity in the lunar environment, refine definition of moon's gravitational field, and serve as a target for tracking operations by Manned Space Flight Network stations. Lunar Orbiter program was managed by LaRC under OSSA direction. Tracking and communications were the responsibility of JPL-operated Deep Space Network. (NASA Proj Off; NASA Release 67–12; AP, *W Post*, 2/5/67, A6; AP, *W Star*, 2/5/67, A8; AP, *NYT*, 2/6/67, 12)

- Decision to use 100% oxygen atmosphere in U.S. spacecraft had not been made as a shortcut to compensate for U.S.S.R.'s superior booster capability, NASC Executive Secretary Dr. Edward C. Welsh told *Interavia Air Letter*. "There were other factors involved," including weight and efficient use of oxygen-consuming fuel cells. Dr. Welsh emphasized there was ". . . no basis at all for a change in [Apollo program] policy" because of Jan. 27 tragedy at KSC and predicted U.S. still had 50–50 chance to land first man on the moon. (UPI, *W Star*, 2/5/67, A8)
- Dr. Donald E. Guss, project scientist for GSFC's Solar Particle Intensity and Composition Experiment, died in Cleveland after a heart attack. (*W Post*, 2/8/67)

February 5: Sounding rocket launched by ESRO from northern Sweden to study aurora borealis reached 116-km (72-mi) altitude. (Reuters, *NYT*, 2/6/67, 3)

- John R. Biggs, former director of NASA Office—Downey at North American Aviation's Downey, Calif., plant, became Deputy Executive Secretary in the Office of the Administrator, NASA Hq. (NASA Ann, 2/6/67)
- 764 students would begin work toward doctorate degrees in space sciences during 1967–68 academic year at 152 institutions as part of NASA's Predoctoral Training program to meet increasing need for highly trained scientists and engineers. (NASA Release 67–19)

February 6: U.S.S.R. sought one-year postponement of meeting of U.N. Conference on Exploration and Peaceful Uses of Outer Space, scheduled for Sept. 11–23 in Vienna. Soviet Academy of Sciences member V. A. Kotelnikov told preparatory committee meeting in New York that U.S.S.R. needed more time for preparation of scientific data and papers. Request caused considerable surprise among U.N. delegates because U.S.S.R. had been chief supporter of resolution, passed by General Assembly Dec. 16, 1966, to hold meeting in September. (*NYT*, 2/7/67, 30; Estabrook, *W Post*, 2/8/67)

- Japan launched three-stage Lambda 3H rocket from Uchinoura Space Center, Kyushu Island, to 1,337-mi (2,142-km) altitude—record altitude for Japanese rocket. The 9.5-ton vehicle carried instruments to obtain nine types of scientific data on the Van Allen belts and the ionosphere; all equipment functioned normally except electronic thermometer. The 29-min flight ended when rocket's 3rd stage impacted 1,400 mi downrange in the Pacific. (AP, *NYT*, 2/7/67, 27)
- President Johnson presented 1966 National Medal of Science to 11 scientists and engineers at White House ceremony and praised their efforts to make new discoveries: "Today, our enormous investment in science and research is our evidence of our faith that science can not only make man richer—but science can make man better." Awards were announced Dec. 24, 1966. (*PD*, 2/13/67, 194–5)
- FAA met with 11 U.S. airlines and one leasing company to discuss financing of SST. Following conference, FAA Administrator William F. McKee announced that airline representatives had "indicated general agreement with the concept of assisting with financing of the prototype development phase." (FAA Release 67-11)
- NASA appointed M/G John D. Stevenson (USAF, Ret.) Director of Mission Operations, OMSF, succeeding Everett E. Christensen who returned to Lockheed Missiles and Space Co. General Stevenson had been serving as special assistant to Dr. George E. Mueller, NASA Associate Administrator for Manned Space Flight. (NASA Release 67-23)
- USAF would not use pure oxygen atmosphere on its manned space flights, partly because of the fire hazard, announced L/Col. John W. Ord, Director of Crew Test Evaluation at USAF School of Aerospace Medicine. Ord said Manned Orbiting Laboratory (MOL) program would use mixture of oxygen and helium because it was lightweight and conducted heat away from the spacecraft. (Lockett, *W Post*, 2/7/67)

February 6–9: "Lectures in Aerospace Medicine" conference was held at USAF School of Aerospace Medicine (SAM), Brooks AFB, Tex.

Dr. Hubertus Strughold, chief scientist at SAM, recommended that astronauts preparing for long-duration interplanetary space flights undergo "prophylactic surgery" to prevent appendicitis or gall bladder attacks and prophylactic dental work to guard against abscessed teeth.

Dr. Alfred C. Koestler, head of Holloman AFB's Altered Atmosphere Pressure Laboratory, reported that 18 chimpanzees had survived rapid decompression to the near vacuum of 150,000 ft for 3½ min without any noticeable residual effects following four-hour recovery period. Tests offered hope that an astronaut outside his spacecraft could be rescued if his spacesuit were to spring a leak.

Bends, the painful decompression sickness that cripples deep-sea divers, would imperil astronauts leaving a spacecraft with "mixed-gas" atmosphere to work in space, L/Col. Robert C. McIver suggested. His conclusion was based on series of experiments in which 74 healthy men underwent a total of 388 exposures to decompression; one fourth developed symptoms of bends. (USAF Proj Off; AFSC Release 274.66; W Star, 2/8/67, A8; W Post, 2/10/67, A4; WSJ, 2/15/67, 1; UPI, NYT, 2/24/67, 53)

February 7: U.S.S.R. launched *Cosmos CXL* into orbit with 241-km (150-mi) apogee, 170-km (106-mi) perigee, 88.48-min period, and 51.7° inclination. All equipment functioned satisfactorily. Satellite reentered Feb. 9. (*Tech Wk*, 2/13/67, 10; GSFC *SSR*, 2/15/67)

- NASA's *Surveyor I* spacecraft, which softlanded on the moon June 2, 1966, was still functioning after transmitting over 11,000 photos to earth, JPL's scientist Leonard Jaffe said in Cal Tech lecture. *Surveyor I* mission had been officially terminated July 21, 1966, after spacecraft's battery suffered a sudden dip in voltage. (*LA Herald-Examiner*, 2/7/67)

- President Johnson submitted space law treaty (see Dec. 8, 1966) to Senate and urged that it be ratified promptly "in appropriate commemoration of the Senate's own role in charting the course that the world now seems willing to follow....

  "The resources of this planet are already taxed to support human existence. Now and even more each day, as the family of man increases so rapidly, fertile soil, clear water, clean air and a safe atmosphere all become more precious to men and nations than the metals and jewels of ages past.... The future leaves no option. Responsible men must push forward in the exploration of space, near and far. Their voyages must be made in peace for purposes of peace on earth. This Treaty is a step—a first step, but a long step—toward assuring the peace essential for the longer journey." (*PD*, 2/13/67, 196–9)

- Charges that haste to meet 1970 deadline to land first man on the moon had contributed to Jan. 27 Apollo tragedy were completely "unfounded," Dr. George E. Mueller, NASA Associate Administrator for Manned Space Flight, told closed hearing of Senate Committee on Aeronautical and Space Sciences in testimony released Feb. 11. ". . . the Apollo program is . . . the longest R.&D. program that we have undertaken in this Nation.... It has been paced at a deliberate pace so that it would in fact reasonably economically, but certainly safely, arrive at a set of equipment capable of carrying out the mission. . . ." Dr. Charles A. Berry, MSC Director of Medical Research and Operations, said that a total of 20,756 hrs had been accumulated testing the use of 100% oxygen in NASA-sponsored spacecraft and spacesuit evaluation programs. Before Jan. 27 accident, three fires had occurred in pure-oxygen ground simulators: in all cases fires had been traced to electrical systems not used in spacecraft and crew members had recovered from burns. NASA witnesses recounted long history of precautions to avoid spacecraft fires, including insulation, circuit design and circuit protection for electrical wires, and fire-testing of over 400 materials. NASA Deputy Administrator Dr. Robert C. Seamans, Jr., suggested that exact source of ignition might never be pinpointed: "It is therefore possible that we may fly the Apollo without having been able to establish the cause of the fire. Even in that case, I believe that the care and skill with which the board and program office are conducting their investigations and review will provide the necessary assurance that such an accident cannot be repeated. . . ." (Testimony)

- Board chaired by KSC Director Dr. Kurt H. Debus convened at MSFC to investigate Jan. 20 explosion of Saturn V 3rd stage at Douglas Aircraft

Co.'s Sacramento plant. (*Birmingham News*, 2/7/67)
- Edward H. White II, one of the three Apollo astronauts who died in Jan. 27 flash fire at KSC, was named to receive AIAA's 1967 Haley Astronautics Award. Selection had been made in early January. AIAA cited White for his historic walk in space and other contributions, undertaken "at great personal risk in the advancement of space flight" during June 3–7, 1965, *Gemini IV* mission. Award would be presented to his widow. (AP, *W Post*, 2/8/67, A6; UPI, *NYT*, 2/9/67, 9)
- U.S. and U.S.S.R. had begun talks aimed at mutual agreement not to deploy antimissile missile systems, Deputy Secretary of Defense Cyrus Vance told Senate Foreign Relations Committee's Disarmament Subcommittee. (AP, *P Inq*, 2/8/67, 3)
- First live global telecast linking North America, Europe, Asia, Australia, and Africa would be carried on National Educational Television (NET) June 25, NET Vice President William Kobin announced. Designed as a live documentary to enable man to see earth as a single entity in space and time, two-hour telecast would examine such major problems as hunger and overpopulation. It would be transmitted to more than 30 nations via four comsats: ComSatCorp's *Early Bird I* and *Intelsat II–B* and two of U.S.S.R.'s *Molniya I* comsats. (Gent, *NYT*, 2/8/67, 63)
- Eleven cosmonauts had been killed in accidents since first Soviet space flight in 1960 according to CIA report prepared for the White House several weeks before Jan. 27 Apollo tragedy, *Northern Virginia Sun* reported. CIA claimed five had been killed when their spacecraft failed to enter orbit and six had died in series of mishaps ranging from boosters exploding during countdowns to training incidents involving helicopter crashes. (Allen, Scott, *Northern Virginia Sun*, 2/7/67)

February 8: Third French satellite, *Diademe I* (D–1C), was launched from Hammaguir Range with three-stage Diamant booster. Orbital parameters were 1,341-km (833-mi) apogee, 418 km (260-mi) lower than nominal; 584-km (363-mi) perigee; 104-min period; and 40° inclination. Reason for low apogee had not yet been determined. Designed to reflect laser beams fired simultaneously from ground stations in France, Greece, and Algeria, 50-lb satellite would make geodetic measurements. It also carried ultrastable oscillator for Doppler effect experiment.

First French satellite, A–1, was launched from Hammaguir Range Nov. 26, 1965; second, *Diapason I* (D–1A), was launched Feb. 17, 1966. Fourth and last satellite in Diamant series was scheduled for launch Feb. 19. (Hess, *NYT*, 2/9/67, 15; *SBD*, 2/9/67, 218; *Av Wk*, 2/13/67, 41)
- *Cosmos CXLI* was launched by U.S.S.R. into orbit with 345-km (214-mi) apogee; 210-km (130-mi) perigee; 89.8-min period; and 72.9° inclination. All equipment functioned normally. Satellite reentered Feb. 16. (UPI, *NYT*, 2/9/67; GSFC *SSR*, 2/28/67)
- USAF launched unidentified satellite employing Thor-Burner II combination from Vandenberg AFB. (UPI, *W News*, 2/8/67)
- *Explorer XXIX* (GEOS–A) spacecraft, launched Nov. 6, 1965, was rendered useless as a geodetic instrument because of failure of range transponder (SECOR). These geodetic investigations had produced important results: for the first time, side-by-side intercomparisons had been made between different tracking systems to show that no significant "hidden" errors existed; improved calibrations had been obtained for NASA Range and Range Rate and MOTS-Minitrack tracking networks; laser tracking sys-

tems had been perfected and were ready for operational use; and gravity field definition had been improved to allow more accurate orbit determinations and the computation of orbits of any given accuracy with a smaller amount of tracking data. (NASA Proj Off)

- NASA turned *Essa IV* meteorological satellite over to ESSA for operation in accordance with NASA–Dept. of Commerce agreement. *Essa IV* was launched by NASA Jan. 26 from WTR. (NASA Release 67–10)

- *The Washington Post* praised space law treaty transmitted to Senate by President Johnson Feb. 7: "The new treaty is an eminently sensible limitation on the military exploitation of outer space. The great powers must continue their efforts until the problems of proliferation also have been met. And then they must go on to search for a means of thermonuclear disarmament that will rescue mankind from the threat of annihilation." (*W Post*, 2/8/67)

- White House had received a letter from an unnamed scientist suggesting that it conduct its own investigation into Jan. 27 accident at KSC which killed three Apollo astronauts, Karl Abraham reported in Philadelphia *Evening Bulletin*. Scientist, who was not identified, felt NASA had not sufficiently heeded warnings of fire hazards presented by use of pure oxygen atmosphere. (Abraham, P *EB*, 2/8/67)

February 8–28: NASA's *Lunar Orbiter III*, launched from ETR Feb. 4, became third U.S. spacecraft to enter lunar orbit; four days later it was successfully transferred to final close-in orbit for photography. Orbital parameters: apolune, 1,145 mi (1,847 km); perilune, 34 mi (55 km); inclination, 20.93°; and period, 3 hrs 29 min. Spacecraft performed 357 attitude changes, responded to 2,866 commands, and recorded no micrometeoroid hits. A total of 211 medium- and high-resolution photos were taken before mission's photo acquisition phase was terminated Feb. 24 because of a malfunction in the priority readout system. Readout would be completed March 6. Among photos transmitted, described by JPL officials as of "excellent quality," was a picture of Hygius Rille, a deep gorge similar to the Grand Canyon. The rille—not considered by NASA as a potential landing site—could not be seen distinctly through earth-based telescopes, but through *Lunar Orbiter III*'s cameras it appeared clearly as a deep steep-walled ditch with one large central crater and numerous smaller ones. Pictures taken of the Sea of Tranquility confirmed that the site was probably smooth enough for manned landings: they showed a thin scattering of craters, some as small as three feet in diameter, and only a few ridges steep enough to upset a landing spacecraft. (NASA Proj Off; AP, *NYT*, 2/16/67; *W Star*, 2/21/67, A4; UPI, *W Post*, 2/23/67, A3)

February 9: NASA Javelin sounding rocket launched from Ft. Churchill carried Rice Univ.-instrumented payload to 467-mi (752-km) altitude to measure ratio of fluxes of auroral zone precipitated protons, and alpha particles, fluxes and spectra of low-energy protons and electrons, and energy spectrum of protons with energies of 100 to 500 kev. Rocket and instrumentation performed satisfactorily. (NASA Rpt SRL)

- Feb. 15 launch of third satellite in ComSatCorp's INTELSAT II series was postponed by NASA until mid-March to study recent failure in a Scout launch vehicle's FW–4S motor. INTELSAT-II comsats were launched by Thrust-Augmented Improved Delta boosters which used motors similar to FW–4S. (NASA Release 67–24)

February 10: Three Apollo astronauts died in Jan. 27 flash fire at KSC because "some engineer, technician, or inspector had gotten careless," former Astronaut Donald K. Slayton, MSC Director of Flight Crew Operations, told employees at General Motors AC Electronics Div., Milwaukee, where Apollo guidance and navigation equipment was manufactured. Slayton urged workers: "Take a good sharp look at your job. Are you giving everything you've got? Now you know somebody's life depends on it." An MSC official later reported that Slayton's speech had not been intended for publication. Slayton emphasized that he was not trying to affix blame for the Apollo accident, but was just trying to make workers ultraconscious of the importance of their jobs. (UPI, W *Star*, 2/10/67; UPI, *WJT*, 2/11/67)

- *Time* magazine considered the question "Why Should Man Go To The Moon?" and concluded: "The moon itself may not be a particularly valuable piece of real estate. But neither is a flight to the moon an end in itself; the moon is no more than a way station on a route that scientists have only begun to map. And there is no doubt that man is going to make the trip some day. . . .

  "The moon is a challenge that the U.S. has already taken, a milestone that U.S. astronauts are already looking beyond. . . . The real object is for the U.S. to develop the capability of voyaging confidently to the limits of man's imagination and ingenuity. The value of such voyages will always be unpredictable. But the history of the human race, said famed Norwegian Explorer Fridtjof Nansen, 'is a continual struggle from darkness toward light. It is therefore to no purpose to discuss the uses of knowledge; man wants to know, and when he ceases to do so, he is no longer man.' " (*Time*, 2/10/67)

- Astronaut Donn F. Eisele, in Columbus, Ohio, to receive a "Governor's Award" for increasing the prestige of the state, told AP he was opposed to "an unrealistic, arbitrary" chronological date for reaching the moon. He favored a "technical" date, which he explained to mean "as soon as we're ready, when we have performed sufficient missions to be ready, when the spacecraft is ready." (AP, *NYT*, 2/12/67, 31)

February 11: President Johnson's Science Advisory Committee (PSAC), in report on *The Space Program in the Post-Apollo Period*, rejected the idea of selecting a single major goal as focal point for Nation's post-Apollo program and urged instead a "balanced program based on the expectation of eventual manned planetary exploration." PSAC recommended program "integrating manned and unmanned efforts" toward five major objectives: (1) a "limited but important" extension of Apollo manned spaceflight program; (2) a "strongly upgraded program" of unmanned planetary probes aimed at landing instruments on Mars and Venus by early 1970s; (3) a program of technology development and qualification of man for long-duration space flight in anticipation of manned planetary exploration; (4) "vigorous exploitation" of applied space technology for both military and socio-economic purposes; and (5) "exploitation of our capability to carry out complex technical operations in near-Earth orbit (and on the Moon) for the advance of science, particularly astronomy."

Report proposed one or two manned explorations of the moon annually after first manned lunar landing, together with unmanned flights to parts of moon inaccessible to Apollo spacecraft. NASA was cited for "absence of integrated . . . planning" of planetary exploration in the 1970s and urged to immediately study "the relative effectiveness of

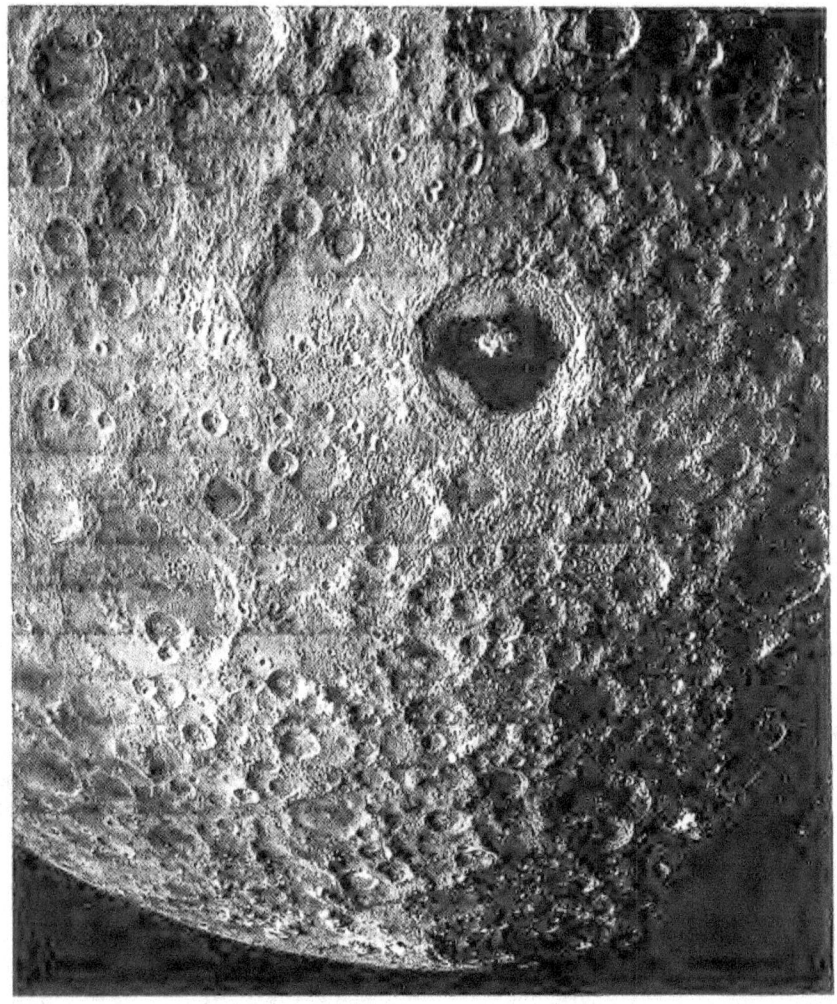

February 8–28: A southward-looking oblique of the backside of the moon taken by *Lunar Orbiter III* on February 19, from an altitude of 900 mi. The large crater toward the center is Tsiolkovsky, 150 mi in diameter. On the right, a much enlarged view of part of the rim and floor of the Tsiolkovsky crater taken by *Lunar Orbiter III*'s high-resolution camera.

man in planetary flyby and orbiter missions to Mars and Venus." Committee endorsed NASA's proposal to use spent 2nd stage of Uprated Saturn I booster as orbital workshop for manned missions up to 56 days but recommended "careful study of the suitability, cost, and availability of Titan III/MOL system for biomedical studies of man for periods up to 60 days."

Astronomy from earth satellite was "focus of . . . [committee's] recommendations for major pure science component of . . . [Nation's]

program." It proposed a program leading to "establishment in earth orbit of a number of astronomical facilities, which by the end of . . . the 1970s will constitute an orbiting astronomical observatory" capable of: (1) exploring full-range of spectrum not accessible from ground; (2) direct scientific control by ground-based astronomers; and (3) extended useful lifetime by servicing in orbit using trained personnel.

Committee urged that "far more intensive effort" be conducted to examine applicability of satellite technology to missions of all Federal agencies. "Such studies should take into account not only technological possibilities but the benefits . . . which might be derived from satellite technology as compared with other methods." (Text)

- MSC Apollo Program Manager Dr. Joseph F. Shea and MSC Director of

Flight Crew Operations Donald K. Slayton had each made a last minute decision against joining Apollo I astronauts in tragic Jan. 27 flight rehearsal. MSC Deputy Director Dr. George M. Low acknowledged that Shea, who had frequently taken part in astronaut flight rehearsals either as a stand-in for a crew member or as an observer, had decided not to participate when technicians were unable to find a communications headset for him to wear. In addition, both men had finally decided that entering the spacecraft would have been "contrary to good test procedural discipline . . . [and] highly irregular." (UPI, *NYT*, 2/13/67; *W Post*, 2/11/67; Hines, *W Star*, 2/11/67, 1)

- Sen. William Proxmire (D-Wis.) urged that Congress cut $9 billion from President Johnson's FY 1968 budget to avoid tax increase. Interviewed on "Youth Wants to Know" television program, Proxmire suggested that $5–7 billion be cut from public works, $1 billion from space, and $1 billion from U.S. forces in Europe. (AP, *W Post*, 2/12/67)

February 13: U.N. Committee on Peaceful Uses of Outer Space adopted Soviet suggestion for one-year postponement of U.N. Conference on Exploration and Peaceful Uses of Outer Space, scheduled for Sept. 11–23 in Vienna. General Assembly would vote on postponement in April. (*NYT*, 2/15/67)

- Dr. Wilmot N. Hess, Chief of GSFC's Theoretical Div., had been appointed director of MSC's newly organized Science and Applications Directorate. He would assume his new post in the near future. (MSC Release 67–8)
- Soviet failure to launch probe toward Mars during January Mars "window" supported reports that U.S.S.R. space program had "run into budget troubles," *Newsweek* reported. (*Newsweek*, 2/13/67)
- USAF successfully launched experimental aerodynamically shaped warhead from Vandenberg AFB aboard an Atlas ICBM to test reentry characteristics. (AP, *St. Louis PD*, 2/14/67, 17)
- Peter V. Dementyev, U.S.S.R. minister of aviation industry, had been awarded the Order of Lenin "for his great services to the Soviet state in developing the aviation industry," *Aviation Week* reported. (*Av Wk*, 2/13/67, 23)
- U.S. civilian-manned satellite-tracking ship *Sword Knot* was still lying off port waiting for U.S. clearance to enter Durban, South Africa, harbor, AP reported. South African government had granted ship permission to enter port only if crew members agreed to obey country's race segregation laws. (AP, *C Trib*, 2/14/67)

February 14: Successful simulation of a meteor was achieved when tiny manmade iron pellet fired by Trailblazer launch vehicle reentered earth's atmosphere at unprecedented speed of 38,180 mph and landed on target. Launched from NASA Wallops Station, six-stage Trailblazer reached 190-mi (306-km) altitude before firing one-gram pellet on downward leg of ballistic trajectory. Artificial meteor was visible for two seconds. Experiment was part of LaRC-Smithsonian Astrophysical Observatory program to study luminosity and ionization of simulated meteors, thereby increasing scientists' knowledge of natural meteoroids. (LaRC Release)

- U.S.S.R. launched *Cosmos CXLII* into orbit with 1,362-km (846-mi) apogee, 214-km (113-mi) perigee, 100.3-min period, and 48.4° inclination. All equipment functioned normally. Satellite reentered Jul. 6. (*Tech Wk*, 2/20/67, 13; GSFC *SSR*, 7/15/67)
- NASA Deputy Administrator Dr. Robert C. Seamans, Jr., submitted second interim report to NASA Administrator James E. Webb on Apollo 204

Review Board investigation of Jan. 27 flash fire in which three Apollo I astronauts died. Board had established 21 panels to conduct inquiry, and Dr. Seamans said he was satisfied that the procedures they were following were "well worked out. When this work is completed, it will give us as clear a view as can be obtained from the evidence."

Source of ignition had not yet been determined, but physical evidence thus far examined indicated fire had had considerable variation and density and might have had more than one phase. "One hypothesis, supported by the cabin pressure history, assumes a small, low-grade fire whose heat was at first largely absorbed by the spacecraft structure and that was burning at the time of the first crew report; that fire may have continued for as long as ten seconds. A more intense fire may have then developed, causing the rapid increase in cabin pressure. This fire was probably then extinguished by the depletion of oxygen." Additional information on accident had been identified and was being analyzed. It was now clear that all three spacesuits had been burned through, Astronaut Grissom's receiving the most exposure and Astronaut Chaffee's, the least. Dr. Seamans said disassembly of spacecraft was proceeding with great care, and spacecraft was expected to be removed to industrial area at KSC by Feb. 17. He concluded: "It is important to note that no single spacecraft element is touched or removed for analysis without full Board approval and evaluation of its possible effect on any of the other on-going studies or analyses." (NASA Release 67–28)

- NASA selected Boeing Co. and Douglas Aircraft Co. for negotiations on cost-plus-fixed-fee contracts for 32- to 36-mo coordinated research programs to minimize jet aircraft noise [see Feb. 1]. Boeing would receive about $7.5 million and Douglas $3 million. (NASA Release 67–26)

February 15: France launched *Diademe II* (D–1D) geodetic satellite from Hammaguir Range with three-stage Diamant booster into orbit of 1,897-km (1,179-mi) apogee, 592-km (368-mi) perigee, 110.4-min period, and 39.9° inclination. Planned apogee was 1,500 km (932 mi), with 635-km (395-mi) perigee. Companion to *Diademe I* (D–1C), launched Feb. 8, satellite was last to be orbited from Hammaguir Range, which would be shut down March 31. All future launches would be from new bases near Bordeaux and in French Guiana. Scientific mission of Diademe series was geodesy by three different means for cross-checks: Doppler effect experiment, laser experiment measuring distance to satellite from three ground stations, and photography of satellites against star background for direction-finding. (*W News*, 2/16/67; *Av Wk*, 2/20/67, 25; Wetmore, *Av Wk*, 3/27/67, 50–58)

- NASA's *Biosatellite I*, carrying more than 10 million tiny living organisms, reentered earth's atmosphere and apparently landed "in the vicinity of Australia." On Feb. 22 NASA announced that search for capsule had been canceled.

    Launched Dec. 14, 1966, from ETR, *Biosatellite I* had been scheduled to reenter earth's atmosphere by parachute and be recovered Dec. 17, but it remained in orbit after retrorocket failed. It was first of three spacecraft in NASA's Biosatellite program to study physiological effects of weightlessness and radiation on organisms. (NASA Release 67–25; Reistrup, *W Post*, 2/14/67, A4; 2/23/67, A3; AP, *P Inq*. 2/16/67, 3)

- GAO report found LeRC innocent of alleged irregularities in construction of $25-million tank chamber at Plum Brook Station but was critical of some of NASA's managerial practices. GAO said it had been advised by NASA "that closer construction management controls had been established. ..." Allegations, brought by Plum Brook employee and referred to GAO by Rep. Charles A. Mosher (R-Ohio), had been denied by LeRC Director Dr. Abe Silverstein Jan. 19. (Text; Cleveland *Plain Dealer*, 2/15/67)

- Boeing Co. would spend $165 million and General Electric Co. $65 million in Phase III (prototype construction) of SST program, M/G J. C. Maxwell (USAF), FAA Director of Supersonic Transport Development, told American Marketing Assn., in Los Angeles. Total cost of Phase III work was estimated at $1.1 billion; most of the difference would be supplied by Government and the small remainder by several U.S. airlines. Boeing and GE contracts would not be effective until President Johnson's authorization to begin Phase III. (Reddig, Jr., W *Star*, 2/16/67, A17; *WSJ*, 2/16/67, 4)

February 16: Sen. Lee Metcalf (D-Mont.) urged on Senate floor ratification of space law treaty submitted to Senate by President Johnson Feb. 7: "So now, in the first flush of triumph over the obstacles of space, is the time to leave behind our terrestrial ambitions, disputes, and claims of sovereignty, and look to new ground rules for peace and cooperation in space. . . . The rivalries, the contentions, and the conflicts that have been our earthly heritage may be abandoned as we jointly and cooperatively launch ourselves into this great adventure in space. It may be that as a result we will achieve a greater cooperative spirit on earth and as a result international confidence and international understanding will grow out of extraterrestrial achievements." (*CR*, 2/16/67, S2072–3)

- NASA Aerobee 150 sounding rocket launched from Ft. Churchill, Canada, carried seven experiments to 100-mi (160-km) altitude measuring: spectral emission line intensity during an aurora as function of altitude; visible emission features of an aurora; and primary and secondary electron streams. Except for high voltage arcing, instrument performance was satisfactory; rocket performance was good. (NASA Rpt SRL)

February 17: President Johnson, in letter transmitting Sixth Annual Report of the Arms Control and Disarmament Agency to Congress, said: "In 1966 a significant link was added to the still slender chain of arms control agreements—a treaty banning weapons of mass destruction in outer space and on celestial bodies. Its significance will grow as our mastery of space grows, and our children will remark the wisdom of this agreement to a greater degree than the present state of our own knowledge quite permits today. . . ." (*PD*, 2/20/67, 274)

February 18: Dr. J. Robert Oppenheimer, "father of the atomic bomb" and recipient of AEC's 1963 Enrico Fermi Award for his "outstanding contributions to theoretical physics and his scientific and administrative leadership," died at age 62. (*NYT*, 2/20/67, 36M)

February 20: NASA Associate Administrator for Manned Space Flight Dr. George E. Mueller stressed importance of manned spaceflight program in speech before IEEE meeting in Washington, D.C.: This is a particularly appropriate date for us to review our manned space flight programs. It was just five years ago today that John Glenn became the first American to orbit the earth in his Mercury spacecraft, Friendship 7. Since that time, we have made noteworthy strides in the manned exploration of space. . . . In the Mercury and the Gemini Programs, we learned much about man's capabilities in space flight. . . .

"We will learn much more from the Apollo missions. By 1970, we will have provided the capability to explore space out to 250,000 miles from earth and to conduct manned operations and experiments on flights of up to two weeks duration. The Saturn I and Saturn V boosters will have injected up to 140 tons of payload per launch into near-earth orbit. The Saturn V will have sent 48 tons to the vicinity of the moon. The Apollo spacecraft will have sustained a three-man crew for two weeks in a two-compartment, maneuverable vehicle and will have landed two men on the moon and returned them with samples of lunar material, to earth. The U.S. manned space flight programs will have logged more than 500 man days in space, during which data and experience will have been acquired from approximately 100 in-flight experiments." (Text)

- Former Astronaut John H. Glenn, speaking at National Space Club luncheon in Washington, D.C., criticized the press for uninformed discussion of the Jan. 27 Apollo tragedy: "I deplore comment in the press recently regarding the accident. The press has a responsibility not to comment in areas about which it is not informed." Speaking on the fifth anniversary of his space flight in *Friendship 7*, Glenn noted that the deaths of three Apollo astronauts were part of the price man must pay in his quest for knowledge: "Man tries something and sometimes he succeeds and sometimes he fails. It is very regrettable. We hate to lose good friends, but we think they went in a good cause." (Sehlstedt, Jr., B *Sun*, 2/21/67, 8; AP, *W Post*, 2/21/67; *NYT*, 2/22/67, 50)

During week of Feb. 20: Princeton Univ. announced development of accelerator capable of raising electrically neutral atoms and molecules to energy levels never before obtained in controlled experiments. Utilizing supersonic jets of mixed gases in high-vacuum container, accelerator could generate molecular beams of neutral particles with kinetic energies of 10 or more electron volts. Particles at this energy level were necessary to many common gas-phase chemical reactions. Dr. John B. Fenn, professor who directed NSF-sponsored project, said accelerator would aid studies in meteorology and aerospace technology. (*NYT*, 2/26/67, 59)

February 21: Reps. Samuel S. Stratton (D–N.Y.) and Durward G. Hall (R–Mo.), members of House Armed Services Committee, released report charging that insufficient safeguards were at least partly responsible for deaths of three Apollo astronauts in Jan. 27 fire at KSC and two airmen in Jan. 31 fire at Brooks AFB. "Our inquiry suggests that [fire prevention] procedures, not only at Brooks, but also at Cape Kennedy, were in fact inadequate," they told newsmen. Congressmen noted what they considered to be questionable procedures in safeguarding against fires in space capsules with 100% oxygen atmosphere: (1) absence of closed-circuit television monitoring of Apollo spacecraft or Brooks chamber; (2) absence of automatic or manually activated fire extinguishing system onboard either capsule; (3) use of sponges by airmen although sponge burns up to 80 times faster in pure oxygen; and (4) use of foam couches by astronauts although foam burns 20–30 times faster in pure oxygen. The congressmen said they had begun their probe because of "an interest in the two fires as they bear on the safety of armed services personnel." Future efforts would concentrate on Brooks AFB fire. (O'Toole, *W Post*, 2/22/67, A4)

- MSC awarded Allis-Chalmers Manufacturing Co. a $3.4-million, cost-plus-fixed-fee contract for development and testing of an improved fuel cell power system for Apollo Applications (AA) program. (NASA Release 67-33)

February 22: USAF launched unidentified satellite WTR using Thor-Agena D booster; satellite reentered Mar. 11. (*Pres Rep 1967*; GSFC *SSR*, 3/15/67)

- Illustrations of the predictive ability of the contemporary engineer were made by ARC's Simulation Sciences Div. chief, George A. Rathert, Jr., at Moffett Field Scholarship Night Banquet for Bay Area Joint Engineers Council: ". . . FAA pilots were experiencing certain characteristics of the supersonic transport on simulators at the Ames Research Center 15 months before the FAA asked for bids on the design studies. You will recall Astronaut Gordon Cooper having to manually control the attitude of his space ship during a reentry retrofiring to successfully complete his mission. Engineering test pilots at Ames were studying problems of coping with manual control of a simulated spacecraft reentering with partly failed control systems more than nine years ago." (Text)
- Soviet antimissile defense system could not prevent all enemy missiles from reaching their targets, Marshal Andrey A. Grechko, acting Soviet Defense Minister, and Marshal Vasily I. Chuikov, head of Soviet civil defense program, wrote in *Izvestia* article commemorating Soviet Armed Forces Day. Their assertion contradicted statement by Gen. Pavel A. Kurochkin, head of Frunze Military Academy, at Feb. 20 news conference: "If enemy missiles fly, they will not arrive in Moscow." (Anderson, *NYT*, 2/23/67, 1)
- France would launch 330-lb Saros comsat—138 lbs heavier than ComSatCorp's INTELSAT II comsats—in 1970 from new French Guiana launch site using ELDO's Europa booster, Reuters reported. Saros would provide direct telephone, radio, and television communications between France and French-speaking countries. (Reuters, *W Post*, 2/23/67, A3)

February 23: GSFC selected Radiation, Inc., for negotiations on a $1-million, cost-plus-fixed-fee contract to develop Versatile Information Processor (VIP), an advanced telemetry system for the Nimbus D meteorological satellite. System would process, format, and store in the spacecraft scientific data obtained from experiments and monitor temperatures, pressures, power supply, and control system. (NASA Release 67–34)

February 24: Astrobee 1500 sounding rocket launched from NASA Wallops Station in third flight test carried 110-lb instrument package to 1,495-mi (3,007-km) altitude before impacting 690 mi downrange in the Atlantic. No recovery was attempted. The 11,600-lb solid-fueled rocket—most powerful sounding rocket in U.S.—transmitted 30 min of data on flight characteristics. From preliminary examination of data test appeared successful, but scientific data on cosmic noise effects were lost because rocket's spin rate was too high for proper deployment of experimental antennas. Astrobee 1500 was being developed by NASA "to provide a research rocket capable of carrying heavy scientific payloads to high altitudes, with a relative ease-of-handling. . . ." First launch attempt (April 8, 1963) failed 16 sec after launch because of a nose-fairing failure; second attempt (Oct. 21, 1964) was successful. (WS Release 67–5)

- USAF launched an unidentified satellite from WTR using Titan III–B booster; satellite reentered Mar. 6. (*Pres Rep 1967*; GSFC *SSR*, 3/15/67)
- Phoebus 1B reactor was operated for 45 min—30 min at 1,500-mw design power—in test conducted by AEC–NASA Space Nuclear Propulsion Office at NRDS, Jackass Flats, Nev. Primary purpose of test, part of Rover

program, was to study high-power-operation effects on reactor—including test conditions to be found in 5,000-mw Phoebus 2, slated for test late this year. Phoebus series would provide advanced technology for solid-core graphite rocket reactors, eventually providing 200,000–250,000 lbs of thrust. (Joint AEC–NASA Release K–46)

- NSF awarded over $28 million in grants as part of expanded program "aimed at accelerating the output of students with advanced degrees in science, mathematics and engineering." Made under NSF's graduate traineeship program, grants would enable 206 institutions to appoint 5,077 students for full-time graduate study beginning in fall 1967 and 896 graduate teaching assistants for full-time study beginning in summer 1967. (AP, *NYT*, 2/25/67, 10)
- "High priests of science"—notably space scientists—were making excessive demands on Federal budget with possible dangers to remainder of the scientific community, Ralph Lapp, nuclear science consultant, told Purdue Univ. audience. He urged "a unified science budget which reflects a 'fair share' for all scientific fields." (AP, B *Sun*, 2/25/67)

February 25: Astrobee 1500 sounding rocket was launched from Wallops Island with radiometer payload to obtain spectral observations in high frequency within the near-earth (up to 2,500 km) environment. Some of the hardware would be for eventual use on proposed Pilgrim satellite. Peak altitude was 1,475 mi (2,380 km). Rocket performance was satisfactory. Improper antenna deployment probably caused loss of most scientific data. (NASA Rpt SRL)

- NASA Administrator James E. Webb issued statement based on third interim report by Deputy Administrator Dr. Robert C. Seamans, Jr., on work of Apollo 204 Review Board. Webb observed: (1) "The risk of fire that could not be controlled or from which escape could not be made was considerably greater than was recognized when the procedures for the conduct of the test were established. Our experience with pure oxygen atmosphere included not only the successful Mercury and Gemini flights but a number of instances where a clearly positive source of ignition did not result in a fire. . . ." (2) Apollo spacecraft had been equipped with such items as Velcro adhesive pads to hold frequently used equipment and nylon netting to catch dropped objects. "While most of these [items] were constructed of low-combustion-potential material, they were not arranged as to provide barriers to the spread of a fire. Tests conducted . . . since the accident have shown that an oxygen fire in the capsule will spread along the surface of Velcro and along the edges of nylon netting much faster than through the material itself. (3) Soldered joints in piping carrying both oxygen and fluids were melted away, with resultant leakage contributing to the spread of the fire. . . ." (4) Capsule burst in such a way that flames traveled over and around astronauts' couches toward rupture. "Under these conditions, and with just a few seconds of time available, the astronauts could not reach the hatch and open it. (5) This fire indicates that a number of items related to the design and performance of the environmental control unit will require the most careful examination and may require redesign. . . ." Webb announced that Senate Committee on Aeronautical and Space Sciences would hold open hearing on Review Board's preliminary findings Feb. 27 and that House Committee on Science and Astronautics' Subcommittee on NASA Oversight would conduct full investigation of accident after completion of Board's inquiry, expected by late March.

Dr. Seamans said in his report that an "electrical malfunction" was regarded as "most likely source of ignition" of the fire but the "possibility exists that no single source will ever be pinpointed." Board was still considering chemical reactions in onboard materials, spontaneous combustion of certain materials, and possible electrical phenomena. Evidence indicated fire had had three distinct phases: originating in left side of spacecraft, it had burned several seconds unnoticed, then spread becoming more intense and causing cabin pressure to rise rapidly as atmosphere became heated. Fifteen seconds after crew first reported fire, cabin ruptured. During second phase, gases and flames flowed past astronauts' couches and through hole moving from left to right. With cabin's oxygen quickly reduced by rush of flames and gas outside, fire continued as localized flame, smoking heavily, in third phase. Dr. Seamans said Board noted: "The experience in flight and in tests prior to the accident had suggested that the probability of a spacecraft fire was low. Continued alertness to the possibility of fire had become dulled by previous ground experience and six years of successful manned missions. . . . Potential ignition sources inside the spacecraft had been treated so as to be considered safe; neither the crews nor the test and development personnel felt the risk of spacecraft fire to be high. . . ." Dr. Seamans emphasized that Board did not recommend changing pure oxygen system or planned cabin pressure for space flight, but did urge that "trade-offs between one- and two-gas atmospheres be re-evaluated . . . [and] pressurized oxygen no longer be used in prelaunch operations." It also recommended: (1) combustible materials be replaced where possible with nonflammable materials; (2) nonmetallic materials be arranged to maintain fire breaks; (3) systems for oxygen and liquid combustibles be made fire resistant; (4) full flammability tests be conducted with mockup of new configuration; (5) more rapidly and easily operated hatch be installed; and (6) on-the-pad emergency procedures be revised to recognize possibility of cabin fire. (NASA Release 67-38)

- NAS had received 923 applications for 20–30 positions in NASA's astronaut training program. Academy was expected to forward NASA a list of 60–75 recommended candidates in March for final selection. (Sehlstedt, Jr., B Sun, 2/26/67)
- Europe was weak educationally and should greatly expand institutions of higher learning to close a "technological gap" that was widening in favor of U.S., Secretary of Defense Robert S. McNamara told students at Millsaps College, Jackson, Miss. "Europe's weakness is seriously crippling its growth. It is weak in general education, it is weak in its technical education, and it is particularly weak in its managerial education." (W Post, 2/25/67, 11)
- Electrically powered cars, trucks, and buses, with limited ranges and speeds, could be produced with existing technology, and their general usage would greatly reduce air pollution, Federal Power Commission (FPC) concluded in report to Senate Commerce Committee. Committee Chairman Sen. Warren G. Magnuson (D-Wash.) said he would hold hearings in March on the report which "hopefully will produce quick response from industry." (Ayres, Jr., NYT, 2/26/67, 69)
- Quasars, which radiate as much light and energy as a galaxy, appear to have a diameter only 1% as large as galaxies, scientists from U.K.'s Royal Astronomical Observatory and Royal Radar Establishment re-

ported in *Nature*. Theory was based on measurements of radio signals of 30 quasars, obtained by using two radiotelescopes 80 mi apart in combination as if at extreme ends of single instrument. Data indicating quasars are smaller than previously believed compounded the problem of explaining their enormous quantities of energy. (*W Post*, 2/26/67)

February 26: U.S. astronauts would one day encounter "other living things" in outer space, Dr. Kurt H. Debus, Director of KSC, told Rollins College audience. "This prospect cannot be dismissed as metaphysical speculation. . . . It is much more a mathematical certainty than were the early theories expounded by scientists and philosophers whose observations and discoveries made possible many of our activities today." (UPI, *NYT*, 2/27/67, 12)

• Vice President Humphrey commissioned into service USN's $130-million Atlantic undersea test center, which he predicted might become the Cape Kennedy of oceanography. Speaking at West Palm Beach, the Vice President said: "This center has been established primarily for purposes of national security. But the knowledge we gain here will help us in other ways." It would contribute to man's knowledge of tides and provide information on untapped mineral resources and unutilized food sources. "We may learn how to divert and even prevent hurricanes," Mr. Humphrey said. "But perhaps the greatest promise of all lies in food from the sea." (AP, *NYT*, 2/27/67, 46)

February 27: Cosmos CXLIII was launched by U.S.S.R. into orbit with 302-km (188-mi) apogee, 204-km (127-mi) perigee, 89.5-min period, and 65° inclination. All equipment was functioning normally. Satellite reentered Mar. 7. (*Tech Wk*, 3/6/67, 15; GSFC *SSR*, 3/15/67)

• NASA's previous approach to preventing spacecraft fires by "preventing their ignition" had been inadequate, Dr. George E. Mueller, NASA Associate Administrator for Manned Space Flight, testified at open hearing held by Senate Committee on Aeronautical and Space Sciences on Jan. 27 Apollo tragedy. Also present were NASA Administrator James E. Webb, NASA Deputy Administrator Dr. Robert C. Seamans, Jr., and Dr. Charles A. Berry, MSC Director of Medical Research and Operations. NASA's future approach would seek to: (1) minimize possibility of ignition, recognizing "that this possibility will always exist"; (2) limit chance of fire's propagating once it has started; and (3) minimize consequences of fire to crew. Dr. Mueller outlined an extensive program of testing, redesign, and procedural changes, which included escape hatch for Apollo spacecraft that could be opened in two seconds; search for new, less flammable materials for cabin interior and spacesuits; and revised emergency procedures. All improvements and changes would be incorporated into an advanced version—Block II—of Apollo spacecraft. No more Block I spacecraft like Apollo 204 would be used for manned space flights. Dr. Mueller said no estimate of fire's eventual cost in time or money would be made until completion of Apollo 204 Review Board's investigation, but he indicated that it would be at least 8 to 10 months before next manned Apollo flight. Dr. Mueller testified that launch rehearsal during which fire had occurred had not been classified as "hazardous" because seven years of experience with Mercury and Gemini programs had never indicated such tests were particularly dangerous. Even if the test had been categorized as hazardous, he said, it would have still taken 90 sec for crew to open the double hatch and complete their egress. Crew died within 20 sec after detection of fire. Webb added that the handle for equalizing pressure, necessary precondition to opening hatch, had not been actuated. Dr. Berry said carbon monoxide had asphyxiated astronauts. He confirmed that bodies had had "some thermal burns of second and third degree" but "not of sufficient magnitude to cause death." Dr. Berry warned that "we are never going to be able to completely eliminate the risk of fire as long as oxygen is available." Dr. Mueller said "an atmosphere which will support life will also support combustion. Therefore the first answer . . . must be fire prevention in terms of strict control of both potential ignition sources and combustible materials." During the questioning session, Sen. Walter F. Mondale (D-Minn.) inquired about a 1965 report by Apollo Program Manager M/G Samuel Phillips (USAF). Dr. Mueller said that he did "not recall such a report," and when asked to supply a copy, Webb said NASA would try and identify the specific report and would make it available to the Comptroller General "under any request that the Committee . . . would make to him." (Testimony; Transcript, 3–4, 24, 41–2, 61, 70, *Apollo Accident,* Part 2, 65–66, 75, 86, 97, 108, 115, 125 ff., and 130 ff.)

- NASA Administrator James E. Webb presented the Distinguished Service Medal—NASA's highest award—to former NASA General Counsel Walter D. Sohier in special Washington, D.C., ceremony. Sohier was cited especially for his efforts on the space law treaty, patent policy, and NASA-university-industry relations. (NASA Release 67–39)

- ERC announced appointment of James M. Bayne, former chief of MSC's Facilities Program Office, as Project Construction Office chief. Bayne would supervise construction of ERC's multi-million dollar permanent site, already under way, at Kendall Square, Cambridge, Mass. (ERC Release 67–3)

February 28: U.S.S.R. launched the meteorological satellite *Cosmos CXLIV* into orbit with 635-km (395-mi) apogee, 585-km (364-mi) perigee, 96.9-min period, and 81.20° inclination. All equipment was functioning normally. (*Tech Wk,* 3/6/67, 15; *Ten Years of Space Research in the U.S.S.R.,* USS–T Trans, vol. 5, No. 5, 1967)

- President Johnson sent $149.8-million FY 1968 budget amendment to Congress which included $91 million for development of Rover nuclear-powered rocket engine. In his message, the President stated: "*Government scientists and engineers have now concluded that a much more powerful nuclear engine would have distinct advantages over rockets now in use. I am recommending that . . .* [U.S.] *move ahead with the development of a nuclear engine having a potential of 200,000 to 250,000 pounds of thrust.

  "Such an engine could be used for a new and much more powerful third stage for our Saturn V launch vehicle. Because of its high efficiency, it could: [1] permit us almost to double the weight of the present payload of the Saturn V vehicle, [2] increase our ability to maneuver spacecraft, and [3] be used in future manned landings and explorations of far distant planets. . . .*" (*PD,* 3/6/67, 334–5)

- President Johnson's FY 1968 budget recommendations marked "a turning point in the Nation's space program," NASA Administrator James E. Webb testified before the House Committee on Science and Astronautics. Webb said President Johnson had "placed before the Congress a program which says that this Nation must go forward in space and not call it quits at the end of this decade." He summarized recommendations for

(1) completion of Apollo program; (2) development of capabilities for long-duration manned space flight in Apollo Applications (AA) program; (3) development of "practical applications of our space know-how" using capabilities of AA program and "advanced unmanned systems like the ATS-4"; (4) planetary exploration through Voyager program; (5) development of Nerva II nuclear rocket engine; and (6) increase in aeronautical research efforts to deal with "emerging problems of civil as well as military aviation" [see Jan. 25].

Webb announced that NASA had established Voyager program management responsibilities "designed to make the best possible utilization of the proved capabilities" of JPL "and of our field centers and laboratories." Voyager program, since 1964 under NASA Hq. Division of Lunar and Planetary Programs, would become a separate division in OSSA with Oran W. Nicks as Director and Donald P. Hearth, Deputy Director. JPL would be responsible for surface laboratory, tracking, and mission operations systems; MSFC, for orbiting spacecraft and Saturn V launch vehicle systems; and LaRC, for landing capsule bus system.

Discussing the NASA program, Webb said: "I should be very clear on the subject of risks. In moving ahead to utilize the resources made available to us, we have had to take technical risks in the development of our equipment and in establishing our schedules. These have included risks that a particular design or line of development would not succeed in meeting the specifications; risks that schedules might not be met; and that we could not recover from a serious setback because we did not have parallel or back-up developments. But we have not knowingly accepted a higher level of risks in order to meet our manned flight schedules. In our specifications, trade-off studies, test criteria, or mission plans, we have taken no risks to the lives or safety of the astronauts that we could find a way to avoid."

Webb urged that decisions on Voyager and AA programs and on Nerva II be made this year to avoid acceleration of the "dispersal of the know-how that we have built up in government, industry, and universities that has already begun...."

Questioned by Rep. James G. Fulton (R-Pa.) on whether U.S. or U.S.S.R. led in space technology, Webb replied that U.S.S.R. would be ahead in large spacecraft until "we get the NERVA as a third stage for the Saturn V which will roughly double the capability of the Saturn V for Earth-escape missions." While U.S.S.R. may lead in planetary exploration, he said, U.S. was superior in "the very sophisticated use of microminiaturization and many new techniques of handling data." (Testimony; Transcript, 14–15, 28–9, 34; NASA Release 67–40; *NYT*, 3/1/67, 29)

- NASA awarded Carco Electronics Co. a $692,000 contract to design, fabricate, install, and check out a dynamic test stand for prototype Apollo Telescope Mount (ATM) control moment gyro system. Installation at LaRC was scheduled for completion by spring 1968. (NASA Release 67–41)
- NASA Associate Administrator for Space Science and Applications Dr. Homer E. Newell stressed importance of rocketsonde in speech at AIAA Sounding Rocket Vehicle Technology Specialist Conference in Williamsburg, Va. "There is still a lot of important research to be done between the maximum heights of balloons and the minimum heights of satellites. The sounding rocket gives us the means for carrying out such investigations. [It] is valuable for checking out equipment, testing the design

of an experiment, conducting an exploratory survey, probing for early discoveries, and in general providing an inexpensive means to lay the groundwork for more expensive satellite and space probe experiments to follow." Important uses for sounding rockets included: detection of x-ray sources; generation of artificial plasma clouds in space; solar observations; investigation of eclipses; study of atmospheric and ionospheric phenomena; and collection of micrometeoroids. Dr. Newell concluded: "We at NASA regard the sounding rocket program as a vital part of our total national space effort. It will continue to receive our strong support." (Text)

- NASA had presented monetary awards totaling $4,200 to six inventors, all employed by industrial contractors, for advancements in space technology: Charles R. Peek and Lewis E. Boodley, RCA—$1,000 for new type of interconnector for satellite solar cells; G. Richard Blair, Hughes Aircraft Co.—$1,000 for passive heat protective coating used on Surveyor spacecraft; Kenneth A. Ruddock and Robert G. Rempel, Spectra-Physics, Inc.—$1,000 for magnetometer to measure earth's magnetic field more precisely; and Paul A. Jensen, Hughes—$1,200 for low-noise antenna feed system used in all NASA Deep Space Network stations. (NASA Release 67–42)

During February: Space flights lasting one month or more should pose no new respiratory problems for astronauts, but additional research was necessary to ensure against respiratory problems that could arise from prolonged weightlessness and an artificial environment, NRC Space Science Board study concluded. Only serious effects of weightlessness on respiratory system anticipated were possibility that astronauts might inhale dust and other solid particles floating in the spacecraft and that muscles controlling the lungs might weaken because work was easier in weightless environment. Areas recommended for further study included: toleration limits for carbon dioxide in case spacecraft's absorption system malfunctioned; problems of infection during long missions; possibility of intentionally filling astronauts' lungs with a liquid to prevent collapse of blood vessels and tissues during acceleration greater than 15 g; and basic medical training and use of drugs for self-medication by crew during weightlessness. Committee also recommended efforts to improve methods to detect and extinguish spacecraft fires, and suggested that a two-gas system with nitrogen or helium be used until effective methods of fire control in 100% oxygen were developed.

Report, to be released shortly by NAS, was the product of a June 26–July 9, 1966, conference held at Woods Hole, Mass., at OART request. (NAS–NRC–NAE *News Report,* 2/67)

- Soviet Academy of Sciences, using results obtained from *Luna III* and *Zond III* probes, had handed over to International Commission on Lunar Nomenclature list of objects on other side of moon and their names—list containing 153 objects of invisible hemisphere and two areas of visible hemisphere. Names would honor great scientists of the past and renowned contemporaries from many countries of the world. For two areas of visible hemisphere, it was proposed to call region of first impact of *Luna II* (1959), the Bay of Lunik; and region of first nondestructive lunar landing of *Luna IX* apparatus (1966), Bay of Lunar Landing. (*S/F,* 2/67, 38–9)

- Initial reaction by the press to Jan. 27 flash fire in which three Apollo astronauts died:

Business Week: "After 16 space flights, in which 19 astronauts returned from their missions safely, many had forgotten the inescapable peril of exploration. No previous frontier has ever been crossed without the loss of life; it was not to be expected that space—the most perilous frontier of them all—could be conquered without sacrifice. None of this lessens the grief that every man in NASA feels for fallen comrades. But it is the nature of test pilots . . . to accept disaster not as defeat but as a call for intensified effort and dedication." (*Bus Wk*, 2/4/67)

Washington *Evening Star:* "Now that the initial shock . . . is beginning to wear off, the second guessers are . . . wondering whether we should be going to the moon at all. . . . From any rational point of view, the only thing to do is to carry on. The Apollo program necessarily will mark time until the true cause of the spacecraft disaster is located and corrected. It is conceivable that the delay could cause the Dec. 31, 1969, deadline to be missed. It is entirely possible that the Russians may get to the moon ahead of us. But so what? The world will not end nor the United States vanish, if the Russians beat us to the moon. . . . It would be senseless to stop now. Fortunately, it probably would also be impossible." (Hines, W *Star*, 2/2/67)

Editor & Publisher: "The recent disaster at Cape Kennedy indicates we are no less reluctant to face up to disaster than anyone else. Not only were our reporters not prepared for it—none of them were there for the simulated flight test—but neither were our space officials . . . [because] the full import of the tragedy was fed out piecemeal. . . . The lessons for NASA officials and for newsmen are obvious. Space travel is fraught with dangers in every test, every maneuver, every piece of equipment, but we became complacent about them because of our successes. NASA's record of dealing with the press up to now has been pretty good and if it hasn't already developed a 'disaster plan' for such accidents it soon will. News media have learned that nothing in the space program should be considered 'routine' and that front row seat should never be left unoccupied at any time." (*Editor & Publisher*, 2/11/67)

Some press reaction to the accident later became increasingly critical of NASA:

Washington *Evening Star:* "The implication of the NASA story was that no fire extinguishing apparatus or quick escape mechanism—neither of which was present on the Apollo 1—could have saved the men. The implication of the truth is that if the spacecraft had been equipped with a means of blowing open the hatch, Grissom, White and Chaffee might have had a chance. NASA's motive in attempting to distort the truth may have been to protect the families of the astronauts from undue mental anguish. But it also served the function of protecting NASA by heading off criticism of the agency . . . by deliberately misleading the press, Congress and the public." (W *Star*, 2/9/67)

New York Times: "The most basic issue . . . is not, as NASA has tended to emphasize, the source of the spark that started the fire. It is rather the question of why so much was permitted to hinge on the assumption that there would be no fire. . . .

"The answer is only secondarily, as a NASA official has suggested, that 'alertness . . . had become dulled' by previous successes. . . . More fundamental was the willingness to gamble for the sake of winning the race to the moon." (*NYT*, 2/28/67, 34)

- Congressional actions in response to Jan. 27 flash fire at KSC in which three Apollo astronauts died:

    Sen. Harrison A. Williams, Jr. (D-N.J.), urged Senate to ratify space law treaty [see Dec. 8, 1966] as a worthy tribute to the astronauts. (*CR*, 2/3/67, S1449–50; NASA LAR VI/16)

    Sen. Spessard L. Holland (D-Fla.) and Rep. Edward J. Gurney (R-Fla.) introduced joint resolution (S.J. Res. 30) to establish a Commission "to formulate plans for a memorial to astronauts who lose their lives in the line of duty in the U.S. space program." Astronauts Memorial commission would be composed of five Presidential appointees—one NASA representative, one DOD representative, and three civilians. (*CR*, 2/7/67, S1657; NASA LAR VI/17)

    Sen. Vance Hartke (D-Ind.) praised efforts of Mitchell, Ind., residents to erect a memorial to Apollo Astronaut Gus Grissom. "The memorial will preserve the memory of that record for future generations who might otherwise lose sight of the spectacular achievements of the few individuals who made the first forays into space." (*CR*, 2/16/67, S2071)

- Lift-drag space vehicle had been gaining more attention, *Space/Aeronautics* reported; reasons were said to be: "X–15 hypersonic tests, unmanned lifting-body reentry, manned subsonic glides and wind tunnel research." For manned spacecraft, atmospheric lift would put an end to worldwide recovery alerts as in Mercury/Gemini experience. (*S/A*, 2/67, 52)

March 1967

March 1: Lewis Research Center awarded two one-year award-fee contracts totaling $22,500,000—$17,665,000 to General Dynamics Corp. and $5,171,000 to Honeywell, Inc.—for management and sustaining engineering services on Centaur launch vehicle (NASA Release 67–43)
- Dr. Louis B. Arnoldi, former Command Surgeon of Air Force Logistics Command, had been named NASA Director of Occupational Medicine, succeeding Dr. Harold H. Stoddard, who resigned. (NASA Release 67–45)
- Special NATO council began study to define and analyze technology gap separating U.S. and Western Europe. Project, initiated by Italian Foreign Minister Amintore Fanfani, would reactivate cooperation within NATO and lead to cooperative programs to close the gap. (Mooney, *NYT*, 3/2/67, 12)

March 1–2: *Lunar Orbiter III* continued transmitting photos of the lunar surface [see Feb. 8–28]. On March 2—during 149th orbit—final photographic readout was interrupted when film failed to advance in the camera system. Efforts to clear the jammed film and continue readout were unsuccessful. Photographic data had been received on all but 29 of 211 frames, including six of the 12 primary Apollo landing sites and 29 of 31 secondary areas. NASA Hq. program and project officials and LaRC engineers would make an on-site analysis of telemetry time histories and engineering tapes at JPL to determine exact cause of failure. (NASA Proj Off; NASA Release 67–46)
- NASA Deputy Administrator Dr. Robert C. Seamans, Jr., testified in support of NASA's FY 1968 budget authorization before the House Committee on Science and Astronautics. He reviewed 1966 accomplishments and discussed future space-flight activities: "We completed 22 missions during 1966 successfully out of 29 attempts, or a 76% success. This record is somewhat lower than the previous year, but it is marked by qualitative advances greater than any before.

    "The FY 1968 request represents the first true post-Apollo decisions. . . . The combination of the 1971 Mariner flights and the initiation of a continuing, major Voyager program" represented effective translations into action of President's Science Advisory Committee's recommendation that planetary exploration be given a high priority in the post-Apollo period. "During FY 1968 we will be decreasing somewhat our application of manpower effort on the manned space flight programs with a considerable shift in effort from the Apollo to the Apollo Applications part of that activity. We will be increasing our total effort in space science and applications programs, particularly in the build-up of the Voyager program activity . . . [and devoting] more effort to our advanced research and technology activities. . . ."

    Emphasizing that he was "not prepared nor willing to state that a

March 1–2

manned lunar landing cannot be achieved before 1970," Dr. Seamans admitted there was a high probability that Apollo program would have to operate on a delayed schedule. "An important contributor to . . . success will be adequate resources and the flexibility to apply them when needed. . . . We cannot say today that the $2.6 billion requested for Apollo is not enough [but] we can say it is not too much. . . . It cannot be reduced and still accomplish the objectives we have defined." (Testimony; Transcript, 118)

March 2: Two Aerobee 150 sounding rockets carrying instrumented payloads were successfully launched by NASA from WSMR. First rocket carried an ultraviolet stellar spectrometer and input telescope to 97.5-mi (156.8-km) altitude to measure spectral irradiance of five early stars in 1,100 to 4,000 Å wavelength. Second, carrying detectors to observe sky in the infrared, reached 98.5-mi (158-km) altitude. Rockets and instrumentation performed satisfactorily. (NASA Rpt SRL)

- President Johnson told White House press conference that Soviet Premier Aleksey Kosygin had responded to his Jan. 27 letter: "This reply confirmed the willingness of the Soviet Government to discuss means of limiting the arms race in offensive and defensive nuclear missiles.

  "This exchange of views is expected to lead to further discussions of this subject in Moscow and with our allies. It is my hope that a means can be found to achieve constructive results. . . ." Asked if he viewed Senate passage of the consular treaty, the space treaty, the East-West trade treaty, and a nonproliferation treaty as "one movement," the President responded: "I think they are all very desirable moves in the national interest of the United States.

  ". . . we are exploring, with every means at our command, every possible way of relieving tensions in the world and promoting peace in the world." (*PD*, 3/6/67, 353–4)

- Informed sources said U.S.S.R. had resumed sending meteorological data to U.S. from an orbiting satellite following a four-month pause, AP reported. Sent over a special teletype "cold line" established in 1964 between Moscow and Washington, D.C., for exchange of weather information, data were apparently transmitted from *Cosmos CXLIV*, launched Feb. 28. National Environmental Satellite Center later confirmed that it was receiving data from *Cosmos CXLIV*. (AP, *NYT*, 3/5/67, 31; AP, B *Sun*, 3/9/67)

- GSFC selected GE Missile/Space Div. to negotiate a $1.7-million contract to develop, integrate, and test attitude control system for Nimbus D meteorological spacecraft. (GSFC Release G–11–67)

- MSFC would design and build inhouse a multiple docking adapter (Mda) for use in Apollo Applications (AA) payload cluster scheduled for launch in 1968–69 [see Jan. 26]. Preliminary designs called for a 10-ft-dia, 15-ft-long cylinder surrounded by five 36-in-dia tunnels with docking collars and sealing hatches for orbital docking. (MSFC Release 67–42)

- Purdue Univ. had approved memorials for alumni Virgil I. Grissom and Roger B. Chaffee, who died in Jan. 27 Apollo spacecraft fire. Memorials included naming two campus buildings for astronauts, providing full four-year scholarships for the two children of each man, and establishing astronauts' scholarship fund for worthy students. (AP, *NYT*, 3/3/67)

- General Dynamics Corp. Convair Div. engineers said NASA had tentatively selected its double delta wing design for a hydrogen-fueled hypersonic

aircraft of the future. Aircraft would weigh 500,000 lbs and cruise at six times the speed of sound. (AP, *NYT*, 3/2/67)
- It was essential to have a system of international law to protect health of extraterrestrial life, Soviet scientist Gennadi Zhukov told Tass in an interview. The space law treaty signed Jan. 27 was only the beginning of space law as a science, Zhukov said; additional legal controls would be needed to ban potentially harmful experiments, prevent man's thoughtless use of planetary mineral resources, protect any existing forms of extraterrestrial life, and establish permanent orbital inspection stations around earth. (Reuters, *NYT*, 3/2/67, 20)
- U.S. should develop a 5,000-mph commercial aircraft and a 8,000-mph bomber, Dr. Antonio Ferri, Director of New York Univ.'s Guggenheim Aerospace Laboratories, told AIAA luncheon meeting in Washington, D.C. "Civilization has grown with the speed of moving around. The history of civilization shows that gradually you want to get in contact with people further away." Dr. Ferri noted that jet aircraft had made Europe accessible, and the next generation of aircraft would bring other parts of the world within equally easy reach. He concluded that the technology and the "tremendous opportunities" of hypersonic aircraft were near at hand, but that necessary decisions and funds were still lacking. (Clark, *NYT*, 3/3/67, 16)
- AEC–NASA Space Nuclear Propulsion Office awarded $64.5-million contract extension through September 1967 to Aerojet Corp. to continue technology research being performed for SNPO. (AEC Proj Off)

March 3: U.S.S.R. launched *Cosmos CXLV* into orbit with 2,135-km (1,327-mi) apogee, 220-km (137-mi) perigee, 108.6-min period, and 48.4° inclination. Equipment was functioning normally. (*Tech Wk*, 3/20/67, 13)
- NASA would delay scheduled 1967 launch of second Orbiting Astronomical Observatory (OAO) until late 1968 to incorporate changes made necessary after failure of *Oao I*, NASA Associate Administrator for Space Science and Applications Dr. Homer E. Newell said in testimony before the House Committee on Science and Astronautics' Subcommittee on Space Science and Applications. Third OAO would be launched in 1969 and fourth in 1970 to continue studies in ultraviolet, x-ray, and gamma-ray spectral regions.

    Dr. Newell explained that "the intensive investigations by NASA of *Oao–I* program development, test, and operations history have resulted in design and program modifications directed toward increasing the reliability and the operating efficiency of the OAO system in order to insure the success of the second OAO missions. . . . Although the remaining three approved missions have been delayed . . . the scientific goals of these missions will be accomplished . . . [and OAO] will continue to be the backbone of the NASA space astronomy program. . . ." (Testimony)
- Successful 15-sec captive firing of ground test version of Saturn V 1st stage (S–IC) was conducted at MTF by Boeing Co., prime contractor, to prove operational readiness of new S–IC test stand and support facilities and of the Boeing test team. Stage was powered by five Rocketdyne F–1 kerosene- and liquid-oxygen-fueled engines capable of developing 7.5 million lbs thrust. (*Marshall Star*, 3/8/67, 1)
- Lockheed Aircraft Corp. was evaluating under a $65,000 NASA contract the usefulness of slush hydrogen for three proposed long-distance vehicles: Uprated Saturn I 2nd stage (S–IVB) as a lunar logistics vehicle to land

payloads or manned modules on lunar surface; cryogenic service module for manned Apollo flights; and "space tanker" to be filled with hydrogen and used for inflight orbit refueling by other space vehicles. (Lockheed Release M3367–11)

- Rep. Chet Holifield (D-Calif.) challenged President Johnson's budget priorities and attacked Administration plans to develop a nuclear rocket system for interplanetary space exploration. He told newsmen that it would be a much greater tragedy to limit school aid than to have a Russian cosmonaut welcome U.S. astronauts to Mars. ". . . We can't be like the Roman Empire. They gave their people circuses when they needed bread. And that empire disappeared." (UPI, *W Post*, 3/4/67, A2)

- Washington *Evening Star* writer William Hines criticized Apollo 204 Review Board's failure to consult Walter C. Williams, operating head of Project Mercury from 1958–63, in investigation of Jan. 27 flash fire. "NASA has no ready answer . . . to questions about why Williams' name was omitted from the consultant panel. His standing in the space business is such that it could hardly have been due to chance oversight." Hines said Williams had left NASA in April 1964, reportedly after "disagreement with superiors over management policies. His relations with headquarters officials had been strained for six months prior to his departure, ever since a final report on Project Mercury was issued . . . [in which] Williams castigated industrial contractors, saying they had given poor attention to quality control on hardware for manned missions. . . ." (Hines, W *Star*, 3/8/67, A5)

- AFSC was conducting five-month study of ways to convert excess Minuteman I ICBM's to launch vehicles, thereby saving millions of dollars. Results of study, to be completed by May, would determine feasibility of flight-test program in which two converted Minuteman boosters would launch instrumented payloads into circular orbits from ETR. Preliminary studies had already shown that converted Minuteman booster could place large payloads into 575-mi circular orbit, or smaller payloads into an elliptical orbit 172 mi by 57,500 mi. (AFSC Release 20–67)

March 4: Press should find ways to better meet readers' needs for science news, James B. Lemert, Assistant Professor of Journalism at Southern Illinois Univ., suggested in *Editor & Publisher:* "There has been an overwhelming explosion of scientific knowledge, and the press has an increasingly important challenge . . . to report these developments so that the people who wouldn't know a cathode from a microbe—and who shouldn't be expected to know—can understand.

"Public understanding is important, because most of the money for science comes from public funds. Further, the long-term social and political effects of scientific achievements should be watched closely. Sometimes they are not evident until it is almost too late." (*Editor & Publisher*, 3/4/67)

March 5: USAF SV–5D lifting body was successfully launched from WTR by Atlas booster on reentry mission and telemetered "excellent" performance data to earth before losing its flotation gear and sinking in the Pacific. Released by Atlas booster at desired orbital altitude, the 900-lb vehicle became first spacecraft to perform preprogrammed cross-range maneuvers when it glided away from a straight line and then returned to original flight path. Purpose of flight—second of four in USAF's Precision Recovery Including Maneuvering Entry (Prime) program—was to further test theory that a vehicle could be designed to operate like an orbital spacecraft and be maneuvered like an aircraft when it reentered and traveled through the atmosphere. First SV-5D vehicle, launched Dec. 21, 1966, performed successfully but was lost when parachute malfunctioned. (AFSSD Release 77.67; *Av Wk*, 3/13/67, 23; Miles, *LA Times*, 3/6/67)

- Senate Committee on Aeronautical and Space Sciences staff report released by Committee Chairman Sen. Clinton P. Anderson (D-N.Mex.) disputed theory that any nation would have a military advantage with orbiting nuclear weapons. Report, prepared "to provide information for . . . legislative evaluation" on the space law treaty [see Dec. 8, 1966] submitted for Senate approval Feb. 7, noted: "Majority military opinion holds that orbital bombardment is not an effective strategic weapon as compared to land-based ballistic missiles. An effort to use space-based nuclear weapons would have the effect of a strategic warning, thus placing an aggressor in the position of being open to retaliation by strategic weapons." (Text)
- Two- to 12-day delivery time for letters between U.S. and Vietnam could be appreciably shortened with Satellite Mail (S-Mail) system, National Environmental Satellite Center's J. Gordon Vaeth suggested in letter to *New York Times* editor. S-Mail letters would be electronically scanned and converted to radio signals that would be sent by mid-Pacific comsat relay to Vietnam where they would be reconverted and reproduced on a recorder. (*NYT*, 3/5/67, E9)

March 6: Distance between earth and moon had been calculated at 250,000 mi with 20,500-mi variation by JPL scientists Dr. J. Derral Mulholland and William L. Sjogren. New lunar values, which had an accuracy of less than 50 ft compared to previous discrepancy of one mile, were calculated from theoretical work by astronomer Dr. W. J. Eckert from spaceflight data; verified by ranging measurements of *Lunar Orbiter I* and *II* spacecraft; and used to revise a 50-yr table of lunar motion required for plotting spacecraft trajectories, determining launch times, and scientific research. (JPL Release 436)

- Florida radio report that Apollo spacecraft 017 had so many manufacturing defects that NASA was returning it to North American Aviation, Inc.'s (NAA) Downey, Calif., plant was denied by NASA. KSC spokesman acknowledged that about 1,000 "squawks" had been found in spacecraft—40 of which would be corrected—but said that at the present time there was no indication it would be returned to NAA. Spacecraft was scheduled for unmanned suborbital flight in May. (*W Star*, 3/17/67, A4)
- NASC Executive Secretary Dr. Edward C. Welsh stressed at AIAA meeting in Baltimore the importance of space exploration as a contributor to peace: ". . . space activities can be a substitute for aggression, a bridge to understanding between the nations, and a major tool of arms control and disarmament. Once man accepts the challenge of exploring the great universe around us, he can more readily rise above the terrifying grip of parochial hatreds and fears. Space exploration helps satisfy man's unquenchable thirst for knowledge—a basic characteristic which sets him apart from other creatures. Compared to the business of coercing and killing his own kind, the challenge posed by the prospect of crossing the established frontiers of the normal physical world is not only infinitely to be preferred, but its potential is limitless." (Text)
- Advanced solid rocket motor capable of being fired or pulsed on command had been developed at the Air Force Rocket Propulsion Laboratory

(AFRPL) by Lockheed Propulsion Co. under two-year, $2-million R&D program. Motor had been selected to power USAF's short-range attack missile (SRAM). (AFRPL Release)

- DC-8 Super 63, world's largest commercial jet aircraft, was displayed by Douglas Aircraft Co. in Long Beach, Calif. Scheduled to fly within 30 days, aircraft had 187.4-ft-long fuselage, 148.4-ft wingspan, 251-passenger seating capacity, and 4,600-mi range. (DAC Release 67-36)
- *Milwaukee Journal* editorial on space law treaty signed Jan. 27: "What's left unsaid in the new space treaty may be as important to world peace as what the treaty says. The treaty . . . prohibits stationing nuclear or other weapons of mass destruction in space . . . [but] makes no reference to use of spacecraft for military reconnaissance.

    "Spy satellites already in orbit can detect activity on earth in unbelieving detail. Even more sophisticated satellites are being developed. . . . Ground inspection, a barrier to nuclear control agreements, is therefore less important and airplane reconnaissance, which has strained Soviet-American relations, is unnecessary. . . .

    "Reconnaissance satellites orbiting above may thus reduce the chance of war below." (*MJ*, 3/6/67, 10)
- Washington *Evening Star* commented on Senate Committee on Aeronautical and Space Sciences' open hearing Feb. 27 on the Apollo fire: "The nation's space officials evidently have managed to keep their heads in the midst of the emotional chaos that followed the tragic accident at Cape Kennedy. They have resisted pressures from semi-informed sources condemning the pure oxygen environment that supported all Mercury and Gemini astronauts in space. And there is mounting evidence that the arbitrary 1970 goal for a manned lunar landing is no longer a major consideration in the space program . . . [and] that the space planners are determined that they will no longer be prompted to act because of any arbitrary timetable or out of a competitive desire to beat the Soviets to the moon.

    "This is as it should be. This nation should move outward into space with all deliberate speed. And right now, the emphasis should be heavy on that qualifying word." (W *Star*, 3/6/67, A11)
- New York Univ. School of Engineering and Science announced establishment of a new chair in aerospace science endowed by $500,000 grant from the Vincent Astor Foundation. Dr. Antonio Ferri, Director of New York Univ.'s Guggenheim Aerospace Laboratories, was named to the chair. (*NYT*, 3/7/67, 29)
- Australia might agree to operate first joint tracking station with U.S.S.R. in Western Australia if U.S. approved, *Newsweek* speculated. U.S. currently operated three stations in Australia, and two more were under construction; U.S.S.R. had to maintain contact with its satellites via picket ships in the Pacific. (*Newsweek*, 3/6/67, 18)
- 238,154 airmen certificates—a new annual record and a 32% increase over 1965 total—were issued by FAA in 1966. (FAA Release 67-18)

March 6-7: Merits of nuclear rocket propulsion were discussed by William Hines in series of two articles in Washington *Evening Star*: "If Congress approves the [Nerva] project . . . the nation's space horizons will be broadened considerably in the 1980's. Manned flights to Mars and Venus and unmanned missions to even more distant planets could be a reality 15 years hence." Although U.S. had not adopted a manned planetary landing program, Hines explained another advantage of a nuclear rocket—it could be used for continued exploration of the moon after first Apollo landings. "Using a non-nuclear Saturn 5, astronauts are restricted to landing areas in a narrow strip along the moon's equator. A nuclear upper stage would permit a manned spacecraft to land anywhere on the surface of the moon." Other possible uses included manned flyby missions to Mars and Venus; unmanned flights of large payloads to Jupiter, Saturn, and Mercury, or within a few million miles of sun; and manned orbital flights to Mars and Venus. (Hines, W Star, 3/6/67, 1; 3/7/67)

March 7: Secretary of State Dean Rusk, accompanied by U.S. Ambassador to the U.N. Arthur J. Goldberg, urged prompt Senate ratification of the space law treaty [see Dec. 8, 1966] at Senate Foreign Relations Committee hearings: "In my view, the interests and security of the United States would be advanced by its ratification. . . . The conclusion of this Treaty, we feel, augurs well for the possibility of finding areas of common interest and agreement with the Soviet Union on other significant issues—especially in those fields in which there are genuine common interests affecting all mankind."

In response to charges by Sen. Albert Gore (D-Tenn.) that some of the provisions were too "vague and even fuzzy," Ambassador Goldberg explained that Senate reservations were "unnecessary" because the treaty was "so crystal clear." Reservations entered by the Senate, he warned, might invite other countries to also enter reservations and thus "impair the value of the treaty." (Testimony; Finney, *NYT*, 3/8/67, 20)

- Impact of Jan. 27 Apollo accident on Apollo program schedule was discussed by NASA Associate Administrator for Manned Space Flight Dr. George E. Mueller, testifying before the House Committee on Science and Astronautics: "We will include in our schedules the time necessary to conduct a thorough program of reverification and requalification of changes . . . based on what we have learned from the AS-204 accident. . . . Our policy of full qualification prior to manned flight will remain unchanged.

"I do believe we still have a reasonable possibility of meeting the major milestones for the Apollo program which were established in 1963. In particular, although the probability is lowered, I believe we will be able to land men on the moon and return them safely to earth before 1970."

Asked about a manned lunar landing in 1968, NASA Administrator James E. Webb, who accompanied Dr. Mueller, replied that it was "not impossible, but almost impossible." (Testimony; Transcript, 286)

- Vesta rocket launched by France from Hammaguir Range carried macaque monkey Martine to 233-km (145-mi) altitude in experiment to measure effects of extreme altitude and weightlessness on reactions of the monkey, who had been trained to push a button in response to a red light, for possible application to problems of manned space flight. Monkey was successfully recovered when she landed by parachute 98 mi from launch site after 21-min flight and was reported to be in good health. (*NYT*, 3/8/67, 24; W Star, 3/7/67)

- Secretary of Agriculture Orville L. Freeman and six other Government officials visited MSFC. NASA Administrator James E. Webb had invited cabinet members and representatives of other agencies to visit NASA centers to become more familiar with NASA activities. (MSFC Release 67-45)

March 7

- U.S.S.R. would fly a prototype of her 1,550-mph Tu-144 supersonic transport to the Paris International Air Show May 26, according to U.S. intelligence sources, UPI reported. If reports proved accurate, U.S.S.R. would become first country to fly a supersonic transport—about one year before scheduled flight of Anglo-French Concorde and several years before U.S. SST. (UPI, *NYT*, 3/7/67, 69)
- Canadian government should establish a national space agency and develop an extensive space program to avoid losing its control over domestic communications to U.S., report to new Science Council of Canada by a four-man panel of scientists and engineers concluded. Report asserted satellite communications were best means of binding Canada's five time zones and bringing advances to isolated northern communities. It advocated that Canada "make or buy" her own satellite launching facilities; centralize her fragmented space program for coordinated planning; renegotiate cooperative agreement for sounding rocket launches with U.S. so that Canada could gradually assume complete control of project; and set up a nationwide comsat system within five years. (Walz, *NYT*, 3/8/67, 21)
- United Aircraft Corp. vice president and chief scientist Perry W. Pratt received the American Society of Mechanical Engineers' Gas Turbine Award for "his distinguished leadership and technical contributions in the development of aircraft turbojet and turbofan engines used widely throughout the world." (United Release)

March 8: NASA's *Oso III* (OSO E) Orbiting Solar Observatory was successfully launched from ETR with three-stage Thor-Delta booster to study the sun and its influence on earth's atmosphere. Orbital parameters: apogee, 354 mi (570 km); perigee, 336 mi (540 km); period, 96 min; and inclination, 33°. Third of eight spacecraft in NASA's OSO program to provide direct observation of the sun, *Oso III* weighed 627 lbs, carried nine experiments, and had two main sections: the wheel (lower) section provided stability by gyroscopic spinning and housed the telemetry-command equipment, batteries, electronic controls, gas spin-control arms, and seven experiment packages; the sail (upper) section was oriented toward the sun and contained solar cells and two solar pointing experiments.

Primary purpose of *Oso III* was to obtain high-resolution spectral data in the range 8Å–1300Å from pointed solar experiments during major portion of one solar rotation and adequate operational support of spacecraft subsystems to carry out acquisition of these scientific data. Spacecraft's experiments were designed to continue and extend work of *Oso I* (launched March 7, 1963) and *Oso II* (launched Feb. 3, 1965) spacecraft by collecting data on solar x-rays, gamma rays, ultraviolet radiation, and other solar activity. Project management officials at GSFC tested or turned on all nine experiments and reported that satellite spin rate, power level, charge rate, and temperatures were normal. Both tape recorders were operating. Last experiment, the high-voltage portion of GSFC's x-ray spectrometer, was successfully turned on March 14. (NASA Proj Off; NASA Releases 67-32, 67-52)

- First contract for purchase from NASA of launch services for a foreign satellite was signed by NASA and ESRO in accordance with Dec. 30, 1966, Memorandum of Understanding on reimbursable launchings. Under contract, ESRO would purchase a Delta launch vehicle from NASA to launch its Heos-A interplanetary physics satellite from ETR in late 1968. The

March 8: NASA's *Oso III* being prepared for spin balance tests before its successful launch into orbit to study the sun and solar influence on the earth's atmosphere.

230-lb satellite, carrying eight experiments prepared by 10 scientific groups in six European countries, would be launched into 120,000-mi-altitude orbit to investigate interplanetary magnetic field and study solar- and cosmic-ray particles outside the magnetosphere. NASA would be reimbursed about $4 million. (NASA Release 67–48)

- NASA Nike-Tomahawk sounding rocket was launched from Churchill Research Range, Canada, and carried GSFC-instrumented payload to 141-mi (227-km) altitude to obtain data on charged particle fluxes and investigate distribution of electric fields in ionosphere during auroral displays. Rocket performance was satisfactory, but only partial data were received from electric field experiment. (NASA Rpt SRL)
- The House adopted H.R. 312 authorizing the Committee on Science and Astronautics, during the 90th Congress, to conduct studies, investigations, and inquiries on astronautical, aeronautical, and other scientific R&D. Rep. James G. Fulton (R-Pa.) urged House to authorize an Inspector General in NASA to evaluate complaints and control launch operations. (NASA LAR VI/30)
- NASA Associate Administrator for Space Science and Applications Dr. Homer E. Newell, accompanied by his deputy Edgar M. Cortright and OSSA Director of Space Applications Programs Leonard Jaffe, stressed

importance of space science in testimony before the House Committee on Science and Astronautics. "The space research effort has been abundantly fruitful in answering first and second generation questions about our space environment, and in turning up a whole new generation of fundamental and important questions, and potentially fruitful practical applications. To answer these new questions, and to continue advancing in this important field, it is time to select new missions to replace old ones."

Past OSSA successes and proposed 1968 program were outlined by Cortright: "During the past two years (1965 and 1966) 23 of the 28 OSSA missions were successful, including 7 out of 8 deep space missions to the Moon and into interplanetary space. In addition, we successfully launched 6 out of 6 payloads for other organizations. Further, the Atlas Centaur development program was successfully completed and Centaur was used to launch Surveyors I and II. The 1966 series of dramatic successes with second and third generation automated spacecraft demonstrated striking advances in the utility, durability, versatility, and effectiveness of this equipment."

The 1968 program—including Voyager, Mariner Mars '71, Sunblazer, Nimbus (E and F), ATS, and Voice Broadcast Satellite—had been carefully planned to meet U.S.'s most urgent minimum future needs, Cortright stressed. ". . . this is a critical year in the development of our Country's program to explore space and to apply space technology to the benefit of man. . . . Those new missions which we have selected to pursue . . . constitute the most important of a long list of exciting opportunities for space exploration and practical application." (Testimony; Transcript)

March 8-9: First of five two-day symposiums for Federal supervisors on equal opportunity problems was held at ARC. Part of drive in San Francisco area against minority discrimination within the Executive Branch of the Government, symposium was sponsored by ARC; Palo Alto VA Hospital: Menlo Park USGS; Moffett Field Naval Air Station; and Sunnyvale Naval Plant Representative's Office. (ARC Release 67–2)

March 9: President Johnson told White House news conference he was "very hopeful" of landing men on the moon by 1970, but did not think "there is any guarantee that we will at all." The President noted that a manned lunar landing was "a very difficult undertaking. I think it has been a very close question since the original target date was set." (PD, 3/13/67, 419)

- Australian government announced it would begin new search for NASA's *Biosatellite I*, carrying more than 10 million tiny living organisms, believed to have landed in western Australia Feb. 15. Initial search had been canceled by NASA Feb. 22. (Reuters, C Trib, 3/10/67)
- Report on spaceflight emergencies and spaceflight safety, released by House Committee on Science and Astronautics' Subcommittee on NASA Oversight, made several specific recommendations: (1) NASA and USAF continue to devote intensive study effort to spaceflight safety and report periodically to the committee; (2) NASA and USAF establish joint working-level committee to eliminate duplication of effort, assure compatibility in equipment features, conduct joint accident reviews, and achieve a coordinated effort; (3) each agency establish a separate flight safety group to monitor system designs for specific missions to ensure close attention to the problem of flight safety; (4) development

of possible space rescue or escape capabilities be considered in the design and development of all future manned spacecraft; and (5) NASA prepare a long-range "plan for meeting future spaceflight emergencies"—including program identification and long-range funding requirements—by the end of 1967. (Text)

- NASA Associate Administrator for Advanced Research and Technology Dr. Mac C. Adams outlined NASA's supporting role in SST development at hearings before the House Committee on Science and Astronautics: "In 1965 NASA provided the FAA with the services of 50 technical people for six weeks for the evaluation of industry-proposed airframe and engine designs. During final evaluations last fall, 75 professionals were provided for two months and two major facilities at Langley and Ames were operated 10 hours a day and 6 days a week during September. We are assisting FAA contractors by making available to them facilities such as wind tunnels and simulators and have helped solve specific problems growing out of detail design. The contractors used about 4100 hours of NASA wind tunnel time for each of the past two years."

  NASA was carrying a major research responsibility as a member of a Federal team on the SST sonic boom program—including definition of sonic boom signatures and effects of sonic boom on structures, Dr. Adams said, and was also continuing its own study of materials suitable for SST structures. "In addition to supporting the FAA on the present SST, we are continuing our general research program aimed at an improved second-generation supersonic transport." (Testimony)

- Development of a military base in space or on the moon would be the least expensive way to protect U.S. cities, Rep. William Jennings Bryan Dorn (D-S.C.) suggested in speech on the House floor. He said that the refinement of U.S. capabilities to destroy, on a moment's notice, any potential enemy anywhere in the world would be far superior to development of an antimissile missile system for the protection of selected cities only. In addition, stations in space that could control the weather and current of the sea might well "deter a madman dictator" and a spacecraft capable of raining atomic warheads could ensure peace. (*CR*, 3/9/67, H2422-3)

- Rep. Charles E. Goodell (R-N.Y.) introduced legislation to House authorizing the President to "designate a day in January of each year as a day of remembrance" for Apollo Astronauts Virgil I. Grissom, Edward H. White II, and Roger B. Chaffee, who died in Jan. 27 fire at KSC. (*CR*, 3/9/67, H2460)

- *Wall Street Journal* editorial challenged theory presented in RAND Corp.-Brookings Institution publication, *Technology, Economic Growth and Public Policy*, that massive Federal effort was necessary to supplement U.S. technology development "in nondefense areas, on projects that will add to the nation's store of industrial technology and further spur economic growth. . . .

  "Improved technology is vital to the nation, and Federal help sometimes is either desirable or unavoidable. But history indicates that the Government's efforts usually should come, not as a matter of bureaucratic routine, but rather as a last resort." (*WSJ*, 3/9/67)

March 9–13: Series of four NASA Nike-Apache sounding rockets launched from Thumba Equatorial Rocket Launching Station (TERLS) carried India's Physical Research Laboratory experiment to 114-mi (183-km) altitude to measure atmosphere winds above 62 mi (100 km) using sodium vapor payload, and electron density using Langmuir probe. In two cases no wind data were collected due to nonejection of sodium vapor; Langmuir probe experiment performed satisfactorily. (NASA Rpt SRL)

March 10: *Cosmos CXLVI* was launched by U.S.S.R. into orbit with 310-km (193-mi) apogee, 190-km (118-mi) perigee, 89.2-min period, and 51.5° inclination. Equipment functioned normally. Satellite reentered Mar. 18. (*Tech Wk*, 3/20/67, 13; GSFC *SSR*, 3/31/67)

- NASA had awarded Western Electric Co. a $1,453,000 contract for operation of ground services at KSC connected with Delta launch vehicle guidance system. (NASA Release 67-53)
- Director of Laboratories (DOL) agency was established within AFSC Hq. "to provide policy and technical direction" to all phases of its eight laboratories' programs and monitor their operations. (AFSC Release 50.67)

March 11: NASA officials dedicated at Corpus Christi, Tex., a 30-ft-dia dish antenna which would be used to receive signals from Apollo spacecraft. Operational responsibility for Corpus Christi tracking station, previously used for Gemini support, had been transferred from DOD to NASA in 1966. (UPI, *NYT*, 3/13/67, 60)

- The Vatican had authorized the Roman Catholic apostolic delegate in U.K. to sign the International Treaty on Peaceful Uses of Outer Space, AP reported. Unanimously approved by the U.N. General Assembly Dec. 8, 1966, treaty was signed simultaneously by representatives of 62 nations Jan. 27 in Washington, D.C., London, and Moscow, and was kept open for additional signatures in those capitals. (AP, *P Inq*, 3/12/67)

March 12: Apollo program was six months behind schedule because of Jan. 27 accident, but time could be made up later, MSFC Director Dr. Wernher von Braun told Stuttgart, West Germany, audience. In Stuttgart to receive gold Boelsche medal for scientific achievement, Dr. von Braun said: "We have in no way lost our chance to fulfill the mission given to us: the landing of a man on the moon in this decade. I admit that the time has become tight. If further snags develop, this decade may run out . . . [but] we are firmly resolved not to take any undue risks." (AP, B *Sun*, 3/13/67; UPI, *C Trib*, 3/13/67)

- Technology gap was beginning to strain U.S.-European relations, Henry R. Lieberman wrote in the *New York Times*. Western Europeans, he said, expressed fear that "a widening gulf threatens to leave them further and further behind as the United States assumes command of the future by its grip on the 'high technologies'—computers, microelectronics, space, communications, nuclear energy, aircraft."

  Secretary of Defense Robert S. McNamara had suggested that Europeans' complaint "is that we are so surpassing them in industrial development that we will eventually create a technological imperialism."

  Noting that Europeans' frustration and resentment might cause increased nationalism, Lieberman cited recent warning by Vice President Humphrey: "Unless we are careful, our concept of the Atlantic partnership can be eroded by the fear and concern about the power of the American capital and technology." (Lieberman, *NYT*, 3/12/67, 1)

- U.S.S.R. launched *Cosmos CXLVII* into orbit with 317-km (197-mi) apogee, 198-km (123-mi) perigee, 89.5-min period, and 65° inclination. Equipment functioned normally. Satellite reentered Mar. 21 (UPI, *NYT*, 3/14/67, 3; GSFC *SSR*, 3/31/67)

- At the insistence of Sen. Albert Gore (D-Tenn.), the Senate Foreign Relations Committee agreed to qualify in its forthcoming report its interpretation of two clauses in the space law treaty [see March 7]. One clause questioned said benefits of space exploration "shall be the province of all mankind"; the other said that each nation would be liable for damages to another caused by its space vehicles. Committee specified that (1) nothing would "diminish or alter the right of the United States to determine how it shares the benefits and results of its space activities"; and (2) damage "pertains only to physical, non-electronic damage." Senator Gore had suggested that if there were two competing U.S. and Soviet space communications networks, one nation might sue the other for damage from electronic interference of one system on the other. (*W Post*, 3/14/67, A7; Finney, *NYT*, 3/14/67, 20)
- NASA was negotiating nine-month, $275,000 contract with Planning Research Corp. to identify, analyze, and evaluate potential economic returns from possible space station activities in mid-1970's. Earth-oriented application areas such as natural resources, meteorology, and communications would be considered as they might affect both the national and world economy. Univ. of Michigan would be major subcontractor to Planning Research in technical feasibility of remote sensing. Contract would be managed by OMSF Advanced Manned Missions Program. (NASA Release 67-54)
- NASA had agreed to fly four DOD experiments on Apollo Applications (AA) missions to support USAF's Manned Orbiting Laboratory (MOL), *Aviation Week* reported; more would be added. Experiments selected to date included: study of an inflatable elastic air lock; use of alternate restraints to determine standard workshop technique in weightlessness; evaluation of suit donning and sleep stations; and integration of multipurpose equipment maintenance. (NASA Proj Off; *Av Wk*, 3/13/67, 18)
- Advanced Defense Satellite Communications Project (ADSCP) currently being revised by DOD was expected to reach capacity of 100–200 voice grade circuits, according to Secretary of Defense Robert S. McNamara, *Aviation Week* reported. In letter to Rep. Chet Holifield (D-Calif.) McNamara said that capacity expansion was being made in accordance with recommendations by House Government Operations Committee's Subcommittee on Military Operations. (Committee Off.; *Av Wk*, 3/13/67, 12)
- *Technology Week* editorial criticized *Los Angeles Times* editor Marvin Miles' accusation that Jan. 27 Apollo accident was caused in large part by NASA's shortsightedness and that NASA was trying to hide its negligence.

    "Mr. Miles discusses the hatch hazard and NASA's failure to provide an explosive hatch and its failure to develop a fire suppression system. Then comes the ringing finale . . . 'The nation should be told the whole truth.'

    "It is our impression that the agency is trying valiantly to come up with just such information and we don't understand the implication of a cover-up."

    Editorial noted that in the past NASA had often been criticized for being too stringent and that despite all precautions, "something somewhere went wrong. . . .

    "We think the agency's standard of conduct has been high. Its statement before Congress was a frank and honest appraisal of the fact that, despite very strong efforts, somehow it had failed to take the steps to protect against the unlikely chain of circumstances that led to disaster." (Coughlin, *Tech Wk*, 3/13/67, 50)

- Dr. Marshall L. Rosenbluth, Univ. of California at San Diego, was selected to receive Princeton Univ. Institute for Advanced Study's $5,000 Albert Einstein Award for 1967 for his "outstanding contributions in the area of plasma physics." Selection of Dr. Rosenbluth, who helped develop the hydrogen bomb, was announced by Dr. Carl Kaysen, Director of the Institute. (AP, *W Post*, 3/14/67, A9)
- Final report of New England Assembly on Nuclear Proliferation endorsed a nonproliferation treaty as "a modest but progressive stride toward nuclear control," and urged that U.S. cancel Project Plowshare program of peaceful nuclear explosions, the *Washington Post* reported. Assembly concluded: "No present or immediate economic gains are apparent that are worth the political costs. We recommend that implementation of this program be stopped and not resumed unless its economic value is proven and suitable international controls agreed to." (McElheny, *W Post*, 3/13/67, A8)

March 14: Lunar Orbiter III photos of *Surveyor I* resting on the moon—first pictures ever taken of an identifiable outline of a man-made object that had been landed on the moon—were released by NASA. Photos were taken Feb. 22 at 30-mi altitude from slant range of 82 mi. (*Marshall Star*, 3/31/67, 2)

- NASA Aerobee 150A sounding rocket launched from NASA Wallops Station carried 190-lb payload with AFCRL solar radiation experiment to 128-mi (206-km) altitude and impacted 60 mi downrange in the Atlantic. Protective nose cone was ejected during ascent, but payload did not separate from rocket. NASA later reported that desired results were not fully achieved because of partial failure in the spectrometer.

  Prime mission objective was simultaneous measurement of solar extreme ultraviolet radiation (EUV) in upper atmosphere with Aerobee's photoelectric spectrometer and similar instrument on NASA's *Oso III* Orbiting Solar Observatory, launched from ETR March 8. Data from two would be correlated to verify satisfactory operation of *Oso III*'s instrument. Secondary experiment on Aerobee 150A involved measurements of ion and electron concentrations and changes in radiation along rocket's trajectory. (NASA Release 67-56; WS Release 67-7)
- President Johnson presented National Space Club's 1967 Goddard Memorial Trophy to Rep. George P. Miller (D–Calif.), Chairman of House Committee on Science and Astronautics, in special White House ceremony. The Trophy, founded in 1958 to recognize outstanding national space leadership, was presented to President Johnson in 1966. (Off of Rep. Miller; *CR*, 3/9/67, H2408-9)
- NASA–Smithsonian Institution agreement providing for transfer of NASA's historical artifacts to the Smithsonian for display in National Air and Space Museum became effective. Agreement also specified that Smithsonian would establish a Committee on NASA Artifacts, which would include the NASA Administrator or his designee, to make curatorial decisions on aeronautical and space artifacts "with regard to their significance, preservation, and ultimate disposition." (NMI 1052.85)
- NASA Associate Administrator for Manned Space Flight Dr. George E. Mueller, testifying before House Committee on Science and Astronautics' Subcommittee on Manned Space Flight, was asked about recent press reports that the Apollo program and the Apollo spacecraft had

serious deficiencies. "I don't recall the Apollo program in an overall sense as being in difficulty," Mueller replied. "I sincerely believe that the basic Apollo spacecraft design is sound and that the reason we can accommodate the changes that we are anticipating within our budget is that the design is basically sound and the changes that are required are going to be relatively nominal." Dr. Mueller denied reports that NASA was considering changing from North American Aviation, Inc., as prime Apollo spacecraft contractor. (Transcript, 32, 38)

- Dr. Mac C. Adams, NASA Associate Administrator for Advanced Research and Technology, and his Deputy Dr. Alfred J. Eggers testified on NASA FY 1968 authorization bill before House Committee on Science and Astronautics' Subcommittee on Advanced Research and Technology. U.S. space program was entering a period when NASA would utilize existing capabilities, Dr. Adams said, and it was therefore particularly important for OART to "emphasize new technology so that new systems will be available for the next major thrust in aeronautics and space." He mentioned highlights of the program to develop new technology: (1) Nerva nuclear rocket engine; (2) experimental research related to hypersonic propulsion; (3) advanced cryogenic engine; (4) improved life-support systems; and (5) increased effort in electronics problems because "electronics systems may well be the limiting factor in our future exploration."

Dr. Eggers noted that R&D budget request in aeronautics for FY 1968 nearly doubled that for 1967 and said the expansion reflected "recognition of the rapidly growing importance of air transportation and the need for increased R&D to provide required advances in the associated technology. Aeronautics is already the order of a $24 billion a year industry and it is netting the nation better than $1 billion per annum in favorable balance of payments. Accordingly, our proposed major strengthening of aeronautics R&D will be to the great commercial as well as military benefit of the nation in the future." (Transcript, 8, 13; Testimony)

- Acting Under Secretary of Commerce Dr. J. Herbert Hollomon, in testimony before Senate Commerce Committee and Committee on Public Works' Subcommittee on Air and Water Pollution, recommended against direct Federal support of an electric car "at this time." He urged instead a program to reward the "best designs of low-pollutant vehicles by basing Government purchase on performance competition. . . . If electric vehicles can compete successfully . . . a market could be made available on which production facilities and schedules could be based. . . ." (Testimony, 9, 11)

March 14–15: AAS–NSC Fifth Goddard Memorial Symposium, "The Voyage to the Planets," was held in Washington, D.C.

Arthur C. Clarke, author and past president of British Interplanetary Society, chided critics of space program expenditures because "'there is so much to do on earth.' . . . There was plenty to do in Europe when Columbus left—there's *still* plenty to do there. But the opening up of the new world did more to revive the stagnant European culture and economy than an internal action could possibly have done.

". . . I welcome the so-called space race, despite the multiplication of cost and human effort that it involves. If we weren't racing, we wouldn't be moving quickly enough. . . ." (Text)

Mariner spacecraft would be increasingly useful in returning scientific data from the planets, JPL Director Dr. William H. Pickering predicted. He noted launch opportunities during 1970's in which Mariner-type spacecraft could be flown to more than one planet in series of gravity-assist missions and said 1970–6 period offered several favorable opportunities for closeup study of the comets by Mariner-type spacecraft carrying up to 150 lbs of scientific instruments. (Text)

GSFC Director Dr. John F. Clark, discussing possible first mission configuration for a Galactic Jupiter probe, said an initial flight could probably be launched during a 20-day launch window centered on first week of March 1972. Planetary encounter would occur between 17 and 20 mos after launch. (Text)

Sterilization of lander and surface laboratory constituted "the most difficult technological task" of NASA's Voyager program to explore Mars and Venus with automated spacecraft during 1970's, Edgar M. Cortright, NASA Deputy Associate Administrator for Space Science and Applications, told the meeting. The most difficult scientific task was developing an effective integrated set of biological experiments. (Text)

Exploration of the planets would contribute to, not detract from, man's ability to meet problems on earth, Dr. Charles Sheldon II, Library of Congress, suggested. "Also one has the right to hope that even if elements of past history repeat in new forms, mankind with a shrinking world and shrinking solar system, and with a greater awareness of his place within the scheme of things will divert his restless energies and struggles to overcoming the space frontiers rather than wasting his talents and his treasures in fratricidal war. . . ." (Text)

U.S. had achieved world leadership in the technology of space exploration, but that leadership was relative and by no means conclusive, Rep. George P. Miller (D-Calif.) asserted. "The challenge is no less great now than it was in 1957 when *Sputnik I* jolted us out of the narcosis of complacency. To our peril, we cannot allow the transient pressures and strains of American life to vitiate our determination to maintain that leadership. At the same time we cannot ignore or overlook any opportunity to reach a rapport and an understanding with Russia that will be of material assistance in making the exploration of space for peaceful purposes truly a monumental achievement of all mankind." (Text)

March 15: NASA and West Germany's Federal Ministry for Scientific Research (BMWF) signed Memorandum of Understanding for study of electric fields in the upper atmosphere. NASA would furnish five Nike-Apache sounding rockets for launch from ESRO facilities at Kiruna, Sweden, beginning in April. Rockets would release barium clouds, particularly in the auroras, to be observed from the ground. BMWF would provide chemical payloads and ground observation equipment and would conduct launchings in cooperation with ESRO and Swedish Space Research Committee (SSRC). Both NASA and BMWF would bear cost of their respective responsibilities. Results would be made available to world scientific community. (NASA Release 67-57)

- NASA and USAF announced agreement to transfer overall management of XB-70 supersonic aircraft research program to NASA Flight Research Center—effective March 25—and to utilize XB-70 "at a reduced rate." Under new arrangement, which updated May 1965 agreement, XB-70 research program would continue as joint NASA–USAF effort, with AFFTC mission support and DOD pilot participation, aircraft support, and research and test projects. Next flight of XB-70, which had been undergoing maintenance, modification, and installation of additional instrumentation since Jan. 31 in preparation for new program, was scheduled for late March or early April. (NASA Release 67-59; FRC Release 5-67)

- Changes in NASA staff alignment: Harold B. Finger, Manager of NASA-AEC Space Nuclear Propulsion Office (SNPO) since its formation in 1960, was named to new NASA position of Associate Administrator for Organization and Management. He was succeeded in SNPO by his deputy, Milton Klein. Reporting to Associate Administrator Finger would be Assistant Administrator for Administration William E. Lilly; Assistant Administrator for Industry Affairs Dr. Bernhardt L. Dorman; Assistant Administrator for Technology Utilization Dr. Richard L. Lesher; and former LaRC Assistant Director Francis B. Smith in new post of Assistant Administrator for University Affairs. University Affairs Office would integrate, coordinate, and guide NASA's relationships with colleges and universities; manage the Sustaining University Program; and assume other duties previously assigned to OSSA's Office of Grants and Research Contracts. Dr. Thomas L. K. Smull, formerly Director of Grants and Research Contracts, was appointed Special Assistant to the Administrator on NASA-NAS and NASA-NAE relationships.

    Other changes: DeMarquis D. Wyatt became Assistant Administrator for Program Plans and Analysis and David Williamson was appointed as his deputy. Budget and programming functions previously under Wyatt were transferred to the Office of Administration, where they would be integrated into a NASA-wide system for resources management, including programming, budgeting, personnel management, and financial reporting. Breene M. Kerr, Assistant Administrator for Policy Analysis, resigned to return to private industry but would continue to serve as part-time consultant to NASA; Gen. Jacob M. Smart (USAF, Ret.) succeeded Kerr in addition to continuing to serve as Special Assistant to the Administrator; William E. Lilly, Director of Program Control, OMSF, became Assistant Administrator for Administration, replacing General Smart; and John Biggs, Deputy Executive Secretary, became Executive Secretary, replacing Col. L. W. Vogel (USA), who returned to his previous assignment as Executive Officer. (NASA Releases 67-47, 67-49, 67-50)

- NASA Associate Administrator for Space Science and Applications Dr. Homer E. Newell, testifying before House Committee on Science and Astronautics' Subcommittee on Space Science and Applications, said the scientific community saw a need to develop a spacecraft to make measurements at low altitudes—about 80 mi—to investigate transition between the meteorology of the lower atmosphere and the aeronomy of the upper. In the lower atmosphere the constituents are thoroughly mixed; in the upper, they tend to separate by gravity with the heaviest below and the lightest above. "The importance of this transition region has been revealed by the [sounding] rocket program, but only satellites can provide the global measurements needed to establish the lower boundary conditions for the upper atmosphere as a whole," Newell said. "Such a mission would require the development of a spacecraft with a restartable engine in order to achieve an adequate lifetime." (Testimony)

- Sen. Clinton P. Anderson (D-N.Mex.), on behalf of himself and Sen. Margaret Chase Smith (R-Me.), introduced to Senate new $5.1-billion FY 1968 NASA authorization bill (S. 1296) to reflect President Johnson's

Feb. 28 request for funds to develop nuclear rocket technology. Original NASA authorization bill for $5.05 billion (S. 781) had been introduced Jan. 31. (*CR*, 3/15/67, S3781)

- Vice President Hubert H. Humphrey told the National Space Club's 1967 Goddard Memorial Dinner that U.S. space program would continue to move forward despite the Jan. 27 Apollo tragedy. "On behalf of President Johnson, I assure you . . . that we have not lost heart—that we will not falter—that we shall move forward with renewed determination in our exploration of space." The Vice President said the first obligation was to "come as close as humanly possible to determining the cause of the Apollo 204 accident, so that we can make sure it will never happen again." He said he was impressed with the "tough-minded way the Review Board is going about its work." (Text; AP, *NYT*, 3/3/67, 31)

- Milton B. Ames, Jr., Director of Space Vehicles, and Dr. Walton L. Jones, Director of Biotechnology and Human Research Div., NASA Office of Advance Research and Technology, testified before House Committee on Science and Astronautics' Subcommittee on Advanced Research and Technology. Ames said land recovery of manned spacecraft was one of OART's major objectives; NASA had enlarged research plans substantially to acquire "without delay" the detailed technology necessary to "permit the confident application of steerable flexible-winged parachutes to the recovery of manned spacecraft." Earliest major potential application would probably be to Apollo, to provide for land recovery in post-lunar-landing period of the early 1970's. "Such a capability would greatly extend the operational flexibility and utility of the Apollo system and . . . contribute to the feasibility of refurbishment and reuse."

    Dr. Jones said increased possibility of aspiration of particulate matter had been identified by NRC Space Science Board as a potentially serious effect of prolonged weightlessness. "In zero gravity conditions, dust particles do not settle out. This could lead to deposits in the respiratory tract which could have important implications in the production of pulmonary disease." Jones said a new instrument to measure and record the number and size of dust particles in a spacecraft was scheduled for an Apollo flight. (Testimony)

- Satellite reconnaissance alone justified spending 10 times what U.S. had already spent on its space program, President Johnson told a group of Southern educators and Government officials in Nashville. Because of this reconnaissance "I know how many missiles the enemy has." Before making his remarks on space, Mr. Johnson said he did not wish to be quoted on that section of his talk, but more than 100 persons not bound by press rules heard the statements. (Clark, *NYT*, 3/17/67, 13; Loory, *W Post*, 3/18/67)

- NASA had successfully completed series of ground tests to qualify improved graphite nozzle insert in FW-4S solid rocket motor which had malfunctioned during launch of Scout booster in January. Malfunction had caused postponement of Feb. 15 launch of ComSatCorp's Intelsat II-C, because FW-4 was 3rd stage of its booster, Thrust-Augmented Improved Delta. (NASA Release 67-60)

- MSC had awarded Bell Aerosystems Co. a nine-month, $5.9-million fixed-price contract to provide three Lunar Landing Training Vehicles (LLTV) for astronauts to practice simulated lunar landings. Capable of simulat-

ing ⅙ gravity environment of the moon, LLTV would be towed to about 4,000-ft altitude and released; pilot could damp out ⅚ of gravity force on vehicle using jet engines; remaining ⅙ gravity would be supported by rocket motors simulating lunar landing conditions. (NASA Release 67–58)
- An electric rocket engine capable of operating off low solar cell voltage, which could be used for satellite station-keeping or attitude control, had been successfully tested for more than 400 hrs at LRC. (LRC Release 67–11)
- U.S. should move its technology beyond the "building blocks" to break out of the "mental straitjacket which is confining the Nation to a stalemate of defeatism," recommended a policy statement adopted by Air Force Assn., at meeting in San Francisco. Asserting that U.S.S.R. was "far ahead" of U.S. in a critical area of nuclear knowledge, AFA called for a "sweeping re-examination of national strategy" to assess the impact of technology "on the world balance of power." In a specific resolution, AFA urged a full-scale U.S. SST program, claiming that U.S.S.R. had been secretly flying a supersonic aircraft prototype for "more than one year." (Davies, *NYT*, 3/16/67, 10; *W Post*, 3/17/67, A4)

  Dr. Harold Agnew, Weapons Div. Leader at AEC's Los Alamos Scientific Laboratory, said U.S. reluctance to deploy an antiballistic missile system reflected a new balance-of-power concept which was characterized by weapons judged on their contributions to world stability rather than their superiority. This philosophy was dangerous because it stifled innovation and placed the Nation continually on the defensive. "Whether we use our full technological advantage should be determined solely on whether it is to our advantage politically, militarily and economically as far as our adversary is concerned. I believe it has been a mistake to allow the fleeting opinions of every nation that has a vote in the U.N. to influence materially our actions. The inhibition of the use of our full technological advantage has hurt our technological future as well as prolonging whatever confrontation is at hand. The basis of advanced technology is innovation, and nothing is more stifling to innovation than seeing one's product not used or ruled out of consideration on flimsy premises involving public or world opinion. . . ." (*Av Wk*, 3/27/67, 11)
- The possibility of using communications satellites to help expedite border-crossing formalities on international flights was considered by Air Transport Assn.'s Facilitation Committee meeting in Ponce, Puerto Rico. It was suggested that computers, working with numerical codes and aided by comsats, could file necessary papers on an aircraft's cargo and passengers from point of departure, thereby facilitating clearances on arrival in the U.S. (*NYT*, 3/15/67, 78M)

March 16: *Cosmos CXLVIII* was successfully launched by U.S.S.R. Orbital parameters: apogee, 436 km (271 mi); perigee, 275 km (171 mi); period, 91.3 min; and inclination, 71°. Equipment performed satisfactorily. Satellite reentered May 7. (Sehlstedt, *B Sun*, 3/28/67; GSFC SSR, 5/15/67)
- *Cosmos CXLIV* satellite, which began transmitting meteorological data to Washington, D.C., March 2, was equipped with an infrared camera for night photography of clouds and ice fields and two television cameras for daylight photography of weather patterns, *Izvestia* reported. Launched Feb. 28, satellite was powered by two wing-like solar batteries

that turned automatically to trap solar rays. (Reuters, *NYT*, 3/17/67, 13)

- NASA Nike-Apache sounding rocket launched from NASA Wallops Station carried 96-lb payload with joint U.S.-India ionosphere experiment to 88-mi (142-km) altitude before impacting 100 mi downrange in the Atlantic. Purpose of flight was to prove out instrumentation and test new antenna design. Experiment, which measured electron and ion densities and Lyman-alpha flux in lower ionosphere with a riometer—cosmic radio noise absorption device—and a Langmuir probe, was being developed for series of cooperative NASA–INCOSPAR launches from Thumba range, India. (NASA Release 67–61; WS Release 67–8)

- NASA Aerobee 150 sounding rocket launched from WSMR carried NRL-instrumented payload to 120-mi (193-km) to obtain spectra from a stellar ultraviolet image converter spectrograph and stellar photometric data from two ultraviolet photon counters. Rocket and instrumentation performed satisfactorily. (NASA Rpt SRL)

- 110 scientists from U.S. and six other countries were selected by NASA Associate Administrator for Space Science and Applications Dr. Homer E. Newell—upon recommendation of Space Science Steering Committee—to conduct experiments with first lunar samples returned by U.S. Apollo astronauts. Principal U.S. investigators represented 21 universities, two industrial firms, three private institutions, and 10 Government laboratories. Nine scientific institutions were represented in U.K.; three in West Germany; and one each in Canada, Japan, Finland, and Switzerland. Most investigations would be conducted in scientists' own laboratories, but time-critical experiments would be carried out in MSC's Lunar Receiving Laboratory during quarantine period—anticipated to last 30 days. Plans called for a total of 122 experiments on approximately 50 lbs of lunar material to study composition of lunar surface and search for evidence of its origin. Four major investigative areas were: (1) mineralogy and petrology; (2) chemical and isotope analysis; (3) physical properties; and (4) biochemical and organic analysis. (NASA Release 67–55)

- Francis J. Sullivan and Dr. Hermann H. Kurzweg of NASA Office of Advanced Research and Technology testified before House Committee on Science and Astronautics' Subcommittee on Advanced Research and Technology. Sullivan, Director of Electronics and Control Div., said that the development of technology for high-resolution, 60- to 120-in space telescopes for optical astronomy, communications, and other potential applications was a major OART effort. "Success in this research will provide the capability to study celestial objects which cannot be viewed from the ground because of atmospheric attenuation and will provide vastly improved resolution over that attainable through the earth's atmosphere."

  Sullivan said because environmental, structural, and control problems of large mirrors in space were almost insurmountable with conventional fabrication and alignment, unique approaches were being studied. One was 24-in-dia, three-segment mirror feasibility model in which distortion sensor, located in front of primary mirror, determined which segments were out of alignment and to what degree. Electronic circuitry then processed and used the alignment data to realign mirror surface segments with actuators located behind each segment.

  Dr. Kurzweg, Director of Research, said the JPL had developed a new mathematical approach for calculating energy requirements for

Mars vehicles which provided, for the first time, a complete day-by-day listing of the energies necessary for a round trip. Calculations had been completed for launching during the entire 1970–80 decade. "The calculated trajectories went from a circular parking orbit about the Earth to the vicinity of Mars, into a circular parking orbit about Mars for a stay time of 7 days, out of the circular parking orbit about Mars to the vicinity of the Earth, and into a circular parking orbit about the Earth. It was assumed that the one-way flight times for both the departure and return trajectories are the same." Dr. Kurzweg said information was very important for "fast trade-off calculations for future advanced Mars missions. . . ." (Testimony)

- The J–2 liquid rocket engine was restarted twice in simulated altitude tests conducted by MSFC at AEDC. Shut down after an initial 30 sec burn, the engine was restarted and run for 5 sec, then started and stopped again. Total burn time was 70 sec. (*Tech Wk*, 3/27/67, 24)
- Thiokol Chemical Corp. presented to Smithsonian Institution's National Air and Space Museum historic liquid-propellant rocket engines made during World War II. Gifts included primitive rocket test stand known as ARS Test Stand No. 2, used by members of original American Rocket Society (now AIAA), on Dec. 10, 1938, to test first U.S. successful, fully regeneratively cooled rocket engine. Other Thiokol gifts included World War II liquid-propellant jet assist takeoff (JATO) units for aircraft, and engines for Lark surface-to-air missile, USAF MX–774 missile, and Viking research rocket. (Smithsonian Institution Release)
- Vice President and Mrs. Humphrey, accompanied by NASC Executive Secretary Dr. Edward C. Welsh, would visit MSFC and U.S. Army Missile Command (AMC) at Redstone Arsenal March 22–23, MSFC and AMC announced. Trip was later postponed because Mr. Humphrey felt he should remain in Washington, D.C., to greet President Johnson on his return from a Pacific trip. (MFSC Release; *Marshall Star*, 3/22/67, 1)
- Request for stricter controls to prevent "instrument abuse" at KSC was made by R. L. Wilkinson, chief of Measurement Systems Div. at KSC, in memorandum to NASA employees and contractor officials. An increasing number of instruments were requiring repairs because of tampering, "simple carelessness," and apparent thievery, Wilkinson noted. He said he had instructed technicians in his instrument calibration shop to report all "cases of obvious instrument abuse or missing rechargeable batteries"; reports would be "reviewed and forwarded through appropriate management channels for corrective action." (Hines, W *Star*, 3/27/67, A1, A8)

March 17: Honeysuckle Creek Tracking Station near Canberra, Australia—newest member of NASA's 16-station network to support manned Apollo missions—was dedicated in ceremonies attended by Australian Prime Minister Henry E. Holt, NASA Deputy Administrator Dr. Robert C. Seamans, Jr., and Edmond C. Buckley, NASA Associate Administrator for Tracking and Data Acquisition. Vice President Hubert H. Humphrey cabled congratulations to Prime Minister Holt: "The establishment of this tracking station is another link in the long chain of cooperative efforts in space between . . . Australia and the United States.

"In its support of the Apollo program, this new station will play a vital role in communications with our astronauts in preparation for their landing on the Moon and return to Earth. . . ." (NASA Release 67–65)

March 17

- President Johnson, in letter transmitting ComSatCorp's 1966 Annual Report to Congress, said: "Accomplishments of the past year under the Communications Satellite Act of 1962 have brought mankind to the threshold of a full-time global communications service to which all nations of the world may have equal access, from which all nations of the world may derive their share of the benefits.

  "Our Space Technology is opening new doorways to World Peace. Within the grasp of the world's peoples is the potential for completely new, heretofore unimagined ways of peaceful cooperation for expanding world trade, for enhancing educational opportunities, for uplifting the spirit and enriching the lives of people everywhere. . . ." (*PD*, 3/20/67, 494)

- U.S. officials reported that *Cosmos CXLVI*, launched March 10, had split into three parts: two were spacecraft and one was final stage of launch vehicle, Evert Clark said in *New York Times*. Exact date of separation was not known. Some officials speculated that flight might have been in preparation for a manned flight on April 12—sixth anniversary of first manned space flight, by Cosmonaut Yuri Gagarin. (Clark, *NYT*, 3/18/67, 11)

- Three Athena missiles were successfuly launched within 4½ hrs from Green River, Utah, to WSMR target area as part of continuing USAF study of missile atmospheric reentry. (AP, *NYT*, 3/19/67, 57; *Tech Wk*, 3/27/67, 28)

- NASA Javelin sounding rocket launched from Churchill Research Range carried a Rice Univ.-instrumented payload to 493-mi (794-km) to determine spatial and temporal variations and transit times of auroral particle fluxes and perform pitch angle studies including backscattered particles. Rocket and instrumentation performed satisfactorily. (NASA Rpt. SRL)

- U.S.S.R. *Molniya 1-2* comsat, launched Oct. 14, 1965, reentered. (GSFC *SSR*, 3/31/67; *W Post*, 4/4/67)

- Paper model of "possible lifting re-entry vehicle design" had flown at 17 times the speed of sound and withstood 4,000° F temperature and pressure equal to 177,000-ft altitude in test to check out recently repaired wind tunnel at LTV Aerospace Corp. in Dallas. Designed by LTV scientist Charles J. Stalmach, model emerged after ¼-sec test with only the nose and leading wing edges slightly blackened. (UPI, *NYT*, 3/17/67)

- President Johnson's Science Advisory Committee's (PSAC) report, *The Space Program in the Post-Apollo Period* [see Feb. 11], was criticized by Dr. Philip H. Abelson in *Science*. Report did not fully discuss the comparative value of space and nonspace activities and "quickly ducked" the difficult problem of priorities, Dr. Abelson said. Nowhere was there an evaluation of the chances of finding extraterrestrial life. "Another deficiency is the lack of a full discussion of the role of man in deep space exploration. To date, manned missions have contributed little scientifically. The unmanned missions have had a cost effectiveness for scientific achievement perhaps 100 times that of the manned flights. Nevertheless, the report implicitly calls for a major role for man in the post Apollo program. . . ." (Abelson, *Science*, 3/17/67)

March 19: Dr. Rem Khokhlov, Univ. of Moscow, predicted at Symposium of Modern Optics in New York that within one year "or a little more" combined Soviet and U.S. techniques could make it possible to tune laser light to any wavelength from shortest wavelengths of ultraviolet across entire spectrum of visible and infrared light to the radio wavelengths used for radar. He described recent Soviet developments in the field and noted that U.S. and U.S.S.R. research was proceeding along parallel lines: both used certain crystals to alter wavelengths of light produced by a laser. Dr. Khokhlov believed tunable laser beam could be used in photochemistry to produce new materials and drugs and would have broad applications in basic physics. (Sullivan, *NYT*, 3/20/67, 33)

March 20: NASA would use AS-204 booster—an Uprated Saturn I—to launch first Apollo Lunar Module (LM) on unmanned flight in summer 1967 because it was last booster equipped with full research and development instrumentation. Original plans had been to use AS-206 booster. Undamaged during Jan. 27 fire, AS-204 launch vehicle would be moved from Launch Complex 34 to Launch Complex 37, which was equipped for unmanned launches of the LM. AS-206 booster would be stored for use in a subsequent mission. Purpose of first LM mission was to obtain data on ascent and descent propulsion systems, including a restart of the descent system; verify LM structure; and evaluate staging. (NASA Release 67-67)

- NASA Associate Administrator for Manned Space Flight Dr. George E. Mueller, in testimony before House Committee on Science and Astronautics' Subcommittee on Manned Space Flight, presented profile of a typical manned Mars mission. "We would launch as we approach the target planet, a series of [unmanned] probes, one would be an orbiter, another would be a Mars surface sample return probe.

  "There would also be some Lander probes which would provide us with information about the atmosphere and the surface on a continuing basis. All of this would be done in a matter of a few days while one was in the vicinity of the planet.

  "We would also be carrying out observations of the local environment and conducting an extensive solar observation program and a planetary observation program from various vantage points one attains from an orbit that is far removed from earth." Asked about the alternative of launching an instrumented spacecraft into a long elliptical orbit with two focal points—the earth and Mars—and manning it from earth, Dr. Mueller replied: "The major problem associated with that is that we are talking about a . . . telescope at least forty inches in diameter. It is a quite difficult task and we are reaching a level of complexity when man becomes an essential part of the system. . . ." (Transcript, 180–1)

- In the year 2000, space satellites would be identifying soil, assessing crop yield, and determining the extent of disease, drought, or insects, Secretary of Agriculture Orville L. Freeman predicted at National Assn. of Science Teachers' convention in Detroit. "While the farmers of tomorrow study reports in their air-conditioned offices, relieved at last of the physical drudgery and occupational anxiety so traditionally theirs . . . these shining space satellites, equipped with the most sophisticated remote sensing instruments will be supplying the information needed to make key decisions.

  "Through information gathered by the satellites, the Government will be able to make accurate predictions to guide marketing and distribution of farm products to avoid waste and local shortages and surpluses." (Text; AP, *NYT*, 3/21/67, 34)

- MSFC awarded Bendix Corp. $7.4 million cost-plus-award-fee contract for development and production of Apollo Telescope Mount (ATM) pointing control system (PCS). Bendix was scheduled to produce three units by August. ATM system would permit Apollo astronauts to point a telescope to selected regions of the sun during period of maximum solar flare activities beginning in late 1968. MSFC had awarded American Optical Co. $740,460 contract to build a dynamic simulator for use in developing the PCS. (MSFC Releases 67-54, 67-58; NASA Release 67-66)

- Tokyo Univ.'s Institute of Space and Aeronautical Science would receive $11 million of Japan's $17.5-million space budget for FY 1967 (April 1967–March 1968), *Technology Week* reported. The Institute would launch 33 vehicles, including a 150-lb scientific satellite; a Mu 4S rocket; and three Lambda 4S vehicles. Remaining funds would go to the Transport Ministry, Meteorological Agency, and Ministry of Posts and Telecommunications. (*Tech Wk*, 3/20/67, 13)

March 21: U.S.S.R. launched *Cosmos CXLIX* into orbit with 297-km (185-mi) apogee, 248-km (154-mi) perigee, 89.8-min period, and 48.4° inclination. Equipment functioned normally. Satellite reentered April 7. (*Tech Wk*, 3/27/67, 28; GSFC *SSR*, 4/15/67)

- Dr. Homer E. Newell, NASA Associate Administrator for Space Science and Applications, defended the Mariner/Mars 1971 mission in testimony before House Committee on Science and Astronautics' Subcommittee on Space Science and Applications. Responding to criticism by Committee Chairman Rep. Joseph E. Karth (D-Minn.) that information from the mission would arrive too late to contribute to the design of the 1973 Voyager spacecraft, Dr. Newell said: ". . . it is very important to recognize . . . that there are three areas of design involved. One is the hardware design. Two is software design, and three is design of the scientific mission. . . . The Mars 71 probe will be fully in time for us to use in the software design, in other words, setting the profile of the mission to get the spacecraft into the atmosphere and onto the surface, and will be fully involved in continuing design of the scientific experiments and observations. . . ." (Transcript, 433–5)

- NASA–Cal Tech cost-plus-fixed-fee contract providing for operation of JPL had been extended for two years through Dec. 31, 1968, without major change. Since October 1958, Cal Tech had staffed and operated JPL; NASA owned the property, facilities, and equipment. (NASA Release 67-68)

- Soviet Vostok spacecraft were equipped with quick-opening explosive hatches and jet-propelled ejection seats—equipment might have saved U.S. Apollo astronauts Jan. 27, Cosmonaut Valentina Nikolayeva-Tereshkova said in *Aviatsiya i Kosmonavtika.* (UPI, *NYT*, 3/23/67)

- ComSatCorp's annual report to its shareholders summarized accomplishments of 1966: "Satellite service brought a new era of communications to the Pacific area. Live transpacific television and other traffic was transmitted by the first of the INTELSAT II series of satellites even though the satellite did not achieve synchronous orbit as intended. A second INTELSAT II satellite was successfully emplaced in synchronous orbit and began regular, 24-hour commercial service over the Pacific, handling telephone, teletype, data, facsimile and television. . . ." Report also revealed that Intelsat's Interim Committee had selected synchronous orbit configuration for basic global system, substantially reducing the number of satellites necessary for global coverage and the complexity of earth stations. Selection was made "after high-quality capability of synchronous satellites was established by the performance of Early Bird." (Text)

- 50-lb. radio backpack that could fix a soldier's position precisely by signals from a satellite had been developed by Johns Hopkins' Applied Physics Laboratory under sponsorship of USN Air Systems Command. Backpack would represent first dry land application of USN navigation satellites to military problems. In most cases, packs would be used in pairs: a forward fire controller would receive satellite signal and relay it to another soldier equipped with a backpack at an artillery base who also would have read satellite signal. Two readings would be transmitted to computer which would instantly tell each man his precise position. Navy sources predicted mass production models of backpack, working with existing satellites, could be in combat use in Vietnam soon. (Clark, *NYT*, 3/22/67, 16)

March 22: NASA launched ComSatCorp's *Intelsat II–C* (nicknamed *"Atlantic II"*) comsat from ETR by three-stage Thrust-Augmented Improved Delta booster into elliptical transfer orbit in preparation for geostationary orbit. Transfer orbit had 22,115-mi (37,200-km) apogee, 183-mi 294-km) perigee, 660.56-min period, and 27° inclination. On March 25, ComSatCorp fired apogee motor transferring satellite into geostationary orbit over the Atlantic.

Scheduled to enter commercial service April 6, *Intelsat II–C* would join *Early Bird I*, world's first commercial comsat, in providing service across the Atlantic and in fulfilling Project Apollo communications requirements. It was third satellite in ComSatCorp's Intelsat II program to place two comsats in synchronous orbit—one over the Pacific, one over the Atlantic. First satellite—*Intelsat II–A*—was launched by NASA Oct. 27, 1966, but failed to achieve synchronous orbit because apogee motor malfunctioned; second—*Intelsat II–B* was launched Jan. 11, achieved synchronous orbit over the Pacific, and entered commercial service Jan. 27. Intelsat II comsats, larger and improved versions of *Early Bird I* (Intelsat I) satellite, were capable of handling television and data transmissions of up to 240 voice channels. (ComSatCorp PIO; ComSatCorp Release)

- *Cosmos CL* was successfully launched by U.S.S.R. into orbit with 373-km (232-mi) apogee, 206-km (128-mi) perigee, 90.1-min period, and 65.7° inclination. Equipment performed satisfactorily. Satellite reentered Mar. 30. (AP, *NYT*, 3/24/67; GSFC *SSR*, 3/31/67)
- NASA Aerobee 150A sounding rocket launched from NASA Wallops Station carried AFCRL solar radiation experiment to 142-mi (229-km) altitude in test to complement March 14 mission which was not entirely successful. Primary objective was simultaneous measurement of solar extreme ultraviolet radiation (EUV) in upper atmosphere with Aerobee's photoelectric spectrometer and similar instrument on NASA's *Oso III* Orbiting Solar Observatory, launched March 8. (WS Release 67-11)
- Maj. Michael J. Adams (USAF) flew X–15 No. 1 to 3,818 mph (mach 5.68) and 135,000-ft altitude during nine-min flight at Edwards AFB to: (1) provide pilot experience at high altitudes; (2) record stabilizer flow angle of attack; (3) test electrical loads; (4) check out 3rd landing skid; (4) check out pressure attitude indicator; (5) record data from X–15 sonic boom; and (6) check out ablative coatings on stabilizer. (NASA X–15 Proj Off)
- An electrical arc caused by a break in insulation on a wire was "the most probable cause" of the Jan. 31 flash fire at Brooks AFB in which two airmen died, USAF investigative board concluded. Board said an elec-

trical cord powering a work lamp outside oxygen-filled pressure chamber apparently developed a break in its plastic insulation, allowing wire to create an electrical arc against the aluminum floor. (AP, *NYT*, 3/23/67, 31)

- Noise problems arising from commercial jet aircraft operations, and the Government's attack on these problems were described by OST Director Dr. Donald F. Hornig in appearance before House Committee on Science and Astronautics' Advanced Research and Technology Subcommittee [see Feb. 1]. He stated: "At the request of the Chairman of the President's Advisory Committee on Supersonic Transport, Secretary McNamara, I assumed responsibility for defining and directing an interagency sonic boom program. Its goal is to determine the potential sources of difficulty or additional costs associated with commercial operation of supersonic transports." OST, said Hornig, had framed "action program" with FAA, NASA, and Depts. of Commerce and of Housing and Urban Development—responding to President Johnson's Transportation Message of Mar. 2, 1966. He added: ". . . NASA's aeronautics research program in the sonic boom problem areas is both broadly conceived and well defined to provide technical support to the FAA and to the aerospace industry in the national SST Development Program." (Text)

- NASA Associate Administrator for Space Science and Applications Dr. Homer E. Newell told House Committee on Science and Astronautics' Subcommittee on Space Science and Applications that NASA did not know whether U.S.S.R. was adequately sterilizing its interplanetary spacecraft. "If we find . . . that . . . [Soviet] sterilization procedure is many orders of magnitude less good than ours, then we will be able to relax in our standards simply because there will be no value in our going through an expensive process only to find that they have already contaminated the planet.

  "However, we do not know at the present time, and there is reason to believe that their procedures may well turn out to be adequate." (Transcript, 446)

- North American Aviation, Inc., and Rockwell-Standard Corp. announced agreement on proposal to merge into new corporation called North American Rockwell Corp. Proposal was subject to approval by directors and stockholders of each company. (*WSJ*, 3/23/67)

- A record 45-million takeoffs and landings were recorded in 1966 by FAA's 304 airport traffic control towers. Figure represented 12% increase over 1965. (FAA Release 67–24)

- Dr. Eugene G. Fubini, an IBM vice president, told IEEE meeting in New York that too many scientists and engineers—as prime movers of change—were not sufficiently attentive to the results of their discoveries and inventions. "Historically, scientists and engineers have underestimated or ignored the social and moral consequences of their work. Don't continue to make that mistake," he warned.

  Dr. Charles H. Townes, MIT provost, received IEEE's 1967 Medal of Honor for his "significant contributions in the fields of quantum electronics which have led to the maser and the laser." (Lieberman, *NYT*, 3/26/67, F1; *Av Wk*, 3/13/67, 97)

March 23: NASA had discarded crew assignments for three manned Apollo missions originally scheduled for 1967 but postponed shortly after Jan. 27 fire, MSC Director of Flight Crew Operations Donald K. Slayton told the press at MSC. "We don't have any crews assigned officially to any missions at the present time. There's not much point in worrying about crew assignments until we know what we're going to do." Backup crew for AS-204 mission had been Astronauts Walter M. Schirra, Jr., Donn F. Eisele, and Walter Cunningham; prime crew for AS-205/208 mission had been James A. McDivitt, David Scott, and Russell Schweickart; for AS-503, Frank Borman, Michael Collins, and William Anders. (UPI, W Star, 3/24/67, A2; AP, W Post, 3/24/67, A6; Wilford, NYT, 3/24/67, 1, 15)

- U.S. officials were studying possibility of live television coverage via *Intelsat II-C* comsat of Latin American summit conference in Punta del Este, Uruguay, April 12-14, which President Johnson was scheduled to attend, Benjamin Welles reported in *New York Times*. Live coverage would require air-lifting a portable ITT ground station to conference site because necessary ground stations for picking up satellite signals did not yet exist in Latin America. Use of ComSatCorp's *Intelsat II-C*, launched by NASA March 22, would not only enhance interest in meeting, officials said, but would also dramatize U.S. technological and political support for Latin America. (Welles, NYT, 3/24/67)

- LaRC Director Dr. Floyd L. Thompson had been selected as one of 10 public servants to receive National Civil Service League's 1967 Career Service Award for his pioneering efforts as an engineer, scientist, and administrator in advancing the science of aeronautics and space flight in the U.S. Award would be presented to Dr. Thompson April 21 at Annual Career Service Awards banquet in Washington, D.C. (NASA Release 67-69)

- NASA Wallops Station employee James C. McConnell had been awarded $1,000 for his invention "A Method of Plating Copper on Aluminum" by NASA Inventions and Contributions Board. Award was part of agency-wide program to encourage and reward NASA employees for meritorious inventions and contributions.

    A second Wallops employee—Joseph W. McAllister—had been named one of the 1967 Economy Champions under President Johnson's national economy program for his "Device to Salvage Power Line Guy Anchors." An economy champion was a Government employee whose achievement or adopted suggestion during FY 1967 had shown first-year measurable benefits of $10,000 or more. (WS Releases 67-9, 67-10)

- Rep. Dante B. Fascell (D-Fla.) introduced in the House a bill (H.R. 7798) to strike commemorative medals in honor of Astronauts Virgil I. Grissom, Edward H. White II, and Roger B. Chaffee who died in the Jan. 27 Apollo flash fire. (CR, 3/23/67, H3301)

- U.S. Ambassador to the U.S.S.R. Llewellyn E. Thompson and Soviet Foreign Minister Andrey A. Gromyko met briefly in Moscow for "preliminary discussions" aimed at mutual agreement not to deploy antimissile missile systems, U.S. Embassy announced. (AP, W Post, 3/23/67, A12)

- Shortage of U.S.-manufactured high-reliability components because of the demands of the Vietnam war was one of the reasons Canada was postponing the launch of her Isis-A satellite from fall 1967 to 1968, Dr. John H. Chapman, Deputy Superintendent of the Defence Research Telecommunications Establishment, told *Toronto Globe & Mail*. Isis-A would measure characteristics of the ionosphere. (*Toronto Globe & Mail*, 3/24/67, 29)

- A correlation between earthquakes and the forces exerted on the earth by the sun and moon was reported by Dr. Alan Ryall, Director of Univ. of

Nevada seismological station. Dr. Ryall said he could support with scientific data the theory that where the earth's crust is fractured and weakened, all that is necessary to cause earthquakes might be the pulling force or gravity of the sun and moon. (*W Post*, 3/24/67)
- U.N. 17-nation disarmament conference in Geneva recessed until May 19 at U.S. request. (Cook, *W Post*, 3/24/67, A12)
- ComSatCorp awarded two four-month study contracts—one to Lockheed Missiles and Space Co. for $272,000 and one to Hughes Aircraft Co. for $299,484—for research on advanced high-capacity synchronous satellites. Lockheed had proposed an inertia-wheel stabilization concept and Hughes a spin-stabilized approach for stabilizing a spacecraft in synchronous orbit. ComSatCorp had advised FCC Jan. 19 of plans to award contracts. (ComSatCorp Release)
- U.K.'s Royal Aeronautical Society had invited foreigners to enter an eight-year-old annual competition to stay aloft in a muscle-operated craft over a figure-eight course with two turning points half a mile apart, Reuters reported. Society had also increased prize from $10,000 to $28,000. (Reuters, *NYT*, 3/23/67, 5)
- U.S. astronauts were honored with a special award at American Personnel and Guidance Assn. convention in Dallas. Accepted by Charles Duke on behalf of his fellow astronauts, the award cited the "exemplary model astronauts display to American youth in intellectual achievement, self-discipline, personal dedication and physical fitness." (UPI, *W Star*, 3/24/67, A6)
- ERC Deputy Director Dr. Albert J. Kelley had resigned to accept an appointment as Dean of Boston College's College of Business Administration effective June 1, NASA announced. He would continue to act as a special consultant to NASA. (ERC Release 67–5)

March 24: U.S.S.R. launched *Cosmos CLI* into 630-km (391-mi) altitude circular orbit with 97.1-min period and 56° inclination. Equipment was functioning normally. (*W Post*, 3/25/67, A4)
- Apollo 204 Review Board announced that completion of its final report on Jan. 27 fire would be delayed one week until April 8. Dr. Floyd L. Thompson, Director of LaRC and Chairman of the Board, issued another call for any information that might aid the probe. Board also announced that it had completed piece-by-piece disassembly of Apollo I spacecraft, although testing and analysis of spacecraft systems and parts were continuing. (UPI, *W Star*, 3/25/67, A5; *W Post*, 3/25/67, A5)
- U.S. policies for sterilizing planetary spacecraft were needlessly stringent and should be revised to take into account new scientific data and past and continuing Soviet activities, two groups of U.S. scientists suggested in companion articles in *Science*. First, by Drs. N. H. Horowitz and R. P. Sharp of Cal Tech and R. W. Davies of JPL, reviewed sterilization standards set in 1964 by COSPAR and agreed to by both U.S. and U.S.S.R. "In effect the [COSPAR] resolution requires that less than one viable microorganism be contained in each 10,000 spacecraft intended for planetary entry. Attainment of such a goal demands heroic measures, and current U.S. directives call for dry-heat sterilization of the completely assembled lander stage." The authors asserted that COSPAR requirements were based on the outdated assumption that terrestrial microorganisms could thrive on other planets, particularly Mars; present knowledge, however, indicated that Mars is so dry, lacking in oxygen,

and soaked with lethal forms of ultraviolet sunlight that proliferation of earth organisms would be unlikely. They suggested that COSPAR-recommended constraints "could be substantially relaxed without compromising to any significant degree the biological condition of Mars."

Second article, by Dr. Bruce C. Murray, Cal Tech, Merton E. Davies, RAND Corp., and Dr. Philip K. Eckman, JPL, said U.S. policy should concentrate on seeing that "U.S. contribution to plantary contamination will remain significantly less than the Soviet contributions" thereby reducing "significantly the cost and time required to carry out serious scientific investigations of the surfaces of Venus and Mars." The authors suggested that U.S.S.R. had already hit both Mars and Venus with spacecraft not sterilized in accordance with COSPAR standards. Any requirement for U.S. to apply more stringent restrictions "is illogical in that . . . the U.S. would be asked to increase greatly the cost and complexity of its planetary program without achieving any significant reduction in the probability of actual contamination." (*Science*, 3/24/67, 1501-10)

- AFSC announced personnel changes. Effective March 27: (1) M/G Harry L. Evans, Vice Director of Manned Orbiting Laboratory (MOL), would become Assistant to the Commander, AFSC, pending his retirement from USAF May 31; (2) M/G James T. Stewart, Director of Space, Deputy Chief of Staff/Research and Development, USAF Hq., would succeed General Evans; (3) B/G Walter R. Hedrick, Jr., Assistant Deputy Chief of Staff/Systems, AFSC Hq., would succeed General Stewart. Effective summer 1967: (1) B/G Joseph S. Bleymaier, Commander of WTR, would become Deputy Director, MOL, with additional duty as Deputy Commander, Space Systems Div. for MOL; (2) B/G Clifford J. Kronauer, DOD Office of Director of Defense Research and Engineering, would succeed General Bleymaier; (3) M/G Vincent G. Huston, Commander of ETR, would be reassigned to AFSC Hq. as Deputy Chief of Staff for Operations; and (4) M/G David M. Jones, NASA Deputy Associate Administrator for Manned Space Flight, would replace General Huston. (AFSC Releases 63.67, 64.67)

- *Life* magazine disputed critics who advocated cutting back NASA's Apollo program: "We do not share the attitude that it doesn't matter who gets to the moon first, we or the Russians. It does matter. . . . Remember the astonishment, the embarrassment, the apprehension, the loss of prestige we all felt so keenly when *Sputnik I* went up on Oct. 4, 1957? The prestige we then sought to regain was not mere prestige, an empty abstract symbol of superiority. The superiority at stake was real, not symbolic. . . .

   "NASA is now engaged in its customary annual defense of its budget requests. We recommend that its expenditures be kept at the $5 to $6 billion level for now. As we continue to compare NASA's budget with other pressing needs, we must be aware of the challenges and opportunities that arise in space. . . ." (*Life*, 3/24/67)

March 25: U.S.S.R. launched *Cosmos CLII*, its eighth Cosmos satellite during March, to achieve record month of space launchings. Orbital parameters: apogee, 512 km (318 mi); perigee, 283 km (176 mi); period, 92.2 min; inclination, 71°. Equipment functioned normally. Satellite reentered Aug. 5. (UPI, *W Post*, 3/26/67, A3; GSFC *SSR*, 8/15/67)

- Capt. Robert F. Freitag (USN, Ret.), Director of Field Center Development, OMSF, addressed the National Student Conference in Ann Arbor, Mich.:

"... the more dramatic aspects of the space program sometimes tend to overshadow the fundamental purposes of the space effort. Just as the moon had influenced the course of young lovers for centuries, so has the excitement of lunar exploration blinded many to the deeper purposes and far reaching benefits of the nation's overall space program, and its deep impact on present and future generations.

"... a dynamic space program will ... bring rewards in knowledge, wonders, and resources far surpassing the investment it requires in money, materials, and brainpower. We do not know all that we shall learn when we are able to place men, instruments, telescopes, and laboratories beyond the envelope of the earth's atmosphere. But we *do* know that we shall know a great deal more than we can possibly *now* know about the reality of which we are a part." (Text)

- Soviet scientist Ruvin Garber revealed that U.S.S.R. had developed a thin thread-shaped tungsten crystal which was the "strongest substance" ever made on earth, UPI reported. Strand was reportedly .0002 cm in diameter and could withstand 230 tons of pressure per sq cm. (UPI, *NYT*, 3/27/67, 67)

March 26: White House Office of Science and Technology (OST) had failed to move effectively in the direction of long-range planning and had been seriously hampered by a "thinly spread" staff, concluded report prepared by Library of Congress Legislative Reference Service and released by House Committee on Government Operations' Subcommittee on Military Operations. Presented as a background document for future congressional action or legislation, report reached no direct findings or conclusions but noted that during four years OST had been in existence its "most conspicuous operating characteristic has been the wide span of issues in which it has become involved.

"But with the office numbering only seventeen to twenty professionals its staff must be thinly spread." Although OST had not moved spontaneously in the direction of long-range planning, report said questions posed by various groups of Congress "at least imply a need for this policy planning, systems analysis capability, especially to deal with issues of transportation, waste management, water resources, and energy development." (Text)

March 27: MSC awarded $1 million in six-month study contracts to General Dynamics, Boeing Co., and McDonnell Co. for space station facilities that could be assembled in orbit early in the 1970s. General Dynamics would study engine room module, a separately launched unit to furnish electrical power and life-support equipment for space station. Boeing would study space station design to include a 38-in optical telescope that would be launched with the space station by a Saturn V booster. McDonnell would study logistics/ferry spacecraft to provide ground landing capability for astronauts. (*Tech Wk*, 3/27/67, 23–4)

- Seven communications companies signed agreement for joint ownership of three existing ComSatCorp-owned earth stations and three new high-capacity stations for expanded international commercial satellite communications. The companies: ComSatCorp; AT&T; Hawaiian Telephone Co., ITT World Communications, Inc.; ITT Cable and Radio, Inc.-Puerto Rico; RCA Communications, Inc.; and Western Union International, Inc. Agreement established a policy-making committee composed of company representatives, provided for sharing of operating costs and capital, and described ComSatCorp's role as manager. It was

filed with FCC in accordance with Dec. 7, 1966, order for joint ownership. (ComSatCorp Release)
- NASA Nike-Tomahawk launched from Barreira de Inferno Range, near Natal, Brazil, carried Univ. of New Hampshire-instrumented experiment to obtain data on neutron intensity at different latitudes, solar x-ray fluxes, Lyman-alpha radiation, and ionospheric electron densities. Rocket and instrumentation performed satisfactorily. (NASA Rpt SRL; *Tech Wk*, 4/10/67, 15)
- William Tier, Deputy Manager of Operations in MFSC's Saturn I/IB Program Office, was named Manager to replace Lee B. James. Tier's former position would be filled by William F. LaHatte. (MSFC Release 67-62)
- *New York Times* editorial on U.S.'s and U.S.S.R.'s competitive programs to land men on moon: "We fail to see that it makes any difference who reaches the moon first or whether the landing takes place in the 1960's or the mid-1970's. Moreover, we see large, unsatisfied needs here on earth—in this country and elsewhere—that could very usefully employ some of the vast resources now being devoted to the Apollo program and its Soviet analogue. A cutback in those resources and their diversion to more pressing needs would make more sense.

    "The moon is not going away, and it can wait a little longer for visitors." (*NYT*, 3/27/67)

March 28: ComSatCorp's *Intelsat II-C* comsat successfully relayed test signals over the Atlantic between ground stations at Andover, Me., and the Canary and Ascension Islands. Satellite was scheduled to enter commercial service April 6. (AP, *NYT*, 3/29/67, 30)
- U.S. Government would receive in royalties more than the estimated $1.3 billion it was contributing to SST development, ATA President Stuart G. Tipton told the Aero Club of Washington (D.C.): "The government investment is to be paid back with the 300th delivery. The manufacturer will thus pay royalties of 4.3 to 6.7 million dollars for each aircraft delivered. . . . Since these royalties are to continue for a 15-year period, they will go on well beyond the 300th delivery, and this phase of royalty payment will produce a return on the government investment comparable to levels normally received by industry." (Text)
- Soviet astronomers from Ukrainian Academy of Sciences had produced a lunar astronomical almanac containing data on the locations of stars as seen from the moon, Science Service reported. Star charts of the sky from earth become outdated after several years, but a star chart for the lunar sky had to be revised after only several days. Precession of earth's axis causes each pole to complete a cone on the celestial sphere every 26,000 yrs; moon's axis has only an 18.6-yr precessional period which makes construction of a star chart of the lunar sky extremely difficult. (SciServ, *NYT*, 3/28/67)
- FAA announced new safety regulation to protect aircraft from off-course or abandoned balloons. Effective April 28, all unmanned free balloons must be equipped with reflective devices to permit tracking by ground radar and at least two independently operated self-destruction devices. (FAA Release 67-28)

March 29: Pan American World Airways cargo jet aircraft approaching Wake Island participated in first public demonstration of a new datalink system for transmitting navigation data from aircraft to ground stations via an orbiting satellite. VHF signals beamed to NASA's *Ats I* satellite by first data-link system ever installed on a commercial aircraft and relayed to Mojave Desert antenna, reached Kennedy International Airport by telephone to fix aircraft's exact position. Experiment was part of a series using satellites to develop more reliable long-range communications and tracking contact with commercial aircraft on transoceanic flights. It was first test of aircraft antenna designed specially for transmitting satellite messages. Pan American officials predicted datalink system could be fully operational by late 1969 and could lead to instant automatic and continuous plotting of flights across vast ocean areas. (Wilford, *NYT*, 3/30/67, 53)

- A report expressing confidence in NASA's management of the Apollo program was released by House Committee on Science and Astronautics' Subcommittee on NASA Oversight. Report was prepared before Jan. 27 Apollo fire, but Subcommittee Chairman Rep. Olin E. Teague (D-Tex.) said the findings of the study were "essentially unchanged. . . . The public may look with confidence to the management, engineers, scientists and technicians of NASA and industry as the Apollo program progresses." Report praised the ability of the NASA-industry team to identify and solve the administrative, technical, and engineering problems "inherent in performing a program contributing major technological progress. . . ." It identified the major Apollo pacing items as the Saturn V 2nd stage, the Lunar Module (LM) for the Saturn I and V, and the Command and Service Module (CSM) for the Saturn I and V. "Any significant delays in attainment of flight capability of these items could affect both program shcedule and cost adversely." Report said that an additional $300 million in NASA's FY 1967 budget would have provided: "(1) increased confidence in meeting unforeseen development problems when encountered; (2) broader latitude in providing system alternatives when engineering difficulties are identified; and (3) an increased assurance of meeting the national goal of a lunar landing in this decade." (Text; Sehlstedt, B *Sun*, 3/20/67)

- JPL had selected Avco Space Systems Div. for negotiations on a six-month, $240,000 contract to study feasibility of launching a flyby spacecraft and atmospheric entry capsule to Venus in 1972. NASA had approved selection of the firm, which would head a team effort with Northrop Systems Laboratories. Study, expected to begin in May, would investigate all aspects of launching by an Atlas-Centaur booster a Mariner-class spacecraft weighing about 1,200 lbs. Near Venus, 100-lb capsule system would separate from flyby portion of spacecraft and enter Venusian atmosphere to take direct measurements during descent to surface. Capsule would not be designed to survive impact on Venus. (NASA Release 67–72)

- United Air Lines 720 jet aircraft dropped 8,000 ft in clear air turbulence near Cheyenne, Wyo., before pilot could bring aircraft under control for an unscheduled landing. One passenger was killed when he released his seat belt and reached for his month-old grandson; other passengers were slightly injured but no one was hospitalized. (UPI, *NYT*, 3/30/67, 41)

- French President Charles de Gaulle launched France's first nuclear-powered submarine, *Le Redoubtable*. The 7,900-ton submarine, which would become fully operational in 1970, would carry 16 missiles that could be fired from submerged positions toward targets up to 1,800 mi away. (*WSJ*, 3/30/67, 1; UPI, W *Star*, 3/29/67, 1)

March 30: President Johnson, signing Executive Order 11340 establishing the Department of Transportation effective April 1, headed by Alan

Boyd, said the new department would "consolidate 35 programs previously dispersed through 7 departments and independent agencies. It will bring together nearly 100,000 employees, and annual expenditures of more than $6 billion."

Tasks of the department, President Johnson said, would be to "modernize and unify our national transportation policy"; "bring greater safety to the travels of all American citizens"; "apply the best of an expanding technology to every mode of transportation"; "strengthen our partnership with private enterprise and State and local governments in meeting America's urgent transportation needs"; and "improve our transportation links with the rest of the world." (PD, 4/3/67, 559-60)

- NASA successfully launched three artificial cloud experiments from NASA Wallops Station. In first experiment Nike-Tomahawk sounding rocket ejected multicolored barium vapor clouds over eastern U.S. at altitudes of 142 and 200 mi (246 and 322 km) to measure electric fields and wind motions in the upper atmosphere by photographing and tracking ionized clouds. Experiment had been scheduled for March 28 but was postponed because of overcast skies.

  In second and third experiments Nike-Apache sounding rockets carried separate triethylborane (Teb) experiments to 100-mi (161-km) altitude to investigate distribution of, and radiation from, atomic oxygen. Teb, a new test material for measuring upper atmosphere wind profiles, was ejected at 90- to 50-mi (145- to 81-km) altitudes and photographed by special photometer as it dispersed. Launches were followed by reported sighting of bright UFO's. (NASA Release 67-74; WS Release 67-12; AP, W Star, 3/31/67, A11)

- NASA Nike-Cajun sounding rocket launched from NASA Wallops Station carried GSFC-instrumented payload to 71.1-mi (114.4-km) altitude to obtain data on temperature, pressure, density, and wind between 35 to 95 km at the transition from wintertime westerly to summertime easterly circulation. Rocket and instrumentation performed satisfactorily. (NASA Rpt SRL; Tech Wk, 4/10/67, 15)

- NASA Nike Tomahawk sounding rocket launched from NASA Wallops Station carried instrumented payload to 149.5-mi (237-km) altitude in a Univ. of Maryland-Geophysics Corp. of America experiment to evaluate capabilities and accuracies of pulse and thermal equalization probes, investigate electron energy distribution in a normal daytime ionosphere, and investigate the use of wing-slope techniques with Langmuir probes. One door covering GCA antennas prematurely ejected causing a complete loss of experimental data and a rocket apogee 4 mi lower than predicted. (NASA Rpt SRL)

- NASA Nike-Tomahawk sounding rocket launched from Wallops Station carried 180-lb Univ. of Maryland payload to 158-mi (254-km) altitude in mission that was coordinated with the overpass of Canadian satellite *Alouette II*. Payload impacted 143 mi downrange in the Atlantic. Primary objective of flight was to evaluate techniques for studying the ionosphere. (WS Release 67-14)

- USAF launched an unidentified satellite from Vandenberg AFB employing a Thor-Agena D booster. (UPI, W Post, 3/31/67, A3)

- Rep. Olin E. Teague (D-Tex.), Chairman of House Committee on Science and Astronautics' Subcommittee on NASA Oversight, told AP he did not think "we will ever know what caused that fire" in which three Apollo astronauts died. Thousands of NASA and industry technicians responsible for Apollo I spacecraft would always wonder whether they contributed to the accident, Teague suggested. Asked if he believed NASA officials had taken unnecessary risks to land a man on the moon by 1970, Teague said: "Maybe we were a little bit careless." He said some of the safety checks probably should have been more thorough, but did not elaborate. (AP, W Star, 3/30/67)

- FRC had awarded 15-mo, $8.9-million contract to North American Aviation, Inc. (NAA), for maintenance and support of XB-70 supersonic research aircraft. As under previous support contracts, NAA would provide manpower, facilities, material, and equipment to support flight operations and would conduct data reduction. Overall management of XB-70 program was assigned to FRC under March 15 NASA–USAF agreement. (NASA Release 67-75)
- Weather photos taken by NASA's *Ats I* satellite in near-synchronous orbit over the Pacific would be relayed to World Meteorological Organization (WMO) conference in Geneva April 3–28 to demonstrate new meteorological capabilities. Transmitted by satellite to Mojave, Calif., photos were traveling by land line to National Environmental Satellite Center in Suitand, Md., where they were being recorded for later transmission by ESSA surface facilities to Geneva. (NASA Release 67-77)
- Small opaque diamonds, formed when graphite-bearing meteorites impacted on earth, had been identified by three General Electric Co. scientists at Schenectady, N.Y., AP reported. Similar to man-made diamonds manufactured by using intense pressure and high temperatures, these diamonds had hexagonal crystal structure, thus differing from cubic structure of diamonds used in rings. Scientists predicted hexagonal diamonds would also be found in meteorites on the lunar surface. (AP, NYT, 3/30/67)

March 30–31: NASA's Space Science Steering Committee's Space Biology Subcommittee met at MSFC. Agenda included, in addition to discussion of subcommittee business, a review of AA program by Leland F. Belew, manager of MSFC Saturn/Apollo Applications Program Office; a presentation on S-IVB space station module study; and a tour of AA payload mockup and simulation facilities. (MSFC Release 67-65)

March 31: Langley Research Center engineers at JPL, after repeated efforts to reactivate *Lunar Orbiter III*'s photo system which failed March 2, had concluded there was virtually no hope for transmission of further photos from the spacecraft. Analysis of events indicated failure had originated with momentary fluctuation in spacecraft's electrical power which disturbed photo system logic by turning the film advance motor memory to forward mode. Cause of fluctuation had not been identified. Motor, following incorrect logic, stalled as it attempted to move film which was being mechanically held in a fixed position while readout was in progress. During stalled condition, motor overheated to failure point. Despite termination of photo transmission, *Lunar Orbiter III* continued to serve as a tracking target, providing practice for NASA Manned Space Flight Network stations, and to record information on radiation and meteoroids in moon's vicinity. (NASA Release 67-76)

- *Explorer XXXII* (AE–B) aeronomy satellite, launched May 25, 1966, had ground support terminated; satellite had suffered complete depressurization [see Jan. 17]. As direct result of loss of internal pressure, battery failure had caused disablement of one electron temperature probe on March 3; ion mass spectrometer and the other electron temperature probe had remained operational and continued to produce quality data until March 22. Last radiofrequency contact had been made on March 26. *Explorer XXXII* mission objectives had been achieved and mission had been declared a success in December 1966. Scientific papers on initial results would be presented at American Geophysical Union meetings April 17–20 in Washington, D.C., and additional results would be reported to COSPAR meetings in London during July. (NASA SP–4007, 192; NASA Proj Off)

- First two of series of 20 Japanese and U.S. sounding rockets were launched from NASA Wallops Station as part of Japanese-U.S. meteorological rocket project to conduct comparative meteorological research between 12-mi (20-km) and 37-mi (60-km) altitudes. Japanese rocket was an MT–135; U.S. rocket was an Arcas. Payloads were ejected at programmed altitudes and descended on parachutes: wind measurements were obtained from radar tracking data; temperature measurements were radioed to ground receiving stations. Objectives of the mission were to compare and cross-calibrate data from various payloads; verify flight and operating characteristics; obtain new information on operation of each rocket system as a whole; and obtain additional data on diurnal cycles of wind and temperature in the stratosphere. Flights were conducted under a Dec. 26, 1966, agreement between NASA, Japanese Science and Technology Agency, and Japanese Meteorological Agency. Results of experiments would be made available to world scientific community. (NASA Release 67–73)

- France's Sahara Missile Proving Grounds at Hammaguir Range closed after 20 yrs, signaling end of operational use of Daimant 1 booster which had launched four French satellites from Hammaguir since Nov. 26, 1965. Next French satellite, 200-lb D–2, would be launched from French Guiana in 1969 with Super Diamant booster. (*NYT*, 2/11/67; *Av Wk*, 3/27/67, 50)

- ComSatCorp requested FCC permission to substantially reduce television and telephone rates between U.S. and Europe effective April 6 when *Intelsat II–C* comsat was scheduled to enter commercial service. New rates, which would cover only the leasing of channels to authorized communications common carriers operating in the Atlantic, would apply both to *Early Bird I* and *Intelsat II–C*. (ComSatCorp Release)

- U.K.'s Royal Artillery Institution donated two 19th-century rockets to the astronautics collection of Smithsonian Institution's National Air and Space Museum. One rocket was made by Sir William Congreve in 1815 and used on the Woolwich Ranges; the other, William Hale's 24-lb rotary rocket, was constructed in 1863. Congreve rockets were used in the U.S. in 1814 while Hale rockets were used in the Mexican War in 1846 and the Civil War. (Smithsonian Release)

- Soviet Defense Minister Rodion Malinovsky, credited with having led Soviet military into the missile age, died in Moscow at age 68, apparently of cancer. (Rosenfeld, *W Post*, 4/1/67, B6)

- U.K. had agreed to buy 40 more F–111 supersonic military aircraft from U.S. at total cost of $785 million, Anthony Lewis reported in *New York Times*. Ten F–111's had been ordered March 31, 1966, with a one-year option to purchase 40 more. Deliveries were expected to begin in 1969. To offset impact on U.K.'s balance of payments, U.S. had agreed to buy $785 million worth of British equipment during the next 10 yrs or help U.K. get orders elsewhere. (Lewis, *NYT*, 4/1/67, 1)

During March: Aerospace praised NASA's contributions to U.S. aeronautical research: "Since the United States started sending experimental vehicles beyond Earth's atmosphere nine years ago aeronautical research has been overshadowed by space exploration as far as the annual budgets of the National Aeronautics and Space Administration are concerned. . . . But dollar comparisons fail to tell the whole story. For its space work, NASA has had to build from scratch a variety of expensive ground facilities and equipment and to make heavy outlays for complex, nonreusable flight hardware. NASA's aeronautical teams spend little on flight equipment and what they do buy is reusable; furthermore, they operate from facilities for the most part built and paid for prior to the Space Age. There is no question but that space has played the dominant role in NASA's programs, but the aeronautical research effort has been far more substantial—and more rewarding—than an examination of budget figures would indicate. In the coming year, assuming Congressional approval of the pending budget, it will be increased." (*Aerospace*, 3/67, 3)

- NAS forwarded to NASA a list of 69 recommended candidates for 20–30 positions in NASA's astronaut training program. A total of 923 applications had been evaluated. NASA was expected to make final selections before Aug. 15. (NAS–NRC–NAE *News Report*, 3/67, 3)
- Hugh L. Dryden Memorial Fund had received nearly $1 million in donations for construction of 700-seat auditorium at NAS Hq. in Washington, D.C., in honor of the late NASA Deputy Administrator. Fund's ultimate goal was $2 million. (NAS–NRC–NAE *News Report*, 3/67, 3)
- Charges that NASA's Apollo program had serious deficiencies were reviewed by William Hines in series of articles in Washington *Evening Star*. A 55-page report submitted to NASA three days before Jan. 27 flash fire by Thomas R. Baron, former North American Aviation, Inc., inspector at KSC Apollo warehouse, charged that improper parts had been installed in Apollo I spacecraft. Careless and unorthodox warehousing practices had caused many parts and materials intended for spacecraft use to lose their identity, Baron told Hines in an interview. "As a result there was probably a bundle of stuff [in Apollo] that they don't have any idea what . . . it is at all." Baron also charged that on at least one occasion, NAA had installed a substandard part with deliberate subterfuge, without the knowledge of NASA. Truth about the part could easily be concealed because NASA had no quality-control men of its own in receiving warehouse, Baron said. He was particularly critical of deviation from test procedures: "When you deviate on your small components, and you deviate on all of them, and you put these in one system and you run a systems test and you deviate again, you've ended up with a system that really isn't worthwhile."

    James Parker, former NASA quality control inspector at KSC, told Hines that NASA quality control supervision on Apollo contractors' work was "spotty" and said he had periodically cited instances of waste and improper expenditures in reports to his KSC superiors. (Hines, W *Star*, 3/15/67, A1, A6; 3/19/67, A7; 3/23/67, 1, A6; 3/26/67, A3)
- ESRO's Kiruna rocket range—ESRANGE—had been officially opened in March, *Space Flight* reported. The Kiruna facility was situated 30 mi north of Kiruna (north of Arctic Circle in Swedish Lapland) and occupied about 300 sq mi of virgin forest and tundra; Kiruna range would be firing base for atmospheric sounding rockets [see Mar. 15]. (*S/F*, 4/67, 114–6)

April 1967

April 1: NASA's March 24 decision to delay issuance of AS–204 accident report until April 8 would not restrict scope of inquiry in public hearings, Rep. Olin E. Teague (D-Tex.), Chairman of House Committee on Science and Astronautics' Subcommittee on NASA Oversight, told Washington *Evening Star* reporter William Hines. Hearings would begin promptly "at 10 o'clock the same day the report is issued," Representative Teague said, and would be held at night if necessary. (Hines, W *Star*, 4/2/67)

- Secretary of Transportation Alan S. Boyd outlined major goals of Dept. of Transportation (DOT) at ceremony in Washington, D.C., marking its first official operating day. DOT would (1) "insist that all forms of transportation become safer and more reliable for the user" and make increased effort to meet transportation needs of the physically handicapped; (2) strive for increased intermodal efficiency through promotion of integrated systems; (3) explore new possibilities to help transportation advance technologically and keep pace with progress of American industry; (4) encourage transportation industry in a selective expansion program for the domestic market; (5) seek ways for transportation to assist economic development of underdeveloped areas; and (6) make more positive contribution to urban development and quality of urban mass transit service. "At the very least, this Department must work to make transportation more efficient, more economical, more expeditious, and more socially responsible. . . . America, today, has the best transportation system of any nation in the world. But it does have its faults. It can be, it must be substantially improved." (Text)
- ComSatCorp requested FCC permission to establish a pilot U.S. domestic comsat system with channels available without charge to educational broadcasters. Urging speedy approval of the plan, ComSatCorp stressed that "differences among private parties as to the best way to provide badly needed domestic satellite service should not be permitted to delay unduly inauguration of such a Program." ComSatCorp proposed two 85-ft send-and-receive antennas—one near New York City and one in California—and 30 smaller receive-only antennas in Pacific and Rocky Mountain time zones; it was prepared to finance and start construction of the system and act as trustee until FCC and Congress made final decision on ownership. System would become operational by late 1968. (ComSatCorp Release)
- NASA Aerobee 150 sounding rocket launched from WSMR carried instrumented payload to 86.5-mi (139.2-km) altitude; it measured spectral irradiance of two early-type stars in wave length intervals 1100 Å to 4000 Å. The other two target stars were pointed at but could not be

observed through opacity of fuel and oxidizer released through control jets. (NASA Rpt SRL)

- Commemorative activities honoring DOT's first official operating day and celebration of Smithsonian Institution's third annual Rite of Spring, sponsored by the Institution, included these events: hourly ascensions in three-man-hot-air balloon by Don Piccard of the famous family of air and underwater explorers; very low speed operation of air cushion vehicle over the Mall (grassy area immediately adjacent to the Institution's site) using Bell Aerosystems Co.'s "Hydroskimmer"; demonstration of Bell's rocket belt to propel a man for short distances through the air; and showing of many models of transportation equipment. (Smithsonian Institution Release)
- Dr. Mervin K. Strickler, Jr., FAA Congressional Fellow, received National Aeronautic Assn.'s 1966 Frank G. Brewer Trophy at National Aerospace Education Council's annual banquet in Tucson, Ariz. Trophy was awarded annually for outstanding contributions to aerospace education of the Nation's youth. (NAA *News*)
- New Jersey Governor Richard J. Hughes proclaimed April 1, 1967, Tiros–Essa Day in New Jersey to mark seventh anniversary of first Tiros launch and to pay tribute to NASA and U.S. space program. (Text)

April 2: One-inch-diameter hollow steel spheres for use in aerospace ball-bearing assemblies were being tested at Air Force Flight Dynamics Laboratory, Wright-Patterson AFB. Hollow spheres were half as heavy and equally as strong as solid ones currently used, had longer life expectancy, and could easily be mass produced. (AFSC Release 44.67)

April 2–5: National Assn. of Broadcasters held 45th annual convention in Chicago. FCC Chairman Rosel H. Hyde strongly urged that U.S. commercial broadcasters support a system of public television and suggested that the broadcasters establish a national committee to aid noncommercial TV: "As diverse as our present commercial broadcasting system is, there is room for innovation, for more diversity, in short, for a viable, supplemental noncommercial service. Such a service could provide a further and differently based competitive spur to the commercial system. It would provide a place where new ideas and experimental techniques could more easily be tested. . . ." (Dallos, *NYT*, 4/5/67, 78)

ComSatCorp Chairman James McCormack termed "nonsense" a charge by the Ford Foundation that ComSatCorp was threatening to become a monopoly: "Comsat, far from being an evil octopus, is a youngster trying very hard to employ private funds and private initiative to . . . maintain the leadership granted to us, under extraordinary complete regulation, to the benefit of the United States and the world more generally. . . ." Restating the corporation's position that there should be one all-purpose domestic communications system, McCormack asserted that ComSatCorp, as a representative of U.S. communications policy, was in best position to establish such a system. "We calculate that a multipurpose system will, from the beginning, provide broadcast transmission services at less cost than will a single-purpose system [and] initial costs will continue to reduce over the years," affording a great deal of flexibility and security. (Text; Dallos, *NYT*, 4/6/67, 67; Reddig, *W Star*, 4/6/67)

April 3: Ford Foundation and AT&T submitted briefs to FCC on domestic comsat system.

Ford Foundation brief sharply criticized ComSatCorp's plan for a domestic comsat system and asked that ComSatCorp's April 1 request for permission to develop an experimental system be denied. "To add to Comsat's authority the exclusive francise for a domestic communications satellite system . . . would give the corporation a world monopoly. One entity would provide all communications satellite services. One company would be responsible for developing all satellite technology. One customer would procure all satellite equipment. A monopoly of this sort would be unprecedented. . . ." Ford suggested instead that the system be developed and initially operated by NASA so that technological knowledge could be gleaned without controversy over ultimate management or ownership. NASA could later be reimbursed by the entity finally chosen to operate the system.

AT&T supported ComSatCorp's proposal, suggesting the system be linked to and integrated with present ground systems. Rejecting Ford Foundation's proposal for a separate nonprofit satellite TV broadcast corporation whose income would support noncommercial TV, AT&T said that such proposals reflected "a parochial disregard for their effects on other satellite systems and terrestrial (land) microwave systems. . . . Installation of such limited systems would . . . jeopardize the existing public communications systems, both domestic and overseas, and would severely inhibit their necessary continued growth." (Text; Gould, *NYT*, 4/3/67, 1; W *Star*, 4/4/67)

- Secretary of Transportation Alan S. Boyd addressed Sixth Annual Editorial Conference of Magazine Publishers Assn. and American Society of Magazine Editors in Washington, D.C.: "Today, April Third, is a memorable day in the history of communication. On this day, one hundred and seven years ago, the Pony Express began. . . . The Pony Express lasted only eighteen months. It was superseded by the transcontinental telegraph. And probably the managers of the Pony Express knew all along that that was going to happen. But they had customers who couldn't wait, who were willing to pay for a faster mail service.

". . . this episode illustrates the driving impatience of the American people for progress. . . . We're doing the same thing today in the aerospace industry. We're building the SST. We already have a grasp of the technology that will supersede the SST, fifteen or twenty years from now. But Americans are not the kind of people who'll passively wait for new developments. . . . They're engaged now in pushing jet aviation technology to its practical limits."(Text)

April 4: U.S.S.R. launched *Cosmos CLIII* into orbit with 291-km (181-mi) apogee; 202-km (126-mi) perigee; 89.3-min period; and 64.6° inclination. Equipment functioned normally. Satellite reentered Apr. 12. (*Tech Wk,* 4/10/67; GSFC *SSR,* 4/15/67)

- NASA Nike-Cajun sounding rocket launched from Point Barrow, Alaska, carried GSFC-instrumented grenade experiment to obtain data on temperature, pressure, density, and wind between 22–59 mi (35–95 km) at the transition from winter-time westerly to summer-time easterly circulation. Rocket and instrumentation performed satisfactorily. (NASA Rpt SRL)
- Astronaut Frank Borman told Cleveland Engineering Society that he and other astronauts hoped U.S. would immediately revitalize the Apollo program: "We'd like to see the crews regrouped and the program under way as soon as possible." (AP, *W Post,* 4/6/67, G2)

April 5: NASA's 815-lb *Ats II* (ATS-A) (Applications Technology Satellite) was launched from ETR by Atlas-Agena D booster on mission to evaluate gravity-gradient system for spacecraft stabilization in 6,900-mi-altitude orbit and to obtain useful data from experiments during first 30 days in orbit. Spacecraft successfully entered elliptical transfer orbit with 6,878-mi (11,069-km) apogee; 120-mi (192.7-km) perigee; 218.3-min period; and 28.4° inclination. Agena engines, which were to have reignited to inject spacecraft into 6,900-mi (11,103-km)-altitude circular orbit, failed to ignite for the second burn; spacecraft separated and, tumbling uncontrollably, remained in transfer orbit.

By April 7, all but one of the scientific experiments had been turned on and operated successfully, but usefulness of data could not be determined because spacecraft's sensors were not compatible with its orbit. Last experiment was turned on April 15. Although it was apparent that mission could not be successfully completed, scientists continued to operate experiments to obtain as much data as possible on spacecraft performance, and ground stations received occasional TV pictures and meteorological photos. Gravity-gradient configuration functions would not be exercised until spacecraft's attitude behavior could be better defined.

Ats II was second in series of five satellites designed to improve spacecraft technology; develop long-life control systems; advance spacecraft communications; and improve long-range weather predictions. Record size—252 ft long with 123-ft booms extended from opposite sides—*Ats II* was first U.S. gravity-gradient satellite to carry meteorological cameras and first with sufficient instrumentation to verify three-axis gravity control system in conjunction with detailed computer program. In addition to gravity-gradient experiment, satellite also carried communications equipment, two Advanced Vidicon Camera System (AVCS) cameras, and eight scientific experiments to measure orbital environment. *Ats I* was successfully launched Dec. 6, 1966, into a 22,300-mi-altitude near-synchronous orbit over the Pacific and was still operating flawlessly. ATS program was managed by GSFC under OSSA direction. (NASA Proj Off; NASA Release 67–71; AP, *NYT*, 3/31/67, 12)

- NASA Aerobee 150, Mod I sounding rocket launched from WSMR carried Lockheed Missiles and Space Co. payload to 99.6-mi (160.3-km) altitude to obtain quantitative measurements of spectrum and intensity of solar x-ray flux in the 2–30 kev interval and to look for x rays from recurrent nova T Pyxides. This was first flight of new Aerobee configuration Mod I. Rocket and instrumentation performed satisfactorily. (NASA Rpt SRL)
- Chairman James McCormack of ComSatCorp, speaking at annual meeting of National Assn. of Broadcasters, in Chicago, made these predictions: "With new satellites, having flexibility to permit working with several earth stations simultaneously and having increased transmission capabilities, and with improved earth stations, we are coming on better days. For a year or so, we will continue to be handicapped by a lack of ground facilities to enable the United States to work efficiently and simultaneously with more than one satellite in either the Atlantic or the Pacific [and] we will continue for a while to be embarrassed by lack of system capacity there, primarily in the earth stations, to handle occasional TV transmissions without buying back other circuits then under annual or monthly lease." He continued: ". . . in a very few years, you will have the potential of continuous, live,

world-wide TV transmission at well under one percent of your total costs."

He went on to explain: "Given the geographic coverage of the satellite, capable of being used directly from a number of points simultaneously, and the rapid growth of all kinds of requirements for service, a single world-wide communications organization was deemed essential from the beginning. The same set of considerations would seem to indicate at least a broadly based international forum for broadcasting, an important and rather unique use of those communications."

Turning to potential world-wide benefits, he said: "The establishment of a world-wide broadcast capability, in being and operating continuously, will enable the almost casual accomplishment of some of the greatest purposes of the United States and many other nations seeking better understanding as one basis for a reasonable peaceful world. I refer in particular of course to cultural exchanges, information and education." (Text)

- NASA rejected industry bids ranging from $13.6–14.9 million for construction of four buildings at NASA's Electronics Research Center in Cambridge, Mass. ERC Director James C. Elms said that difference between bids and $10.4-million Government estimate was too great to be adjusted and that building plans would have to be revised. (McElheny, *Boston Globe*, 4/6/67)
- FRC awarded General Electric Co. a $2,782,272 contract for maintenance of XB–70 aircraft engines and support services. (NASA Release 67–81)

April 6: *Lunar Orbiter III* press briefing was held at NASA Hq.

Dr. Leonard Reiffel, Apollo Program, OMSF, said NASA had selected eight "candidate" locations along the moon's equator as the smoothest and safest possible sites for first Apollo manned landings. Selection had been made after careful screening of photos taken by three Lunar Orbiters and *Surveyor I*. Detailed analysis of photos indicated that safest area might be a broad plain in the southeast part of the Sea of Tranquility—just east of Maskelyne D crater—where the surface was flat and there were few dangerous boulders and no deep craters. NASA officials emphasized, however, that no single site would be selected: each mission would have three possible sites along the equator—one in the east, one in the center, and one in the west—to allow for launch delays.

Harold Masursky, U.S. Geological Survey, described evidence of "seismic erosion" as "one of the major processes on the moon's surface that Lunar Orbiter has given us information about." Several photos showed dunes of soil-like material that appeared to be debris from moonquakes filling old craters. Some of the smoother plains on the moon might be caused in part by this process, Masursky suggested. (Wilford, *NYT*, 4/7/67, 1, 14)

- Rep. Donald Rumsfeld (R-Ill.) introduced in the House H.R. 8145, a bill to establish Aerospace Safety Advisory Panel to "provide the [NASA] Administrator with independent advice and assistance in safety matters, regardless of operational 'commitments' or budgetary considerations. Regardless of the outcome of the current [Apollo accident] investigation . . . the National Aeronautics and Space Administration's . . . safety practices need strengthening. Based on NASA statements and . . . [FY 1968] testimony . . . the area of greatest weakness appears to be in the lack of a truly independent review of NASA's facilities, operations,

and procedures solely from the viewpoint of safety." (*CR*, 4/6/67, H3657; NASA LAR VI/39)
- Col. V. John Lyle (USAF, Ret.), former Assistant to the Comptroller of the Air Force, was made special assistant to NASA Associate Administrator for Manned Space Flight Dr. George E. Mueller. (NASA Ann)
- Reduced television and telephone rates between U.S. and Europe became effective when *Intelsat II–C* comsat entered commercial service. New rates, which covered only the leasing of channels to authorized communications common carriers operating in the Atlantic, applied to both *Early Bird I* and *Intelsat II–C*. Monthly charge for one leased voice channel was reduced from $4,200 for 16-hr daily service to $3,800 for 24-hr service. Rates for one-way black-and-white television in the Atlantic area were $1,100 for first 10 min and $30 for each additional minute per half channel. New rates for color TV reduced existing 150% of black-and-white charge to 125% of black-and-white rate. All rates were based on providing voice-grade or TV half-channel service from a U.S. earth station to an appropriate satellite and did not cover charges for communication links from satellite to an overseas earth station. (ComSatCorp Release)
- Trans World Airlines retired the last of its propeller-driven Constellations and became first major U.S. airline to have an all-jet fleet on domestic as well as overseas routes. (*NYT*, 4/7/67, 62)
- Editorial comment on domestic satellite system controversy [see April 1, 3].
 New York Times: "The Ford Foundation and Communications Satellite Corporation are at odds about how best to set up a domestic satellite system that would carry educational television. Neither side, it appears, has all the answers. Ford is right to question Comsat's demand to run a domestic satellite system . . . [but] Ford's own plan for creating a new educational network is also unsatisfactory. It wants to give the job on a trial basis to the National Aeronautics and Space Administration, which has enough problems with its present over-ambitious objectives and should not have its energies diverted or its resources increased for anything new. Ford's suggestion that savings in communications charges made by the commercial networks should be used to finance educational TV has an inherent disadvantage in that the amount of savings would be inadequate to do a worthwhile job. . . . These differences in approach should not be allowed to hold up development of noncommercial network. The thing to do is to choose what is best from both plans." (*NYT*, 4/6/67, 36)
 Washington Post: "The issue is more than just a squabble over Comsat's request. The Ford Foundation has proposed that the Nation establish more than one domestic satellite system, with one system reserved exclusively for television, both commercial and public. Comsat maintains that one system can do the entire job of communications adequately and will save large sums of money. . . . Given the complexities of the problem, and the need that it be solved, hand-in-hand with the general question of how public television is to be financed, we think Ford's alternative is advisable. . . . NASA has the capabilities to do the job and would be involved, anyway, in putting Comsat's satellite into orbit. This solution would leave all avenues open, for NASA is highly unlikely to become the permanent operator and could recoup its expenditures by selling the equipment and stations to the eventual operator." (*W Post*, 4/6/67)

April 7: Seven communications companies filed joint application with FCC for construction of new earth station near Cayey, Puerto Rico, and the transfer of ownership of three existing comsat stations. Companies—ComSatCorp; AT&T; Hawaiian Telephone Co.; ITT World Communications, Inc. (WorldCom); ITT Cable and Radio, Inc.-Puerto Rico; RCA; and Western Union International (WUI)—submitted application pursuant to March 27 earth-station ownership agreement. (ComSatCorp Release)

- Soviet scientist E. I. Taghiev told seventh World Petroleum Congress in Mexico City that U.S.S.R. planned to drill a 49,500-ft-deep test well to explore earth's crust and expressed regret that U.S. had abandoned its Mohole drilling project in the Pacific. Site of Soviet project was not revealed. (AP, W *Star*, 4/7/67, A14)

April 8: *Cosmos CLIV* was launched by U.S.S.R. into orbit with 232-km (144-mi) apogee, 186-km (116-mi) perigee, 88.5-min period, and 51.6° inclination. Equipment functioned normally. Satellite reentered Apr. 10. (*Tech Wk*, 4/10/67; GSFC *SSR*, 4/15/67)

- Cosmonaut Yuri Gagarin, discussing Soviet space plans in *Ogonyek*, hinted that U.S.S.R. was preparing to launch a large manned space station in the near future. It had been speculated that U.S.S.R.—which had not launched a manned spacecraft since March 18, 1965—would conduct major manned launch in 1967 to commemorate 50th anniversary of the Bolshevik Revolution. Gagarin said there was little value in repeating feats already accomplished: ". . . it is not sensible in the future to put expensive spacecraft into orbit for several days. . . . It is apparent that they will have to stay in orbit a long time. It will be necessary to supply them and change crews with the use of simplified types of space vehicles. . . . The time is not far off when longer and more distant space flights will be made."

    In later Novosti press agency report Soviet Academy of Sciences Chairman Anatoly A. Blagonravov supported Gagarin's predictions and said: "Mankind is entering the 7th year of the age of man's space flight, and we do not doubt that in the not very distant future we will witness new space victories which will exert a growing influence on the life of people." (Anderson, *NYT*, 4/9/67; UPI, W *Star*, 4/9/67, A13; C *Trib*, 4/10/67)

April 9: Most probable cause of Jan. 27 flash fire in which Astronauts Virgil I. Grissom, Edward H. White II, and Roger B. Chaffee died was a sparking short circuit in worn, defective, or poorly insulated wire, Apollo 204 Review Board concluded in its final report. Copies of report were distributed simultaneously to NASA Administrator James E. Webb, congressional committees, and the press. Board was unable to pinpoint exact ignition source, but identified conditions which led to the disaster: (1) a sealed cabin, pressurized with an oxygen atmosphere; (2) an extensive distribution of combustible materials in the cabin; (3) vulnerable wiring carrying spacecraft power; (4) inadequate provisions for the crew to escape; and (5) inadequate provisions for rescue or medical assistance. Having identified these conditions, the Board addressed itself to the question of "how these conditions came to exist. Careful consideration of this question . . . leads . . . to the conclusion that in its devotion to the many difficult problems of space travel, the Apollo team failed to give adequate attention to certain mundane but equally vital questions of crew safety.

"The board's investigation revealed many deficiencies in design and engineering, manufacture and quality control. When these deficiencies are corrected the over-all reliability of the Apollo program will be increased greatly."

Board offered a total of 21 recommendations, including: (1) an "in-depth" review of the entire Apollo life-support system; (2) an investigation of effective ways to control and extinguish spacecraft fires; (3) severe restriction and control of amount and location of combustible materials inside spacecraft; (4) reduction of time required for crew to egress and simplification of escape operations; (5) continuous monitoring of safety of all test operations to assure adequacy of emergency protections; (6) improved ground communications system; (7) continued studies of two-gas cabin atmosphere; and (8) full-scale mockup tests to establish fire safety of reconfigured command module. Board emphasized that its report was not intended as a criticism of NASA's entire manned program: "The board is very concerned that its description of the defects in the Apollo program . . .will be interpreted as an indictment of the entire manned space flight program and a castigation of the many people associated with the program. Nothing is further from the board's mind." (Text; *NYT*, 4/10/67, 1, 28–9; *W Star,* 4/10/67, A1, A6–7; *W Post,* 4/10/67, A1, A8)

- William H. Woodward was promoted from Deputy Director to Director of OART's Space Power and Electric Propulsion Div.—formerly Nuclear Systems and Space Power Div.—succeeding Harold B. Finger, who became NASA Associate Administrator for Organization and Management March 15. Woodward would be responsible for planning, managing, and coordinating advanced research and technology programs in the area of space power, including solar, chemical, nuclear, and electric propulsion. (NASA Release 67–78)
- NASA Office—Downey at North American Aviation's Downey, Calif., plant was disestablished and its functions transferred to MSC and MSFC, (NN 1136)
- Completed section of USN's $80-million radio communications center at Northwest Cape, Australia, was occupied by Navy personnel under Capt. Robert Friedman. Center, which would maintain contact with U.S. nuclear submarines in Pacific and Indian Oceans via very-low-frequency radio, would be fully operational in July. (Durdin, *NYT,* 4/9/67, 3)

April 10: NASA Administrator James E. Webb, accompanied by Apollo 204 Review Board members, NASA Deputy Administrator Robert C. Seamans, Jr., and NASA Associate Administrator for Manned Space Flight Dr. George E. Mueller, testified on Jan. 27 flash fire before House Committee on Science and Astronautics' Subcommittee on NASA Oversight. Webb said the Board had found error, "but it has also found the capability within NASA and the Apollo team of contractors to overcome error." NASA and industry would share their part of the blame "for what we have done or left undone," Webb said, but the committee could have confidence that "NASA and its contractors have the capability to overcome every deficiency required to proceed to successfully fly the Apollo Saturn system and accomplish its objectives." He said Board Chairman Dr. Floyd Thompson had reassured him that the "concepts and the basic design on which the Apollo system is based can be made to work in a reliable manner," and Astronaut/Board member Frank Borman had reported he would be confident to fly the Apollo spacecraft if reported deficiencies were eliminated. Webb concluded: "If any man in this room asks for whom the Apollo bell tolls, it tolls for him and me, as well as for Grissom, White and Chaffee.

"It tolls for every astronaut test pilot who will lose his life in the space simulated vacuum of a test chamber or the real vacuum of space . . . for every astronaut scientist who will lose his life on some lonely hill on the moon or Mars . . . for Government and industrial executives and legislators alike . . . for an open program continuously evaluated by the opinion makers with little time for second thought—operating in the brilliant color and brutal glare of real time, worldwide mass media with the speed of the television, from euphoria to exaggerated detail...."

Asked why six of the eight Board members were NASA employees, Webb replied that they were the best qualified people for the investigation and noted that they were all familiar with the spacecraft or with similar types of investigations. He said that in his opinion the Board had "to a large degree . . . overstated the case." Questioned about charges by the press that many qualified Project Apollo technicians were not called to testify before the Board, Webb responded: "It is easy to make allegations, hard to conduct an investigation, and even harder to make the Apollo Saturn fly at the earliest possible date. I do not believe these allegations are worth your consideration."

Dr. Thompson discussed methods and procedures followed by the Board, and key Board members traced the process of establishing the course of the fire, the "most probable" ignition source, and detailed findings and recommendations. (Testimony; Transcript 10, 15-8)

- Rep. Donald Rumsfeld (R-Ill.) inserted in the *Congressional Record* his statement expressing concern over Apollo 204 Review Board's failure to determine whether Apollo accident was result of a single oversight or "indicative of a basic weakness in the organization of NASA and/or the contractors.... I am deeply disturbed by the report of the Apollo 204 Review Board. My concern is not that the Board was unable to discover the precise cause of the accident during the course of its rigorous technical investigation, but, rather, that the Board failed to examine, or at least report on, the fundamental conditions which permitted the accident to occur and which, moreover, resulted in the tragic deaths of three fine young men. The report's, and apparently the investigation's, shortcomings in this regard leave open the possibility of similar catastrophes in the future." (*CR*, 4/10/67, H3786-7)

- Washington *Evening Star* suggested that House Committee on Science and Astronautics might have been "the victim of a high-level snow job last year when it did research for a 1,200-page report on Project Apollo" because report's conclusions conflicted so greatly with findings of Apollo 204 Review Board. Committee report issued March 29 praised ability of NASA and industry "to identify and solve technical and administrative problems" and stated that the Apollo accident did not "alter the basic conclusions of the report." Board report, by contrast, criticized NASA and NAA for poor management and deficient quality control. (*W Star*, 4/10/67, A7)

- NASA Nike-Cajun sounding rocket launched from Point Barrow, Alaska, carried GSFC-instrumented grenade payload to obtain data on temperature, pressure, density, and wind between 35 to 95 km at transition from winter-time westerly to summer-time easterly circulation. Rocket and instrumentation performed satisfactorily. (NASA Rpt SRL)

April 10

- "The Soviet Anti-Ballistic Missile (ABM) System: a Survey of Soviet Strategic Doctrine, Arguments on Deployment, Intentions and Fulfillment, American Reaction and Implications," Library of Congress document F-239, evaluated the dangers for Soviet ABM deployment. Author, Specialist in Soviet and Eastern Europe Affairs Joseph G. Whelan, summarized present state of affairs, noting Russia's historic preoccupation with the defensive and her clear intent not to accept permanently a second position to the U.S. Whelan quoted from Feb. 23, 1967, speech of Presidential adviser Walt W. Rostow at Univ. of Leeds, Great Britain: "We are all actively trying to find the terms for a non-proliferation agreement; and the emergence of [ABM defense] for Moscow has posed for the United States and the Soviet Union the question of whether the nuclear arms race shall be brought under control or go into a vast and expensive round of escalation on both sides with respect to both offensive and defensive weapons." (Text)
- Soviet space dogs Veterok and Ugolyek, in orbit Feb. 22–March 16, 1966, aboard *Cosmos CX*, had both fathered normal puppies, Tass reported. Effects of prolonged weightlessness [see May 16, 1966] had gradually disappeared, Tass said: "For some time, their movements were languid, and the animals preferred to lie down. But . . . 2½ months after their space voyage, the dogs fully recovered and became cheerful again." (UPI, *NYT*, 4/12/67, 77)
- U.S. defense warning system was "highly vulnerable" to missile attack because of increasing Russian missile strength, report by Institute for Strategic Studies warned. Institute, an unofficial body of military and international affairs' experts from 28 countries, recommended that U.S. establish a new back-up interceptor control against hostile missiles by mid-1970. (*W Post*, 4/11/67, A15)

April 10–13: Aerospace Medical Assn. met in Washington, D.C.

Computer technique which had been used to enhance 12 *Surveyor I* photos of the lunar surface in August 1966 might be applied to diagnostic x rays, Dr. Robert Nathan, JPL scientist who developed the process, suggested in speech over closed-circuit TV broadcast. Process, which separated significant from insignificant detail, could remove distortions and obscurities that often make it difficult for physicians to interpret x-ray films. Dr. Nathan also predicted technique might enable scientists to see the inner details of human chromosomes and gain a new understanding of birth defects. (Randal, *W Star*, 4/11/67, A3)

FAA would appoint 10 radiation biology experts to study the radiation aspects of SST operation, FAA Administrator William F. McKee revealed. "We want to assure ourselves, beyond any doubt, that high-level altitude flights can be made routinely." FAA was also studying new pressurization on equipment to assure passenger safety at 40,000-ft flight levels as well as new escape devices and evacuation methods, McKee said. (*NYT*, 4/13/67, 74)

NASC Executive Secretary Dr. Edward C. Welsh predicted that U.S. would have the best SST in the world. "Despite its timing, I nominate it for the 'clean-up' position in sales, safety, and economy." Dr. Welsh said expenditures for the SST could be justified because it was important to "our international balance of payments, to our technological progress, to our private enterprise system, to improved transportation, as well as to our international prestige." (Text)

Philco-Ford Corp. and General Dynamics Corp. both displayed several monitoring devices which had grown out of technology acquired by NASA in the development of remote sensing equipment for astronauts. The corporations indicated they would attempt to market the devices for use in medical sciences. (*Tech Wk,* 4/17/67, 22)

Dr. Charles F. Gell, chief scientist of Submarine Medical Research Laboratory at U.S. Naval Submarine Medical Center, received Eric J. Liljencrantz Award and $500 honorarium for his contributions to the advancement of aeronautics and space sciences through medical research. (*Av Wk,* 4/24/67, 121)

April 11: Apollo 204 Review Board, accompanied by NASA Administrator James E. Webb, testified before Senate Committee on Aeronautical and Space Sciences. Key Board members summarized the Board's findings and recommendations. Astronaut Frank Borman, asked if he believed U.S. decision to use 100% oxygen atmosphere in spacecraft had been influenced by a desire to achieve first manned lunar landing, replied: "... never since I have been associated with NASA have I ever experienced any decision where a known detriment to crew safety was sacrificed to any operational requirement. And although I am willing to accept risk ... I am not willing personally to accept undue risk and I would not participate in any decision which I thought was expediting a program in an unsafe manner. And in the final analysis the crew is the Review Board because if we do not like the way the spacecraft is configured, we don't have to get in." Borman said if on Jan. 27 he had had the information accumulated by the Board, he would not have entered the Apollo spacecraft because: (1) the test had not been classified as hazardous; (2) the spacecraft contained too many combustibles; (3) the wire that provided the ignition source was vulnerable. Borman said he would be "willing and eager" to enter a 100% oxygen Apollo spacecraft which incorporated Board's recommended improvements. (Testimony; Transcript, 181, 209-10, 258)

- In testimony before the House Committee on Science and Astronautics' Subcommittee on NASA Oversight, North American Aviation President J. Leland Atwood denied that pressures to meet 1970 lunar-landing goal had prompted NAA engineers to take shortcuts in Apollo program. Terming the Apollo 204 Review Board's study "a good and penetrating report," Atwood conceded that deficiencies existed in the Apollo capsule, but disputed Board's charge that "deficiencies existed in Command Module design, workmanship, and quality control. . . ." NAA's Director of Research and Engineering Dr. John F. McCarthy said that although an electrical arc "probably" caused the fire, there were "various other possibilities"—such as Astronaut Gus Grissom accidentally kicking loose a wire that could have later initiated the blaze. Under stern questioning by subcommittee members, however, Dr. McCarthy admitted that this theory was pure speculation.

In response to charge by Rep. William F. Ryan (D-N.Y.) that prevailing attitude prior to fire was not "just a feeling of over-confidence" but "real negligence" by NAA and NASA, Atwood reasserted that the most serious error made by NAA and NASA was classification of the AS-204 test as "non-hazardous." Questioned about the reasoning behind designing to minimize fire hazard, Dr. McCarthy pointed out that NAA had concentrated on eliminating possible ignition sources while it turned out that the amount and placement of flammable materials was crucial.

He characterized this as a "grave error in judgment." Atwood assured the Subcommittee that NAA and NASA had "placed great stress" on the importance of quality and workmanship: "The procedures followed and the depth of inspection which have been employed far surpass those . . . used on any other program. The fact of the accident itself makes it obvious that we must do more than we have done, that we must—guided by the finding of the board—redouble our efforts to approach perfection. Already we have gone over every other design feature of the spacecraft to identify any other area of possible failures . . . [and] we have found none." (NYT, 4/12/67, 34; W Star, 4/12/67, A6)

- *New York Times* commented on Apollo 204 Review Board's final report on Jan. 27 flash fire: "Two months ago, on the basis of the incomplete evidence then available, we suggested that the Apollo catastrophe had taken place because 'many men miscalculated—grossly.' The official report . . . suggests that that conclusion was too kind . . . [since] the report convicts those in charge of Project Apollo of incompetence and negligence.

  "The core of the problem is indicated in the board's fifth finding: 'Those organizations responsible for the planning, conduct and safety of this test failed to identify it as being hazardous. Contingency preparations to permit escape or rescue of the crew from an internal command module fire were not made.'

  "How could those in charge of the test have 'failed to identify it as being hazardous?' The three astronauts had been put into what even a high school chemistry student would know was a potential oxygen incendiary bomb, one needing only a spark to initiate catastrophe. . . . But at Cape Kennedy, in the report's words, 'No procedures for this type of emergency had been established either for the crew or for the spacecraft pad work team.'

  "The incredible complacency thus revealed overshadows even the other disgraceful and disheartening findings of the review board. . . . And as the record of incompetence and careless work—devastating in its impact—is recited in the report, one wonders how many of the previous manned space flights succeeded rather than how the Apollo tragedy occurred. . . ." (NYT, 4/11/67, 44M)

- *Lunar Orbiter III*, launched Feb. 4, had sent full photographic coverage of six primary Apollo sites west of 20° east longitude and partial coverage, including high-resolution photographs, of six remaining sites in east. One objective had been to photograph a landed Surveyor with sufficient resolution to detect the spacecraft. NASA-appointed study group established location of image on Lunar Orbiter photographs that fitted all *Surveyor I* identification criteria. (NASA Proj Off)

- NASA Nike-Cajun sounding rocket launched from Wallops Station carried GSFC-instrumented grenade payload to 78-mi (126-km) altitude to obtain data on temperature, pressure, density, and wind between 22–59 mi (35–95 km) at transition from winter-time westerly to summer-time easterly circulation. Rocket was planned as companion shot with Nike-Cajun launched from Point Barrow, Alaska, April 10. Weather conditions at Wallops Station prevented launch on same day. Rocket and instrumentation performed satisfactorily. (NASA Rpt SRL)

- FAA proposed that current noise abatement rules covering large aircraft be applied also to small turbine-powered aircraft. Many turbine-powered aircraft were not included under the large transport category, but created a higher noise level than piston-type aircraft of comparable size. (FAA Release 67–36)

- *Washington Daily News* editorial warned that U.S. eagerness to sign space law treaty—under consideration in Senate since Feb. 7—might be costly: "Since the risks of Soviet deception are so great, it seems unwise to have accepted the Russian-proposed no-inspection honor system. . . . Since there is no provision for a check on space activity, the Soviet Union could push ahead with a variety of secret military programs in space while publicly attacking the right of the U.S. to an American military defense program in space. . . . Since the U.S. is an open society, any U.S. space defense program would be known. Since the Soviet society is closed, Moscow might well be able to keep its programs secret.

    "The loose way in which this space treaty was written was due in part to the Administration's eagerness to get a treaty in hand. But that looseness could lead the U.S. into a trap." (Cromley, *W News*, 4/11/67, 12)

April 11–13: Witnesses testifying before Senate Commerce Committee expressed conflicting views on proposed Corporation for Public Television.

Ford Foundation consultant Fred W. Friendly said he would oppose any long-range design that relied on Treasury Dept. appropriations to produce news and public affairs programming because such funds would be subject to annual review by Congress. Instead, he suggested, news and public affairs should be underwritten from private funds—preferably out of savings from a non-profit comsat system that would employ income from commercial programs to help meet educational television's overhead.

Dr. James R. Killian, Jr., chairman of Carnegie Commission on Educational Television, urged revision of the proposed legislation so that the Corporation would have continuing responsibility for program selection to ensure freedom "from the centralization and rigidity of a network system" and to protect autonomy of local stations.

NET President John R. White argued that the Corporation's role should be limited to matters of "high policy" and that operational activities, including the choice of programs to be carried on a network, should be subcontracted to other entities. The Corporation, he said, would not be able to make informed decisions until the programs had been submitted for national distribution. (Gould, *NYT*, 4/12/67, 79; 4/13/67, 75; 4/14/67, 67)

April 12: U.S.S.R. marked Cosmonautics Day—anniversary of first manned space flight by Cosmonaut Yuri Gagarin, in 1961—by launching *Cosmos CLV*. Orbital parameters: apogee, 286 km (178 mi); perigee, 203 km (126 mi); period, 89.2 min; and inclination, 51.8°. Satellite reentered April 20. (UPI, *W Star*, 4/12/67, 1; GSFC *SSR*, 4/30/67)

- NASA had been concerned about oxygen fire danger prior to Jan. 27 flash fire and had taken steps to curb the danger, NASA Associate Administrator for Manned Space Flight Dr. George E. Mueller told the House Committee on Science and Astronautics' Subcommittee on NASA Oversight. NASA felt it "had control of the sources of ignition . . . [but] underestimated badly . . . the course such a fire would take. If more thorough testing of flammable materials used in the cockpit had been conducted, the accident might not have happened." He denied emphatically April 11 suggestion by NAA Director of Research and Engineering Dr. John F. McCarthy that Astronaut Gus Grissom could have triggered the fire by kicking loose a wire, and said he generally agreed with the Apollo 204

Review Board's criticisms of deficiencies in project operations, conceding: "We were wrong in the standards we applied to the selection of materials for the spacecraft." Efforts were currently being directed toward eliminating flammable cockpit materials such as nylon and replacing them with glass fiber cloth and Teflon plastic. Although a study was also under way to assess the relative merits of pure oxygen and two-gas life-support systems, Dr. Mueller doubted that present system would be abandoned.

In answer to specific questions concerning escape and rescue, Dr. Charles A. Berry, Director of Medical Research and Operations at MSC, testified that the astronauts had died from asphyxia before the spacecraft hatch was opened and that their burns alone would not have caused death.

Apollo Program Director M/G Samuel C. Phillips told the Committee that his concern "about the load of work on NAA" had prompted an inspection by a NASA team of NAA's Downey, Calif., plant. He had reported to NAA President J. Leland Atwood Dec. 19, 1965, "some of things that if done differently would contribute to the progress of the program," and on April 22, 1966, noted that progress was satisfactory: NAA was "completely responsive to the programs we described ... and a very aggressive effort ... was applied to these problems." He declined to discuss specifics as requested by Rep. William F. Ryan (D-N.Y.) but indicated that NASA would consider releasing his report to the subcommittee [see April 18]. (Wilford, *NYT*, 4/13/67, 52; Reistrup, *W Post*, 4/13/67, A2)

- LaRC engineers successfully completed first of two planned orbit adjustments to place NASA's *Lunar Orbiter II* and *III* spacecraft in favorable position for major lunar eclipse April 24. Adjustment, made by burning velocity-control engine of *Lunar Orbiter III* for five seconds, decreased spacecraft's period by 57 sec to shorten length of time it would be in darkness during eclipse to 1 hr 20 min. Scientists speculated that power system of spacecraft—which experienced one hour of darkness per orbit—would not be able to survive four-hour dark period that would have occurred during eclipse without adjustment. Orbit of *Lunar Orbiter II* would be adjusted April 14. (NASA Release 67-87)
- Chairman of the Joint Chiefs of Staff Gen. Earle G. Wheeler (USA) and Deputy Secretary of Defense Cyrus Vance endorsed the space law treaty at hearings by the Senate Foreign Relations Committee. "The treaty," Vance said, "is one of the finest reflections of the people of our time meeting their duty—their duty to themselves and to future importance. It is a treaty of practical importance. We have studied this problem carefully . . . [and] looked at the implications for weapons development programs and at verification consideration, and we have concluded that this treaty will enhance our national security." General Wheeler added that he had "no military objection" to the treaty and believed it could "potentially keep the vast new frontier of space free of obstacles to international peace and order." (W *Star*, 4/12/67, A18; *NYT*, 4/13/67, 52)
- At NASA request, NAS would conduct Space Applications Summer Study at Woods Hole, Mass., during summer 1967 and 1968 to determine how orbiting satellites could best be used for practical benefits on earth. Participants would review present U.S. programs and plans for peaceful applications of earth satellites; consider and evaluate probable future uses of satellites in a variety of areas; and provide advice on nature and scope of R&D programs necessary to achieve the possibilities. The

1967 session would "provide first judgment on . . . most immediately useful applications of satellites." The 1968 session would reach final recommendations on future satellite uses. Lehigh Univ. President W. Deming Lewis was named chairman of the Central Review Committee, which would oversee the study. Work would be conducted within NRC's Engineering Div. with Laser Inc., vice-president Eugene T. Booth serving as study's executive director. (NAS-NRC Release)

- Many building blocks of life, "and possibly life itself," may exist on Jupiter, ARC scientists Dr. Cyril S. Ponnamperuma and Fritz Woeller suggested at American Chemical Society meeting in Miami. Using electrical discharges as an energy source the scientists had simulated activity of Jupiter's upper atmosphere and produced numerous building blocks common to life on earth. They concluded that energy transfer and resulting chemical processes in Jupiter's turbulent atmosphere—similar to processes believed to have produced life on earth—"may produce organic chemicals such as the forerunner compounds of amino acids and of the living cell nucleus." (NASA Release 67-79; ARC Release 67-4)

- Karl G. Harr, Jr., President of Aerospace Industries Assn., speaking before Aerospace Security Analysts in New York City, stated: "The aerospace industry today is the nation's largest industrial employer accounting for in excess of 1,250,000 jobs [and] is the nation's largest manufacturing exporter, in excess of $1.5 billion [and] employs more than one-fifth of the nation's privately employed scientists and engineers." He reported that 1966 sales approached $24 billion. Harr added: ". . . the industry as a whole has never been healthier or sounder." This soundness he attributed to ". . . stability of the government segment of the market [and] recent and projected growth of civil aviation of all kinds [and] the beginnings of successful application of aerospace capabilities to a wide range of non-aerospace activities. . . ." (Text)

- Hall of Aerospace Art would become "an integral part" of the new National Air and Space Museum, Smithsonian Institution announced. Permanent collection already included paintings and drawings of NASA manned space flight projects by Norman Rockwell and Chesley Bonestell and 18th and 19th century prints illustrating balloon flight. (Smithsonian Institution Release)

- Prof. G. A. Tokaty of The City University, London, speaking to Royal Aeronautical Society's Astronautics and Guided Flight Section meeting, in London, said: ". . . the Soviets have been (from about 1962) . . . working on a rocket system of 7 to 9 million lb thrust. Evidence of a general nature suggests that they are making good progress and have now reached an advanced stage in achieving a payload-carrying capability of the order of 400,000 to 550,000 lb. . . ." He added: "In the USSR, space science and technology have become an impressive source of new ideas, theories, industrial methods, especially in electro-radio-electronic instrument design [and] all these revolutionised the Soviet precision industry and accelerated the development of the national economy as a whole." (Text)

- MSFC Director Dr. Wernher von Braun presented Henry Ricketts a $1,625 NASA suggestion award—largest sum ever awarded for an employee suggestion—for new method of removing coating from printed circuit boards. Method shortened removal time from 15 min to two minutes and reduced costs by $571,750 the first year. (MSFC Release 67-80)

- USAF successfully launched Titan II ICBM from Vandenberg AFB toward target in the Pacific. (UPI, *W Post*, 4/13/67, A22)
- Javelin sounding rocket launched from Churchill Research Range, Canada, reached 496-mi (798-km) altitude in Southwest Center for Advanced Studies/Defence Research Telecommunications Establishment experiment to determine in conjunction with an overpass of an Isis satellite: vertical distribution of flux and energy spectra of soft electrons during a mild ionospheric disturbance; vertical distribution of thermal ion and electron densities and temperature of various species of ions; and flight qualification of a soft-electron spectrometer to be flown on ISIS–A spacecraft. Rocket and instrumentation performed satisfactorily. (NASA Rpt SRL)
- Number of mid-air aircraft collisions could be reduced "if pilots accept the idea that the plane can be quicker than the eye," Lockheed Aircraft Co. aviation psychologist Dr. Earl J. Ends said in a Washington, D.C., interview. According to his calculations based on ideal conditions: "A pilot flying 600 miles an hour will travel 88 feet from the time his eye spots an oncoming plane until the message is relayed to the brain. By the time the brain interprets the message, the plane has flown 920 feet. When the pilot decides to climb, descend or turn, he has already covered 2,680 feet. And by the time he carries out his decision the plane has covered 4,792 feet. Put another way, one of the pilots would have to see the other plane 9,548 feet away before he could change course to avoid a collision. (UPI, *NYT*, 4/13/67, 74)

April 13: NASA Deputy Administrator Dr. Robert C. Seamans, Jr., and Associate Administrator for Manned Space Flight Dr. George E. Mueller, testifying before the Senate Committee on Aeronautical and Space Sciences, absolved North American Aviation, Inc., of direct blame for the Jan. 27 Apollo fire but criticized contractor workmanship. NAA "has not always been sufficiently dedicated to engineering design or workmanship" in the Apollo program and "did not address itself properly to the training and supervision of its personnel and the adequate inspection of the work that was done," Dr. Seamans charged. As a result of NASA's demand for numerous management reforms in late 1965, NAA had made "marked improvement . . . in the last year and a half," Dr. Seamans noted, but he stressed that "it should not have been necessary for the Government to take this kind of strenuous action."

Dr. Seamans told the Committee that NASA would implement Apollo 204 Review Board's recommendations and build stronger discipline into technical requirements, schedules and management while placing the Apollo program "back on a firm operating schedule." He asserted that nothing in the Board's report suggested that the fundamental design or management structure was faulty. Basic revelation of the accident was fallibility, he stated. Dr. Mueller pointed out that the flammability and placement of cabin materials had been misjudged and compared this flaw in engineering judgment to those which resulted in catastrophic accidents in the course of aircraft development. (Testimony; Wilford, *NYT*, 4/14/67, 10; Abramson, *W Post*, 4/14/67, A2; *W Star*, 4/14/67, A7)

- President Johnson called for more Inter-American cooperation at conference of Latin American leaders in Punta del Este, Uruguay. U.S., he said, would aid Latin American development by (1) helping to develop a new common market; (2) contributing additional funds to Inter-American

Bank for "special operations" such as construction of comsat earth stations; (3) exploring possibility of temporary preferential tariff; (4) offering agricultural and educational assistance; (5) increasing food program for pre-school and school-age children and operating a nutrition demonstration project; (6) establishing Alliance for Progress centers at U.S. colleges and universities; and (7) aiding development of science and technology, especially in areas of educational broadcasting, marine science, and atomic energy. (*PD*, 4/24/67, 636)

- USAF launched unidentified satellite from WTR using Scout booster. (*Pres Rep 1967*)
- Japan's third attempt to orbit a satellite was unsuccessful when 3rd stage of Lambda 4S-3 failed to ignite at Uchinoura Range. In first attempt, Sept. 26, 1966, rocket went off course and missed orbital trajectory. Second attempt, Dec. 21, 1966, was unsuccessful because of 4th-stage ignition failure. (*W Post*, 4/13/67, A17; *SBD*, 4/14/67, 264)
- Second of three phases of tests to evaluate use of heavy-lift helicopters for Apollo crew and spacecraft recovery in KSC launch areas was being conducted by NASA and DOD at Eglin AFB. Tests would develop operational equipment, techniques, and procedures to be used at KSC where standard recovery equipment would be ineffective in case of a launch escape system abort during liftoff because of the surrounding marsh, beach, and surf. (MSC Release 67-16)
- USAF and USN issued statements supporting development of F-111 fighter aircraft as an all-purpose military aircraft.

    Secretary of the Air Force Harold Brown: "The F-111A is designed to provide an all-weather strike potential superior to that of any tactical aircraft in the world today. It will fly faster, farther and carry a greater payload, nuclear or conventional, with an unexcelled capability to penetrate sophisticated enemy defenses. . . . We are confident from the level of performance already displayed in the very extensive flight tests performed on development aircraft that the operational F-111A will provide these vitally needed capabilities, and will meet the military requirements established for the aircraft at the inception of the program."

    Secretary of the Navy Paul H. Nitze: ". . . development of a system as complex as the F-111B/Phoenix has produced sizeable engineering problems. I continue to be impressed with the manner in which they have been met and overcome. I have confidence that changes proposed by the Navy, approved by the Secretary of Defense and now being implemented, will make the F-111B suitable for operation from our first-line attack carriers and will give it the capabilities to perform its mission." (*CR*, 4/17/67, S5300)
- U.S. bombers were being guided to targets in North Vietnam by daily meteorological photos received from Essa and Nimbus satellites, USAF announced. Detailed photos, monitored by USAF's weather stations in Saigon and Ugon, Thailand, permitted USAF to divert aircraft to areas that were unexpectedly free from clouds. Since equipment for reception was relatively inexpensive, USAF officials speculated that North Vietnamese might also be receiving the photos and using them for defense planning. (AP, *NYT*, 4/14/67, 3)
- A 2,000-mph USAF SR-71 reconnaissance aircraft on routine training flight from Beale AFB, Calif., crashed and burned near Las Vegas, N. Mex.—60 mi from site of Jan. 25, 1966, crash of another SR-71. Crew escaped major injury. Aircraft had been built for USAF by Lockheed as successor

to the U-2 and had been undergoing tests since December 1964. (UPI, *NYT*, 4/15/67, 13; AP, *W Post*, 4/14/67, 9)

April 14: LaRC engineers successfully completed second of two planned orbit adjustments to place NASA's *Lunar Orbiter II* and *III* spacecraft in favorable position for major solar eclipse April 24, by firing *Lunar Orbiter II*'s velocity-control engine for three seconds and shortening spacecraft's period by 65 sec. Similar maneuver had been accomplished with *Lunar Orbiter III* April 12. (NASA Release 67-92)

- Launch of ESRO's scientific satellite Esro II from WTR was postponed from April 27 until late May because of tape recorder malfunction. Satellite was first of two scheduled for launch by NASA under cooperative agreement with ESRO. (NASA Release 67-91)

- MSFC had awarded one-year, cost-plus-incentive-award-fee contract extensions totaling more than $57 million for engineering, fabrication, and institutional support services for Saturn launch vehicle program. Recipients: Sperry Rand Corp., $12,695,727; Vitro Corp., $5,344,159; Brown Engineering Co., $12,350,140; Spaco, Inc., $5,971,638; Northrop Corp., $3,905,000; Hayes International Corp., $4,969,277; Management Services, Inc., $5,560,941; Rust Engineering Co., $599,090; and RCA Service Co., $5,749,907. (MSFC Release 67-81)

- MSC awarded Owens-Corning Fiberglas Corp. a $97,000, cost-plus-fixed-fee study contract to evaluate Beta fiber, a noncombustible fabric, for possible use in Apollo spacecraft and crew systems equipment. (NASA Release 67-90)

- Informed sources speculated that U.S.S.R. had been conducting heat-shield reentry tests with Cosmos satellites since November 1966 in preparation for manned flights to moon, the *New York Times* reported. Spacecraft on missions from moon would reenter earth's atmosphere at about 25,000 mph and heat to 5,000–6,000° F—too hot for heat shields currently used on earth-orbiting missions. (*NYT*, 4/15/67, 13)

- Employment in the aerospace industry would reach 1,384,000 by September 1967—a 3.4% increase over September 1966—Aerospace Industries Assn. predicted. Forecast, based on survey of 60 companies representing 80% of entire aerospace industry, cited continuing demand for civilian aircraft; stable levels of missile and space sales; and increases in nonaerospace activities as reasons for steady growth in number of employees. (AIA Release 67-17)

- Former Astronaut John H. Glenn, Jr., had signed contract to appear in at least eight Wolper Productions' TV documentaries on great explorations in history, the *New York Times* reported. Series, beginning in January 1968, would include Henry M. Stanley's African search for David Livingstone, Captain James Cook's exploration of the South Pacific, and Roald Amundsen's expedition to the South Pole. (Gent, *NYT*, 4/14/67, 67)

- NASA and USAF had completed loan agreement whereby USAF had loaned to NASA's FRC three F-104 aircraft to "support NASA projects." Agreement would terminate Dec. 31, 1968, unless extended by mutual arrangements. Logistical support of aircraft was included. (NMI-1052.89)

April 16: Astronaut Frank Borman objected to making anyone a scapegoat for Jan. 27 Apollo flash fire and said everyone connected with the program must be blamed. Interviewed on ABC television network with Chairman of House Science and Astronautics Committee's Subcommittee on NASA Oversight Rep. Olin E. Teague (D-Tex.), Borman said he had

"never . . . seen safety sacrificed for anything" in U.S. space program. Both men suggested that the greatest safety problem was tendency toward complacency about dangers inherent in tests. (W *Star*, 4/17/67, A2)
- Delegates from U.S.S.R. and Communist-bloc countries, who met in Moscow April 3–15, had drafted a "joint satellite and rocket launching program" and had agreed to set up a satellite communications program "open to all countries willing to join," Tass announced. No further details were released. (UPI, *NYT*, 4/17/67, 10)
- U.S. technological lead might be lost because of increasing Vietnam war costs, L/G Ira C. Eaker (USAF, Ret.) wrote in the *Newport News Press*. He cited examples of this "dangerous possibility": (1) defrayment of war costs at the expense of other military programs such as maintenance of strategic forces and development of new weapons; (2) deferment and possible cancellation of SST program which "now appears to be stalled on dead center"; and (3) cutting of space program budget and jeopardizing of its goals. "Needed now in this emergency is a people alert and aware of their hazard and determined to maintain the world's highest living standard the way they won it, with technological leadership." (Eaker, *Newport News Press*, 4/16/67, 21)

April 17: NASA's *Surveyor III* (Surveyor C) was successfully launched from ETR by Atlas-Centaur (AC–12) booster on 65-hr lunar intercept trajectory. Primary mission for the 2,283-lb spacecraft was to perform a soft-landing east of the *Surveyor I* landing site within the Apollo zone and to obtain post-landing TV pictures of the lunar surface. As secondary mission spacecraft would obtain data on lunar surface bearing strength, radar reflectivity, and thermal properties and photograph use of a surface sampler to manipulate the lunar surface [see April 19].

At 34:53 GET *Surveyor III* with two-burn Centaur—being used for the first time on a Surveyor mission—separated from Atlas and ignited to reach 100-mi-altitude circular parking orbit where it coasted for 22 min; Centaur then reignited to escape earth orbit and boost spacecraft toward the moon. *Surveyor III* separated from Centaur, deployed its solar panels, and locked on the sun and the star Canopus. Based on a 248-mi target miss estimate, a 4.2-mi midcourse maneuver was conducted shortly after midnight April 18 to assure precise landing on target in the Ocean of Storms.

Surveyor III was third in series of seven spacecraft designed to prove out design, develop technology of lunar soft-landing, obtain post-landing TV pictures of lunar surface, and provide basic scientific and engineering data in support of Project Apollo. Surveyor program was directed by OSSA Lunar and Planetary Programs Div.; project management was assigned to JPL; Atlas-Centaur launch vehicle was managed by Lewis Research Center; and prime contractor for spacecraft development and design was Hughes Aircraft Co. (NASA Proj Off; NASA Release 67–85)
- Astronaut M. Scott Carpenter would never be able to fly in space again because of an arm injury suffered in 1964 motorcycle accident, the Washington *Evening Star* reported. Surgery performed six months ago had failed to restore full mobility of the arm. (AP, W *Star*, 4/17/67, A6)
- Soviet scientists had succeeded in raising the temperature in lower atmosphere by as much as 14.4° F by dissipating clouds with carbon dioxide and other chemicals, Soviet Hydrometeorological Service director Dr.

Yevgeny K. Federov told World Meteorological Congress in Geneva. Dispersal of "tens of thousands of square kilometers" of clouds could serve as "a mechanism for triggering off the modification of more massive meteorological processes," he said. (*NYT*, 4/18/67, 26)

- The press praised Apollo 204 Review Board's investigation of Jan. 27 flash fire.

Technology Week: "The harsh, scathing report of the Apollo 204 Review Board has effectively laid to rest charges that the board was assembled to conduct a whitewash of the January 27 fire, which took the lives of three astronauts. The board should be commended for the vigor with which it undertook to establish the cause of the fire, the energy with which it attacked what it regards as dangerous practices and procedures and the directness with which it distributes the blame. . . . Perhaps the question that will linger longest about the January tragedy is the part played by the pressure of the lunar landing deadlines. We personally believe that such target dates play a large and important factor in motivating the space program and keeping it moving at a healthy pace. But they should not be so sacred as to invoke carelessness. . . . Without waiting for another accident, it might be well to undertake the same sort of review of practices and procedures throughout the major elements of Apollo that has just taken place in regard to the spacecraft. . . . The Apollo 204 Review Board has set a good example of what can be accomplished." (Coughlin, *Tech Wk*, 4/17/67, 50)

Aviation Week: "Its report is technically incisive and sharply critical of NASA and the private contractors manufacturing Apollo hardware. It is as thorough and technically solid as any other investigation of aerospace disaster in the past three decades. It disproves earlier charges that NASA is incapable of investigating its own technical faults. In fact, it would have been impossible to assemble another group sufficiently knowledgeable about the Apollo system from any other source to complete a useful investigation within this time span.

". . . there are none of the pat answers, so often demanded by legislators and harping critics, to the dangerous and complex problems of probing the unknown. Apollo is man's most daring foray in his history on this planet. If all the problems involved in a lunar voyage were so obvious and the solutions so simple as many critics seem to think, we could all buy tickets for a lunar excursion tomorrow." (Hotz, *Av Wk*, 4/17/67, 21)

- A satellite capable of broadcasting directly to the home, bypassing local stations and national networks, might be technically feasible within five years, ComSatCorp President Dr. Joseph V. Charyk testified before FCC. Economic feasibility, however, might require more than five years, he said. Dr. Charyk appeared as a witness for the Justice Dept., which was challenging the proposed acquisition of ABC by ITT. (Mintz, *W Post*, 4/18/67, A2)

- NASA Administrator James E. Webb told the Senate Committee on Aeronautical and Space Sciences that NASA would present, within two weeks, detailed estimates of how much the Jan. 27 Apollo accident had cost; what changes in spacecraft would be made to reduce fire hazards; and the earliest date for the first manned Apollo mission. He declined to disclose his thoughts concerning canceling NASA's contract with North American Aviation, Inc., because of failures to meet development schedules and stay within cost estimates: "They did not do all the things they should have done, but I can say the same thing about us at NASA." He stated his conviction that he was obligated to "act responsibly in respecting confidences between himself and contractors." Denying that undue risks were taken in order to speed the program, Webb emphasized that NASA had "never put schedule above safety." Apollo 204 Review Board's study was "a good report," he said, but it should not be construed to be an evaluation of the entire space program: "The board has done an outstanding investigation of one accident but it is not a full and fair appraisal of the space program . . . [and] cannot be taken as a guide to the realities of this program. . . ." In response to questioning Webb stated that, in his judgment, the pacing item in the manned lunar landing program remained the Saturn V booster, and when it was ready for manned flight there would be a spacecraft ready.

Committee Chairman Clinton P. Anderson (D-N.Mex.) requested that NASA witnesses appear again not later than May 9 to detail changes the program, as suggested by Webb. (Wilford, *NYT*, 4/18/67, C26; O'Toole, *W Post*, 4/18/67, A1; *SBD*, 4/18/67, 275)

- Astronauts Donald K. Slayton, Alan B. Shepard, Walter M. Schirra, Jr., Frank Borman, and James A. McDivitt, testifying before the House Committee on Science and Astronautics' Subcommittee on NASA Oversight, expressed confidence in NASA's management of U.S. space program. They noted that there had been only one mishap in the program, compared with 16 successful manned flights under the same management. Since fire had been foreseen as a significant danger in orbit but not on the launch pad before fueling of the launch vehicle, everyone had been "lulled into a sense of false security . . . [and] grossly underestimated the fire potential of the spacecraft," they said. Schirra explained confidence astronauts felt prior to the fire Jan. 27: "The back-up crew [of which Schirra was commander] had left the Cape with the feeling that we had a good spacecraft behind us. . . . I was not at all prepared for the news I received when I arrived at Houston." (Hines, *W Star*, 4/18/67, 1)

April 17–20: Some of the more important preliminary results from NASA's interplanetary spacecraft *Pioneer VI* (launched Dec. 16, 1965) and *VII* (launched Aug. 17, 1966) were presented at American Geophysical Union's 48th annual meeting in Washington, D.C. Spacecraft, which had flown almost 200 million mi in their solar orbits, had provided: (1) better definition of solar atmosphere; (2) additional data on solar wind; and (3) improved measurement of earth's magnetosphere. Report by ARC scientists Dr. John H. Wolfe, R. W. Silva, and D. D. McKibbin indicated that magnetosphere ends about 3.5 million mi from earth going away from the sun; previous estimates had varied from 240,000 mi to 100 million mi. (NASA Release 67-94)

April 18: NASA Administrator James E. Webb submitted summary of M/G Samuel Phillips' (USAF) report on North American Aviation, Inc.'s workmanship (the "Phillips Report") to the Senate Committee on Aeronautical and Space Sciences: "Our main criticism was that the Space and Information Division was overmanned and that the S–II and CSM programs could be done, and done better by fewer people, better organized with particular emphasis on achieving greater competence in key management and technical positions. . . . Work tasks were inadequately defined and scheduled . . . and, largely as a result of these conditions, budget and control were deficient.

"... in each of ... [the] problem areas we identified in December 1965, we found ... improvement in the situation in April 1966. I do not intend, however, to leave the impression that we were completely satisfied with North American's performance. We did determine ... that the corrective actions already taken and those in the process of being implemented in April 1966 were appropriate ones that reflected a responsive reaction by NAA to the criticism we had expressed during our review and provided some measure of confidence for improved future performance. ..."

Testifying in support of NASA's FY 1968 budget, Webb warned against permitting the Jan. 27 Apollo accident to deter U.S. space efforts. It should not, he said, "lead us to lose sight of the continued progress and vigorous efforts the Soviets are making. ... With the world situation what it is, and with clear indications that the Soviets are going ahead with a large-scale, long-term program in space, it is especially important this year for the United States to let the world know that we cannot and will not surrender our hard-won position, even though we will not mount the larger effort required to catch up with them." Describing the proposed budget as "forward looking," Webb conceded that it would not "enable us [NASA] to move toward important goals with as much assurance as we would like." He outlined NASA's planned program and urged support for all future projects: "If we fail to make these decisions in our consideration of the 1968 budget, they will go by default. ... Later when we find we do not have the capabilities we could have had and that others are accomplishing what we might have achieved, it will be too late." (Wood, *W Post*, 4/19/67, A9; Sehlstedt, *B Sun*, 4/19/67, 1)

- NASA Nike-Cajun sounding rocket launched from Point Barrow, Alaska, carried GSFC-instrumented grenade payload to obtain data on temperature, pressure, density, and wind between 22–59 mi (35–95 km) at transition from winter-time westerly to summer-time easterly circulation. Rocket and instrumentation performed satisfactorily. (NASA Rpt SRL)
- Informed sources reported that U.S.S.R.'s Tu–144 supersonic transport would not be ready for the Paris International Air Show in May as UPI reported March 7 but would still fly before the Anglo-French Concorde and U.S. SST, according to AP. First test flight of Tu–144 would be in late 1967. (AP, *NYT*, 4/19/67, 73)

April 19: First manned Apollo flight had been delayed at least one year by Jan. 27 fire at KSC, Dr. George E. Mueller, NASA Associate Administrator for Manned Space Flight, told Senate Committee on Aeronautical and Space Sciences. He said he believed men could still land on the moon by the original 1970 deadline "although the probability is lowered." Dr. Mueller told the Committee that the replacement spacecraft for the first manned Apollo mission, which incorporated "all changes resulting from the findings of the Apollo accident investigation," was on the assembly line at North American Aviation, Inc.'s plant in Downey, Calif. (AP, *B Sun*, 4/20/67)

- USAF SV–5D lifting body vehicle was successfully launched from WTR by Atlas booster on reentry mission and telemetered excellent performance data. Vehicle—third of four in USAF's Precision Recovery Including Maneuvering Entry (Prime) program—was second to perform preprogrammed cross-range maneuvers and first to be successfully recovered. First SV–5D vehicle was launched Dec. 21, 1966, and lost when parachute malfunctioned during reentry; second, launched March 5, sank in Pacific after losing its flotation gear. (*Tech Wk*, 4/24/67, 13; *Av Wk*, 4/24/67, 35)

- NASA Nike-Cajun sounding rocket launched from Wallops Station carried GSFC-instrumented grenade payload to 67-mi (108-km) altitude to obtain data on temperature, pressure, density, and wind between 22–59 mi (35–95 km) at transition from winter-time westerly to summer-time easterly circulation. Rocket and instrumentation performance were only partly successful due to low apogee of rocket, which placed the highest grenade burst at only 42-mi altitude. (NASA Rpt SRL)

- NASA Administrator James E. Webb addressed Edison Electric Institute's 35th Annual Convention in New Orleans: "Our flight projects, both manned and unmanned, depend upon electronic links back to earth where their data are recorded and analyzed, where their actions are controlled, and where the new knowledge is put to use. We must therefore continue to provide and operate the tracking and data acquisition networks and their associated communications. Our technology utilization program will continue to provide effective means for making available the fruits of our technological progress to the scientific and industrial community at large. And, in every area of our program effort, we are continuing the underlying supporting research and technology which make possible the development of experiments, the implementation of ideas, and the advances in technology which culminate in meaningful aeronautical and space projects. . . . Research and development, scientific inquiry and technological advance, are the province of men working here on earth. New knowledge and new tools for its application are major elements of our total national power—political, military, and economic." (Transcript)

- Edgar M. Cortright, NASA Deputy Associate Administrator for Space Science and Applications, testifying on NASA FY 1968 authorization bill (H.R. 6470) before House Committee on Science and Astronautics' Subcommittee on Space Science and Applications, said he had "high confidence that we can sterilize the Voyager system by 1973 to the satisfaction of the biologists." He disagreed with the opinion of Dr. N. H. Horowitz (Cal Tech) that U.S. was adhering to sterilization standards which were exceedingly severe and represented an obstacle to U.S. plans for planetary exploration [see March 24]. To support his view, Cortright presented a summary of the findings of Spacecraft Sterilization Advisory Board, chaired by Dr. Richard Bond, Univ. of Minnesota, which was "examining the statements" made by Dr. Horowitz. Cortright said it was felt that "probably Prof. Horowitz has overestimated the difficulty posed by the current sterilization requirements and underestimated the ability of aerospace companies and government laboratories to cope with it." Most of the components and parts NASA planned to use in Voyager and sterilize were already "demonstrably sterilizable," so there seemed to be no justification in abandoning a goal which the majority of the scientific community feels is legitimate." (Transcript)

- The "Phillips Report"—report by NASA's Apollo Program Director M/G Samuel C. Phillips which "found insufficient competence in key management and technical positions" at North American Aviation, Inc., 16 mo before Apollo accident—was being suppressed by NASA, Rep. William F. Ryan (D-N.Y.) charged in an AP interview. Representative Ryan said he would ask Chairman of House Committee on Science and Astronautics' Subcommittee on NASA Oversight Rep. Olin E. Teague (D-Tex.)

to demand NASA's submission of the full report for examination by the subcommittee. (AP, W *Star*, 4/19/67, A6)

- Letter to editor of *Space Business Daily* by unidentified leader in U.S. space program attacked NASA's use of pure oxygen in life-support systems: ". . . most engineering people in industry, most reputable bio-engineering people in USAF and NASA felt two gas was the way to go. In addition, there has never been an overwhelming engineering reason for a pure oxygen system. . . . The argument of reliability and complexity is ludicrous when you examine the complexity of other systems in the spacecraft and booster. Now it is even more ironic in that, if pure oxygen is kept, all sorts of complicating compensations must be made. The truth is that a few key people in NASA flew in the face of strong industry technical advice against pure oxygen. To add to the problem, Gemini made them cocky and they forgot completely the sleeping monster they had created. Take anyone, regardless of skill and training, show him a fire in pure oxygen, and ask him if he'd agree to live in pure oxygen with that danger in the complex environment of a spacecraft. The answer is painfully obvious. Now we'll make a bad situation worse by jury rigging a system to avoid pure oxygen on the pad—but—we'll still have it in space.

    "NASA and everyone must face it. It was a bad decision compounded by complacency, for example, the hatch design. If the atmosphere were two gas, chances are, survival would have occurred. Let's get guts, fix the basic deficiency and take the delay. That is the only answer and sooner or later it must be faced." (*SBD*, 4/19/67, 282)

- Five communications companies filed joint application with FCC for construction of new comsat earth station near Rowlesburg, W. Va. Companies: ComSatCorp; AT&T; ITT WorldCom; RCA; and WUI. Proposed station, to be completed in 1968, would handle all types of high-quality communications. (ComSatCorp Release 67–31)

- M/G Holger Nelson Toftoy (USA, Ret.), a leading Army ordnance expert who recommended that former German V-2 development team under the leadership of Dr. Wernher von Braun be brought to U.S. at end of World War II, died at Walter Reed Army Medical Center after a long illness. General Toftoy became commander of White Sands Proving Grounds immediately following World War II. In 1952 he became Director of the Ordnance Missiles Laboratory at Redstone Arsenal, Huntsville, Ala., and in 1958, when Redstone was reorganized as the Army Ordnance Missile Command, he was named Deputy Commander under M/G John B. Medaris. General Toftoy subsequently became commander of Aberdeen (Md.) Proving Ground until his retirement in 1960. (UPI, *NYT*, 4/20/67)

April 19–30: NASA's *Surveyor III* became second U.S. spacecraft to soft-land on the moon when it touched down in the Ocean of Storms after 64 hr 59 min flight and began transmitting the first of 5,487 detailed television pictures to JPL Deep Space Facilities, Goldstone, Calif.

Landing sequence began when *Surveyor III* shifted its normal cruising attitude to position main retrorocket. Triggered by radar (later ejected), main retromotor burned and slowed spacecraft to about 250 mph. Vernier engines, which were to have slowed spacecraft to about 3 mph, failed to cut off at 14 ft above lunar surface as planned, and the 620-lb spacecraft made three separate landings—covering distance of 33 ft—before engines shut off 37 sec after initial touchdown.

April 19-30: Lunar material scooped onto *Surveyor III* footpad by surface sampler on the moon was photographed through red filter in spacecraft's TV camera April 26. Color calibration chart in foreground is mounted on the spacecraft's leg.

First pictures transmitted were unclear because of sun's brightness, but later photos were excellent. They revealed that spacecraft was resting in a depression or small crater with 5° slope—which was expected to hinder photography of lunar horizon—and showed impressions made by footpads during multiple landing. Shortly after touchdown a telemetry malfunction falsely indicated major drain on power system, but photos confirmed that spacecraft's batteries were intact and undamaged.

In addition to photographic equipment, including two additional mirrors to expand camera's field of view, spacecraft also carried a shovel-like surface sampler which would scoop soil, move debris, and strike lunar rock with an expandable arm operated by radio signals from earth. On April 21–22 sampler began bearing strength tests and excavation of trenches up to six inches deep. On April 23–24, cooled by two-hour lunar eclipse, *Surveyor III* took 118 excellent color photos of Venus, a star cluster, and earth as it passed between the moon and the sun. On basis of preliminary digging operations and photos, project scientists concluded that lunar soil had consistency similar to wet sand and a bearing strength of about six pounds per square inch—firm enough to support Apollo spacecraft. Cal Tech professor and advisor to JPL on soil mechanics Dr. Ronald F. Scott explained: "The surface is like ordinary soil ... a lot like fine grained sod or damp beach sand, and it behaves in perhaps a disappointingly ordinary way. The area presents no hazardous conditions and looks good for a landing site. An astronaut walking across the surface would not need snowshoes." On May 3 communications with spacecraft were halted temporarily to conserve battery strength throughout two-week lunar night.

Performance of *Surveyor III*, with equipment identical to *Surveyor I* and *Surveyor II* except for soil sampler and two additional mirrors, was near flawless, and expected lifetime was indefinite. *Surveyor I* was launched May 30, 1966, successfully softlanded on moon June 2, 1966, in Ocean of Storms, and transmitted 10,338 pictures to earth. *Surveyor II* was successfully launched Sept. 20, 1966, but failed to softland because of an ignition failure. (NASA Proj. Off; NASA Release 67–85; UPI, W *Star*, 4/23/67, A10; O'Toole, W *Post*, 4/22/67, A4; 4/23/67, A1)

April 20: NASA successfully launched *Essa V* (TOS–C), fifth meteorological satellite in ESSA's Tiros Operational Satellite (TOS) system, from WTR using Thrust-Augmented Thor-Delta booster. Satellite achieved nearly-polar, sun-synchronous orbit with 883-mi (1,421-km) apogee; 840-mi (1,352-km) perigee; 113.5-min period; and 101.9° inclination. Wheel orientation maneuver would be completed during 18th orbit [see April 24], at which time first photos would be programmed and NASA would check out spacecraft before turning its operation over to ESSA.

An advanced version of the cartwheel configuration, 325-lb *Essa V* carried two Advanced Vidicon Camera System (AVCS) cameras which would provide 24-hr global weather coverage. Photos would be stored on magnetic tape until readout by ESSA's Command and Data Acquisition (CDA) stations at Fairbanks, Alaska, and Wallops Island, Va., to supplement *Essa III*—expected to become almost completely inoperable by early June.

ESSA financed, managed, and operated TOS system; GSFC was responsible for procurement, launch, and initial checkout of spacecraft in orbit. *Essa I* was launched Feb. 3, 1966; *Essa II*, Feb. 28, 1966; *Essa III*, Oct. 2, 1966; and *Essa IV*, Jan. 26, 1967. (NASA Proj Off; ESSA Release 67–39)

- NASA Deputy Administrator Dr. Robert C. Seamans, Jr., testifying on NASA's FY 1968 authorization bill (H.R. 6470) before House Committee on Science and Astronautics' Subcommittee on Advanced Research and Technology, urged congressional approval for funding to begin program to develop Nerva II nuclear rocket engine. Estimated cost over 10 yrs would be $2 billion: $500 million would be needed to develop nuclear stage and $1 billion for ground testing. NASA's decision to develop Nerva II was based primarily on "our own national objectives in space; what do we do in the future; now looking ahead to the late '70s and '80s, as a requirement for booster capability in this country. . . . However, we also must consider our capability in the world arena and, of course, specifically today we have to be cognizant the best that we can of the Russian capability and the Russian plans." Dr. Seamans noted that Soviet Proton booster could launch 50,000 lbs into earth orbit, compared to 40,000–41,000-lb capability of Uprated Saturn I. "We have every reason to believe that in the [Soviet] program they recognized the same kind of needs that we recognized for our program and that they are moving on a broad front, including the development of very large boosters." (Transcript, 1131–4)

- NASA Associate Administrator for Space Science and Applications Dr. Homer E. Newell, testifying on NASA's FY 1968 authorization bill (S. 781) before Senate Committee on Aeronautical and Space Sciences, urged congressional approval to begin 10-yr, $2.2-billion Voyager program. "Our concept in developing the Voyager spacecraft is to develop a technology that can be used for exploring the solar system. Although we talk in terms of going to Mars for the first missions, our plan is to modularize the system so that by rearranging and reorganizing pieces, one can go to Venus or one might develop a spacecraft to go out to Jupiter . . . we are planning on developing for planetary exploration the kind of technology that we developed in the Explorer series for exploration of the vicinity of the earth." Dr. Newell said NASA had no proposal for manned exploration of the planets at the present time. (Transcript, 320–2)

- Testimony by NASA and NAA officials at congressional hearings on Apollo accident was loaded with technical information to avoid discussion of mismanagement and responsibility, William Hines charged in the Washington *Evening Star:* "In execution, it was a masterful operation. Experts avalanched data on the committees until . . . [they had] far more technical information than they could possibly handle. . . ." Hines cited statement by Sen. Walter Mondale (D-Minn.) which "got to the very core" of the problem: " 'I am not interested in knowing enough about building spacecraft so that I can build one in my backyard, but I do think I have a responsibility to know enough about questions of management, contractor performance and quality control so I can pass judgment on the question of whether these are fundamental issues that we need to deal with.' "

"If members of the House committee had all kept their eyes on this central issue and refused to be confused by the irrelevant and the trivial, the hearings on that side of the Capitol would have been far more fruitful to date." (Hines, W *Star*, 4/20/67, A6)

- Unconfirmed reports in Moscow indicated that "a spectacular and significant" space venture by U.S.S.R. would occur within one week, according to AP. (AP, W *Star*, 4/20/67, 1)

- Third U.S. Redstone rocket in the "Project Sparta" antimissile-missile research program was fired from Woomera Rocket Range near Canberra, Australia, Supply Minister Denham Henty announced. (AP, *C Sun-Times*, 4/21/67)

- Large antennas on satellites, missiles, aircraft, and household rooftops might soon be replaced by a tiny two- or three-ounce microcircuit device called Subminiature Integrated Antenna (SIA) being developed by USAF. Broad-banded SIA would have several octaves of bandwidth to cover entire VHF and UHF range; allow 50–100 times as much inflow of electrical current for sharper TV pictures; replace larger antennas that sometimes revealed presence of combat troops to the enemy; and reduce weight of future air and space vehicles by 10–500 lbs. (AFSC Release 40.67)

- NATO was considering establishing a $45-million comsat system which would permit simultaneous consultation of its leaders during crisis, William Beecher reported in *New York Times*. U.S. would supply and launch two satellites; costs would be shared by NATO members. Proposal was expected to be formally approved during May. (Beecher, *NYT*, 4/21/67, 11)

April 20–21: Thirteen high school students from Northwestern states presented scientific papers at regional Youth Science Congress at ARC. Congress was sponsored jointly by NASA and National Science Teachers Assn. to "encourage promising students to enter science as a career, and to develop their abilities in science through scientific investigation." (ARC Release 67-6)

April 21: Seven members of the House Committee on Science and Astronautics' Subcommittee on NASA Oversight continued hearings on Jan. 27 Apollo accident at KSC by touring AS–204 launch pad and questioning eyewitnesses.

Donald Nichols, assistant test manager at KSC, told the Subcommittee that the decision to keep the hatch closed during Jan. 27 Apollo rehearsal might have cost the lives of the crew. On Oct. 31, 1966, he said, decision was made to change plans that would have kept the hatch open until the crew had been better trained on emergency procedures. Officials decided that hatch could be locked and emergency escape practice conducted after the test, he said. If test procedures had not been changed, the hatch would have been open or the astronauts not present in the spacecraft.

Thomas R. Baron, former North American Aviation, Inc., quality control inspector, said that Apollo 1 astronauts smelled smoke in their spacecraft 12 min before they died and tried for five minutes to escape. This information, he said, had been supplied by NAA electrical technician Mervin Holmberg.

Holmberg later made a surprise appearance before the Subcommittee and denied Baron's allegation: "I listened to his [Baron's] speculation, but I didn't make any comment" on what caused the fire. "I wasn't even near the accident when it happened." Baron, who had submitted a report to NASA [see During March] prior to Jan. 27 accident charging laxity, poor morale, and bad management on the part of NAA, was sharply criticized by subcommittee members, and his statement was disclaimed. (Abramson, *W Post*, 4/22/67, A4)

- USN F–111B fighter aircraft crashed on takeoff in test flight at Grumman Aircraft Engineering Corp.'s Calverton, N.Y., flight test center. Both Grumman pilots were killed. Adapted for use by USN and USAF, the F–111 (formerly TFX) had been subject of continued controversy between Congress and DOD since Secretary of Defense Robert S. McNamara insisted that the F–111 be used by both services and overruled military leaders' objections to selection of General Dynamics Corp. for construction of the aircraft. Crash was first for USN test model, which had been stripped to remove excess weight; USAF test model (F–111A) crashed in January, killing one crew member. (Wilson, *W Post*, 4/22/67, A5)

- Aquanauts diving to extremely low ocean depths should breathe almost pure hydrogen in their diving suits, Dr. R. W. Brauer of Wrightsville Marine Bio-Medical Laboratory suggested at Federation of American Societies for Experimental Biology in Chicago. Dr. Brauer planned to subject himself to such an atmosphere—98% hydrogen and 2% oxygen—at a pressure equivalent to 700- or 800-ft depth within the next six months. Divers long ago passed the depth limits that were feasible with compressed air and were now approaching the probable safe limits with an atmosphere in which helium was the major ingredient, Dr. Brauer said. Yet oil drilling, prospecting operations, and other undersea activities were pushing divers to further depths which required new types of atmosphere and, possibly, drugs to modify the narcotic effects produced by the high pressure air. (Schmeck, *NYT*, 4/23/67, 60)

- NASA awarded $213,000 contract to Lockheed Missiles and Space Co. to design an orbiting primate spacecraft. (LaRC Release)
- U.S. scientists' interest in military research had declined substantially, Bryce Nelson wrote in *Science*: "Since the beginning of the Second World War, many American scientists have regarded it as a duty to work on military research in times of national emergency and have often done so with enthusiasm. However . . . many scientists now seem to show little feeling of obligation to do military research." Suggested reasons for lack of interest: (1) disagreement with U.S. policy in Vietnam; (2) boredom with defense problems; (3) higher salaries in private industry than in DOD; and (4) increased interest in civilian problems such as urban development, poverty, and transportation. (Nelson, *Science*, 4/21/61, 24)
- The efficiency of Soviet scientific research was deteriorating and the price of getting an idea into production was rising, biologist Alexander Neifakh wrote in *Literaturnaya Gazeta*. Although scientists had doubled every 7 to 10 yrs, published works had doubled only every 13–15 yrs. "This shows that the effectiveness of scientific activity per scientist or per ruble invested in science is consistently falling," Neifakh concluded. (Reuters, *NYT*, 4/22/67, 31)

April 21–23: Editorial comment on NASA *Surveyor III*'s successful softlanding on the moon April 19.

Miami News: "The National Aeronautics and Space Administration badly needed a victory to demonstrate that the moon program is not threatened by technical deficiencies, as the Apollo investigation seems to suggest. The landing of Surveyor III . . . and the transmission of TV picture fills the bill. . . . There is no denying, however, that Apollo is in serious difficulty . . . [and] NASA is creating the impression that it is unwilling to see the investigation carried to a complete conclusion. . . . Surveyor and dozens of similar achievements demonstrate that the overall space program has been a remarkable success. NASA would make a mistake to see the luster of this record dimmed by its failure to cooperate fully in the Apollo investigation." (*Miami News*, 4/21/67)

New York Times: "Inevitably, Surveyor 3 calls attention to the fact that the first stages of lunar study and exploration do not require men, that the moon can be investigated to a considerable extent without incurring either the great risks or high costs of manned lunar flight. From the scientific and technical points of view, Surveyor-type flights are a rational and feasible alternative to Apollo. . . . Manned space exploration has unavoidable dangers, but these are multiplied many times by the pressure of the 1970 deadline that dominates the Apollo project. What Surveyor 3 has done is to remind us that—aside from propaganda and prestige considerations—there is no need to endanger brave men's lives." (*NYT*, 4/23/67, 10B)

Washington Post: "With the success in landing Surveyor III . . .

[NASA] once again exhibited its stunning scientific and technological skills. . . . Surely the time cannot be far off when men will be shuttling from earth to the moon—and far beyond. . . . Bold ventures of mankind are always undertaken in din of criticism, and the manned flight to the moon will be no exception to the rule. Space exploration will continue—not because it excites expectations of an El Dorado or a Fountain of Youth—but because it is an inescapable challenge to human curiosity and imagination." (*W Post*, 4/23/67, C6)

April 23-24: U.S.S.R. successfully launched *Soyuz I* spacecraft with Vladimir M. Komarov, but mission ended 26 hrs after launch when spacecraft failed to reenter properly and crash-landed, killing Cosmonaut Komarov.

Launched April 23 from Baikonur by a "powerful new carrier rocket," *Soyuz I* entered initial orbit "close to the estimated one," Tass announced with unprecedented speed, only one hour 20 min after launch. Orbital parameters: apogee, 224 km (139 mi); perigee, 201 km (125 mi); period, 88.1 min; inclination, 51.4°. Onboard equipment was "functioning normally" and Komarov was "in good health and feeling well." During the flight Tass distributed only a few terse reports in contrast to extensive publicity usually accorded manned missions. Final announcement, which followed 11-hr silence, said accident occurred during 18th orbit after program of test flight had been completed and spacecraft had been successfully braked with retrorockets for reentry. ". . . when the main parachute was opened at an altitude of seven kilometers [4.3 mi], the lines of the parachute, according to preliminary information, got snarled and the spaceship descended at great speed, which resulted in Komarov's death." Tass did not report whether the cosmonaut had a personal parachute to attempt an escape. Western experts believed *Soyuz I* was tumbling or spinning out of control during reentry: when parachutes were deployed, tumbling caused them to tangle among themselves or wrap around spacecraft resulting in crash-landing. Experts said difficulties apparently developed earlier in the flight because Komarov tried to reenter during 15th orbit but was unable to control his spacecraft.

Informed sources had speculated that mission involved the launch of a second spacecraft carrying several cosmonauts that would dock with *Soyuz I* and permit at least one cosmonaut to transfer from second spacecraft to *Soyuz I*. Tass's official announcement said mission would "test the new piloted spaceship, check the ship's system and elements in conditions of space flight, conduct expanded scientific and physical-technical experiments and studies in conditions of space flight, and continue medical and biological studies on the influence of various factors of space flight on the human organisms." Mstislav Keldysh, President of Soviet Academy of Sciences, said on April 26 that *Soyuz I* spacecraft had been tested previously in two to four unmanned flights. Weight, size, and capacity of new spacecraft were not revealed.

Immediately following the accident, which was first known fatality during a space flight, the Kremlin announced it was appointing a Government commission to investigate "all the circumstances" of the cosmonaut's death. Komarov would be cremated and buried in the Kremlin Wall—a final resting place for Soviet heroes—and a monument would be erected on the site of his birthplace in Moscow. (*NYT*, 4/24/67, 1, 27; 4/25/67, 1, 20; 4/27/67, 2; *W Post*, 4/23/67, A1, A16; 4/24/67, A5; 4/25/67, A1, A16; *W Star*, 4/23/67, A1)

April 24: After successful wheel orientation maneuver *Essa V*'s No. 1 meteorological camera was turned on for the first time and transmitted "high quality" cloud cover pictures. Photography, originally scheduled to begin April 22, had been delayed until analysis of spacecraft's telemetry system confirmed that spacecraft and subsystems were working satisfactorily. Operation of *Essa V*, launched by NASA April 20, would be turned over to ESSA in early May. (NASA Release 67-98)

- U.S. leaders expressed sorrow at the death of Cosmonaut Vladimir M. Komarov.

    President Johnson: "The death of Vladimir Komarov is a tragedy in which all nations share. Like three American astronauts who lost their lives recently, this distinguished space pioneer died in the cause of science and in the eternal spirit of human adventure."

    Vice President Humphrey: "It is with great sorrow that we learn of the tragedy involving Vladimir Komarov. The loss of a dedicated life is indeed a heavy price. Yet progress and space development is necessary."

    NASA Administrator James E. Webb: "All of us who have faced the difficulties of understanding and putting to use the forces of nature at the outer edge of Man's knowledge of what is possible in this decade deeply regret the loss of life represented by the death of Cosmonaut Komarov, and extend our sincere sympathy to his family and associates. We feel certain that man will achieve great things in space. Some of these will determine what the men will be able to do on earth.

    "We also feel that at this dawn of the space age, man has the duty to seek cooperation between nations, such as the USSR and the United States on a realistic basis.

    "We at the National Aeronautics and Space Administration want to make every realistic effort. Could the lives already lost have been saved if we had known each other's hopes, aspirations and plans? Or could they have been saved if full cooperation had been the order of the day?"

    U.S. astronauts: "We are very saddened by the loss of Colonel Komarov. We feel comradeship for this test pilot because we have met several of his fellow cosmonauts and we are involved in a pioneering flight effort which is not without hazards." (Text; *PD*, 5/1/67, 660; AP, *NYT*, 4/25/67, 21)

- NASA Aerobee 150 sounding rocket launched from WSMR carried GSFC-instrumented payload to 112-mi (181-km) altitude to observe solar spectrum from 3 Å to 400 Å in conjunction with *Oso III* spectrometer instrument [see Mar. 8]. Rocket and instrumentation performance were satisfactory. (NASA Rpt SRL)

- NASA revised designations for Apollo and AA missions: (1) all Apollo missions would be numbered sequentially in the order flown with the next Apollo mission to be designated Apollo 4, followed by Apollo 5, etc.; (2) AA missions would be designated sequentially as AAP-1, AAP-2, etc. Number designations would not differentiate between manned or unmanned Uprated Saturn I and Saturn V missions. (Text)

- Report on space rescue prepared by RAND Corp. for NASA recommended that all future spacecraft be compartmented with crew spread among separate detachable modules. Each module would carry additional survival equipment for use in space rescue missions in earth orbit, during lunar flights, and manned planetary missions, and would be equipped with a spare Earth Reentry Module (ERM). Study estimated, however, that even with an emergency system available, rescues would be success-

ful in only 80% of space emergencies. (Taylor, *Tech Wk*, 4/24/67, 16)
- MSC awarded MIT a $500,000 contract to study, develop, and test an advanced control guidance and navigation system for long-duration, post-Apollo manned spaceflight missions. (MSC Release 67-17)
- Water might have played an important role in shaping the lunar surface, Dr. Harold C. Urey, Univ. of California physicist, told National Academy of Sciences. As evidence to support his theory of the former existence of "lakes" on the lunar surface, Dr. Urey showed spacecraft photos of the moon's surface in which material had filled in depressions as though it were free-flowing. He also cited carbon-rich meteorites which seemed to be fragments of ancient lake beds. Dr. Urey believed many, if not all, the stony meteorites were fragments dislodged from the moon by impacting objects such as iron meteorites. As further evidence of water action, he noted the presence of meteorites within which numerous founded fragments were embedded which appeared to have been shaped by frictional effects within flowing material. A body much smaller than the moon would not have had sufficient gravity for this flow effect, Dr. Urey argued. (Sullivan, *NYT*, 4/25/67, 23)
- ComSatCorp selected Holmes & Narver, Inc., to provide architectural and engineering services for three new earth stations under two contracts—$361,600 for West Virginia and California facilities, and $194,200 for Puerto Rico station. (ComSatCorp Release 67-32)

April 25: Senate unanimously approved space law treaty, under consideration since Feb. 7, and transmitted it to President Johnson for signing and formal ratification. Treaty would become effective when ratified by U.S., U.K., U.S.S.R., and two other nations. (NASA LAR VI/51; Finney, *NYT*, 4/26/67, 1; *W Star*, 4/26/67, A3)
- Col. Joseph F. Cotton (USAF) and NASA test pilot Fitzhugh Fulton made successful emergency landing of XB-70 No. 1 at Edwards AFB after crew entry door opened on takeoff and main landing gear malfunctioned. (XB-70 Proj Off)
- M/G Benjamin D. Foulois (USA, Ret.), oldest U.S. military pilot and first American to fly in combat, died at age 87 at Andrews AFB, Md. A member of USA since 1898, General Foulois had been first person to operate an Army dirigible; first military observer on a cross-country flight (with Orville Wright); first military man to teach himself to fly; first military test pilot; first man to fly more than 100 mi nonstop; first man to use an aircraft in combat; and first Chief of Staff of the Army Air Corps to be a military aviator. He had received five campaign badges for field services, in addition to the Distinguished Service Medal, the French Legion of Honor, the Crown of Italy, and the Congressional Medal of Recognition. (UPI, *NYT*, 4/26/67, 43; Casey, *W Post*, 4/26/67, B8)
- NASA awarded Douglas Aircraft Co. a one-year, $13.5-million fixed-price-incentive contract extension to provide continued support services for Delta launch vehicle. Contract covered nine launches—seven at ETR, and two at WTR—for 12 mo beginning Jan. 1. (NASA Release 67-99)
- Nearly 100,000 mourners filed past the bier of Cosmonaut Vladimir Komarov in Moscow. Premier Aleksey Kosygin and President Nikolay V. Podgorny briefly joined an honor guard of soldiers near the bier. (Anderson, *NYT*, 4/26/67, 3; UPI, *W Post*, 4/26/67, A16)

April 26: Italy's *San Marco II* (San Marco B) scientific satellite was successfully launched from Mobile Launcher in the Indian Ocean off coast of Kenya, Africa. Four-stage NASA Scout launch vehicle boosted satellite

April 26: Italian scientific satellite *San Marco II* lifts off from platform at sea off coast of Kenya, Africa, on four-stage Scout booster supplied by NASA.

into equatorial orbit with 748-km (465-mi) apogee; 219-km (136-mi) perigee; 94-min period; and 2.9° inclination. The 285-lb satellite—first to be launched from a platform at sea—would measure air density by monitoring spacecraft's drag forces and investigate ionospheric characteristics which interfered with long-range radio transmissions.

San Marco II was second Italian satellite to be launched under May 31, 1962, cooperative agreement between NASA and Italian Space Commission (ISC). First satellite, *San Marco I*, was successfully launched from NASA Wallops Station Dec. 15, 1964. San Marco was a mutual program of NASA and ISC with no exchange of funds. NASA supplied launch vehicle and provided personnel training and tracking and data acquisition services. ISC was responsible for design, fabrication, and testing of payloads and for launching of satellite built by Centro Ricerche Aerospaziali (CRA) of the Univ. of Rome. (NASA Proj Off; NASA Release 67-93; WS Release 67-17)

- NASA test pilot William H. Dana made successful emergency landing of X-15 No. 3 at Edwards AFB after low pressure developed in fuel pump. (AP, B *Sun*, 4/27/67)
- Ashes of Soviet Cosmonaut Col. Vladimir M. Komarov were buried in the Kremlin wall in a military funeral attended by Premier Aleksey N. Kosygin, President Nikolay V. Podgorny, and other high Government leaders and scientists. Cosmonaut Yuri A. Gagarin, in a brief speech, pledged that Soviet cosmonauts would "continue the cause" for which Komarov gave his life. A NASA spokesman said plans to send U.S. Astronauts L. Gordon Cooper and Frank Borman to funeral had been abandoned because U.S.S.R. denied permission, saying it was an internal Soviet affair. (AP, W *Star*, 4/26/67, A24; *NYT*, 4/27/67, 3; UPI, W *Post*, 4/27/67, A4)
- Rep. William F. Ryan (D-N.Y.), in statement on the House floor, charged that NASA refused to release the Phillips Report because "the truth concerning NASA's failure properly to supervise Apollo operations and incredible mismanagement on the part of NASA's major Apollo contractor is highly embarrassing." He said that if NASA failed to officially release the report by April 29, he would make the full text available to the press. "In a democracy no agency should be permitted to withhold critical information for its own protection. On the contrary, the public which, at a tremendous cost is financing NASA's efforts in space, has a right to know how its money is being spent." (*CR*, 4/26/67, H4733)
- *New York Times* praised Senate ratification of the space law treaty April 25 and called for increased international cooperation in space: "Ideally, the probing of space and the planets would be the province of a World Space Organization affiliated with the United Nations, financed by the contributions of all nations desiring to participate, and drawing upon the technical manpower and knowledge of all nations. This international organization could be the owner of the moon and other solar bodies, arranging for the exploitation of resources found there, and using any extraterrestrial profits for the benefit of all men.

    "It will take much time before such an ideal can be reached. But even now the more sober spirit evident in Moscow and Washington opens the door to greater bilateral cooperation. Both nations have technical knowledge in this field that they could exchange to their mutual advantage. Their space tracking and rescue facilities could be coordinated into a single world system; they could agree on a division of labor that would eliminate such duplication as is represented by Surveyor and Lunar Orbiter. The result could easily be major savings in money and increased safety for all astronauts, regardless of nationality." (*NYT*, 4/26/67)
- James Smith McDonnell, founder of McDonnell Aircraft Corp., received

National Academy of Engineering's second Founders Medal at NAE's third Annual Meeting in Washington, D.C. Medal honored "outstanding contributions by an engineer both to his profession and to society." (NAE Release)

- Former NASA Associate Administrator for Advanced Research and Technology Dr. Raymond L. Bisplinghoff was among 45 new members and 10 foreign associates elected to NAS. (*NYT*, 4/27/67, 57)

April 27: U.S.S.R. launched the meteorological satellite *Cosmos CLVI* into orbit with 640-km (398-mi) apogee, 586-km (364-mi) perigee, 96.9-min period, and 81.19° inclination. Initial plane of orbit was shifted relative to plane of meteorological satellite *Cosmos CXLIV* (launched Feb. 28) by 95°. These two satellites would form experimental system called "Meteor." Mutual positions of orbits for satellites had been chosen to permit meteorological observations at intervals of about six hours; system would obtain data on half of the earth's surface during 24-hr period. (GSFC *SSR*, 4/30/67; *Ten Years of Space Research in the U.S.S.R.*, USS–T Trans, vol. 5, no. 5, 1967)

- NASA awarded General Electric Co. and Lockheed Missiles and Space Co. $50,000, fixed-price contracts to conduct four-month parallel studies of a medical laboratory for Apollo Applications (AA) flights. Designated Integrated Medical and Behavioral Laboratory Measurement System, laboratory would permit detailed measurements of human systems and crew functions during space flight. It could be flown as a complete laboratory or as selected group of measurement modules on specific missions. (NASA Release 67–102)

- Nuclear physicist Dr. Edward Teller, speaking at a Washington, D.C., press conference, recommended that U.S. proceed with a ballistic missile defense system. Qualifying his opinion as "inexpert," Dr. Teller noted there were other factors to be considered in any decision for the deployment of a missile defense system. Secretary of Defense Robert S. McNamara opposed the antimissile system on the grounds that it would not withstand a sophisticated assault. (B *Sun*, 4/28/67, 7)

April 28: USAF Titan III–C booster launched from ETR successfully orbited five unmanned satellites: two Vela nuclear detection satellites and three scientific satellites, *Ers–XVIII*, *Ers–XX*, and *Ers–XXVII*. The 508-lb Velas—improved versions of six Vela payloads previously launched—entered initial elliptical orbits with 69,000-mi (96,642-km) apogee and 5,400-mi (8,690-km) perigee. On April 29, orbit of one Vela was circularized at 69,000-mi (96,642-km) altitude; orbit of second would be circularized May 1. Designed to operate at least 18 mo in orbit, Vela satellites were part of DOD's Vela program to monitor space for violations of nuclear test ban treaty. The 14- and 20-lb scientific satellites contained scientific and engineering experiments to (1) investigate effects of vacuum on characteristics of 16 metals, (2) monitor radiation of Van Allen belts, and (3) measure solar radiation. (*NYT*, 4/29/67, 17; *W Post*, 4/29/67, A4; *Pres Rpt 1967*)

- Maj. Michael J. Adams (USAF) flew X–15 No. 1 to 167,000-ft altitude and 3,682 mph (mach 5.32) to provide pilot experience at high altitude and check out horizon scanner, horizontal stabilizer alpha, 3rd landing skid, and electrical loads. (X–15 Proj Off)

- McDonnell Aircraft Corp. and Douglas Aircraft Co. merged to form McDonnell Douglas Corp., pending Justice Dept. approval. (*Av Wk*, 4/24/67, 35)

April 28
- Joint application for authority to construct new comsat earth station near Jamesburg, Calif., was filed with FCC by ComSatCorp, AT&T, ITT WorldCom, RCA, and WUI. (ComSatCorp Release 67–34)

April 29: President Johnson authorized construction of two SST prototypes by Boeing Co. and General Electric Co., who had been continuing development and refinement of aircraft and engine designs since their selection in design competition Dec. 31, 1966. He said he would ask Congress to appropriate $198 million to finance Government's share of project during FY 1968. Construction of first flight model of the 1,800-mph, variable-sweep-wing aircraft was scheduled for completion by 1970. (Frankel, *NYT*, 4/30/67, 1; O'Toole, *W Post*, 4/30/67, A1)

- NASA announced it was conducting series of conferences with Apollo prime contractors and "a number of other companies that may have resources which could be utilized to effect the most rapid feasible recovery from the setback to the Apollo program caused by the Apollo 204 accident." Contractors with whom discussions were being held included: Aerojet-General Corp., Boeing Co., General Electric Co., Lockheed Aircraft Corp., Martin Co., McDonnell Douglas Corp., and North American Aviation, Inc. (NASA Release 67–105)

- NASA launched two Nike-Cajun sounding rockets, one from NASA Wallops Station, the other two hours later from Point Barrow, Alaska, to obtain temperature, pressure, density, and wind data at 22- to 59-mi (35- to 95-km) altitude during transition from winter-time westerly circulation to summer-time easterly circulation. Both rockets and instrumentation performed satisfactorily. (NASA Rpt SRL)

- Rep. William F. Ryan (D-N.Y.) made public the full text of the Phillips Report which had been withheld by NASA. Full report contained no significant information that had not been included in the summary presented by NASA Administrator James E. Webb to Senate Committee on Aeronautical and Space Sciences April 18. (UPI, *NYT*, 4/30/67, 38)

April 30: Launch of first manned MOL had been postponed by USAF from 1968 to 1970 because of "development problems," George Wilson reported in the *Washington Post*. According to USAF officials, the problem was not money but the "meshing of the various parts that go into the Manned Orbiting Laboratory." (Wilson, *W Post*, 4/30/67, C1)

- Vladimir Komarov was twelfth Soviet cosmonaut to die in a spaceflight accident, Julius Epstein, a Stanford Univ. professor, told UPI. He claimed he had conclusive evidence including the names of Soviet cosmonauts who had supposedly died on space missions. Epstein said U.S. policy of not disclosing Soviet space disasters that U.S.S.R. does not publicize was based on the State Department's desire "not to embarrass the Russians." (UPI, *P Inq*, 5/1/67, 1)

During April: The U.S. press commented on the April 24 death of Soviet Cosmonaut Vladimir Komarov.

New York Times: "Men have risked their lives in pioneering since history began—explorers, mountain climbers, the first aviators who conquered the globe's blanket of air, and now the cosmonauts who enter the 'infinite spaces.' That much has to be accepted—but how much? It seems clear that the three Americans who died in the oxygen fire at Cape Kennedy were victims of insufficient care, slipshod work and haste. There is no way of knowing whether similar factors led to the death of Komarov yesterday; but it is a fact that the Russians are aiming at time schedules, hoping to beat the United States to the moon, trying to meet special dates connected with the fiftieth anniversary of their Revolution. The United States still insists on trying to reach the moon by 1970.

"Both nations are duplicating costly and dangerous work. Thus good and brave men die unnecessarily, vast sums are wasted, and without doubt the progress that humanity could make through cooperation in the thrilling quest for knowledge of the universe is being hampered by pride, prestige and the nebulous possibility of strategic gain." (NYT, 4/25/67, 40)

Philadelphia *Evening Bulletin:* ". . . hard on the heels of a test disaster that took the lives of three U.S. Apollo astronauts, tragedy has also struck the Russian space program with the crash of a new giant spacecraft and the death of its pilot, Cosmonaut Vladimir Komarov. . . .

"Had the Russians succeeded in mounting a new space spectacular at this time, it is possible that some of the sobering impact of the Apollo on this country would have been dissipated. Although there have been repeated disclaimers of a space 'race,' the sight of the Soviet Union again forging ahead would surely have fanned anew American competitive instincts. Surely the death of brave men on both sides will reemphasize for both the element of human fallibility and the need for offsetting it with infinite pains and caution in space exploration." (P EB, 4/25/67)

Technology Week: "The Soviet tragedy, by emphasizing the hazards of space flight, has perhaps eased some of the pressure of the National Aeronautics and Space Administration and North American Aviation. There is more awareness now that the exploration of space will not be a series of glittering successes, one after another. There are dangers; there will be further tragedies, other brave men to mourn. While bringing home the care which must always be taken, the Soviet accident also makes it clear that in programs of this magnitude humans are not always infallible. . . ." (Coughlin, *Tech Wk,* 5/1/67, 50)

Aviation Week and Space Technology: "It is fortunate that the Apollo and Soyuz tragedies came this far downstream in the history of manned space flight. If they had occurred in the initial phases of Vostok or Mercury, the hue and cry of technical timidity in both countries might have killed the manned space flight programs before they had a chance to demonstrate their technical feasibility. Viewed against the perspective of Vostok, Mercury, Voskhod and Gemini and more than 2,000 hr of successful space flight logged by crews of both nations, the recent tragedies spotlight development problems but do not raise any fundamental doubts about man's ability to reach the moon. . . ." (Hotz, *Av Wk,* 5/1/67, 11)

• Press commented on Congressional hearings on Apollo accident.

Philadelphia Inquirer: "With both the House and Senate conducting their own investigations of the Apollo tragedy, in the aftermath of the exhaustive inquiry by the National Aeronautics and Space Administration's own review board, there is a danger that what will result is a negative witch hunt instead of constructive criticism of benefit to the space program. Congress should strive to protect the Nation's investment in this critically important and expensive venture and that is not to be accomplished by laboring the deficiencies of Apollo I. Above all, the concern should be for the safety of future space voyagers. It should be to certify that no unnecessary risks will be taken that might invite another lethal disaster." (*P Inq,* 4/12/67)

New York Times: "It is a new, and rather shocking, NASA image that is being projected by the current Congressional investigations into the Apollo tragedy. One would have expected that NASA would make available every possible bit of relevant information. . . . The Space Administration cannot afford even to raise the suspicion that it may be covering up evidence relevant to the deaths of the three astronauts . . . [and] certainly does not want to reinforce the view of cynics who insist that NASA actually stands for 'Never A Straight Answer.' " (*NYT*, 4/19/67, 40)

Washington Post: NASA "has merited the reproach of the investigators of the Apollo disaster and the searching scrutiny of the congressional committees reviewing the program. . . . What must not be overlooked however, are the risks of the space program that cannot be eliminated altogether without abandoning it. . . . When these catastrophes take place, the Government and the people must and should react as they reacted to the Apollo fire, by demanding new and further precautions, even more careful and complete protection. But there will not be much bold venturing in a society that visits a sanguinary fury upon those unfortunate enough to preside over calamity. For in that kind of society, more and more men will discover that the way to avoid getting any reproach is to avoid taking any risks." (*W Post*, 4/19/67)

- Rep. Gerald R. Ford (R-Mich.) in joint dialogue with Rep. Carl Albert (D-Okla.), appearing in *General Electric Forum*, when asked: "What is the present climate in the 90th Congress for other key legislative programs such as space?" stated: "Congress is interested in space and will support a strong space program. Those who are knowledgeable recognize that we have gone past the point of no return in giving full support to the Apollo lunar landing program. But many Congressmen are concerned about our course in the future exploration of space. When you relate the financial demand of another big quantum jump in space technology to our other fiscal burdens, you get the feeling that the 90th Congress will not be willing to embark on another major manned space program beyond the moon that will lead to another commitment of five or ten years or longer." (*GE Forum*, 4/6/67, 11–4)
- NASA awarded Univ. of Arizona a $14,970 research contract for a case history of LaRC from 1917–47. (NASA Release 67–86)
- Dept. of Commerce announced it would construct a multimillion dollar oceanographic laboratory on Florida's Virginia Key similar to West Coast Laboratory in Seattle. East Coast Laboratory would operate three ships and conduct research in geophysics, oceanography, and sea-air interaction. (UPI, *NYT*, 4/25/67, 26)

May 1967

May 1: NASA established the NASA Lunar and Planetary Missions Advisory Board to " . . . develop and review . . . scientific objectives and general strategy for manned and unmanned lunar and planetary missions [and to formulate] guidelines and specific recommendations for the design of missions and for the scientific payloads. . . ." The Board would have a close working relationship with all senior NASA officials involved in lunar and planetary exploration and would work ". . . through the Associate Administrator for Space Science and Applications." (NMI-1156.12)

- Five NASA officials were among 45 young business and Government executives from U.S. and abroad selected to receive MIT's 1967-68 Alfred P. Sloan Fellowships for one year of study at MIT's Sloan School of Management "to accelerate their development into positions of major executive responsibilities. . . ." NASA recipients: R. Bryan Erb, Assistant Chief, Structures and Mechanics Div., MSC; Robert H. Kirby, Jr., Flight Mechanics and Technology Div., LARC; Calvin H. Perrine, Jr., Assistant Chief, Mission Operations Div., Apollo Spacecraft Program Office, MSC; Robert J. Schwinghamer, Technical Assistant to Director of Manufacturing and Engineering Laboratory, MSFC; and William E. Scott, Deputy Chief, Research Div., Office of University Affairs, NASA Hq. (NASA Release 67-104)

- Orbit of second Vela nuclear detection satellite, launched April 28 by USAF Titan III-C booster, was circularized at 69,000-mi altitude. First Vela's orbit had been circularized April 29. Satellites were part of DOD's Vela program to monitor space for violations of nuclear test ban treaty. (AP, *NYT*, 5/2/67, 16)

- Field Enterprises Educational Corp. announced that it would not exercise its option to extend its rights for book and newspaper distribution of the personal stories of NASA astronauts and their families. Present four-year agreement, which paid the astronauts $320,000 annually for exclusive publishing rights and provided each astronaut with $50,000 life insurance policy, would expire Aug. 31. Field Enterprises president Bailey K. Howard said large number of astronauts now, compared with number four years ago, made continuation of the project impractical. (AP, *NYT*, 5/3/67, 13; *US News*, 5/15/67)

- Director of FAA SST Development B/G J. C. Maxwell signed the contracts for construction of two SST prototypes by Boeing Co. and General Electric Co. Industry representatives had signed earlier. President Johnson authorized FAA to proceed with Phase III (prototype construction) of SST program on April 29. (Sehlstedt, B *Sun*, 5/2/67, 1)

- U.S.S.R.'s May Day Parade was transmitted across U.S. for broadcast in Japan, but was not televised on American networks. The 1½-hr program,

May 1

transmitted through the cooperation of ComSatCorp and U.S.S.R., originated live in Moscow and was picked up by *Early Bird I* over the Atlantic; relayed across U.S. by Andover, Me., and Brewster Flat, Wash., ground stations; and transmitted across Pacific via *Intelsat II* for viewing in Tokyo. Japan paid approximately $30,000 for broadcast and transmission services. (Birger, *M News*, 5/5/67; ComSatCorp)

- Washington *Evening Star* editorial praised Senate's unanimous approval of the space law treaty April 25: "The action lends timely support to President Johnson's policy of seeking to promote a spirit of detente through various types of agreements designed to build bridges between our country and the Communist East.

    "... the new space treaty is at best only a modest step in the direction of improved Soviet-American relations. Still, it is a step forward, and a good one. Anybody who objects to it on security grounds can rest assured: It will not interfere in the slightest with our country's ability to detect Russian cheating between us and the stars if such cheating actually takes place." (W *Star*, 5/1/67)

- Haste might have contributed to the April 24 crash-landing of *Soyuz I* in which Cosmonaut Vladimir Komarov died, Edmund Stevens suggested in Washington *Evening Star*. He speculated that Soviet leaders, influenced by the goal of orbiting men around the moon by Nov. 7—50th anniversary of the Bolshevik Revolution—had launched *Soyuz I* before it was fully man-rated. (Stevens, W *Star*, 5/1/67, A1, A6)

May 1–3: NASA's *Surveyor III* spacecraft continued photography and digging operations on lunar surface. On May 3, communications with spacecraft were halted to conserve battery power during two-week lunar night. Photos taken during mission since April 19 soft-landing on the moon totaled 6,315. (NASA Proj Off)

- American Astronautical Society's (AAS) 13th Annual Convention, "Commercial Utilization of Space," was held in Dallas.

    MSFC Director Dr. Wernher von Braun stressed the importance of space exploration to man's general welfare: "The space program is providing us with new and valuable scientific information about our environment in this universe. The space budget cannot be justified, however, solely on the basis of expanding scientific knowledge, even though the discoveries may be regarded priceless by one segment of society. Space technology must continue and expand its contributions to the satisfaction of the needs and desires of society in general if it is to become a permanent outlet for man's creativity and energies." Dr. von Braun later suggested to the press that the lunar module (LM) might not be ready for the Apollo lunar mission as soon as the other launch vehicle and spacecraft elements. In that event, he said, NASA might decide to "fly the lunar mission [as a manned circumlunar flight], but just not activate one phase of it." (Text; *Av Wk*, 5/8/67, 17)

    Future potentials of navigation satellites were outlined by Eugene Ehrlich, Navigation and Traffic Control Program Chief, OSSA. Navigation satellites could provide: (1) high-accuracy position determination service to craft, people, and shore stations; (2) communications service to pilots, traffic controllers, and passengers; (3) air traffic control and maritime coordination; (4) SST radiation warning; (5) search and rescue aid during emergencies; (6) weather routing for aircraft and ships; (7) iceberg warning; and (8) data on migratory habits of marine, land, and sea life. (Text)

George S. Trimble, Director of Advanced Planning, OMSF, called for increased industry responsibility in space exploration. NASA builds "the roads into space. It is beholden on . . . industry to make certain that we build the proper ones for . . . future use. . . . As we look forward to missions of ever-increasing complexity, it is important that we continue to insure the most effective accomplishment of the most worthwhile experiments that this country can devise." (Text)

Rep. Joseph E. Karth (D-Minn.), describing industry's attempts to capitalize on non-Government space business opportunities as "substantially insufficient," urged industry to submit imaginative proposals for commercial development of space and to solicit Congress' support for the effort. (*Av Wk*, 5/15/67, 67)

A more realistic basis for determining space transportation costs should be devised to reflect actual use of the object in space, KSC Director Dr. Kurt H. Debus suggested. Although cost reductions were being studied, prospective commercial users of space should expect to spend about 10% of their overall program costs for launch operations, he said. He emphasized that more substantial savings could be realized in other areas such as reusable boosters and payload; increased life cycle of payload; reserviceable payload; and booster and spacecraft propulsion. (Text; *Av Wk*, 5/15/67, 69)

NASA Assistant Administrator for International Affairs Arnold W. Frutkin noted that space program had to defend itself continuously to enlist international support: "In ten years of space development, no nation has been harmed in any way. We must point to this record constantly in order to improve the atmosphere and wisdom of international consideration . . . [and] emphasize the expansion of benefits rather than restrictions upon progress. The record makes clear that we can address the more complicated questions of international organization on an evolutionary basis with the advantage of growing experience. The needs of the developing world and the magnificent record of space applications so far combine to show that the issues of international organization are secondary to the task of accelerating benefits for the world." (Text)

MIT Head of Dept. of Aeronautics and Astronautics Dr. Raymond L. Bisplinghoff and NASA Assistant Associate Administrator for Advanced Research and Technology John L. Sloop discussed selection of technology and optimum conditions for its development. They pointed out how such development ". . . triggered vast areas of accomplishment in [its] wake [and] produced a sustained and long-term stimulus to education [and provided] influence [that is] one of spirit rather than power. . . ."

Turning to discussion of impact of space program on engineering practices, they cited a major stimulus which was produced by ". . . the extreme environment of outer space combined with the requirements for low weight, small size, and exceptional reliability." They pointed to the new stimulus to "imagination and creativity" among engineers. Among important reasons for having a space program they listed satisfaction of society's needs; developing technology should serve as ". . . a solution to an existing or foreseeable social problem."

Calling attention to the international arena of technological development, the speakers urged "a balance of technologies in our favor [which] will . . . require wisdom in the selection of new technologies . . . so vital to our national well-being and survival that [we] must, to the greatest extent of our ability, use imagination, good judgment, and energy in their planning and execution." (Text)

TRW Systems Group President Dr. Ruben F. Mettler suggested that space environmental characteristics could be useful and, in some ways, unique for certain manufacturing operations. He stated: "An environment so widely different from that to which we are accustomed suggests we [determine] whether the new features can be used for making conventional products of improved quality or performance or at lower cost, whether they can support new processes previously considered impractical, or if they can lead to new products of such characteristics that they create new and currently unidentified markets." Mettler suggested these environmental characteristics for space operations which would be of interest to "possible specialized manufacturing operations": low temperatures, approaching absolute zero with use of space cryostat; high temperatures with use of solar concentrator; near-perfect vacuum of infinite extent; weightlessness in orbit; low gravitational forces on moon; and radiation environment. (Text)

Awards for achievement in 1966 presented at convention included: (1) AAS Space Flight Award to MSC Director Dr. Robert R. Gilruth for "distinguished contributions to aeronautical and space research . . . direction of continuing investigation of man's capabilities in space, and . . . active participation in the scientific community for the public good"; (2) AAS Flight Achievement Award to Astronauts Charles Conrad and Richard F. Gordon for "their new space flight altitude record and the world's first one-orbit rendezvous on Gemini XI"; (3) Melbourne Boynton Award to Col. William K. Douglas (USAF), Assistant Deputy Chief of Staff, Bioastronautics and Medicine, AFSC, for exceptional accomplishments as the first flight surgeon in charge of astronaut medical operations; (4) W. Randolph Lovelace II Award to Dr. Robert M. Page, former Director of Research, Naval Research Laboratory, for pioneering work in rocket astronomy and contributions to radar development; and (5) AAS Fellowships to MSC Director of Medical Research and Operations Dr. Charles A. Berry "for providing exceptional medical support for the NASA manned spacecraft programs," and to MSC Director of Engineering and Development Maxime A. Faget for "contributions to the basic conceptual design of the Mercury spacecraft and . . . subsequent engineering efforts" on Mercury, Gemini, and Apollo Projects. (MSC Release 67-15)

- North American Aviation, Inc. (NAA), NASA's prime Apollo spacecraft contractor, announced major management changes: (1) William B. Bergen, former President of Martin Marietta Co., who joined NAA April 7, replaced Harrison A. Storms, Jr., as President of Space Div.—formerly Space and Information Systems Div.; (2) Ralph H. Ruud, corporate vice president in charge of manufacturing, replaced William Snelling as executive vice president of Space Div. Snelling assumed new post of assistant vice president; (3) Bastian Hello, former Martin Marietta Co. executive, assumed new post of vice president in charge of Space Div.'s launch operations at KSC; and (4) Paul R. Vogt, former vice president of engineering at NAA's Rocketdyne Div., assumed new post of assistant to President of Space Div. in charge of quality control. (Clark, *NYT*, 5/2/67, 1, 53; UPI, W *Star*, 5/2/67, A13; O'Toole, *W Post*, 5/2/67, A4; Wilford, *NYT*, 5/4/67, 9)

May 2: NASA Assistant Administrator for Policy Analysis Gen. Jacob E. Smart (USAF, Ret.) received National Aerospace Services Assn.'s Quarter Century award for "outstanding contributions to aerospace advancement" and continuing leadership and service in NASA. (Natl Aero Serv Assn; *Av Wk*, 5/8/67, 13)

- Turbulence-measuring device developed at U.S. Weather Bureau's National Storms Laboratory in Norman, Okla., might be used by U.S. airports in 1968 to guide commercial airline pilots through severe and moderate turbulence areas, Martin Waldron reported in *New York Times*. Device measured severity of thunderstorms by analyzing and averaging radar pulses reflected from cells within the storm and then registered cloud structure on a radar scope. Data were relayed to pilots by flight controllers on the ground. Chief of U.S. Weather Bureau's radar systems section Stewart Bigler said a refined version of the device had been installed at Washington, D.C., National Airport in January and would be tested in June. If it performed satisfactorily, similar devices would be installed in other U.S. airports in 1968. (Waldron, *NYT*, 5/4/67, 51)

- *Washington Post* editorial criticized suggestion by nuclear physicist Dr. Edward Teller that nuclear explosions be used to provide data about moon's interior: "Speaking for a multitude of laymen, we want our moon left intact. If not many farmers are left who do their planting by the times of the moon, at least all of us like to remember moonlit nights when we were not thinking of nuclear fission. . . . If Dr. Teller wants to dig another Isthmian Canal with 'a small nuclear explosive,' we might consider that within the range of our needs. But let our moon alone. We look forward once again to sitting in the moonlight on a summer's night and smelling honeysuckle without the thought that a chunk of moon might blast off suddenly into another orbit." (*W Post*, 5/2/67)

May 3: NASA's Lunar Orbiter Incentive Evaluation Board awarded Boeing Co. a $1,918,725 bonus—maximum amount permitted under May 1964 contract provisions—for the *Lunar Orbiter II* mission. Spacecraft had obtained 98% of planned prime site photography after it entered lunar orbit Nov. 10, 1966; equipment failure resulted in loss of other 2%. Because *Lunar Orbiter II* also took excellent secondary photos—including two million square miles of the hidden side of the moon—Board ruled that usefulness of photographic data merited maximum award. (NASA Release 67-112)

- Dr. Mac C. Adams, NASA Associate Administrator for Advanced Research and Technology, was featured speaker at MIT conference sponsored by Electronics Industries Assn. to acquaint aerospace industry with NASA's requirements for electronic systems on future space missions. He outlined objectives of NASA's electronics research program: (1) stabilization systems to maintain orientation in space environment and to point telescopes and other experiments at celestial bodies with precise accuracy; (2) communications systems to translate data acquired in space activities over millions of miles and at rates comparable to exchange of information on earth; (3) communications and navigation satellites to meet increasing demands in commercial air transportation; (4) data-handling systems to quickly and efficiently store, catalog, analyze, and edit data produced by space exploration; (5) instrumentation systems to detect and measure environmental characteristics over

broad ranges of temperature, pressure, and density; and (6) components and technology which could operate reliably for long periods of time despite extremes in radiation and temperature. (Text; NASA Release 67-103)
- Charles W. Harper, Director of OART's Aeronautics Div., was appointed to new position of NASA Deputy Associate Administrator for Aeronautics. Aeronautics Div., renamed Aeronautical Vehicles Div., would be directed by A. J. Evans. (NASA Release 67-108)
- Confidential report submitted to NASA by General Electric Co. cited numerous serious workmanship flaws in Apollo spacecraft No. 17, scheduled to be launched by Saturn V booster on an unmanned mission no earlier than August, *New York Times* reported. Flaws included damaged parts, corroded valves, leaky pipes, three small holes in heat shield that "could have catastrophic implications" during reentry, and more than 1,300 "discrepancies" in the 20 mi of electrical wiring. Report was prepared by GE's Apollo Support Dept. under terms of a 1962 contract with NASA to conduct computer-assisted checks of all systems in Apollo spacecraft prior to launch. George C. White, Jr., NASA Director of Reliability and Quality, OMSF, told *New York Times* reporter John Wilford in a telephone interview that report was "a working document that, in effect, summarizes known problems" for KSC supervisors and should not be taken as "an alert of really big problems." NASA officials said many of the flaws cited in the report had already been corrected; others had yet to be changed. (Wilford, *NYT*, 5/3/67, 1, 2)

May 4: NASA's *Lunar Orbiter IV* (Lunar Orbiter D) unmanned spacecraft was successfully launched by Atlas-Agena D booster from ETR on mission to photograph lunar surface [see May 8–June 1].

Agena 2nd stage fired to boost 850-lb spacecraft into 100-mi (161-km)-altitude parking orbit, reignited after 21-min coast period, injecting spacecraft on 89-hr translunar trajectory, and separated. On schedule *Lunar Orbiter IV* deployed its four solar panels and two antennas, locked its five sensors on the sun, and fixed its star tracker on Canopus. At 16:45 GMT midcourse maneuver with 53.8-sec engine burn was successfully conducted to slow spacecraft's speed and alter its aim point slightly.

Primary objectives of NASA's *Lunar Orbiter IV* mission, fourth in series of five, were (1) to place three-axis stabilized spacecraft into high-inclination lunar orbit; (2) to obtain broad systematic photographic survey of lunar surface; (3) to improve knowledge of the moon; and (4) to provide basis for selecting sites for more detailed scientific study by subsequent orbital and landing missions. Full photographic flight plan would require more than 200 camera-pointing maneuvers, compared to 50 for *Lunar Orbiter III*. Photos would cover more than 80% of the moon's front face and more than 90% of the hidden side. Spacecraft would also monitor micrometeoroids and radiation intensity in the lunar environment, refine definition of moon's gravitational field, and serve as a target for tracking operations by Manned Space Flight Network stations. Lunar Orbiter program was managed by LaRC under OSSA direction. Tracking and communications were the responsibility of JPL-operated Deep Space Network. (NASA Proj Off; NASA Release 67-101)
- North American Aviation, Inc., President J. Leland Atwood assured the Senate Committee on Aeronautical and Space Sciences that NAA was

making numerous technical, inspection, and management changes that would enable U.S. to "effectively accomplish the lunar mission in this decade." He accepted NAA's share of responsibility for Jan. 27 Apollo fire but urged the Committee to view the deficiencies cited by Apollo 204 Review Board [see April 9] in the perspective of the standards involved. "... it must be recognized that in space work the standards are and must be extremely high. Literally perfection is the goal." Noting that many of the problems in NAA's management of the Apollo program had resulted from "a rapid build-up in manpower ... encountered in almost all large, complex developmental programs," Atwood explained that the survey of NAA workmanship summarized in the Phillips Report [see April 18] had been conducted during time when these problems were reaching their peak. Thus the report did not truly represent overall NAA performance "but reflected the fact that the [Apollo] mission criteria had been evolving towards their final form." Atwood stressed, however, that General Phillips' recommendations had been accepted and corrected to NASA's "general satisfaction" by April 1966.

To further improve Apollo program progress, Atwood said, he and NASA Administrator James E. Webb had reviewed "the overall working relationship" between NAA and NASA. As a result of this review efforts would be made to (1) improve coordination of management effectiveness and program control; (2) establish clearer and more specific objectives and measurement techniques for program performance; (3) conduct more periodic surveys and audits of program progress and performance; and (4) include sharper incentives and penalties in contract relationships. Only major changes in Apollo spacecraft would be incorporation of a quick-opening hatch and installation of additional insulation for interior wiring and tubing. (Hines, W *Star*, 5/4/67, 2; Wilford, *NYT*, 5/5/67, 19; *SBD*, 5/5/67, 33–4)

- Rep. William F. Ryan (D-N.Y.) recommended to the House that President Johnson appoint a high-level commission to fully investigate NASA and to determine whether NASA "has the competence to carry out the space program and to recommend changes in the agency to insure that our space program is efficiently and safely administered." Asserting that Americans "must have complete confidence in the administration of the space program" in order to give it unqualified support, he insisted that only a commission of non-NASA officials could thoroughly evaluate NASA. "From the [Apollo accident] testimony and subsequent developments it is clear that the entire management capability and supervisory competence of NASA is in question.

    "... NASA is not competently administering one of the largest Federal programs ... [and] has been extremely reluctant to be candid about its operations. There is every reason to believe that NASA will not examine itself sufficiently to form the basis for the changes that are needed to insure that there are no further catastrophes." (*CR*, 5/4/67, H5109–13)

- Former Soviet Premier Nikita Khrushchev made at least two offers in the early 1960's to show to the U.S. photos of U.S. military bases taken by orbiting Soviet satellites "in return for certain favors," Dr. Charles S. Sheldon, formerly an NASC staff member and currently at the Library of Congress, told an AIAA meeting in Atlanta. He declined to specify what favors Khrushchev wanted. Dr. Sheldon said frequent launchings of satellites into low circular orbits indicated strongly that U.S.S.R. was observing U.S. from space: "I think there's little doubt they are in the observation business." It was difficult to ascertain the status of the Soviet manned space program, he said, but U.S.S.R. might feel ready to attempt a manned orbit of the moon within the next year. Concerning reports that several cosmonauts had died during space flights, Dr. Sheldon said U.S.S.R. had definitely not lost any men in orbit and probably had not lost any men in the space program before the April 24 death of Cosmonaut Vladimir Komarov. He acknowledged the death of a man in a parachute accident and the death of a Soviet test pilot, but said he was not certain these deaths were involved in the space program. (AP, B *Sun,* 5/5/67)

- NASA Nike-Cajun sounding rocket launched from Wallops Station carried GSFC-instrumented grenade payload to 72-mi (115-km) altitude to obtain data on temperature, pressure, density, and wind between 22–59 mi (35–95 km) at transition from winter-time westerly to summer-time easterly circulation. Rocket and instrumentation performed satisfactorily. (NASA Rpt SRL)
- NASA Aerobee 150 sounding rocket launched from WSMR carried GSFC-instrumented payload to 109-mi (176-km) altitude to study special UV radiation of star Zeta Ophiuchi. Rocket and instrumentation performance was satisfactory; no spectra were obtained because a thin film of oil-like substance covered grating and mirrors. Source of substance was under investigation. (NASA Rpt SRL)
- Total estimated cost for USAF's Manned Orbiting Laboratory (MOL) had risen to $2.2 billion, according to March testimony by Secretary of Defense Robert S. McNamara released by House Armed Services' Appropriations Subcommittee. Estimate represented $700-million increase over total cost of $1.5 billion predicted by President Johnson in his Aug. 25, 1965, announcement approving DOD development of MOL. (Text; Getler, *Tech Wk,* 5/8/67, 14)

May 5: U.K.'s *Ariel III* (UK–E) scientific satellite was successfully launched from WTR by NASA four-stage Scout booster into orbit with 373-mi (600-km) apogee, 306-mi (492-km) perigee, 95.6-min period, and 80° inclination. The 198-lb satellite, which had a one-year design lifetime, carried five experiments for the investigation of earth's atmosphere; four were operating at liftoff and the fifth was turned on after third orbit. All five were functioning normally.

Third in a series of U.S.–U.K. cooperative space projects, *Ariel III* was designed to supplement and extend atmospheric and ionospheric investigations conducted by *Ariel I* (launched April 26, 1962) and *Ariel II* (launched March 27, 1964). Primary NASA mission objectives were to place satellite into planned orbit and provide tracking and telemetry support. NASA supplied Scout booster, conducted launch, and provided tracking and data acquisition services with STADAN facilities, under overall management of GSFC. Major portion of the technical effort on *Ariel III*, including design and fabrication of spacecraft and five experiments, was accomplished in U.K. under management of Science Research Council's Space Research Management Unit (SRMU). (NASA Proj Off; NASA Releases 67–96, 67–115)

- U.S. should measure its return on space technology in terms of such things as the development of communications, meteorological, and simulation systems, rather than in dollars, NASA Administrator James E. Webb said in the Second Annual Cortez A. M. Ewing Lecture at the Oklahoma Center for Continuing Education in Norman. "The marginal

May 5: Engineers check the United Kingdom's *Ariel III* before successful launch by NASA from WTR for U.K. Science Research Council, to investigate atmosphere and ionosphere.

return on the investment in the space program is greater than in any other area. . . . One must take into account the value of the services rendered."

Asserting that U.S. investment in the space program served as "a message to the world that we do not expect to be behind the world in the area of effective use of energy," Webb noted the success of U.S. space efforts. "We have accomplished in a decade what most informed sources in 1957 thought would take several decades."

Webb presented the Cortez A. M. Ewing Foundation with several photographs of the moon taken by Surveyor spacecraft, two atlases of the moon made from photos taken by *Rangers VIII* and *IX*, and color photos of India and Saudi Arabia taken by Gemini astronauts. (*Norman Transcript*, 5/7/67, 3; Howard, *Oklahoma Daily*, 5/9/67)

May 5

- NASA Aerobee 150 sounding rocket launched from WSMR carried GSFC-instrumented payload to 98-mi (159-km) altitude to measure spectral irradiance of Venus. Rocket and instrumentation performed satisfactorily, but telescope and camera separated from recovery body during reentry and were damaged at impact. (NASA Rpt SRL)
- NASA Nike-Apache sounding rocket launched from Churchill Research Range, Canada, carried Southwest Center for Advanced Studies experiment to 85-mi (136-km) altitude to: provide test of instrumentation; prove out feasibility of payload recovery and reuse; and provide geophysical observations of auroral zone under either quiet or disturbed conditions. All experimental objectives were met except payload recovery. (NASA Rpt SRL)
- Dr. Heinz von Diringshofen, a German pioneer in flight and space medicine, died in Frankfurt, Germany, at age 67. Among the first to study the effects of weightlessness in vertical flight, Dr. Diringshofen recorded reactions of humans in weightless periods of up to eight seconds. He also developed systems currently used in the training of U.S. astronauts, including a centrifuge that produced 17 g's pressure. (NYT, 5/9/67, 43)
- MSC awarded Rocket Research Corp. a $405,000 contract to deliver by February 1968 two hand-held EVA maneuvering units and two extra propellant tanks for future manned spaceflight missions. (MSC Release 67-20)
- Highest known temperature at which a material becomes superconducting—losing all resistance to electric current—was reported in *Science* by Bell Telephone Laboratories scientist Dr. Bernd T. Matthias. Conducting experiments under USAF contract, Dr. Matthias and six colleagues discovered a composition of niobium, germanium, and aluminum that became superconductive at $-434°$ F. Discovery was expected to lead to construction of "super magnets" which could be used to shield manned spacecraft from high-energy particles in space and development of "super-efficient" electric motors and electronic computers requiring virtually no power. (*Science*, 5/5/67, 645–6; *NYT*, 5/6/67, 62; AP, W *Star*, 5/5/67, A9)
- Wilbur L. Pritchard, Group Director, Communications Satellite Systems, Aerospace Corp., was appointed director of ComSatCorp's new laboratories, scheduled for completion in Montgomery County, Md., by 1969. (ComSatCorp Release 67-35)
- NASA had funded 1st flight unit of new $9-million Small Astronomy Satellite–A (SAS-A) project to map stellar x-ray sources. Spacecraft, scheduled to be launched in 1969 by four-stage Scout launch vehicle, would be placed into 330-mi-altitude circular orbit where it would measure position, strength, and time variation of all detectable x-ray sources. Project was under GSFC management. (NASA Release 67-111)
- Alumni Assn. of Western High School, Washington, D.C., would honor alumnus Edward H. White II, one of three Apollo astronauts who died in Jan. 27 accident, with a memorial, which would include furnishings, decorations, and possibly a space exhibit for school's guidance library and counseling rooms. (W *Star*, 5/7/67, B1)

May 7: Busts of Wilbur and Orville Wright were unveiled in the Hall of Fame for Great Americans at New York Univ.'s Bronx campus. They were among 93 people named to the Hall of Fame since its opening in 1900. (*NYT*, 4/2/67, 24)

During week of May 7: Stevens Institute of Technology, Hoboken, N.J., announced plans for what was believed to be the Nation's first academic center for the study of cryogenics—a branch of physics related to the production and effects of very low temperatures. Dr. John G. Daunt was named to direct the new center. Stevens President Dr. Jess H. Davis said center would have two major objectives: "to help meet the pressing need for scientists and engineers who have substantial training and experience in this developing science," and "to conduct basic research, which seeks to extend man's knowledge of cryogenics into areas as yet unexplored." (Bird, *NYT*, 5/14/67, 63)

May 8: Maj. William J. Knight (USAF) flew X-15 No. 2 to 96,000-ft altitude and 3,239 mph (mach 4.8) in flight at Edwards AFB to check stabilization and control with dummy ramjet and check out thermalcouple system. Flight was successful except for break of dummy ramjet chute cable during landing sequence. (X-15 Proj Off)

- NASA turned over *Essa V* meteorological satellite to ESSA for operation in accordance with NASA-Dept. of Commerce agreement. *Essa V* was launched from WTR by NASA April 20. (NASA Release 67-114)
- ARC-sponsored study of Direct Lift Control (DLC)—system to help reduce aircraft noise and improve handling—began at Oakland International Airport with specially-equipped Boeing 707 aircraft. System consisted of applying various methods, such as retracting spoilers, modifying main flaps, and controlling boundary layer, to increase aircraft's lift directly. Tests would continue through July 1. (ARC Release 67-9)
- *Newsweek* reported that Soviet diplomats in Washington, D.C., admitted to East European colleagues that *Soyuz I* mission had been launched one week early in an attempt to offset the impact of Svetlana Alliluyeva's press conference. Mrs. Alliluyeva, daughter of the late Joseph Stalin, defected to the U.S. in April. U.S.S.R. reportedly denied, however, that rescheduling had any effect on the April 24 crash-landing which resulted in Cosmonaut Komarov's death. (*Newsweek*, 5/8/67)
- U.S.S.R. was completing construction of several ground stations for use with second-generation Molniya comsats, *Aviation Week* reported. First Molniya II, expected to be launched in late 1967, would be capable of transmitting to several stations simultaneously; Molniya I's could reach only one at a time. (*Av Wk*, 5/8/67, 13)
- "Despite the high cost and dangers in exploring space, tragically dramatized by the deaths of three U.S. astronauts and a Russian cosmonaut in recent months, Americans overwhelmingly approve of the U.S. effort to land men on the moon," the *National Observer* reported. Survey in 12 metropolitan areas showed that 77% of people questioned supported the lunar landing program; 17% opposed it; and 6% were undecided. There was less enthusiasm, however, for manned planetary exploration programs: 57% favored manned planetary expeditions; 31% opposed them; and 12% were undecided. (*Natl Obs*, 5/8/67)
- Earth had been cooling off rapidly since 1950 because of accumulation of atmospheric dust, Univ. of Wisconsin meteorologist Reid A. Bryson told UPI. Solid particles clogging paths of sun's warming rays had caused earth to experience an average loss in temperature from $2/3°$ to $1/2°$, he said. (*St. Louis Post-Dispatch*, 5/9/67)

May 8–June 1: NASA's *Lunar Orbiter IV* became fourth U.S. spacecraft to circle moon when it entered lunar orbit following successful deboost maneuver which reduced its speed by 1,475 mph, permitting lunar capture. Orbital parameters: apolune, 3,797 mi (6,111 km); perilune,

1,681 mi (2,706 km); period, 72 hr 1 min; and inclination, 85.5°. Launched from ETR May 4, spacecraft performed 670 attitude changes, responded to 7,067 commands, and recorded two micrometeoroid hits. Spacecraft performed normally except for five-day period when camera thermal door operated improperly, causing overexposure and fogging. Problem was solved, however, and 193 medium- and high-resolution photos—including coverage of 99% of moon's front face—were taken with almost a hundredfold increase in discernible detail for most of the area covered. Final readout, underway since mission's photo-acquisition phase was terminated May 26 because of an encoder failure, was completed June 1.

By its telephoto coverage of eastern limb areas, *Lunar Orbiter IV* provided the basis for extending the cartographic grid system established for front-face mapping around to the hidden side, so that features there could be precisely located for future mission planning and operations. Initial photos, taken at 2,176-mi altitude, showed broad panorama of moon's South Pole region—which had never before been photographed—with typical rugged terrain of craters and ridges. Closer, wide-angle pictures of this area revealed a 200- by 10-mi crevice which scientists speculated had been formed by volcanic action. Other photos transmitted included excellent pictures of the Alpine Valley and Mare Orientale, a large and relatively young impact crater whose center lay beyond the western rim of the visible face of the moon. (NASA Proj Off; NASA Releases 67-127, 67-143; *W Post*, 5/12/67, A10; 5/13/67; AP, *NYT*, 5/28/67, 50)

During week of May 8: A sandbox full of hundreds of thousands of bullet holes accurately simulates the crater-pocked lunar surface, U.S. Geological Survey reported. Dr. Henry J. Moore, one of Survey's astrogeologists who fired projectiles of differing calibers into a sandtable at ARC, told UPI that results of experiment supported theory that moon's craters were, for the most part, created by the impact of meteorites. (UPI, *NYT*, 5/14/67, 29)

May 9: NASA Administrator James E. Webb, Deputy Administrator Dr. Robert C. Seamans, Jr., and Associate Administrator for Manned Space Flight Dr. George E. Mueller testified on Apollo project reprogramming before the Senate Committee on Aeronautical and Space Sciences. First "man-rated" Apollo Block II spacecraft would be delivered to KSC in late 1967 and launched three months later with Walter M. Schirra, Walter Cunningham, and Donn F. Eisele as prime crew, Webb said. Back-up crew would be Thomas P. Stafford, John W. Young, and Eugene A. Cernan. Spacecraft would still have 100%-oxygen atmosphere, but use of noncombustible and fire-resistant materials would minimize fire hazard, he asserted. "We are confident that the results of the first Block II Apollo flights will justify moving rapidly to follow-on flights, thus overcoming some of the effects of the present delay . . . [so that] we can carry out this plan within the funds now available for fiscal year 1967 and in NASA's budget for fiscal year 1968." In order to "assure maximum progress" and reduce pressures on North American Aviation, Inc. (NAA), prime Apollo spacecraft contractor, NASA was realigning its contracting arrangements, Webb said. Contracting changes included: negotiation of a new strong-incentive contract with NAA for manufacture, test, and delivery of standard Block II spacecraft; extension of Boeing Co.'s present contract responsibility to include integration of command

and service module and lunar module with the Saturn booster system; selection of a third contractor to make necessary modifications to standard Apollo spacecraft produced by NAA; and consideration of offers by other aerospace companies to provide contractor assistance in Apollo systems management and check-out and test procedures.

Dr. Seamans described impact of Jan. 27 fire on Apollo schedule and budget. Lunar landing before 1970 "remains possible," he said. "The impact . . . has been to reduce that probability, not eliminate it. . . . After the first manned Block II flight on the uprated Saturn I in 1968, we plan three to four manned Saturn V missions . . . [and] in 1969 . . . up to six Saturn V missions." Resulting $75 million of additional costs through FY 1968—for materials and equipment changes, flammability testing, launch facility modifications, and spacecraft delivery rescheduling—would be absorbed within the total budget plan currently before Congress.

NASA's responses to recommendations by Apollo 204 Review Board were outlined by Dr. Mueller: (1) bulk combustibles and other flammable materials had been replaced with flame-resistant materials, and necessary flammables stored in fireproof containers; (2) emergency procedures would be reviewed and exercised prior to each hazardous operation, and a quick-opening hatch installed for simplified egress from capsule; (3) new NASA Office of Flight Safety would be established under OMSF to review and evaluate safety provisions and monitor test operations; (4) all emergency equipment had been reviewed and additions or substitutions—including installation of water fire extinguishers—made where necessary; (5) detailed design review of spacecraft communications system had been conducted and minor changes made to assure reliable operation; (6) full-scale mockup tests had been conducted to assess fire hazard and problems of gas detection and control; and (7) management and organizational changes had been instituted to clarify responsibilities and improve coordination between NASA centers and contractors.

In response to questions by Sen. Margaret C. Smith (R-Me.) regarding rating of companies in the 1961 Apollo spacecraft contract competition, Webb said that Martin Co. had been rated first, on the basis of its technical proposal, by the source evaluation board. However he, Dr. Dryden, and Dr. Seamans—considering other factors such as lower cost estimate, outstanding company performance record, overall capability offered, and experience with X-15 experimental aircraft—chose NAA as prime candidate.

Senator Smith expressed concern that the Committee was not kept informed of serious problems revealed in the Phillips Report and asked whether NASA followed any guidelines to determine when situations should be brought to the attention of the Committee. Webb replied that no such guidelines existed and indicated that he thought it unnecessary to detail the problems to the Committee since he thought that NASA could control them. Several other committee members also expressed their desire to be informed of any problems approaching the magnitude of those surrounding the Phillips Report. When Sen. Walter F. Mondale (D-Minn.) continued to criticize NASA's reluctance to make the report public, Webb offered to discuss it, in complete context, in executive session, and repeated his objection to releasing such evaluations of

May 9

- contractor performance out of context. (Testimony; *SBD*, 5/10/67, 57–66; O'Toole, *W Post*, 5/10/67, A1)
- NASA Nike-Cajun sounding rocket with grenade payload was launched from Point Barrow, Alaska, to obtain temperature, pressure, density, and wind data between 22–59 mi (35–95 km) at transition from winter-time westerly circulation to summer-time easterly circulation. Rocket and instrumentation performed well. (NASA Rpt SRL)
- Aerobee 150 sounding rocket launched by NASA from WSMR carried Naval Research Laboratory payload containing two coronagraphs to 113-mi (180.8-km) altitude. Primary purpose of flight was to launch externally occulted coronagraphs during period when moon was very close to the sun. Rocket also carried a photographic spectroheliograph, a photographic extreme ultraviolet heliograph, and an ion chamber to measure total solar flux at Lyman-alpha. Rocket and instrumentation performed satisfactorily. (NASA Rpt SRL)
- NASA successfully conducted fifth test of a rocket-launched parachute as part of an advanced technology effort to investigate possible parachute landing systems for Voyager program. Two-stage Honest John-Nike rocket launched from WSMR ejected ringsail parachute at 130,000-ft altitude, where earth atmosphere resembles that of Mars. Parachute descended to earth with 200-lb payload carrying camera and instruments to record deployment of parachute and its characteristics in flight. Test was part of LaRC's planetary entry parachute program. (NASA Release 67–120)
- Second Bell Aerosystems Co. X–22A V/STOL research aircraft successfully underwent 18-min demonstration flight at Niagara Falls International Airport. During test, which increased aircraft's total accumulated flight time to nine hours, X–22A achieved speeds of up to 130 mph, made several vertical and short takeoffs and landings, and executed series of fast-climb-out turns. Designed and built by Bell under USN-administered program for Tri-Service V/STOL research project, aircraft would undergo 18-mo flight-test program at Bell and would then be delivered to Naval Air Test Center, Patuxent, Md., for further evaluation. First X–22A aircraft had been damaged beyond practical repair after crash-landing in summer 1966. (*Tech Wk*, 5/15/67, 18)
- USAF launched two unidentified satellites from WTR using Long Tank Thrust-Augmented-Thor-Agena D booster as first stage. One satellite reentered July 13. (*Pres Rep 1967*)
- ComSatCorp President Joseph V. Charyk, speaking at ComSatCorp's fourth annual meeting in Washington, D.C., said a shortage of ground stations was delaying development of a complete international comsat network. "A fourth satellite of the Intelsat–II series exists and could be launched this fall. . . . Unfortunately, appropriate earth station facilities do not exist and even after approval of the necessary applications . . . [by FCC], operational availability will still be 14 to 18 months away." Because of this lag in number of earth stations and because a satellite failure could cause severe service interruption, ComSatCorp was seriously considering "the desirability of placing the fourth satellite of the Intelsat–II series as a spare in orbit this fall and . . . [seeking] a limited solution to the high-traffic demands by interim utilization of this satellite through a transportable earth station in Hawaii." (Text; Smith, *NYT*, 5/10/67, 63)
- In February and March testimony released by Senate Foreign Relations

Committee's Subcommittee on Disarmament, Director of Defense Research and Engineering Dr. John S. Foster, Jr., indicated that U.S. was developing a new type of nuclear warhead designed to destroy enemy missiles with high-intensity x-rays. X-ray method, which would be used with three-stage Spartan missile, could destroy enemy missile by (1) disrupting avionic circuits used for guidance and detonation; (2) heating warhead to such high temperatures that it would ablate prematurely and burn up during reentry; or (3) triggering the warhead and causing premature detonation. (Committee Off; *Av Wk,* 5/15/67, 22; UPI, *P Inq,* 5/11/67, 3; Finney, *NYT,* 5/10/67, 1)

- L/G Nikolay P. Kamanin, military commander of Soviet manned spaceflight program, hinted in *Zemlya o Vselennaya* that U.S.S.R. would use Soyuz-type spacecraft for manned lunar landing. He made no direct mention of *Soyuz I* spacecraft [see April 23–24]—believed to be heaviest Soviet space vehicle ever launched—but said "payloads of more than 15 tons have already been put into orbit." These payloads, he said, could soft-land on moon from lunar orbit. (AP, *B Sun,* 5/10/67)
- End of the "space race" between U.S. and U.S.S.R. was nowhere in sight, Holmes Alexander speculated in the Philadelphia *Evening Bulletin:* "As matters stand today, the space contest is pretty much a race between the Communist and capitalistic systems. . . . Percentagewise the Soviet government is investing 2.5 times as much of its gross national product in the moon dash as we are. Moneywise, the investment is about equal. But the Russian economy, being much more 'managed' than ours, permits the ordering of what our economists call an 'internal brain drain.' Russian consumer industries are raided in order to feed the USSR space programs, whereas . . . NASA must compete . . . [with other agencies].

"It is wishful thinking to believe that the Russians will make any substantive concessions to ease the cost of the space race. As in Vietnam, we are locked in an expensive contest against Communism. Nobody has yet found a way to the negotiation table." (Alexander, P *EB,* 5/9/67)

May 10: M2–F2 lifting body vehicle crashed on landing at Edwards AFB about three minutes after air-launch from B–52 aircraft, injuring NASA test pilot Bruce A. Peterson. Vehicle, on mission to evaluate effects of reduction in automatic damping for roll and yaw, turned over several times after touchdown and was heavily damaged. Although exact cause of crash was unknown, one unidentified NASA official speculated that pilot had been distracted by a helicopter flying nearby. NASA immediately set up a committee to investigate crash and said that powered M2–F2 flights scheduled for June would be delayed indefinitely until after repairs and conclusion of investigation. (*Av Wk,* 5/15/67, 34; *Tech Wk,* 5/15/67, 14; *W Post,* 5/11/67, C21; West, *LA Times,* 5/12/67)

- NASA Administrator James E. Webb, Deputy Administrator Dr. Robert C. Seamans, Jr., and Associate Administrator for Manned Space Flight Dr. George E. Mueller appeared before the House Committee on Science and Astronautics' Subcommittee on NASA Oversight to submit Apollo project reprogramming plans presented to the Senate May 9. Dr. Seamans assured the Subcommittee that schedules could be met and urged that Congress continue to support NASA: "We have confidence in the ability of our technical and administrative teams, both in government and industry, to meet the challenge before us. But that is not sufficient; the Congress, representing the people of the Nation, must

May 10: Prime crew for first manned Apollo mission—shown here preparing for mission simulator tests at North American Aviation plant—are, left to right, Donn F. Eisele, senior pilot; Walter M. Schirra, Jr., command pilot; and Walter Cunningham, pilot.

feel and show this same confidence if the work and the sacrifice of the past are to have meaning and the objectives of manned space exploration are to be achieved."

Several subcommittee members, however, condemned NASA for withholding facts about cost and management problems with North American Aviation, Inc. (NAA), and indicated that they would be more critical of NASA in the future. In response to charges by Rep. William F. Ryan (D-N.Y.) that Webb had "an obsession with secrecy," Webb asserted that NASA had been exceptionally cooperative: "No program of this magnitude has ever been operated with greater public scrutiny. No agency has ever given its committees more information on what it is trying to do." He offered to elaborate on the Phillips Report [see April 18] in closed session but declined to discuss it publicly, insisting that further public discussion would destroy close relationship between NASA and its prime Apollo spacecraft contractor. (Testimony; Wilford, *NYT*, 5/11/67, 26; O'Toole, *W Post*, 5/11/67. A9)

- Prime and backup crews for NASA's first manned Apollo mission [see May 9] told a news conference at North American Aviation, Inc.'s Downey, Calif., plant that they were "very impressed" with the design changes being made in the Apollo spacecraft as a result of the Jan. 27 flash fire. Major changes included: (1) a new hinged hatch that could be opened outward in about two seconds by activating a springed lever; (2)

replacement of many highly flammable materials such as nylon netting; (3) sheathing of open wiring with hard panels; and (4) flame-resistant spacesuits. In addition, spacecraft would be pressurized with nitrogen rather than oxygen on launching, and in orbit—where oxygen is indispensable—pressure would be reduced one third to 5 psi.

Questioned about his attitudes toward the program since the Jan. 27 Apollo accident, Command Pilot Walter M. Schirra replied: ". . . we went through a very agonizing self-appraisal; we suffered through three months of recrimination; we've culminated these three months with some very firm ideas." These ideas, Schirra said, were being phased into an orderly program that "will continue—with orderly haste . . . [and] we are not going to run rampant on schedules." The crews anticipated that Apollo 101 spacecraft would be delivered to KSC in December and that first manned mission would be flown in first quarter of 1968. Primary mission objective would be to "exercise the vehicle in earth orbit," Schirra said. Asked if the April 24 death of Soviet Cosmonaut Vladimir Komarov had had any adverse effects on plans for earth landings in the Apollo program, Astronaut Thomas P. Stafford replied: "The earth landing system that consists of the three main parachutes and the two drogue chutes had been qualified completely. . . . We still have some earth landing drops to be made. But right now as far as the crew is concerned, either the parachutes in this vehicle are qualified or we don't fly." (Transcript; *Newsweek*, 5/22/67, 94; Hill, *NYT*, 5/11/67, 27)

- NASA awarded Douglas Aircraft Co. a $4,665,000 modification to an existing contract for reliability and verification testing of Saturn S-IVB stage components. (NASA Release 67-118)
- Small fire in KSC building housing an astronaut training simulator damaged $1,200 worth of equipment, AP reported. Fire occurred in a rack of gear which provided electricity to the equipment, but which was not connected to the spacecraft simulator. NASA officials said no astronauts were inside simulator when fire broke out but even if they had been there would have been no damage. NASA issued no official announcement on the fire. (AP, *NYT*, 5/13/67, 15)
- Fifteen outstanding young scientists received Air Force Office of Scientific Research (AFOSR) Postdoctoral Research Awards "for advanced study and fundamental research in the sciences." Selections, made by a board of NAS-appointed senior scientists, were based on "demonstrated competence and creativity in original research and on the scientific merit of the proposed postdoctoral investigation." (NAS-NRC Release)
- DOD awarded General Dynamics Corp. a $1.8-billion fixed-price-incentive contract for production of 493 F-111 aircraft: 331 F-111As for USAF; 64 F-111 strategic bombers for SAC; 24 F-111Bs for USN; 24 F-111Cs for Royal Australian Air Force; and 50 F-111Ks for British Royal Air Force. Contract included only costs of basic airframes and minor electronic equipment; additional costs for engines, major electronic equipment, and modifications had not yet been fully determined and were not included. There was speculation that congressional opposition to adaptation of F-111 (formerly TFX) as an all-purpose military aircraft—heightened by large increases in estimated costs and by January and April crashes of USAF and USN test models—might delay appropriation of production funds. (DOD Release 427-67; Wilson, *W Post*, 5/11/67, H11; Sheehan, *NYT*, 5/11/67, 1)

May 11: President Johnson submitted to Congress NASA's 15th Semiannual Report, covering January to June 1966, and praised the continued progress of U.S. space program: "The achievements reported here reflect not only our progress in space flight, but also new steps taken toward the real objective of all our efforts in space—the application of new knowledge to bettering the lives of all people. Already, we see dramatic examples of success in the satellites which have improved our weather forecasts and navigation, and which are extending radio and television communication to the farthermost regions of the earth. . . .

"The United States space program, as reflected in this report, continues to exemplify our Nation's conviction that the road to peace, progress, and abundance is through continued cooperation among all nations." (*PD*, 5/15/67, 729–30)

- To clarify apparently conflicting statements presented during Congressional testimony, NASA Administrator James E. Webb issued statement regarding selection of North American Aviation, Inc., as prime Apollo spacecraft contractor. In separate appearances before Senate Committee on Aeronautical and Space Sciences, Webb had indicated (1) that NAA was the unanimous choice and (2) that Martin Co. had been recommended by the NASA Source Evaluation Board in 1961. Webb explained: "Dr. Robert Gilruth, Dr. Robert C. Seamans, and Dr. Hugh L. Dryden . . . were unanimous in their judgment that of the five companies submitting proposals . . . [NAA] offered the greatest experience in developing high-performance manned flight systems and the lowest cost.

    "In the selection . . . the work of the Source Evaluation Board was not rejected or discarded. It was used as the basis for a more extensive and detailed examination of all pertinent factors . . . necessary to determine whether the facts then available formed an adequate basis for our selection of a contractor. We decided in the affirmative and then proceeded to select the contractor the facts indicated offered the most to the government." (NASA Release 67–122)

- Subcommittee on Manned Space Flight, in its report to House Committee on Science and Astronautics, recommended that $46.8 million be cut from $3.4 billion requested by NASA for FY 1968 manned space flight program. Recommendation, which included $25-million reduction in Apollo program, was approved by full committee May 16. (*W Post*, 5/12/67, A13)

- Glennan Space Engineering Building in Cleveland at Case Institute of Technology was dedicated. NASA Administrator James E. Webb said: "We need more men like NASA's first administrator . . . Dr. Keith Glennan . . . men who are willing . . . to seek out new facts, new concepts of reality as to the current human situation and [who] have the courage to base new and untried political, professional, industrial and technological ventures on information they test and learn to trust."

    Recognizing the value of institutions who provide graduate education, he said: "There is little further room to doubt that a high level of basic research, an intellectual activity that must be closely associated with graduate education, is indispensable to an advancing front of scientific knowledge. . . ."

    Webb referred to a June 1957 speech of J. S. McDonnell of St. Louis who had made predictions to graduating students at Rolla's Missouri School of Mines. Prediction No. 1 had a satellite circling earth and moon in 12 years. Webb said accomplishment was completed in three months for earth and in two years for moon. Prediction No. 2 had a satellite circling earth and Mars in 23 years. Webb said this was accomplished in seven years. Prediction No. 3 had manned space flight in earth orbit in 33 years. Webb pointed out this was done in four years.

Webb continued: "If we look backward, we can easily see that it took 37 years to conquer the earth's oceans . . . by the Caravelle sailing vessel [and] after that 37 years, the mind of man was never the same again. Control of the earth's ocean avenues by England through the maintenance of a favorable balance of ocean technology was a predominant force for stability in the affairs of men for over 400 years [but] no clear favorable balance of air technology was held by any one nation long enough to create world stability. It was, in fact, a predominant force in the affairs of men for less than 60 years. Space technology will be a dominant force far longer than was the case for those of ocean and air. Not limited as was the ship to the water or the airplane to the air, or by international boundaries, or to the fuel that its tanks can carry, spacecraft are today unlimited tools to explore and measure the environment within which the earth itself moves and has its being." (Text)

- NASA Nike-Cajun sounding rocket launched from Wallops Station carried GSFC-instrumented payload to 44-mi (68-km) altitude to develop experimental atmospheric composition profiles in mesospheric region and to measure ozone distribution in regions of 12–40 mi (20–65 km). Because of malfunction in parachute system only two minutes of ozone data were acquired. (NASA Rpt SRL)

- Naval Research Laboratory scientists reported the Milky Way galaxy was not capable of producing numerous new stars; instead, its star-producing days were virtually over, based on data gathered from rocket-borne instrument, a Far-Ultraviolet Image Intensifier Spectrograph [see Mar. 16]. Mission of spectrograph was to measure amount of invisible, gaseous, molecular hydrogen in interstellar space along the Milky Way; should bountiful amounts of hydrogen be indicated by the spectrograph, proof of Milky Way's capability to produce numerous new stars would have been obtained. Instead, spectrograph measurements had found no such evidence. (AP, *NYT*, 5/14/67)

- NASA awarded 28 12-mo, $2,400 predoctoral training grants to Univ. of Pittsburgh for public administration and to Georgia Institute of Technology, Univ. of Kansas, Cornell Univ., Purdue Univ., and Stanford Univ. for engineering design. (NASA Special Releases)

- U.S. industry would spend $16.6 billion for R&D in 1967—double the amount spent in 1957, McGraw-Hill, Inc., reported in its annual economic survey. Federal Government had financed 54% or $8.4 billion of the $15.5 billion spent for research in 1966. Survey estimated that 1970 expenditures would reach $21 billion. It noted that more than half of the R&D was concentrated in two industries—aerospace, and electrical machinery and communications—both of which were heavily involved in defense and space exploration. (Koshetz, *NYT*, 5/12/67, 67)

- Civic groups such as the Lions, Kiwanians, and Rotarians could make significant contributions to UFO investigations by equipping police squad cars with cameras, Dr. J. Allen Hynek, USAF consultant on UFOs and Chairman of Northwestern Univ.'s Astronomy Dept., told the MIT Club meeting in Washington, D.C. Policemen, whom Dr. Hynek characterized in general as good observers, often sighted UFOs but were not believed, he said: "For some reason we think that a policeman whose testimony could be sufficient to send a man to the electric chair is simply not to be believed when he makes a UFO report." Charging that Americans had "neglected looking at flying saucers scientifically too long." Dr. Hynek suggested that a "two-dimensional" assessment which considered both the "strangeness" of the event and the credibility of the witness be undertaken by physical and social scientists. (Hines, W Star, 5/12/67, A5)

- France had informed shipping companies and airlines that a zone of 185 km (115 mi) around Mururoa Atoll in the South Pacific would be considered dangerous beginning June 1 because of imminent atomic tests. (AP, W Star, 5/11/67, A1)

- U.K. was considering developing jointly with France a new generation of nuclear missiles to replace U.S.-made Polaris and Poseidon missiles, Karl E. Meyer reported in the *Washington Post*. British Prime Minister Harold Wilson had hinted in telephone interview that U.K. would soon announce its decision not to acquire additional U.S. missiles because of economic, political, and strategic considerations, Meyer said. *Development of new missiles with France would be more beneficial to U.K.* than purchasing U.S. missiles, Meyer suggested, because (1) it would prove that British were "European-minded" in defense policy and were weakening their ties with U.S.; and (2) it might persuade France to accept British application for admission to European Common Market. (Meyer, W Post, 5/11/67, A28)

May 12: U.S.S.R. launched *Cosmos CLVII* into orbit with 296-km (184-mi) apogee, 202-km (126-mi) perigee, 89.4-min period, and 51.3° inclination. Equipment functioned normally. Satellite reentered May 20. (UPI, NYT, 5/14/67; GSFC SSR, 5/31/67)

- Memorandum of understanding on participation of NASA with DOT in R&D and testing of high-speed ground transportation became effective. DOT would plan, manage, and conduct tracked air cushion vehicle (TACV) program and keep NASA informed, especially of areas in which NASA participation and facilities would be most valuable. NASA would assist by performing at LaRC aerodynamic research on possible high-speed ground vehicle configurations and on various interference effects. (NMI–1052.88)

- Sen. Clinton P. Anderson (D-N. Mex.), Chairman of Senate Committee on Aeronautical and Space Sciences, issued statement on Jan. 27 Apollo accident, commending "the diligence, integrity, and patent independence" of the Apollo 204 Review Board which would serve as a model for future inquiries. Noting the "admitted mood of overconfidence" exhibited by NASA and North American Aviation, he urged that a "renewal of chain successes . . . never again be permitted to encourage indulgence in such a potentially dangerous attitude.

". . . the Apollo accident may well move the date for an American landing on the moon beyond the reach of this decade. That would be regrettable. But there is nothing sacred about the goal. When set in 1961 it was a goal which was technically achievable. A target date was, and still is, essential to efficient planning and to maintaining a vigorous and competent organization. . . . The accident has taken a toll in morale and in momentum within the program . . . [but] that momentum will be regained and NASA will emerge stronger. I intend to support NASA in its requests for manpower and funds to get on with the important job." (Text)

- Sensors aboard two Vela nuclear detection satellites, launched April 28 by USAF Titan III-C booster, were turned on, and large amounts of data were being received. Satellites were part of DOD's Vela program to monitor space for violations of nuclear test ban treaty. (AP, *NYT*, 5/14/67)

May 13: Manned expeditions to every part of the solar system would be achieved by the end of the 20th century, Vice President Humphrey predicted in his opening statement to a student panel at North Dakota State Univ. in Fargo. He envisioned "the development of a whole family of earth orbiting stations, manned and supplied by regular ferry services; the launching of unmanned probes to every part of our solar system, and probably manned expeditions as well." (UPI, *NYT*, 5/14/67)

- Australia would be able to maintain its prominence in radio-astronomy with a new 150-in optical telescope to be constructed at Siding Spring, New South Wales, the *New York Times* reported. Construction of the $11,760,000 telescope, a joint British and Australian project, was expected to be completed in six years to provide Australian scientists—who currently had to send findings to Mt. Palomar (Calif.) Observatory for verification—with a suitable size optical telescope to check their observations. (*NYT*, 5/14/67, 37)

- L/G James H. Doolittle (AFRes., Ret.) received the Sixth Annual Thomas D. White National Defense Award at Air Force Academy ceremonies in Denver. An aviation pioneer and Medal of Honor holder, General Doolittle was cited for making "outstanding contributions to the national defense and security . . . [and giving] a lifetime of leadership to military and civil aviation." (*Denver Post*, 5/14/67, 32)

May 14: U.S.S.R. was training more scientists, faster, than any other country in the world, Dr. G. M. Dobrov, Kiev Academy of Science professor, told a London meeting of scientists from 14 countries. During the last 50 yrs, the number of scientists in the world had doubled every 15 yrs: U.S. had doubled its personnel every 10 yrs; U.S.S.R., every seven years. Dr. Dobrov said U.S.S.R. had 10,000 professional scientists in 1917, compared to 700,000 in 1967. He suggested that women be encouraged to enter scientific professions to enlarge resources of potential manpower. (Reuters, *NYT*, 5/14/67, 85)

- A 59-lb, $350,000 computer used on *Gemini V* mission had been reconditioned and was being used by USN in tests of new airborne weapons system designed to locate and destroy enemy radar, AP reported. USN declined to name the aircraft or reveal where the tests were being conducted. (AP, *Denver Post*, 5/14/67)

May 14–18: Aviation-Space Writers' Assn. met in Las Vegas.

Dr. Mac C. Adams, NASA Associate Administrator for Advanced Research and Technology, discussed NASA's attempts to solve the increasing problem of aircraft noise. Efforts included: (1) active participation in the interagency task force established by President Johnson to study aircraft noise [see Feb. 1], (2) study of methods to reduce noise of conventional aircraft engines, (3) investigation of the use of steeper descent angles during approach to landing and steeper climb on takeoff to reduce noise by moving the source farther from the ground, and (4) research on a new turbine engine with low noise as a basic requirement. (Text; *N News*, 5/16/67)

Former AFSC Commander Gen. Bernard A. Schriever (USAF, Ret.) charged that the Johnson Administration was not "pushing technology" and suggested that $400–700 million more should have been spent annually on exploratory and advanced development during the past four or five years. He was particularly critical of the "arbitrary separation of space activities into peaceful and military. . . . It simply serves to emphasize, by comparison, the straightforwardness of the Soviet program. Their single-purpose program seeks only to attain a versatile technological superiority which, once attained, will serve the political purpose they choose regardless of what kind of a space program we have or how we categorize it. . . ." (Text)

Anglo-French Concorde supersonic aircraft, scheduled to enter commercial service in 1971, would pioneer three markets, monopolizing the first two until "pushed away" by U.S. SST, scheduled to enter passenger service in 1974–5, W. J. Jakimiuk, president of Sud-Aviation, predicted. Concorde would dominate the North Atlantic market where SSTs would first be used; however, "as soon as the U.S. SST penetration of the North Atlantic pushes away the Concordes, they will be introduced on the major other markets, where a position of monopoly again will be found," Jakimiuk said. When U.S. SST "pushes" Concorde away from these routes, it would then move to a new series of routes where traffic density would be so low that "there will be no SST competition." (*NYT*, 5/21/67, 88)

Director of FAA SST Development B/G J. C. Maxwell, referring to report by Institute for Defense Analyses (IDA) which suggested that SST program could seriously weaken the Nation's balance of payments [see May 16], said that "our balance of payments will continue to benefit if American and foreign airlines continue to buy and operate American aircraft in the upcoming supersonic area." General Maxwell said FAA estimated that 1990 market for SSTs, if operations were limited by sonic boom considerations, would be 500 aircraft, improving balance of payments position by $17 billion. If sonic boom did not limit operations, 1,200 aircraft would be sold, improving balance of payments position by $32 billion. (Text; Lardner, *W Post*, 5/17/67, C1)

Dr. John S. Foster, Jr., Director of Defense Research and Engineering, responded to criticism that U.S. was losing its strategic superiority. "For many years, the Soviet Union apparently had been following our lead in every important strategic system technical development. . . . This is still the case. We are following their activities with great care. We see no evidence that our planned strategic capabilities will be endangered by recent Soviet technological actions. . . ." Dr. Foster announced that first MOL would be launched by USAF in 1970, confirming rumors that program was two years behind schedule. (*Av Wk*, 5/22/67, 11)

Awards presented included: (1) Monsanto Aviation Safety Award for the "most significant and lasting contribution to air safety" in 1966 to FAA Deputy Administrator David D. Thomas, who was recognized as "the principal architect of the U.S. air traffic control system"; and (2) 1967 Robert S. Ball Memorial Award for distinguished and meritorious aerospace writing to Howard Benedict, AP senior aerospace writer, for series of articles from Aug. 8–Sept. 19, 1966, on *Gemini XI* mission. (*Editor & Publisher*, 5/13/67; FAA Release 67–7)

May 15: *Cosmos CLVIII* was launched by U.S.S.R. into circular orbit with with 850-km (528-mi) altitude, 100.7-min period, and 74° inclination. Equipment was functioning normally. (AP, *NYT*, 5/16/67)

• NASA announced that *Oso III* Orbiting Solar Observatory, launched from

KSC March 8, had not only achieved but exceeded its primary and secondary objectives. Primary objective was to obtain high-resolution spectral data from pointed solar experiments during major portion of one solar rotation and adequate operational support of spacecraft subsystems to carry out acquisition of these scientific data. Secondary objective was to obtain useful data from the non-pointed experiments, and to obtain data from pointed experiments during more than one solar rotation. *Oso III* had operated satisfactorily for over two solar rotations and was continuing to acquire useful scientific data from both pointed and wheel experiments. (NASA Proj Off)

- NASA Nike-Cajun sounding rocket, launched from Point Barrow, Alaska, carried GSFC-instrumented grenade payload to obtain data on temperature, pressure, density, and wind between 22–59 mi (35–95 km) at transition from winter-time westerly to summer-time easterly circulation. Rocket and instrumentation performed well. (NASA Rpt SRL)
- NASA Administrator James E. Webb, at White Sulphur Springs 86th Annual Meeting of Proprietary Association, presented social implications of the Nation's space program. He asked: "If our future is now more dependent than ever on human intelligence, and if these activities made possible by the rocket engine are adding large increments of grist for the human intelligence mill, and trained workers to use the result, how much more important is it today than ever before for leaders in an industry like yours to understand those institutions of organized society that will foster, encourage, and make proper use of the vast powers the human mind is deriving from its understanding and use of the forces, materials and forms and activities of life which are provided by nature in this universe?" He added: "The comprehension of space exploration and its implications for mankind is a tremendous human intellectual enterprise . . . and our capability to use it effectively is only limited by our ability to systematically order it into viable patterns." (Text)
- NASA's *Nimbus II* meteorological satellite successfully completed one year of operation and was still transmitting useful data. Spacecraft had taken more than one million weather photos since its launch from WTR May 15, 1966. (AP, B *Sun*, 5/15/67)
- David S. Gabriel, former Program Manager for the Advanced Agena at Bell Aerosystems Co., was named Deputy Manager of AEC–NASA Space Nuclear Propulsion Office (SNPO). (NASA Ann; NASA Release 67-131)
- Technique for predicting solar flares by the positions of the planets was described by Rex Pay in *Technology Week*. Predictions were made by calculating time rate of change of the gravitational field at sun's surface, taking into account local areas of persistent solar activity and their alignment with respect to earth. "Principal influences on rate of change of the gravitational field at the solar surface are the planets. Although their gravitational effects are many orders of magnitude less than that of the Sun itself, the time rate of change of the resultant planetary field vector appears to have some triggering effect on the release of solar flares. It appears, therefore, that solar flares can be predicted from the positions of the planets." Dr. Richard Head, ERC, told *Technology Week* that technique had been used to predict intense proton storm in late August 1966 which occurred shortly after NASA's *Lunar Orbiter I* had completed its photographic mission. (*Tech Wk*, 5/15/67, 35)
- Scientists at National Radio Astronomy Observatory (NRAO), Green Bank, W. Va., and Naval Research Laboratory successfully demonstrated new

May 15

technique to obtain high-resolution pictures of quasars. Radiotelescopes at both locations were focused simultaneously on a quasar 240 trillion mi away, and radio signals beamed to earth by the quasar were recorded on magnetic tape at both centers; processed by a computer; and matched to form a single, clearer tape. NRAO Director Dr. David Heeschen speculated that facilities even farther away could be combined for sharper pictures: "Using this technique, there's no reason why we can't link up two telescopes 1,000 miles apart and improve our resolution 100 times over what we had." (O'Toole, *W Post*, 5/25/67, D18)

- U.S.S.R. planned to display two new jet passenger aircraft at Paris International Air and Space Show in late May, Tass announced. First was 24-passenger Yak-40, scheduled to replace propeller aircraft on Soviet secondary air routes; second was 164-passenger Tu-154, scheduled to be main medium-range Soviet passenger aircraft. (AP, *C Trib*, 5/16/67, 3)

- Washington *Evening Star* columnist Richard Wilson compared "cloud of suspicion over large undertakings" in both DOD and NASA programs: "An identical parallel is now disclosed between the award of the multi-billion dollar TFX fighter-bomber contract and the award of the very large contract for the Apollo capsule. In both instances the recommendation of the experts was rejected and a handful of public officials chose the contractor they favored. . . . In both instances many other facts and expert opinions were suppressed.

  "Contrary to the findings of expert boards and committees and of the Joint Chiefs of Staff, the TFX contract was awarded to a company with good political connections which was temporarily in financial trouble. Contrary to the findings of a group of 200 space experts, the Apollo contract was awarded to North American Aviation, strongly represented in Washington's political life, instead of to the Martin Company.

  "Congress and the public need to know how these huge undertakings have been handled because there will be many more of them in the future." (Wilson, *W Star*, 5/15/67)

May 16: House Committee on Science and Astronautics voted to report $4.9 billion NASA authorization bill (H.R. 10340)—$107.8 million less than $5.1 billion NASA had requested for FY 1968. Committee reductions included $46.8 million from manned space flight; $35.3 million from space science and applications; and $25.6 million from advanced research and technology. Report was filed with the House June 6. (O'Toole, *W Post*, 5/17/67, A1, A6; Hines, *W Star*, 5/17/67, A9; *Av Wk*, 5/22/67, 24)

- Sen. Stephen M. Young (D-Ohio), speaking on the Senate floor, urged U.S. and U.S.S.R. to cooperate in the exploration of outer space. "Let us proceed to inquire together seeking to effect a treaty with officials of the Soviet Union for the joint exploration of outer space for peaceful purposes, including efforts for a joint lunar landing, sharing the cost on a 50–50 basis. The tremendous expense would be shared equally by our two great nations. Also, if we were no longer engaged in a race with the Soviet Union for space achievements, the likelihood of tragedies resulting from too much haste—such as both nations have recently witnessed—would be greatly diminished.

  "This would save the taxpayers of our country at least $1 billion next year and billions of dollars in future years. Even more important, it would be a great advance toward permanent peace in the world." (*CR*, 5/16/17, S6903)

- Twenty-sixth meeting of Interim Communications Satellite Committee (ICSC)—17-member governing body of International Telecommunications Satellite Consortium (INTELSAT)—accepted invitation of Japan to meet in Tokyo, May 18–23. ComSatCorp, representing U.S. in INTELSAT and acting as its manager, announced that international participation had expanded from 13 countries in 1964 (when interim agreements were opened for signature) to 56 countries that now account for more than 95% of world's commercial communications. Tokyo meeting would be first to be held in Asia, and was second time ICSC had convened away from headquarters offices in Washington, D.C., Paris having been site of meeting in July 1965.
- ComSatCorp announcement summarized commercial satellite communications in operation around the world: *Early Bird I* (INTELSAT I series) launched April 6, 1965, and *Atlantic II* (INTELSAT II series) launched March 22, 1967, linking North America and Europe; and *Lani Bird II* (INTELSAT II series) launched Jan. 11, 1967, linking U.S. and Hawaii, Japan, and Australia. Fabrication had been started in 1966 on advanced satellites in INTELSAT III series that would have five times the communications capability of *Early Bird I*. ComSatCorp reported that these latest satellites would be launched beginning in 1968 to expand communications system to global capability. As to ground facilities, 14 earth stations were now operating in countries with more than 40 planned for by the end of 1969. (ComSatCorp Release)
- SST program recommended by the Johnson Administration could seriously weaken the Nation's balance of payments, Institute for Defense Analyses (IDA) concluded in report prepared for and released by FAA. Report said gains from foreign sales of SST would be more than offset by a large increase in U.S. tourist spending abroad as well as by a reduction in sales of subsonic U.S. aircraft. IDA study was one of four reports commissioned by FAA for $633,000 but withheld pending President Johnson's decision to proceed with Phase III (prototype construction) of SST program [see April 29].

    New York Times later commented: "The very fact that there can be no certainty about the sales prospects for the SST or its impact on the balance of payments argues for a deliberate and conservative approach in financing it. Statements based on hope and supersalesmanship are considerably less convincing than the hard facts of supersonic economics. The SST is supposed to be a commercial venture. It should meet commercial standards of investment." (Lardner, *W Post*, 5/17/67, C1; *NYT*, 5/21/67, E12)

May 17: NASA Argo D-4 (Javelin) sounding rocket launched from NASA Wallops Station carried Syracuse Univ. Research Corp. payload containing vacuum-ion chamber to 625-mi (1,006-km) altitude on flight to measure ion densities in the upper atmosphere. Telemetry signal was received for 18 min. (NASA Rpt SRL)

- NASA test pilot William H. Dana flew X-15 No. 1 to 3,205 mph (mach 4.84) and 71,000-ft altitude in flight at Edwards AFB. Purposes of test: (1) PCM system checkout; (2) measurement of coldwall heat transfer; (3) measurement of step panel heat transfer; (4) boost guidance checkout; (5) energy management checkout; (6) check of tip-pod accelerometer; (7) sonic boom study; and (8) study of horizontal tail loads. (X-15 Proj Off)
- U.S.S.R. successfully launched two Cosmos satellites: *Cosmos CLIX* entered

orbit with 60,600-km (37,655-mi) apogee, 380-km (236-mi) perigee, 19-hr 33-min period, and 51° inclination; *Cosmos CLX* entered orbit with 205-km (127-mi) apogee, 142-km (88-mi) perigee, and 49.6° inclination. Equipment on both satellites functioned normally. *Cosmos CLX* reentered May 18. (UPI, W *Star*, 5/18/67, C18; GSFC *SSR*, 5/31/67)

- NASA's *Surveyor III* spacecraft, resting on the moon, would not be reactivated until about May 23 despite the end of the two-week lunar night, NASA announced. Officials said they wanted spacecraft to warm up gradually before resuming communications. (UPI, *NYT*, 5/18/67, 22)

- Congress, the Administration, and the press had failed to put the U.S. space program—particularly the Apollo fire—into proper perspective, Rep. Joseph E. Karth (D-Minn.), Chairman of the House Committee on Science and Astronautics' Subcommittee on Space Science and Applications, told National Space Club meeting. "Despite reports and statements to the contrary, neither the Apollo nor the rest of the space program is falling apart. Despite reports to the contrary, NASA has revealed and discussed its errors and those of the contractor to an unprecedented degree. In fact, I have never heard more candid admissions of failures and responsibilities. Yet, these errors must be weighed in the overall balance with the tremendous accomplishments of NASA and its contractors during the past nine years. It is unfair and super-bad-judgment to measure this present 'low water' mark with any other yardstick. . . ." (Text; Clark, *NYT*, 5/18/67, 30)

- Soviet space program would be seriously delayed because of the April 24 accident which killed Cosmonaut Vladimir Komarov, Soviet Cosmonaut Yuri Gagarin disclosed in *Komsomolskaya Pravda*. "Flights of spaceships of the type of *Soyuz I* will be possible only after a complete determination of the causes of the accident, their elimination and then test flights. This, of course, will take time." Gagarin confirmed that *Soyuz I* mission had been first of a series of flights planned for summer 1967. Concerning rumors that Komarov had survived the crash landing of *Soyuz I*, he said: "How we all wish that Volodya Komarov was alive. . . . But one must look facts in the face." (*NYT*, 5/18/67, 31; AP, *W Post*, 5/18/67, A26)

- Michigan's Governor George Romney said U.S. exhibition at Expo '67 in Montreal was "a discredit to the Nation and the people of this Nation. . . . The only thing inside the exhibit that I could see that had any merit were the space capsules." (UPI, *W Post*, 5/18/67)

May 18: USAF launched unidentified satellite from WTR using Scout booster. (*Pres Rep 1967*)

- NASA personnel changes: (1) George H. Hage, former Engineering Manager for Boeing Co.'s Lunar Orbiter Program, had been appointed Deputy Associate Administrator for Space Science and Applications (Engineering), replacing Robert F. Garbarini, who resigned effective April 27. Hage would be responsible for maintaining the highest possible standards of engineering excellence and discipline, systems performance and project management on the spacecraft and launch vehicles in the OSSA program. (2) Bernard Moritz, former NASA Deputy Assistant Administrator for Industry Affairs, had been appointed Assistant Administrator for Special Contracts Negotiation and Review. Moritz would be responsible for negotiation and review of certain assigned contracts involving new policy or administrative arrangements requiring consideration or decision by the Administrator or Deputy Administrator. (3) M/G John

May 17: United States exhibits spacecraft at Expo '67 in Montreal: Lunar Orbiter (top left), Surveyor (center), and Apollo Command Module suspended from parachutes (right).

G. Shinkle (USA, Ret.), Apollo Program Manager at KSC, resigned for "personal reasons." Dr. Kurt H. Debus, KSC Director, commented: "I deeply regret the decision of General Shinkle to leave the program. Throughout his association here, his experience has been of major assistance in our contribution to the Apollo program." (NASA Releases 67-128, 67-129; AP, *NYT*, 5/20/67, 2)

- NASA announced it would negotiate three-month $500,000 fixed-price con-

May 18

tracts with Martin Marietta Corp. and McDonnell Aircraft Astronautics Co. for preliminary design studies (Phase B) of planetary entry capsules for the Voyager program. Phase B contractor efforts were the responsibility of JPL in conjunction with LaRC. Responsibility for system design and hardware development subsequent to Phase B would be divided between JPL and LaRC with JPL responsible for surface laboratory and LaRC responsible for capsule bus system, which included entry and landing equipment. (NASA Release 67-126)

- NASC Executive Secretary Dr. Edward C. Welsh, delivering the Steinmetz Memorial Lecture at Union College, Schenectady, N.Y., emphasized the contributions of the U.S. space program to national security. "Surely, we must all realize that the more certain potential aggressors are that we are informed and alert, the less is the likelihood of attack and the greater is the likelihood of peace. Space competence contributes significantly to our state of alertness and awareness. . . . We are not developing space weapons systems which might threaten the security of other countries, but rather we are developing insurance against the violation of international treaties as well as against surprise attacks. Those who want to live within the reasonable limitations of international law and who desire and are willing to work for world peace surely should welcome a program which lessens the risks of world conflict. Our space activity constitutes such a program." (Text)

May 19: U.N. space law treaty had been ratified by the Presidium of the Supreme Soviet, Tass announced. Presidium acted on behalf of the Supreme Soviet when the full body was not in session. Ratified by U.S. Senate April 25, treaty would become effective when U.S., U.S.S.R., U.K., and two other nations had filed ratification papers.

Washington *Evening Star* later commented: ". . . [the space law treaty] may not be one of the most significant agreements of our time, but it is surely the most far-reaching, encompassing nothing less than the infinities between us and the stars. It is an agreement, moreover, that puts America and Russia on the same side. It can be helpful as an example of what the two of us can do together—down here on the troubled earth—to promote a good and enduring peace." (*NYT*, 5/20/67, 1, 5; AP, *W Star*, 5/20/67, A5; *W Post*, 5/20/67, 1; *W Star*, 5/22/67, A16)

- Press briefing on Esro II satellite and NASA International Cooperative Programs was held at NASA Hq.

Arnold W. Frutkin, NASA Assistant Administrator for International Affairs, discussed the areas of cooperation between NASA and ESRO, emphasizing the "unprecedented data exchange agreement. . . . I don't think there is anything quite as sophisticated or as extensive in the international field. Under this agreement . . . ESRO assumes responsibility to collect all of the European technical report literature relevant to space activities, indexes and abstracts it, reduces it to a standard microfiche or microcard format which is compatible with our own system, and provides that material to us. We incorporate it in a single combined publication every two weeks, listing all of the U.S. and European literature in the field so that it is available to both sides. . . ."

Brian M. Walker, ESRO Project Coordinator for Esro II, announced that next ESRO satellite—Esro I—would be launched in late 1967 to study the ionosphere. Esro II was scheduled for launch from WTR May 29. (Transcript)

- USAF awarded two fixed-price incentive-fee contracts to continue development of the MOL: (1) Douglas Aircraft Corp., contractor for the laboratory vehicle, received $674,703,744; and (2) McDonnell Aircraft Corp., contractor for Gemini B spacecraft, received $180,469,000. (DOD Release 464-67)
- A street and school in Warsaw, Poland, had been named for Vladimir Komarov, Soviet cosmonaut who died April 24 when *Soyuz I* spacecraft crash-landed, AP reported. (AP, *NYT*, 5/19/67, 4)

May 20: Air Transport Assn. (ATA) issued statement urging FAA to expedite its program for automatic equipment to improve the control of air traffic. Statement noted that by 1970 airlines would have more than two thirds of their share of the equipment installed compared to FAA, which would have only one third of its ground stations suitably equipped. ATA urged that FAA temporarily abandon plans for a complex and costly system and proceed instead with less elaborate equipment. (AP, *NYT*, 5/22/67, 25)

May 21: Astronauts Walter M. Schirra, Jr., Thomas P. Stafford, and Frank Borman were interviewed on NBC-TV's "Meet the Press." Borman said he had "nothing but admiration for the way NASA had conducted its investigation of the Jan. 27 Apollo accident and that, far from losing confidence in the agency, he had gained confidence in it. Asked if he believed that some other group should also investigate the fire, he replied: "Absolutely not. I don't understand the desire for another investigation." Schirra, asked if he believed it were necessary to send men to the moon when unmanned spacecraft were obtaining valuable information, responded: "It is only natural that man must go out there and see for himself." The computer in unmanned flights is only a product of man's mind, and, "if something different occurs, man is best equipped to make the move." (AP, *Newsday*, 5/22/67)

- One remaining "lightweight" prototype of USN's F-111B fighter aircraft had been grounded pending investigation board's determination of cause of April 21 accident in which an F-111B, which had been stripped of excess weight, crashed on takeoff. "You can't lose that high a proportion of your total assets and not have it hurt . . . a lot," Capt. K. C. Childers, deputy director of F-111B-Phoenix program, told Washington *Evening Star*. He emphasized, however, that three other USN models of the aircraft were still flying, and two of them were being used in tests of the Phoenix missile that formed an integral part of the F-111B weapon system. (W *Star*, 5/21/67, A7)

May 22: Cosmos CLXI was launched by U.S.S.R. into orbit with 343-km (213-mi) apogee, 205-km (127-mi) perigee, 89.8-min period, and 65.7° inclination. Equipment functioned normally. Satellite reentered May 30. (AP, *W Post*, 5/23/67, A9; GSFC *SSR*, 5/31/67)

- USAF launched two unidentified satellites by Atlas-Agena D boosters from WTR. One satellite reentered on May 27; other reentered May 30. (UPI, *W Post*, 5/23/67, A9; *Pres Rep 1967*; GSFC *SSR*, 5/31/67)
- France and Quebec had agreed on joint use of a comsat for transmission of French radio and television programs to the French-speaking areas of Canada, Quebec's Premier Daniel Johnson disclosed at a Paris press conference. (*NYT*, 5/23/67, 77)
- The time had to come to end the "belabored post-mortems of the Apollo fire" and to restore confidence in the Apollo program, William Coughlin suggested in *Technology Week*. "NASA certainly has its faults, par-

ticularly in the quality control area. But we believe that the overall record of the Apollo 204 probe offers some inspiration for confidence in NASA management. It takes a great deal of integrity to give an accident review board full freedom and support when the outcome can only be critical of those setting up the board—and NASA provided that freedom and support. It then takes courage to stand up and disagree with some of the findings of the board—and NASA did that. The Phillips report, damaging as it was, also was testimony that NASA had been on top of the problems at North American and was making progress in solving them. . . .

"Confidence in Apollo must be restored and the program allowed to move ahead, without undue restriction, on its new lunar schedule.

"Anything less will diminish the sacrifice made by three brave men." (Coughlin, *Tech Wk*, 5/22/67, 58)

- Cosmos, Minn.—a community of 560—had given all its streets astronomical names such as Mars, Jupiter, Pegasus, and Comet. (AP, *NYT*, 5/22/67)
- Dr. J. Herbert Hollomon, Acting Under Secretary of Commerce and Assistant Secretary of Commerce for Science and Technology, had been appointed President of Oklahoma Univ., effective Sept. 1. (AP, *NYT*, 5/23/67, 32)
- USAF F-111A jet aircraft was flown from Loring AFB, Me., to Le Bourget Airfield, France, for display at the Paris International Air and Space Show. (UPI, *NYT*, 5/23/67, 77)

May 22–23: Vice President Hubert H. Humphrey, accompanied by NASC Executive Secretary Dr. Edward C. Welsh, toured MSFC and U.S. Army Missile Command (AMC) at Redstone Arsenal. During his visit, the Vice President watched a full-duration static-firing of an Uprated Saturn I 1st stage which lasted 145.6 sec. (*Marshall Star*, 5/24/67, 1)

May 23: Science press briefing on *Surveyor III* was held at NASA Headquarters.

Dr. Ronald F. Scott, Cal Tech, showed a movie composed of a time-lapse sequence of still pictures of the surface sampler digging a small lunar trench. The device also picked up a small rock in left side of scoop and tried unsuccessfully to crush it, exerting between 200–300 psi. Dr. Scott concluded that rock examined was at least harder than sandstone, which would have broken under impact of about 200 lbs. *Surveyor III* also confirmed that moon's surface could support manned lunar landing. Consistency of surface was found to be comparable to wet sand with bearing strength between three and eight psi. Dr. Scott said of an exploring astronaut: "He'll leave a definite footprint, that's all."

Dr. Eugene Shoemaker, U.S. Geological Survey, said a series of 50 color photos taken by *Surveyor III*'s cameras using color filters indicated that lunar surface was gray. "The gray varies in shade from a pale to a very dark gray, but it appears to be basically all gray." He said this color probably resulted from lack of moisture and of air on moon. Minerals on earth were brightly colored because of higher state of oxidation possible under earth environment. Dr. Shoemaker noted that some of the lunar material photographed by *Surveyor III* was very dark, suggesting that it had been coated with a darkening substance which he nicknamed "lunar varnish."

Among photos displayed were two pictures of the eclipse in which earth was depicted as a dark circle sitting between spacecraft and sun. Earth appeared to have partial halo of deep sunset color in one photo and orange areas connected by a faint purple halo in the other. (O'Toole, *W Post*, 5/24/67, A12; Clark, *NYT*, 5/24/67, 50; Hines, *W Star*, 5/25/67, A13)

- NASA Administrator James E. Webb, speaking at colloquium on "Urban Government in the Decade Ahead," at Fels Institute of Local and State Government, Pennsylvania Univ., discussed urban American goals and the urban society: "What can one say about the use of new scientific approaches for governing such a complex [society]?"

  Referring to earth-orbiting spacecraft achieving dynamic equilibrium in their mastery of space, he asked if the "ability of the human mind" could translate this concept of dynamic equilibrium into an entirely new way of thinking. Webb stated: "Here I believe we do have something to learn from the perspective the earth satellite gives us [and] if we continue to expose this kind of overview to the largest possible number of able minds, there can be little doubt that new knowledge, new capability, new perception of worthwhile and quite specific goals will emerge." He added: ". . . I believe that what the space frontier opened up in 1957 is the kind of frontier on which pioneers can experiment and learn to use many facets of modern technology here on earth—here in America's cities. In urban America, as elsewhere, men need the constant development of new concepts of reality and the inspiration that comes from a chance to better know this city and this universe in which we live." (Text)

- First SV-5D lifting body vehicle ever recovered from a suborbital space flight was displayed by USAF at Pentagon press briefing. Seven feet long and weighing 860 lbs, it was heat-blackened and slightly charred on its outer surface from 3,000° F reentry heat. Col. Curtis L. Scoville, AFSC, said vehicle, launched April 19 by an Atlas booster, had achieved an orbital speed of 17,000 mph and reentered in a maneuvering descent that took it "hundreds of miles" off its straight-line trajectory. A USAF aircraft recovered it in mid-air during final parachute descent. Mission was third successful flight in USAF's Precision Recovery Including Maneuvering Entry (PRIME) program; first two vehicles were not recovered, but program was considered so successful that planned fourth flight had been canceled. (Schmeck, *NYT*, 5/24/67, 50)

- Soviet Institute of Oceanology's Marine Electronics Laboratory was developing a television-equipped robot to probe the ocean floor at depths of four–six km (2½–four mi), Science Service reported. Controlled electronically through a specially made cable, robot would carry a manipulator to pick up objects sighted by TV camera. Engineers hoped to eventually develop a robot that could be controlled entirely by ultrasonic signals transmitted through the water. (Sci-Serv, *NYT*, 5/23/67, 50)

May 24: NASA successfully launched *Explorer XXXIV* Interplanetary Monitoring Platform (IMP-F) from WTR by three-stage Thrust-Augmented Improved Delta booster into near-perfect polar orbit. Orbital parameters: apogee, 133,131 mi (214,348 km); perigee, 150 mi (241 km); period, 106 hrs; inclination, 67.2°. Preliminary data indicated that all spacecraft systems and the 10 of its 11 experiments which had been turned on were operating normally. Final experiment would be turned on by ground command June 1.

Explorer XXXIV would make measurements of solar and galactic cosmic rays within and at the boundary of earth's atmosphere and interplanetary space during period of maximum solar activity. Primary mission objectives were to: (1) place spacecraft into orbit with an apogee of approximately 92,000 mi, or greater, to investigate the region between the magnetosheath and the shock front; and (2) to obtain scientific data on the cislunar environment from a number of its experiments. *Explorer XXXIV* was ten times more complex than any of four previous satellites launched in NASA's Interplanetary Monitoring Platform (IMP) series, two of which—*Explorer XXVIII* (launched May 29, 1965) and *Explorer XXXIII* (launched July 1, 1966)—were still operating and providing scientific data. IMP series was managed by GSFC under OSSA direction. (NASA Proj Off; NASA Release 67–21; UPI, *NYT*, 5/26/67, 9; *WSJ*, 5/25/67, 1)

- President Johnson, in a White House ceremony, presented M/G James W. Humphreys, Jr., of the Air Force Medical Corps, the Distinguished Service Medal for helping to develop "a national medical program for the Vietnamese people." General Humphreys would become Director of Space Medicine, OMSF, June 1. (*PD*, 5/29/67, 778; *W Post*, 5/25/67)

- NASA announced that 2nd stage of the first Saturn V booster would be dismantled at KSC to check for any "hairline" cracks. Decision was made after similar cracks had been found in an identical stage of the vehicle at North American Aviation, Inc.'s Seal Beach, Calif., plant. Additional checks were not expected to delay first Saturn V flight—an unmanned mission scheduled for mid-August—by "more than a week or so." (NASA Release 67–132)

- Ronnie J. Lagoe, 17-yr-old high school senior whose $\frac{1}{7}$-scale model of NASA's Surveyor spacecraft had made five successful soft-landings on earth from varying heights, was introduced at a New York news conference by Dr. Allen E. Puckett, Executive Vice President of Hughes Aircraft Co., manufacturer of Surveyor spacecraft. Lagoe built his 2-lb model—designated Surveyor-R ("for revised")—with wood strips, foam rubber, bathtub calk, glue, ping-pong balls, cork fishing floats, a battery, and four tiny rockets. Total cost: $28.95. He described the 85-ft flight sequence following Surveyor-R's launch from the top of a water tower: "After retro-fire, a marked decrease in spacecraft velocity was noted until retro-burnout occurred at 20 feet and the verniers burned alone. Complete counteraction of gravity was achieved at two feet, after which vernier shut-off occurred and the craft dropped to the surface." Hughes presented Lagoe with a summer job at KSC, a free trip to its Culver City, Calif., plant, and a $5,000 college scholarship. (Wilford, *NYT*, 5/25/67, 43)

- 1967 Robert J. Collier Trophy was presented to James S. McDonnell, founder of McDonnell Aircraft Corp., by Vice President Humphrey in ceremony at Smithsonian Institution. Making the award on behalf of President Johnson, Mr. Humphrey said McDonnell had been selected because of his leadership in aeronautics and astronautics as demonstrated by the performance of the McDonnell Gemini spacecraft and the F–4 Phantom II jet fighter. Trophy was presented annually by *Look* and National Aeronautic Assn. "for the greatest achievement in aeronautics or astronautics in America, with respect to improving the performance, efficiency, or safety of air or space vehicles." (NAA *News*; AP, *NYT*, 5/25/67, 81)

- The creation of a European Institute of Science and Technology to help narrow the technological gap between U.S. and Western Europe was proposed at a Deauville, France, conference of 70 U.S. and European businessmen, scientists, academicians, and public officials. Institute, which would be supported initially by private funds, would offer one-year postgraduate course. Faculty and student body would be international. Conference concluded that such an institute would be "a considerable stimulant for the various European educational systems and a rich source of trained engineers." Meeting was sponsored by North Atlantic Assembly's Science and Technology Committee and Univ. of Pennsylvania's Foreign Policy Research Institute. (Mooney, *NYT*, 5/30/67, 25)

May 25: U.S.S.R. successfully launched *Molniya 1–5* comsat into orbit to relay television signals from the Soviet Far East to Moscow and on to Paris. Orbital parameters: apogee, 31,658 km (19,672 mi); perigee, 459 km (285 mi); period, 11 hr 55 min; and inclination, 64.8°. Tass said primary goal of the Molniya I series of comsats was "further checking and experimental exploration of a system of long-distance two-way television, telephone, and radio communications." *Molniya 1–5* comsat was believed to be replacement for *Molniya 1–2* comsat which had reentered prematurely March 17. (UPI, *NYT*, 5/26/67, 3)

- Report by NRC Space Science Board on the concept of orbiting large reflecting mirrors concluded that there was no overwhelming evidence that scientific damage would result from the deployment of a single reflector system. It recommended, however, that such a satellite not be considered in the future unless the ability to destroy it by ground signals were an inherent part of the design and unless detailed studies of its effects on ecology, biology, and astronomy were previously conducted and made public. It said it could see no scientific merit for such a satellite system commensurate with its cost to the public and its nuisance to science. Report was prepared by Board's Committee on Potential Contamination and Interference from Satellites as a result of NASA's 1966 announcement that it had asked five aerospace companies to study feasibility of orbiting large reflecting mirrors that could illuminate land masses at night. Dr. Donald F. Hornig, Special Assistant to the President for Science and Technology, confirmed in a letter to NAS President Dr. Frederick Seitz that the Government no longer had plans for such a project. (NAS–NRC–NAE *News Report*, 5/67, 2)

- National Academy of Sciences Panel on Applied Science and Technological Progress had concluded that the "most important invention in the pursuit of modern . . . applied science is the big mission-oriented industrial or Government laboratory," and had underlined importance of a symbiosis between mission-oriented institution and university. Panel's report, forwarded to House Committee on Science and Astronautics, also recognized Government's special responsibility for integrity and sufficiency of man's environment and for dealing with national and regional social questions. It urged adaptability of large, interdisciplinary Government laboratories to new national problem areas. (Text)

- JPL announced that 24-hr efforts—begun May 23—to reactivate NASA's *Surveyor III* spacecraft on the moon after two-week lunar night were continuing. Officials said they were not concerned about spacecraft's temporary failure to respond, noting that it had taken seven days to reestablish contact with its predecessor *Surveyor I.* (UPI, *NYT*, 5/27/67, 2)

May 26–31: 27th Paris International Air and Space Show at Le Bourget Airport was opened by French President Charles de Gaulle, who toured the French, British, U.S., and U.S.S.R. exhibits. More than 450 aerospace companies representing 16 countries participated in the show. Exhibits in U.S. pavilion traced the progress of aviation during the 40 yrs since May 20–21, 1927, when Charles A. Lindbergh flew from New York to Paris, becoming the first man to fly the Atlantic non-stop alone. Included were: (1) a replica of Lindbergh's aircraft, "The Spirit of St. Louis"; (2) Douglas Aircraft Co.'s DC-8 Super 63, world's largest commercial jet aircraft; (3) SV–5J, first manned jet lifting body vehicle; (4) USAF F–111A swing-wing jet fighter; and (5) models of NASA's Lunar Orbiter, Gemini, and Apollo spacecraft. Soviet exhibit featured first public display of Vostok booster, similar to the one used to launch Cosmonaut Yuri Gagarin on world's first manned space flight April 12, 1961.

On May 29, Cosmonaut Pavel Belyayev entered the Apollo spacecraft during a visit to the U.S. pavilion and discussed the instruments, levers, and buttons inside the spacecraft with Astronaut David Scott. Belyayev was first Soviet cosmonaut ever to have entered a U.S. spacecraft.

On May 31, it was announced that two USAF rescue and recovery helicopters had left New York enroute to Paris, attempting to achieve the first nonstop helicopter crossing of the Atlantic. Helicopters would refuel in flight about eight times during flight, which was expected to take 28–32 hrs. USAF said purpose of the crossing was to demonstrate "the long-range capability of the HH–3E [helicopter] for employment in long range recovery of personnel who may be forced down at sea a great distance from land." Helicopters landed safely at Le Bourget Airport June 1. (Long, *NYT*, 5/21/67, 88; UPI, *NYT*, 5/27/67, 50M; *W Post*, 5/27/67, D20; *W Post*, 5/30/67, A1)

May 27: Venus is devoid of water and, therefore, of life, Dr. Gerard P. Kuiper, Univ. of Arizona Lunar and Planetary Laboratory, told *New York Times* reporter Richard D. Lyons in a telephone interview. "We now know that Venus is a dead planet. The astronomical literature is full of references to water clouds and ice crystals. These two things are definitely wrong. The complete absence of water means that the chemistry of Venus is totally different from the chemistry here." Practical absence of water in the Venus atmosphere provided new information on the constitution of the planet, Dr. Kuiper said: "It is probable now that the Venus clouds are dust and that a substantial fraction of the high surface temperature must be attributed to internal heat reaching the planet's surface through a thin solid crust." Dr. Kuiper's conclusions were based on astronomical observations made May 13 from a specially fitted NASA-owned Convair 990 jet aircraft which flew at 37,000-ft altitude over Canada. His theory conflicted with one set forth in 1966 by Johns Hopkins Univ. astrophysicists Drs. William Plummer and John Strong that surface temperatures of Venus were low enough to support life in certain areas. (Lyons, *NYT*, 5/28/67, 21)

- The frontier of space is limited only by man's "ability to maintain individual freedom and yet join many minds in concerted action," NASA Administrator James E. Webb said at the Celebration of The

Prelude to Independence in Williamsburg, Va. "It is the essential requirement for each of us to do what we can as individuals and as groups to make our system of representative government work, and then work better. More and more this means that each of us must find a way to reach through complexity and organized prejudice to trusted sources of information and organized facts. We must not do less than to make sure we understand the fundamentals in the many important disciplines of human activity. Without this understanding of at least the fundamentals . . . today's citizen cannot play the role only he can play in representative society, cannot bridge the gap from the old to the new, from one discipline to another. Without this capability, a citizen today cannot be a fully effective participant in a free society. Indeed, without this basic understanding on a large scale at all levels our nation is likely to forfeit the capability of collectively responsible action, and with it, the basis for an effective free society. . . ." (Text)

- NASA Administrator James E. Webb had been encountering a "storm of criticism" since the Jan. 27 Apollo accident, but "odds are he'll see his job through," *Business Week* concluded in an article summarizing Webb's career. It noted some of Webb's accomplishments as NASA Administrator: ". . . Webb has steered steadily growing budgets—from $1-billion in 1961 to this year's $5-billion—among the Congressional reefs. He runs a NASA organization of nearly 35,000 people in 16 facilities that stretch from Boston to Cape Kennedy, from Houston to the West Coast—and a worldwide network of tracking stations. At the same time, he has had to deal directly or indirectly with nearly 21,000 contracting companies, which employ about 400,000 people. . . ." Article concluded that Webb had been "bruised, not beaten" by the controversy surrounding the Apollo accident. ". . . odds-makers in Washington are betting that when the smoke clears Webb will emerge still at the helm of NASA. Further, they think NASA still may meet its deadline of a moon landing before 1970." (*B Wk*, 5/27/67, 71-7)

May 28: Dept. of Agriculture scientists, in April testimony released by House Committee on Appropriations' Subcommittee on Dept. of Agriculture, predicted that the application of remote sensing techniques developed in conjunction with U.S. space program, could revolutionize agriculture. They requested $340,000 for FY 1968 to adapt equipment to agricultural requirements. Equipment, currently flown in aircraft, was used to identify accurately and within minutes the soil conditions and moisture, crop yield, and incidence and severity of diseases. Dr. G. W. Irving, Jr., Agriculture Administrator of Agricultural Research Service, predicted: "Once we are in operation from satellite-type spacecraft, the possibilities for securing vast amounts of useful information within extremely short time periods are almost limitless." (AP, B *Sun*, 5/29/67)

- Some of the developments in the U.S. space program which had benefited the field of medicine were cited by John G. Rogers in *Parade* magazine. "A TV camera so tiny it can be dropped down your throat to study your ulcers . . . an ultra-sensitive device to detect the now undetectable first faint muscle quivers of dread Parkinson's disease . . . a heart examination table that escapes all foreign vibrations by suspending the patient on a sheet of air . . . a versatile electronic system that can keep close watch on 128 intensive-care patients at one time and 'shout' an alarm if one needs instant attention." Rogers also noted "magic glasses" which enable paralysis victims, by moving their eyeballs, to regulate electronically many activities necessary to daily life. (Rogers, *Parade*, 5/28/67, 14–5)

May 29: Esro II, first satellite designed, developed, and constructed by ESRO under July 8, 1964, NASA–ESRO agreement, was launched from WTR by four-stage Scout booster but did not achieve planned polar orbit. NASA officials were studying telemetry in an attempt to determine cause of Scout's failure.

Primary NASA mission objectives were to place *Esro II* in planned orbit and to provide tracking and telemetry support. 153-lb satellite carried seven experiments for solar astronomy and cosmic ray studies representing six different organizations from U.K., France, and the Netherlands. ESRO was responsible for delivery of satellite to launch site, for equipment and personnel necessary to mate spacecraft to launch vehicle, and for testing spacecraft. (NASA Proj Off; UPI, *NYT*, 5/31/67, 1)

- GAO had begun "detailed examinations into certain specific areas of the Apollo program," Assistant Comptroller General Frank H. Weitzel wrote in a letter to Rep. William F. Ryan (D-N.Y.), who had inquired about GAO reviews of NASA–NAA contracts. Examinations were begun "on the basis of information developed during" a preliminary review, started in October 1966, which had covered NASA Hq., "several NASA centers," and plants of NAA and Grumman Aircraft Engineering Corp. (Hines, *W Star*, 6/1/67, C16)

- LaRC selected Northrop Ventura Co. to negotiate $3-million contract to conduct research flight-test program using an all-flexible parawing. Experiments would be performed at various scales with remotely controlled unmanned vehicles to establish a body of parawing technology which could be potentially adapted to manned spacecraft recovery systems. NASA's goal was to extend recovery capabilities for the AA Program command module (CM) to include land landings in the early 1970's. Northrop, relying on design criteria gained in earlier LaRC parawing research programs, would evaluate parawings in the 200–600-lb and 5,000-lb payload capacities, including flight tests, before designing 15,000-lb payload system required for the Apollo CM. (NASA Release 67–134)

- Edward R. Mathews, chief of KSC Saturn systems office, was named acting Apollo Program Manager at KSC, replacing M/G John G. Shinkle (USA, Ret.), who resigned May 18. (AP, *NYT*, 5/30/67, 44)

- AFSC awarded General Electric Co. $110,020,000 cost-plus-incentive-fee contract for experiment integration work on the MOL. (DOD Release 497–67)

- U.S.S.R. would launch test series of "carrier rockets of space objects" into the Pacific between May 30 and June 30 "to further explore outer space and accumulate experimental data," Tass announced. Aircraft and ships were warned not to enter 129-km (80-mi)-wide target area about 161 km (100 mi) from Jarvis Island, a U.S. possession, between noon and midnight local time each day. (UPI, *NYT*, 5/30/67, 2)

- Dr. Peter Franken, Deputy Director of Advanced Research Projects Agency (ARPA), was appointed Acting Director of ARPA, succeeding Dr. Charles M. Herzfeld, who resigned to take an executive position with ITT. (DOD Release 493–67)

- Paul C. Aebersold, a pioneer nuclear physicist who worked on the Manhattan Project which developed the first atomic bomb, jumped to his death from the top of an apartment building in Chevy Chase, Md. Aebersold had retired as Director of AEC's Office of Isotopes Development in 1964. (AP, *NYT*, 5/31/67, 18)

May 30: House Committee on Appropriations' Subcommittee on Dept. of Defense Appropriations released testimony on U.S. missile strategy by DOD officials at closed hearing March 20. Dr. Charles M. Herzfeld, who resigned as Director of Advanced Research Projects Agency (ARPA) May 26, testified that advanced missile systems based in space rather than on land were being studied. He recalled that DOD had studied, until 1964, the "Bambi" system designed to use satellites to locate and destroy enemy missiles shortly after they had been launched. Work had been stopped because it was "much too costly," but "we think the time is getting ripe again to look at the whole question because the costs of putting things in orbit have gone down dramatically, the reliability of space engineering has gone up dramatically so that the overall cost of the system ought to come down significantly."

Dr. John S. Foster, Jr., Director of Defense Research and Engineering, revealed that repairs had had to be made on all Minuteman ICBMs because of missile's failure to fire in three tests during October 1966. He emphasized, however, that all missiles were operational at the time of his testimony and said overall readiness of the Minuteman and Polaris missile forces was "excellent." (Beecher, *NYT*, 5/31/67, 1, 28; *W Star*, 5/31/67, A6)

May 31: NASA completed membership of its Lunar and Planetary Missions Board and expanded its scope of activities to include: (1) assistance in the "planning and conduct of all manned and unmanned missions to explore the Moon and planets," and (2) development of "scientific objectives and general strategy for such missions including specific recommendations for mission design and scientific payloads." Board would report to NASA Administrator James E. Webb through Associate Administrator for Space Science and Applications Dr. Homer E. Newell. (NASA Release 67-133)

- USAF launched 8 satellites by Thor-Agena D booster from Vandenberg AFB. This solar-cell-powered, 85-lb, satellite experiment, designed for developing satellite navigation technique as easy to use as celestial navigation and more accurate, would assist pilot or navigator of ship or aircraft who could make direct, simultaneous measurements. Signal measurement from two or more satellites with known positions to determine location of ship or aircraft would make use of signal transmission times and standard navigation techniques. Experiment had been built by Naval Research Laboratory as one to be compatible with Doppler equipment for low-altitude satellites. Naval Air Systems Command had been project sponsor. (UPI, *W Post*, 6/1/67, C4; NRL *News* 5/31/67)

- NASA Nike-Apache sounding rocket launched from WSMR carried instrumented payload to 26-mi (42-km) altitude to gather micrometeoroid particles during a quiet period immediately before the Zeta Perseid/Anetid meteor shower. No scientific data were obtained because "burnthrough" near the igniter headcap caused an explosion. (NASA Rpt SRL)

- NASA had amended its contract with Bendix Corp. for the Apollo Lunar Surface Experiments Package (ALSEP) to add a heat flow experiment substation. $2.4-million amendment increased total value of contract to $23.6 million, plus incentive fees. Heat flow subsystem woud measure net outward flux of heat from moon's interior, providing a comparison of radioactive content of the moon's interior and earth's mantle, a thermal history of moon, a lunar temperature-depth profile, and thermal parameters of the first three meters of moon's crust. (NASA Release 67-135)

- Proposal that Canada immediately begin developing a Canadian-owned, multipurpose communications satellite system to be operational by 1970 was made to Transport Minister John W. Pickersgill by eight major companies of Trans-Canada Telephone System and Canadian Pacific and Canadian National Telecommunications. $80-million system would involve 54 earth stations, the purchase of three comsats, and the launching of two of them into synchronous orbit. Proposal recommended that system be fully integrated with existing networks to carry TV, telephone calls, data, and other services. (Can Press, *NYT*, 6/1/67)
- Two-part report evaluating land use on and around airports and its relation to noise and hazards in the operation of aircraft was released by the FAA [see Feb. 1]. Report, which included surveys of 70 different types of land use on and around 120 U.S. and Canadian airports, pointed out land uses which were compatible with airports and land uses which adversely affected the safety of flight operations. It also considered the problem of aircraft noise. (FAA Release 67-51)
- A national information system for assuring the availability of every significant scientific or technical document, regardless of its origin, was "inevitable," Edward J. Brunenkant, director of AEC's Division of Technical Information, predicted at New York meeting of the Special Libraries Assn. Noting that over 1 million scientific articles were being printed annually in 35,000 journals, Brunenkant conceded that the establishment of a national system would be "extraordinarily complex," but said the need for it was "urgent." (*NYT*, 6/1/67, 33)
- New York's Mayor Robert Lindsay announced plans to construct a $10-million Hall of Science in Queens during the next 14 mos. Structure, which would be connected to science building used during 1964–65 World's Fair, would constitute the second stage of a center for science study and general education. "The Hall of Science when completed . . . will be not only one of the great science museums of the nation and the world, but also a major center for science research and training," the Mayor said. One feature of the new building would be a $1.5-million nuclear reactor donated by AEC. (Knowles, *NYT*, 6/1/67, 33)
- Soviet carrier Aeroflot would begin operating two flights a week from Moscow to Montreal in early June to accommodate "the big inflow of Soviet tourists to the Expo 67 exhibition," Tass announced. (*NYT*, 5/31/67, 66M)

During May: NASA conducted contract negotiations for KSC support services. Federal Electric Co. and RCA Service Co. were selected for competitive negotiations on a five-year, cost-plus-award-fee contract to provide operational communications and instrument services at KSC; Catalytic Construction Co. and Dow Chemical Co. were selected to jointly provide KSC facilities support services under $42-million, four-year, cost-plus-award-fee contract; Trans World Airlines received a $17,006,394 contract extension for installation support services; and

Bechtel Corp. received a $2,039,967 contract extension for specialized facility maintenance, repair, and minor construction services. (NASA Releases 67-106, 67-107, 67-109, 67-119)
- Editorial comment on Apollo accident hearings:

    New York Times: NASA "went to great lengths . . . to assure the Senate Space Committee that all the expensive lessons taught by the fatal January fire will be taken into account before the next try. Yet doubts must persist because the pressurized oxygen system is retained. So is the prime contractor, North American Aviation, whose past errors and deficiencies contributed so much to the debacle. Moreover, NASA has preferred to ignore the pointed questions that have been raised about its own competence to supervise and manage this intricate endeavor." (*NYT*, 5/10/67, 40)

    Washington *Sunday Star:* "Webb and his associates are not solely—perhaps not primarily—to blame for their attitude in time of crisis. In its formative years, the agency was pampered and spoiled by Congress like a favored child. The press willingly cooperated in the creation of NASA's shining image of a superagency staffed by supermen. . . . Inevitably, NASA became the spoiled brat of the federal establishment.

    "But NASA, of course, is no different than any other federal agency. It is composed of wise and foolish men, of dedicated and self-serving public servants, of heroes and of knaves. It is doing an extraordinary—and, in our opinion, a worthwhile—task [and] deserves continued public support." (*W Star,* 5/14/67, C1)

    New York Times: "Now that the public finally knows that North American was picked for the Apollo project by a few high officials—and not by a large group of technical experts as originally imagined—the question of why it was picked becomes even more intriguing than before. The cynics, of course, have always believed that the award went to North American as the result of a battle among influence peddlers, including Bobby Baker, former Secretary to the Senate Democrats. The cynics may be wrong, but the public cannot know without a full investigation. Congress has the obligation to force out all the facts." (*NYT*, 5/11/67, 44)

- Editorial comment on authorization for SST prototype construction [see April 29]:

    Washington Post: "President Johnson made the right decision in requesting that Congress appropriate Federal funds for the development of the supersonic transport plane. . . . But in proposing what it calls 'a creative partnership between our Government and American industry,' the Administration should provide more adequate protection for the taxpayer who will bear a substantial share of the development costs.

    "Why should not the taxpayers, who are to provide about 25 per cent of the capital required, be permitted to share proportionately in the enormous profits that would be earned by a successful SST? Although it is necessary to launch the SST project with the shortest possible delay, Congress should be concerned with a more adequate protection of the public interest." (*W Post,* 5/1/67)

    World Journal Tribune: "Actually, the basis of the SST project is coldly commercial. At stake is not some vague blue ribbon of achievement but American leadership in the highly profitable business of supplying planes to the world's airlines.

    "SST, by its size, speed and sonic boom, will cause a number of problems which have not yet been solved. But it was inevitable that this country should enter the supersonic race—not for national pride or out of scientific curiosity, but as a matter of good business.

"The international plane market is too good for America to let it go by default." (*WJT*, 5/4/67, 22)

New York Times: "Mr. Johnson acknowledged that the building of an American SST 'carries high technical and financial risks.' Yet in his anxiety to get on with the job he has increased the potential hazards to the taxpayer. . . .

"Congress has a responsibility to establish guidelines to keep the Administration's investment in the SST from getting out of hand. It should see to it that the Government does not write any blank checks and become overly involved in what is supposed to be a commercial venture. And industry, too, should recognize that if it asks Washington to shoulder the costs, then Washington may well end up running the entire operation." (*NYT*, 5/2/67, 43)

- American Helicopter Society, meeting in Washington, D.C., presented the Dr. Alexander Klemin Award to Avco Corp. executive Dr. Anselm Franz for "leading the development of a noteworthy series of gas turbine engines for helicopters and other VTOL aircraft. . . ." (*Av Wk*, 5/29/67, 55)

- NASA's broad program of lunar exploration included, according to *Space/Aeronautics*, "quickening conversion of data and designs into hardware and techniques for . . . planting a coordinated set of instruments on the lunar surface for remote, year-long readout; bringing back those lunar samples for exhaustive, quarantine-flavored analysis in custom labs; going back with wheels, rocket pogos, and sophisticated drills for more extensive exploration; [and] exploiting, eventually, the moon's capacities as a base for astronomy, or as a source for fuel." (*S/A*, 5/67, 68)

June 1967

June 1: U.S.S.R. launched *Cosmos CLXII* into orbit with 280-km (174-mi) apogee, 201-km (125-mi) perigee, 89.2-min period, and 51.8° inclination. Equipment functioned normally. Satellite reentered June 9. (AP, W *Star*, 6/1/67, 1; GSFC *SSR*, 6/15/67)

- NASA Aerobee 150 sounding rocket launched from WSMR carried GSFC-instrumented payload to 111.5-mi (179-km) altitude to measure and record spectra of Spica and Zeta Ophiuchi in extreme ultraviolet range. Rocket and instrumentation performed well. (NASA Rpt SRL)

- NASA reported success of *Lunar Orbiter IV*, launched May 4, noted that final readout of all photographs had been completed. Photographs would provide photo coverage of 99% of the moon's front face and show high resolution photos of polar and limb areas with extensive geological detail. Report said convergent stereoscopic coverage had been obtained of entire Apollo zone of interest; photographic coverage obtained of the back side of the moon was still under review (NASA Proj Off)

- NASA's *Ats II* mission had been adjudged a failure "based upon review of the [spacecraft's] assessed performance . . . in its unplanned elliptical orbit," NASA announced. Satellite, launched from ETR April 5 on mission to evaluate gravity-gradient system for spacecraft stabilization, had remained in transfer orbit, tumbling uncontrollably, after Agena engines failed to ignite for second burn. All experiments were functioning, but possibility of obtaining useful data was poor. (NASA Proj Off)

- NASA personnel changes:

    M/G James W. Humphreys (USAF), recently returned from a Vietnam assignment as Chief of the Public Health Service Div., Agency for International Development (AID), became Director of Space Medicine, OMSF, replacing B/G Jack Bollerud (USAF), who had been Acting Director since the December 1965 death of Dr. W. Randolph Lovelace II. General Bollerud returned to duty with USAF.

    Jerome Lederer, vice president and technical director of Flight Safety Foundation, Inc., and director of Cornell-Guggenheim Aviation Safety Center, was appointed Director of Safety, OMSF. Lederer would develop policies and procedures related to manned spaceflight systems safety; act as focal point for consideration of and decisions on matters related to system and flight safety; and provide continuing review and evaluation of safety provisions.

    M/G Charles R. Roderick, who retired May 31 from USAF where he was serving as Deputy Assistant to the Secretary of Defense for Legislative Affairs, was named Special Assistant to NASA Administrator James E. Webb.

    Harold Sims, retired career foreign service officer currently serving as mayor of Sparta, Tenn., and Gerald J. Lynch, president and chairman of the board of Menasco Manufacturing Co., were sworn in as consultants to NASA Administrator James E. Webb. (NASA Release 67–84, 67–138, 67–139, 67–140, 67–141)

- NASA turned on the last of 11 experiments onboard *Explorer XXXIV* satellite, launched from WTR May 24. Data from experiments, designed to measure solar and galactic rays, were being transmitted, and equipment continued to operate satisfactorily. (NASA Release 67–144)
- Dr. Albert J. Kelly's resignation as ERC Deputy Director became effective. Now dean of Boston College's School of Business Administration, he would continue to serve as a part-time consultant to NASA. (*Tech Wk*, 3/27/67, 13)
- First nonstop helicopter crossing of the Atlantic was successfully completed when two USAF HH–3E helicopters landed at Le Bourget Airport, opposite the U.S. pavilion at the Paris International Air and Space Show, following 30-hr 48-min flight from New York [see May 26–31]. Helicopters, which followed the route taken by Charles Lindbergh in 1927 when he became first man to fly the Atlantic nonstop alone, traveled 4,160 mi and made nine refuelings in flight. First flight crew was commanded by Maj. Herbert B. Zehnder (USAF); second by Maj. Donald B. Maurras (USAF). Both crews were submitting claims to Fédération Aéronautique Internationale for new speed records from New York to London and New York to Paris. (UPI, P *EB*, 6/1/67, 1; *NYT*, 6/2/67, 17)
- F–111B was "a very easy plane to fly" with outstanding low-speed qualities, V/Adm. T. F. Connolly, Deputy Chief of Naval Operations (Air), said in April 18 testimony released by the House Committee on Appropriations' Subcommittee on the Dept. of Defense. Admiral Connolly's description of the F–111 (formerly TFX), subject of increasing criticism since April 21 crash of USN test model, was described by Rep. George Andrews (D-Ala.) as "the most complimentary statement [about the F–111] . . . which has been made before this committee."

  Basing his views on his personal inflight observations, Admiral Connolly conceded that there was "a lot of work to do on the airplane," including making configuration changes to improve pilot visibility. He asserted that when difficulties involving engine stall were overcome there should be no major problems and said that aircraft's anticipated weight satisfied USN requirements. (Transcript; AP, *NYT*, 6/2/67, 23)
- Dr. Alexander H. Flax, Assistant Secretary of the Air Force for Research and Development, was awarded National Geographic Society's General Thomas D. White Space Trophy for outstanding contributions to U.S. aerospace progress. First civilian to receive the award since its establishment in 1961, Dr. Flax was cited for "his effective direction of Air Force aerospace research and development programs enhancing national security and advancing space technology." (NGS Release)
- Robert F. Allnutt, assistant general counsel of NASA, was named NASA's Assistant Administrator for Legislative Affairs, effective June 15. (NASA Release 67–137)

June 2: Col. Joseph Cotton (USAF) and NAA test pilot Van Shepard flew XB–70 No. 1 to mach 1.43 and 42,200-ft altitude during two-hour performance evaluation flight at Edwards AFB. (XB–70 Proj Off)
- NASA selected RCA to negotiate $38-million contract to maintain and operate portions of the Satellite Tracking and Data Acquisition Network

(STADAN) at Rosman, N.C., and Fairbanks, Alaska, STADAN sites, and at GSFC control centers for individual spacecraft. Effective Oct. 1 for three-year period with NASA options for two one-year extensions, contract covered actual cost-plus-incentive award based on performance. (NASA Release 67-142)

- GAO, responding to March 14 request by Sen. Margaret Chase Smith (R-Me.), member of Senate Committee on Aeronautical and Space Sciences, had reviewed NASA's Apollo contract with North American Aviation, Inc. (NAA), and published a "Summarization of National Aeronautics and Space Administration Management Review of North American Aviation, Inc., Activities" based on the Phillips Report. According to GAO, the Phillips Report "stated that, at the start of the CSM and S-II programs, key milestones were agreed upon, performance requirements were established and cost plans were developed . . . [but] as the program progressed, there were slippages in key milestone accomplishments, degradation in hardware performance, and increasing costs." In evaluating the probability of NAA's ability to "meet future commitments, the report stated that, for both the S-II and the CSM, significant technical problems still remained which had to be resolved. The review team did not find significant indication of actions underway, for the S-II, to 'build confidence that future progress will be better than past performance.' While some progress was indicated to improve the outlook for the CSM, the report stated that 'there is little confidence that NAA will meet its schedule and performance commitments within the funds available for this portion of the Apollo program. . . .'" (Text)
- Astronaut Edwin E. Aldrin received first honorary membership awarded by International Assn. of Machinists and Aerospace Workers at IAM Aerospace Conference in Houston. Aldrin, considered by the Union to be the first space mechanic, was cited for his work with tools outside *Gemini XII* spacecraft Nov. 13, 1966. Astronaut Alan B. Shepard, Jr., received an Award of Merit for being first American in space. (*Houston Chron*, 6/2/67)

June 3: U.S.S.R.'s next manned space flight would use a Soyuz spacecraft similar to the one which crashed April 24, according to Astronauts Michael Collins and David Scott who had discussed future space plans with Soviet cosmonauts at the Paris International Air and Space Show [see May 26-31]. In UPI interview Collins and Scott said they were "amazed by the similarities between the two [U.S. and U.S.S.R. space] programs." Collins said that Cosmonauts Pavel Belyayev and Konstantin Feoktistov indicated "that there would be several earth orbital flights and then . . . a circumlunar flight." (UPI, *W Post*, 6/4/67, A9)

June 4: USAF launched unidentified satellite with Atlas-Agena D booster from WTR; satellite reentered June 12. (UPI, *NYT*, 6/6/67; *Pres Rep 1967*)

- Dr. Lloyd V. Berkner, a leading scientific pioneer in the U.S. space program, died at age 62. He collapsed, apparently of a heart attack, while attending a Washington, D.C., meeting of the NAS Council, and died the next day. Credited with being the "father" of the International Geophysical Year (IGY), Dr. Berkner was the principal administrator of the U.S. part of the IGY program during its operation in 1957 and 1958. From 1958-1962, he was chairman of NRC's Space Science Board, which advised the Government on the national program of space research. For his "outstanding and pioneering leadership," of SSB, he received NASA's Distinguished Public Service Medal Oct. 7, 1966. He served as President of the Graduate Research Center of the Southwest in Dallas from 1960 until 1965, when he retired because of health. He continued until recently as Chairman of the Board of Trustees, and was a member of the NASA Historical Advisory Committee. A Rear Admiral in the U.S. Navy Reserve, Dr. Berkner would be buried with full military honors in Arlington National Cemetery June 7. (*NYT*, 6/5/67, 41; *W Star*, 6/5/67, B5; *Science*, 6/9/67, 1349)

- Charles F. Bingman, Director of Management Programs, OMSF, became Deputy Director of the Office of Organization and Management Planning. (NASA Ann, 6/9/67)

June 5: Cosmos CLXIII was launched by U.S.S.R. into orbit with 593-km (368-mi) apogee, 255-km (158-mi) perigee, 93.1-min period, and 48.4° inclination. Instruments performed satisfactorily. Satellite reentered Oct. 11. (UPI, *NYT*, 6/6/67, 3; GSFC *SSR*, 10/15/67)

- *The Washington Post* commented on the death of Dr. Lloyd V. Berkner: "In every generation there are a few men whose personal careers reflect with precision the great changes that have come over the country in their lifetime. Lloyd Viel Berkner built his first radio transmitter as a schoolboy in Sleepy Eye, Minnesota, and in 1928 he joined an expedition to the South Pole. Three decades later he emerged as one of the most effective advocates of American exploration in space.

    "In those years American science had become a vast interwoven fabric of academic and Federal laboratories, foundations and private corporations, civilian operations and military operations. Mr. Berkner moved in this world, which he had signally helped to create, not only as a scientist of distinction but as an organizer and manager of extraordinary force and capacity. . . ." (*W Post*, 6/5/67)

- NASA selected J. H. Lawrence Co. for negotiations on a one-year, $2.25-million contract to provide plant maintenance at GSFC. (NASA Release 67–145)

- A manned Apollo spacecraft should be launched to recover *Explorer I*, first U.S. satellite, whose orbit was expected to decay in late 1969, G. Snowden of Melbourne, Australia, suggested in a Letter to the Editor of *Aviation Week*. "When Apollo swings into action, maybe one of the tasks could be a rendezvous and retrieval of man's first satellite. I say first because Sputnik 1 and 2 have decayed, leaving Explorer 1 at the head of the list. A replica can be built at a fraction of the cost of retrieval, but there is nothing like having the original. Not only would it have value as a historic memento of man's venture into space, but also to scientists. I am certain after nearly 10 years in space it would be valuable to them. . . ." (*Av Wk*, 6/5/67, 110)

- Cost of reserving future delivery positions for SST would be $750,000 each, under new policy announced by FAA and Boeing Co., SST airframe contractor. In addition, all payments would be made directly to Boeing Co. and would be used in lieu of Government funds to help finance SST development. Original policy, established in November 1963 and applicable to 113 positions presently assigned, required airlines to deposit with U.S. Treasury a total of $200,000 for each delivery position reserved—$100,000 initially and $100,000 six months after initiation of prototype construction. On February 6, FAA asked 12 U.S. companies on SST delivery list to also contribute $1 million per delivery position to the cost of prototype construction. (FAA Release 67–52)

- Pilot performing with the French Patrol, the aerobatic team of the French Air Force, was killed when his jet aircraft crashed and burst into flames during final aerial display at the Paris International Air and Space Show. Aircraft failed to pull out of a dive, bounced along the airfield, and then exploded about 100 yds from the grandstand, injuring several spectators. (NYT, 6/6/67, 77M)
- France exploded nuclear device from a balloon over Mururoa Atoll in the Pacific. Test was first of four in 1967 series designed to perfect a trigger for the hydrogen bomb France planned to explode in 1968. (UPI, NYT, 6/23/67, 2)

June 6: NASA Aerobee 150 sounding rocket launched from WSMR carried ARC-instrumented payload to 86-mi (137-km) altitude to collect meteoritic debris during peak of meteor shower. Rocket and instrumentation performance was satisfactory. Further examination of the 12 collection modules would be required to determine if micrometeoroid particles were collected. (NASA Rpt SRL)
- NASA Nike-Tomahawk sounding rocket launched from Churchill Research Range carried Univ. of New Hampshire experiment to provide data on the neutron intensity at different altitudes using a neutron detector, solar x-ray fluxes using an x-ray counter, and 2–10 Å Lyman-alpha radiation using a photoionization chamber. Rocket and instrumentation performed satisfactorily. (NASA Rpt SRL)
- House Committee on Science and Astronautics filed $4.9-billion NASA authorization bill (H.R. 10340) with the House and recommended that it be passed without amendment [see May 16]. (NASA LAR VI/64)
- Japan Air Lines jet aircraft carrying American and Japanese scientists from San Francisco to Tokyo successfully maintained radio contact with Mojave, Calif., ground station via NASA's *Ats I* satellite (NASA Proj Off; P *EB*, 6/8/67)
- 37-yr-old Astronaut Edward G. Givens, Jr. (Maj., USAF), was killed near Houston when the car he was driving missed a curve and crashed into an embankment. Two other USAF officers with him were injured, one critically. Givens was the first U.S. astronaut to die while off duty. Six others had died in connection with the space program: three in air crashes and three in the Jan. 27 Apollo fire. He was selected to join the NASA astronaut team April 4, 1966. (MSC *Roundup*, 6/9/67, 1; UPI, NYT, 6/7/67, 30; UPI, *W Post*, 6/7/67, A7)
- MSFC Director Dr. Wernher von Braun was presented the Smithsonian Institution's Langley Medal by Dr. Fred L. Whipple, Director of Smithsonian Astrophysical Observatory, in Washington, D.C., ceremony. Dr. von Braun was cited in "recognition of his creative vision of the practical application of rocket power to space flight leading to the first U.S. satellite, and of his technical leadership in development of the Saturn class of large launch vehicles upon which the Apollo moon flight is based." He was the 13th recipient of the award, which was established in 1908 to commemorate the aeronautical achievements of Samuel Pierpont Langley, third Secretary of the Smithsonian Institution. Other recipients included Orville and Wilbur Wright, Charles A. Lindbergh, Robert H. Goddard (posthumously), Dr. Hugh L. Dryden, and Astronaut Alan B. Shepard, Jr.

In making the presentation, Dr. Whipple said U.S. could have launched the world's first satellite if it had endorsed the proposal which he and Dr. von Braun developed in 1954 to launch small spacecraft from the equator using existing rockets. Government decision to fund, instead, USN's Vanguard project, allowed U.S.S.R. to win "the first space race." (Text; Smithsonian Release; *Marshall Star*, 6/7/67, 1; AP, *NYT*, 6/6/67, 3)

- President Johnson appointed Astronaut James A. Lovell, Jr., Special Consultant to the President for Physical Fitness, to replace Stan Musial who resigned in January. Lovell would assume the new position in addition to his regular duties as an astronaut. (*PD*, 6/12/67, 833–4)

- Chairman of the House Committee on Science and Astronautics George P. Miller (D-Calif.), in a Letter to the Editor of the Washington *Evening Star*, commented on William Hines' May 25 column criticizing Committee member Joseph E. Karth's (D-Minn.) speech before the National Space Club May 17. "The article unfairly alleged that Rep. Joseph E. Karth . . . by speaking out . . . was the 'tool' chosen by the administration to do a public relations job for NASA and the space program. . . .

   "What Mr. Hines failed to say was that Congressman Karth has led a number of hard-hitting congressional space investigations over the years for me. Both NASA and the contractors involved were 'taken to the woodshed.' The Apollo program is no exception. Mr. Karth has been critical—and made it clear that the accident responsibility falls squarely on NASA and the contractor.

   "Very simply, Mr. Karth said that perhaps the Congress, the administration and the press could do a better job of giving the public objective information on the space program. The wide swings from 'everything is great' to 'everything is wrong' must be confusing. . . ." (*W Star*, 6/6/67, A12)

- NASA Associate Administrator for Manned Space Flight Dr. George E. Mueller outlined Apollo Applications (AA) program at AIAA meeting in Washington, D.C. Series of innovations would be made, he said, to permit sufficient reductions in unit costs: (1) reuse of Command Module (CM); (2) addition of land-landing capability which would facilitate CM reuse, permit increase in crew capacity, and, possibly, make use of naval recovery forces unnecessary; (3) "double use" of Uprated Saturn I's 2nd stage as booster during launch phase and as Orbital Workshop in space; (4) repeated use of Orbital Workshop as an embryonic space station; (5) longer-duration flights of one year or more; and (6) use of Apollo flight hardware and physical plant and employment of Apollo program officials and industrial organizations as they become available.

   ". . . this is a program that provides for a detailed measurement of the utility of man in space at a relatively low cost. The measurement is obtained by doing useful things—astronomical observation, extended exploration of the moon and experiments with sensing equipment that can lead to benefits of enormous significance to all mankind." (Text)

- AFSC Director of Communication Satellite Programs Col. M. B. Gibson (USAF), speaking to annual convention of Armed Forces Communications and Electronics Assn. in Washington, D.C., explained management of Initial Defense Communication Satellite Program (IDCSP): ". . . [we] recognized need for the innovation of special management techniques [and] a rigorous, streamlined procedure for Government acceptance of satellites under a high production rate . . . [and] satellites which have a requirement for long-life operation in space. . . ." Gibson concluded: "The thoroughness of satellite acceptance procedures is evident from the performance of the 15 communication satellites in orbit today." (Text; AP, *W Post*, 6/7/67, 10)

- Mrs. Lyndon B. Johnson presented Phoenix, Ariz., Mayor Milton Graham with FAA's Airport Beautification Award for the city's "outstanding community-wide program to make its airport premises a center of culture and beauty." (FAA Release 67-53)

June 7: Launch of the first manned Apollo mission by March 31, 1968, "is a very difficult milestone to achieve, but it is possible," George M. Low, Manager of the Apollo Spacecraft Program at MSC, told the press at MSC. All changes in the Apollo spacecraft necessitated by the Jan. 27 Apollo fire had been planned, he said, but until daily scheduling was completed, no date could be set for the mission. Low noted that Apollo parachute landing system might require requalification because design changes had increased spacecraft's weight by 200-300 lbs, but he indicated that testing, if necessary, could be conducted in summer 1967 without seriously affecting the Apollo schedule. (*Tech Wk,* 6/12/67, 15; *SBD,* 6/8/67, 217)

- Sen. William Proxmire (D-Wis.) warned the Senate that U.S. deficit could reach $30 billion unless drastic cuts were made in Federal spending. He suggested that a $5-billion reduction in proposed FY 1968 spending be made by: (1) postponing public works programs such as highway construction until economic and military demands were lowered; (2) foregoing the "expensive frill" of SST development; (3) curtailing military costs by further reducing U.S. troop strength in Europe; and (4) cutting NASA's post-Apollo funds by over $1 billion by postponing "such projects as a soft-landing on Mars, which has nothing to do with our present goal of a moon landing.

"The situation that we face in fiscal 1968 involves an economy dominated by a heavy Federal budget, no matter how it is measured. If these circumstances materialize, there is no question about the immediate urgency of substantial expenditure reductions. . . . We may need a tax increase too, but first priority by all means should go to the reduction of spending." (Text)

- NASA selected Bendix Field Engineering Co. for negotiations on $21-million contract to provide maintenance and operations services at two U.S. and four foreign Satellite Tracking and Data Acquisition Network (STADAN) sites, and at NASA Communications Center and Network Test and Training Facility, both at GSFC. Effective Oct. 1 for three-year period with NASA options for two one-year extensions, contract covered actual cost-plus-incentive award based on performance. It completed maintenance and operations arrangements for STADAN. (NASA Release 67-148)

- OAR's AFCRL would launch four 812-ft-long balloon systems from Walker AFB in summer 1967 to complete series of five launches conducted by NASA to investigate possible parachute landing systems for Voyager program [see May 9], OAR Commander M/G Ernest A. Pinson announced. Scheduled for June 27, July 20, Aug. 9 and 24, flights would evaluate performance of different types of parachutes at pressures and velocities equivalent to a descent through the Martian atmosphere. Each balloon would carry, for release at high altitude, a simulated Voyager capsule and a parachute designed to softland the capsule on Mars. Five cameras—two on the capsule and three on the parachute—would be used to evaluate each test. First flight in the series was conducted Aug. 30, 1966. (OAR Release 67-17; AP, *NYT,* 6/8/67, 3)

- MSFC had awarded Lockheed Missiles & Space Co. a $334,031 contract to

June 7

study methods of increasing the number of astronauts that could be delivered to the moon with the Saturn V booster and to determine the most promising concepts for efficient delivery of logistic payloads. For the purposes of the study, Lockheed would consider: (1) a hypothetical 1975 mission involving two "product improved" Saturn V boosters—one to launch a modified Apollo spacecraft and one to launch a cargo payload; and (2) a hypothetical 1980–82 mission in which two Uprated Saturn V boosters would launch a six-man spacecraft directly to the moon. (MSFC Release 67-123)

- Widow of NASA X–15 test pilot Joseph A. Walker filed $1-million damage suit for the June 8, 1966, death of her husband when his F–104 fighter aircraft collided with an XB–70 experimental bomber near Barstow, Calif. Aircraft had been flying in tight formation "to allow photographic coverage of aircraft powered by General Electric engines" for publicity purposes. Mrs. Walker's suit, filed in Federal Court in Los Angeles, charged negligence on the part of General Electric Co. and its chief test pilot John M. Fritz, and North American Aviation, Inc., manufacturer of the XB–70. (UPI, *W Post*, 6/9/67, A2)

- Scientists have an increased responsibility to relate the merit of their research to the public "in understandable terms," M/G Ernest A. Pinson, commander of USAF Office of Aerospace Research, suggested at 12th Annual Science Seminar in Albuquerque, N. Mex. "Science has to do a better job in making known its values and its needs. Science has to remember that the past 25 years, when everything came comparatively easy for financial support of research, are gone and that science has to compete with many other high priority sectors of the economy" for operating funds. (OAR Release 67-13)

- Two ten-week programs for college faculty members—administered jointly by Auburn Univ., Univ. of Alabama, and MSFC—opened at MSFC. First program would be summer institute in space-related sciences, sponsored by NASA and American Society for Engineering Education (ASEE), and would be conducted by Univ. of Alabama. (Five other summer institutes, sponsored by NASA and ASEE, would be conducted by NASA centers and nearby colleges at MSC, GSFC, ARC, LaRC, and LeRC.) Second ten-week program, administered by Auburn Univ., would be 1967 NASA Engineering Systems Design Summer Faculty Fellowship Program. (MSFC Release 67-124)

June 8: U.S.S.R. successfully launched *Cosmos CLXIV*. Orbital parameters: apogee, 320 km (199 mi); perigee, 202 km (126 mi); period, 89.5 min; and inclination, 65.7°. Satellite reentered June 14. (AP, *NYT*, 6/9/67, 3; GSFC *SSR*, 6/15/67)

- Adjustments of *Lunar Orbiter IV*'s orbit were successfully conducted June 5 and 8 to place spacecraft on path similar to that intended for Lunar Orbiter V. Adjustments, which lowered apolune to 2,450 mi (3,943 km) and perilune to 48 mi (77 km), were made to enable LaRC engineers "to gather useful tracking and gravitational field experience" in preparation for flight of Lunar Orbiter V. (NASA Release 67-154)

- Memorial services in Houston for Astronaut Edward G. Givens, Jr., were attended by many of his fellow astronauts and their families. Givens would be buried June 9 in Quanah, Tex. (MSC *Roundup*, 6/9/67, 1; AP, *W Post*, 6/9/67, B8)

- Single global time zone would be advantageous with the speed of communications and transportation in the space age, OSSA Director of Space

Applications Programs Leonard Jaffee told the Armed Forces Communications and Electronics Assn. in Washington, D.C. "We are almost approaching the point where the remaining obstacle to global business, aside from the political problems, will be the zonal time differences.

"At the moment time differences are aggravations to the total realization of our new-found communications and transportation capabilities, but not complete deterrents. . . . Today, if it is important enough, we make allowances for and tolerate time differences—tomorrow we may have to do something about [them] . . . because as our business and cultural patterns change in this space era, these time differences will become intolerable.

". . . the space age has already contributed the concept of 'one world-one time.' By creating a common time and acceptance of a day not geared to the rising and setting of the Sun, we may be able to extend this to an acceptance of 'one day' consisting of 24 hours . . . during which we will make optimum use of our roads, our schools, our factories, our communications and transportation facilities . . . [with] 3-shift use . . . [so that] current facilities could handle three times as many people." (Text)

- "Although junior colleges constitute a major segment of higher education in terms of number of institutions (about one-third), their resources for science, in terms of expenditures and manpower, are meager," NSF reported to House Committee on Science and Astronautics' Science, Research, and Development Subcommittee. NSF report, *The Junior College and Education in the Sciences*, stated that junior college "share of expenditures for separately budgeted research and development, in 1963–64, was one-tenth of 1 percent of the total for all colleges and universities; for science 'plant,' 3.7 percent; and for instruction and departmental research, 4.8 percent." (Text)
- FAA Administrator Gen. William F. McKee received Washington, D.C., Air Line Traffic Assn.'s 1967 "Man of the Year" award. (*Av Wk*, 6/19/67, 13)

June 9: USAF launched an experimental reentry vehicle by Atlas-F booster from Vandenberg AFB as part of Advanced Ballistic Re-Entry Systems (ABRES) program. (*Tech Wk*, 6/19/67, 15)
- Ground-based optical astronomy had been throttled by lack of large telescopes, especially in southern latitudes, said Director Horace W. Babcock of Mt. Wilson and Palomar Observatories, who explained that "optical astronomy occupies a central position and will continue to do so." He deplored the fact that, in recent years, vital support had been "woefully small" for construction of ground-based telescopes of large size (up to 200-in class). He cited 1964 NASA-sponsored study by Panel on Astronomical Facilities (chaired by Lick Observatory's Albert E. Whitford) which had recommended $1-million study for construction of "the largest feasible optical reflector," and construction of three 150- to 200-in telescopes, four 60- to 84-in telescopes, and eight 36- to 48-in models. (Babcock, *Science*, 6/9/67, 1317–22)
- Rep. J. Edward Roush (D-Ind.) introduced to the House H.R. 10674, a bill to rename the Indiana Dunes National Lakeshore in honor of Astronaut Virgil I. Grissom who died in Jan. 27 Apollo accident. (NASA LAR VI/65)
- U.S., France, and U.K. had had excellent displays at Paris International Air and Space Show [see May 23–31], but visitors had seemed most

impressed by U.S.S.R. exhibit, *Time* magazine noted. "The greatest Soviet surprise was the launch vehicle that in 1961 sent Pioneer Cosmonaut Yuri Gagarin into orbit in Vostok 1. . . . Visitors at the show flocked to a huge mock-up of the 13.6-ton Proton satellite, [to] . . . a model of their advanced Molniya communications satellite, [and to] . . . the variety of new Russian jet liners.

". . . the U.S. went all out . . . [and] Britain and France also put their best fleet forward. . . . But it was the Russians who stole the show, simply by taking the wraps off space hardware—some of it a decade old—that they had never before displayed in the West. . . ." (*Time*, 6/9/67)

- Decrease in worldwide air temperature since the 1940's might be due to emission of man-made pollutants into the atmosphere, National Center for Air Pollution scientists Robert A. McCormick and John H. Ludwig suggested in *Science*. Increasing amount of solid material made atmosphere more turbid and thus reduced amount of solar radiation reaching earth. Although excessive amount of carbon dioxide in atmosphere as a result of prolonged coal and oil burning should have made earth warmer, the authors noted, its effect had been offset by the turbidity caused by the pollutants. (McCormick, Ludwig, *Science*, 6/9/67, 1358–9)

June 10: LaRC selected Lockheed California Co. for negotiation of a $750,000 contract for a 19-mo R&D program leading to design of a hypersonic aircraft wing structure. Lockheed would "assess the validity of several wing structural concepts . . . and provide the technology required for NASA to determine which combination of concepts is best. . . ." (NASA Release 67-152)

- President Johnson announced that he had accepted "with deepest reluctance and regret" the resignation of Cyrus R. Vance as Deputy Secretary of Defense, effective June 30. Vance would be succeeded by Secretary of the Navy Paul H. Nitze. (*PD*, 6/19/67, 860)

June 11: Recent experiments by USAF scientists had proved that satellites could communicate directly—unaided by relay stations—from opposite sides of earth, AFSC announced. Experiment, in which HF and VHF signals were transmitted between two satellites launched Nov. 3, 1966, confirmed theory that ionosphere resembled a "whispering gallery" in which sound waves traveled along a curved area with only minor volume loss. (AFSC Release 116.67)

- U.S. and Australian scientists would begin extensive survey off coast of western Australia Aug. 8 to test continental drift theory—hypothesis that earth's continents were once part of one or two "supercontinents" that broke apart millions of years ago and have since been constantly drifting in earth's mantle. Survey, which would be conducted as part of a global expedition currently underway by USC&GS *Oceanographer*, would use echo soundings to determine depths and angles of ocean floor and measure seismic waves produced by underwater explosions in an effort to determine whether Australia was once joined with Antarctica, and possibly India. (ESSA Release 67/51)

- A 236-in-wide mirror which would be used by U.S.S.R. in world's largest reflector telescope was nearing completion, Walter Sullivan reported in the *New York Times*. After cooling from a molten state for 1½ years, mirror had been cut to produce concave, parabolic configurations and was currently undergoing final grinding at a factory near Moscow. It would soon be shipped to a new observatory in the northern Caucasus, which would be operated directly by the Soviet Academy of Sciences. Largest reflector telescope in use at present time was 200-in-wide instrument at Mt. Palomar Observatory. (Sullivan, *NYT*, 6/11/67, 15)

June 12: *Venus IV* unmanned probe was successfully launched by U.S.S.R. on a four-month journey toward Venus. Tass announced that the 2,438-lb spacecraft—heaviest Venus probe ever orbited—had been launched into a parking orbit and then injected on a trajectory "close to the prescribed one." All onboard equipment was functioning normally. There was no indication whether spacecraft—launched two days before scheduled launch of NASA's *Mariner V* Venus flyby mission—would attempt a flyby of the planet or a soft-landing. (*NYT*, 6/13/67, C19; AP, *W Post*, 6/13/67, A11; Nordlinger, B *Sun*, 6/13/67)

- *Cosmos CLXV* was launched by U.S.S.R. into orbit with 1,542-km (958-mi) apogee; 211-km (131-mi) perigee; 102.1-min period; and 81.9° inclination. Equipment was functioning normally. (UPI, W *Star*, 6/13/67, A70)

- Rep. William F. Ryan (D-N.Y.), speaking on the House floor, urged Congress to reopen its investigation of NASA, with particular emphasis on management policies and procedures. He said the Phillips Report indicated NASA "has had grave difficulty in enforcing standards of workmanship, implementing safety and inspection procedures and in properly supervising contractor costs and time schedules. The results of more recent reviews of North American and reviews of other contractors have again not been made public.

  "Under such unsatisfactory conditions, accidents and failures can well be expected. Congress should be less concerned with the technical cause of a particular fire and more with the space agency's lack of control over its own program. It is with this in mind that I have regretted the untimely close of the congressional hearings and requested further substantive investigation into the area of NASA management. . . ." (*CR*, 6/12/67, H7033)

- Although ARC was the smallest NASA field center, *Newsweek* said, it was producing "some of the agency's biggest ideas." Summarizing ARC history, the article noted that when center opened in 1940 it had been primarily a site for testing warplane aerodynamics. "Ames still maintains its original wind tunnel—the world's largest—as well as 29 others, [but aeronautics] . . . represents only a quarter of the effort at Ames today. Most of the scientists are otherwise engaged in pure research in such fields as planetology, exobiology . . . and chemical evolution.

  ". . . the center's laboratories enable more than a thousand scientists, engineers and technicians to pursue as wide a variety of problems as their imaginations and the broad guidelines of the space program permit. The problems now range from designing a new hard space suit to studying the effect on mice of reproducing in a 2G environment." (*Newsweek*, 6/12/67)

- ERC awarded MIT a three-year, $7.9-million contract for R&D of high-performance gyroscopes and accelerometers "to guide and control the vehicles planned for the complex aeronautical and space missions of the 1970s and 1980s." It was anticipated that instruments would make possible systems which would navigate advanced supersonic aircraft to intercept runway landing beams without present runway aids; provide highly precise pointing references for satellites; and guide interplanetary vehicles on missions lasting one year or more. (NASA Release 67-155)

June 12

- NASA awarded Aerojet-General Corp. a two-year, $17-million contract extension for R&D work on SNAP-8 nuclear reactor electrical power system. (NASA Release 67-156)
- NASA was consulting with Aerospace Industries Assn. (AIA) and with individual aircraft companies on a proposed research program which would support areas of general aviation, *Technology Week* reported. At the request of several users, NASA had identified a number of technical areas where the application of advanced technology seemed warranted but emphasized that NASA's efforts should be with aircraft designers and it should have no responsibility for the application of technology. (*Tech Wk*, 6/12/67, 3)

June 13: NASA Nike-Apache sounding rocket launched from Wallops Station carried American Science and Engineering experiment to 112-mi (179-km) altitude to measure neutron flux in vicinity of earth using directional neutron detector. Once proven, detector was planned for use in OGO-E program. Rocket and instrumentation performed satisfactorily. (NASA Rpt SRL)

- Twenty-one U.S. astronauts began a week of training at USAF's Tropical Survival School in Panama. Their first meal was iguana thermidor and baked armadillo. (*W News*, 6/13/67; AP, *C Trib*, 6/14/67)
- Donald L. Mallick, a pilot and engineer at FRC, and L/Col. Emil Sturmthal (USAF), chief of AFFTC's bomber section, were selected as pilots in joint NASA–USAF XB-70 research program. (FRC Release 17-67)
- NASA test pilot Milton O. Thompson received AIAA's 1966 Octave Chanute Award in Los Angeles for "his contributions both as an engineer and a pilot in the development and conduct of the manned lifting body flight test program." Award was presented annually to the pilot who had contributed most to the aerospace sciences during the preceding year. (FRC Release 15-67)
- Average U.S. taxpayer would not help finance any space program beyond manned lunar landing unless he could understand it and identify with its purpose, M/G John B. Medaris (USA, Ret.) told AAS Symposium in Huntsville. "Putting a man on the moon was a good objective. People can see the moon. They can understand communication and weather information and possible defense benefits from a moon landing.

    "But after the astronaut puts an American flag on the moon's surface and satisfies the public, what will happen to our program then? We have to have a steady commitment of great resources to keep this program going. We have got to convince the public that there is a link between them and our next objectives . . . [and] give them some sort of objective they can identify with their own welfare."

    Noting that "people recoil in fear from the concept of the endlessness of interstellar space," General Medaris suggested that public would approve of an objective such as establishment of a lunar colony. "We better forget the spectacular and use a public relations approach like the soap companies do and figure out what the public will pay for." (Houtz, *B News*, 6/14/67)
- ITT had signed contract "involving several million dollars" with Indonesian Government to build Indonesia's first satellite communications earth station near Djakarta. Project was subject to obtaining financing under investment guarantees requested from Agency for International Development (AID). ITT predicted that station would be operational within 16 mos. after finances were arranged. (DJNS, *W Star*, 6/13/67, A21)

- Sen. Howard W. Cannon (D-Nev.) inserted in the *Congressional Record* a UPI dispatch stating that U.S.S.R.'s Tu-144 supersonic aircraft would be test-flown "within a few months." Noting that U.S. SST program was "from 3 to 3½ years behind the time when the [Anglo-French] Concorde and the TU-144 will enter into commercial service," Senator Cannon urged the House to quickly approve the $198-million authorization request for SST construction. "The supersonic transport is an important program . . . [and] it is imperative that this country proceed with the construction. . . . This is an arena of industrial competition in which the United States cannot afford to finish second." (*CR,* 6/13/67, S8141-2)
- U.S.S.R. preferred "to wait for technical reasons" until its Il-62 aircraft was ready and thus was postponing first direct Moscow-New York flights until "later this year," U.S. State Dept. announced. Commercial service by Aeroflot and Pan American World Airways had been scheduled to begin in late spring under Nov. 4, 1966, agreement. (AP, *NYT,* 6/14/67, 47)
- Prime Minister Harold Wilson confirmed U.K.'s decision against the purchase of U.S. Poseidon missiles [see May 11]. Asked in the House of Commons if he planned to keep the House informed on any major changes in the British nuclear armory, Wilson replied: "If you are referring to a proposal to replace Polaris by Poseidon, the answer is that the Government has no such intention." (*NYT,* 6/14/67, 11)
- Current FAA R&D programs were discussed by FAA at its first annual "Report to Industry" meeting in Washington, D.C. Meeting was designed to keep aviation public informed of R&D programs which would result in new elements in the National Airspace System. (FAA Release 67-8)
- USAF had developed new transportable "fold away" radar system which could be assembled or dismantled in less than 20 min. Called TPS-44, system would be used in tactical aerospace operations. (AFSC Release 88.67)

June 13-28: Rep. J. Edward Roush (D-Ind.), member of House Committee on Science and Astronautics, attended tri-annual meeting of International Organization for Standardization in Moscow. He urged increased U.S. participation: "If the United States is to take full advantage of world markets, if we are to have a louder voice in the process of establishing international standards, if we are to be persuasive in obtaining standards of our own choosing, and if we are to have immediate access to information relative to standards which are accepted on an international level, then we must be more effective in our participation in the setting of international standards." (Text)

June 14: NASA's *Mariner V* (Mariner E) spacecraft was successfully launched from ETR by Atlas-Agena D booster on four-month, 212-million-mi Venus flyby mission—NASA's only attempt to conduct a Venus flyby during the 1967 launch window. Agena stage separated from Atlas on schedule and ignited to reach 115-mi (185-km)-altitude parking orbit, where it coasted for 13 min before reigniting to inject spacecraft onto transfer trajectory toward Venus. Spacecraft then separated from Agena, deployed its four solar panels, and locked its sensors on the sun and the star Canopus. Midcourse maneuver was successfully conducted June 19 to assure that spacecraft would fly to within 2,500 mi of Venus Oct. 19. [U.S.S.R.'s *Venus IV,* launched June 12, was expected to reach Venus around same time.]

June 14: NASA's *Mariner V* being weight-tested at Cape Kennedy before its successful launch on a four-month mission to perform a close flyby of the planet Venus.

Mariner V, fifth in a series of interplanetary probes, had originally been built as a Mars probe to back up *Mariner IV* but was remodeled to accomplish Venus '67 mission. Primary mission for the 540-lb spacecraft was to fly by Venus to obtain scientific data which would complement and extend data obtained by *Mariner II* pertaining to the origin and nature of Venus and its environment. Spacecraft's secondary mission was to obtain data on interplanetary environment during a period of increased solar activity and provide engineering experience in converting and operating spacecraft designed for flight to Mars into one flown to Venus. *Mariner V* carried automatic sensing devices to record data for derivation of radiation levels, temperatures, pressures, magnetic fields, and atmospheric density near Venus.

First Venus probe, *Mariner I*, was destroyed during launch July 22, 1962; second, *Mariner II*, was successfully launched Aug. 27, 1962, and flew within 21,600 mi of Venus Dec. 14, 1962. *Mariner III*, a Mars probe launched Nov. 4, 1964, failed to achieve desired orbit when shroud remained attached to spacecraft. *Mariner IV* was successfully launched Nov. 28, 1964, and transmitted first close-up photos of Mars in July 1965. Mariner program was directed by OSSA Lunar and Planetary Programs Div.; project management and responsibility for spacecraft, mission operations, and tracking and data acquisition were assigned to JPL; and Atlas-Agena launch vehicle was managed by Lewis Research Center. (NASA Proj Off; NASA Release 67-124)

- NASA announced incorporation of at least seven changes in Apollo Lunar Module (LM) program "to reduce the possibility of fire and to increase crew safety" in accordance with Apollo 204 Review Board findings [see April 9]. Changes, which would increase total weight by 25–125 lbs, included: (1) reassessing combustibility of all nonmetallic materials and substituting or redesigning where necessary; (2) reevaluating material acceptability by intentionally starting fires in LM mockup; (3) modifying water hose and providing a hose nozzle to permit use of water as a fire extinguisher; (4) using a built-in handle, rather than a removable tool, to open hatches; (5) instituting more stringent standards and controls on electrical systems and wiring; (6) adding isolation switches to electrical systems to ensure that no connector is "hot" while being connected or disconnected; and (7) installing TV camera to monitor activity inside LM cabin during ground tests. Changes were being made at Bethpage, N.Y., plant of Grumman Aircraft Engineering Corp., LM prime contractor. (NASA Release 67-159)

- Special tests to determine best way to recover and reuse Saturn V 1st stage (S-IC) were being conducted at MSFC. In tests, a 1/10 scale model of the S-IC was dropped from altitudes of up to 156 ft into a 20-ft-deep water tank to simulate descent of stage through earth's atmosphere and into the ocean. Tests were studying a "soft splash" concept designed to bring back S-IC with a minimum of damage. MSFC engineers said a workable recovery method would save "millions of dollars" in future years. (MSFC Release 67-127)

- Astronaut James A. Lovell, Jr., told the press at MSC that he did not foresee any problems in combining his new assignment as President's Consultant on Physical Fitness [see June 6] with his duties as an astronaut. He said he would devote about one fourth of his time to the fitness program and the rest to keeping "in a state of readiness" for future space flights. "The fact that I am not spending all of my efforts on my new assignment does not mean I consider it any less important than my role in space. In fact, I can't think of any mission more important than that of persuading the American people to keep themselves healthy and fit. These qualities are basic to everything we do. Whether we are trying to reach the moon, or merely running to catch a bus, it helps to be in good physical condition. . . . A healthy population is the backbone of a strong nation." (Text)

- KSC awarded General Electric Co. a $7.7-million, cost-plus-fixed-fee contract modification for additional Apollo engineering support services. Modification, effective through Sept. 30, 1967, increased total value of original contract to $49 million. (NASA Release 67-158)
- Four U.S. universities—Cal Tech, Univ. of California, Univ. of Michigan, and Stanford Univ.—had formed Associates for Radio Astronomy to design, build, and operate world's largest, fully-maneuverable radiotelescope. The 328-ft-dia dish-shaped radiotelescope—78 ft larger than one at Jodrell Bank Experimental Station which was largest in operation at present time—would be constructed northeast of Pasadena in Owens Valley, Calif. (UPI, *St. Louis P-D*, 6/14/67)
- Joint Senate-House Atomic Energy Committee approved Weston, Ill., as site for AEC's 200-billion electron volt (bev) proton accelerator. (UPI, *WSJ*, 6/15/67, 5)
- Secretary of Transportation Alan S. Boyd received National Aviation Club's Award for Achievement in Washington, D.C., for "leadership in the development and establishment of . . . [DOT] and for his long and successful career in . . . air transportation and public utilities." (NAC; *Av Wk*, 6/12/67, 23)
- Aircraft flown by pilots "lacking professional ability"—usually those in general aviation—were responsible for most air accidents, according to study by Cornell Univ.'s Daniel and Florence Guggenheim Aviation Center. Fatalities involving general aviation aircraft in 1966 in U.S. totaled 1,069, compared with a loss of 59 passengers by scheduled domestic airlines, the study noted. "It is apparently futile to try to persuade the majority of pilots to secure and maintain instrument proficiency. . . . It would seem more effective to attack the problem of requiring manufacturers to incorporate better aircraft stability and controllability characteristics. . . ." Other methods suggested to improve air safety included: (1) raising airport standards by improving lighting and increasing landing aids, anti-collision devices, and weather data accuracy; (2) reviewing pilot certification procedures and incorporating more "human factors" into aircraft design; and (3) intensifying methods of crash fire protection and reevaluating emergency evacuation procedures. (*NYT*, 6/15/67, 81)

June 15: Maj. Michael J. Adams (USAF) flew X-15 No. 1 to 3,570 mph (mach 5) and 227,000-ft altitude in test to build up pilot altitude and check out horizon scanner and horizontal stabilizer. (X-15 Proj Off)

- New $1.3-million science laboratory for extensive research on problems of fatigue in aircraft structures and space vehicles was being placed in operation at LaRC. Laboratory would utilize 50 testing devices to simulate loads, temperatures, and other environmental conditions encountered by the materials in the structures of aircraft and space vehicles. (LaRC Release)

June 16: U.S.S.R. launched *Cosmos CLXVI* into orbit with 578-km (359-mi) apogee; 283-km (175-mi) perigee; 92.9-min period; and 48.4° inclination. Instruments performed satisfactorily. Satellite reentered Oct. 25. (UPI, *W News*, 6/16/67; GSFC *SSR*, 10/31/67)

- USAF launched two unidentified satellites with a Thor-Agena D booster from WTR; one satellite reentered July 20. (AP, *KC Times*, 6/17/67; *Pres Rep 1967*)

- NASA, in cooperation with West Germany's Ministry for Scientific Research (BMWF) and Brazilian Space Activities Commission (CNAE), successfully launched an Argo D-4 (Javelin) sounding rocket from Barreira do Inferno range near Natal, Brazil, on ballistic trajectory through Van Allen radiation belts to 650-mi (1,000-km) altitude. Purpose of mission was to flight-test instruments being developed for West German research satellite, scheduled for launch in 1969 to investigate earth's radiation belts. Several German organizations provided scientific experiments; CNAE provided range and launching services; NASA provided sounding rocket, downrange telemetry, and radar. (WS Release 67-23)
- NASA had completed post-launch evaluation of *Essa V* meteorological satellite and adjudged the mission a success. Launched by NASA from WTR April 20, satellite had been turned over to ESSA for operation May 8. (NASA Proj Off)
- NASA and Boeing Co. signed a $20-million letter contract extending the scope of Boeing's work with NASA to include integration of Apollo spacecraft's three modules with Saturn V launch vehicle [see May 9]. Boeing would: (1) assist and support NASA and its three manned spaceflight centers—MSC, MSFC, and KSC—in performance of certain technical tasks for Apollo missions AS-501 through AS-515; and (2) be responsible for supporting Apollo Program Office in integrating Saturn V booster with Apollo Command Module (CM), Service Module (SM), and Lunar Module (LM). Work would be performed under overall direction of OMSF Apollo Program Office. Boeing Co.'s currently contracted Saturn work included engineering, construction, and testing of Saturn V 1st stage; support of assembly and system integration of Saturn V 2nd and 3rd stages with the 1st; and design engineering support of certain ground support equipment at KSC. (NASA Release 67-161)
- NASA approved Phase II of a Dec. 14, 1966, contract and awarded Goodyear Aerospace Corp. $650,000 to design, fabricate, and test a full-scale model of a 30-ft-dia parabolic antenna. Antenna, planned as part of experimental package on proposed second-generation Applications Technology Satellites (ATS F and G), would be used for advanced communications research. (GSFC Release G-32-67)
- The use of ammonia as a fuel for central power plants, turbine-driven buses, trucks, and trains, and possibly private cars, would solve the problem of air and water pollution, Lockheed Aircraft Corp. executive Leon Green, Jr., wrote in *Science*. He suggested that fossil fuels such as gasoline, oil, coal, and natural gas be converted into ammonia rather than burned directly. Heat for the conversion could be provided by nuclear power plants. The economic feasibility of the process would depend on "the economic value of the byproducts of sulfur, carbon dioxide, water and possibly nitrogen," Green said, "and upon the price we are willing to pay for a clean environment." (Green, *Science*, 6/16/67, 1448-50; AP, *NYT*, 6/17/67, 15)

June 17: *Cosmos CLXVII* was successfully launched by U.S.S.R. into orbit with 286-km (178-mi) apogee, 201-km (125-mi) perigee, 89.2-min period, and 51.8° inclination. Satellite reentered June 25. (*SBD,* 6/20/67, 280; GSFC *SSR,* 6/30/67)
- Nation's most powerful 260-in solid-propellant rocket motor, SL-3, developed 5.7 million lbs thrust and burned 1.6 million lbs of propellant in 80-sec test-firing conducted for NASA by Aerojet General at

Homestead, Fla. Only flaw was explosion of a portion of the exhaust nozzle at the end of the test. Purpose of firing, third in NASA's Large Solid Rocket Technology Program, was to test "a large, submerged-type, ablative nozzle; the use of inert slivers to control the burn-out of propellant; elements of a motor failure warning system; and a higher burning rate propellant." First two 260-in motors, SL-1 (fired Sept. 25, 1965) and SL-2 (fired Feb. 23, 1966), both operated about two minutes and produced 3.5 million lbs thrust in tests to check strength of the maraging steel motor case, structural integrity of the case propellant, insulation and ablative nozzle, and the repeated performance of test hardware. Managed by LeRC under OART direction, program was designed to demonstrate feasibility of building and operating solid motors of greater size than those in current use for multistage launch vehicle systems carrying heavy payloads into space. (NASA Release 67-151; *Lewis News*, 6/9/67, 1; Reistrup, *W Post*, 6/18/67, A2; *NYT*, 6/18/67, 6; *Marshall Star*, 6/21/67, 10)

- NASA Administrator James E. Webb received an honorary Doctor of Science in Commerce degree from Drexel Institute of Technology in commencement exercises in Philadelphia. (NASA PAO; *P SB*, 6/18/67)
- Communist China, using ground site, successfully explored her first hydrogen bomb over western region of country, Hsinhua, Chinese Communist press agency, announced. (*NYT*, 6/18/67, 1, 3)

June 18: NASA published *Earth Photographs from Gemini III, IV, and V*, an atlas containing reproductions of 244 high-quality earth photos taken by Gemini astronauts. Pictures, which were taken with hand-held cameras from 100- to 215-mi altitudes as scientific, meteorological, and terrain experiments, showed major natural features and some man-made ones in 50 countries around the world. (NASA SP-129; NASA Release 67-153)

- Citizens League Against the Sonic Boom, an organization with 230 members in 18 states, had started a campaign to stop the construction of a U.S. SST, *New York Times* reported. Harvard Univ. physicist Dr. William A. Shurcliff, director of the campaign, said in a Cambridge, Mass., interview that the League would attack the SST program in Congress. "The amount of money provided thus far by Congress is less than one-quarter of what is needed to get a prototype in the air. The government must put about $1.1 billion together, and to date Congress has only supplied $200-million or $299-million.

    "I think our chances of stopping the program are good." (*NYT*, 6/18/67, 60)

June 19: Tiros VII, oldest U.S. meteorological satellite in operation, completed its fourth year in orbit. Designed with a three- to six-month operational lifetime, satellite had traveled 594,000,000 mi, completed 21,600 earth orbits, and taken over 124,500 photos of 50,000,000,000 sq mi of earth and its cloud cover. NASA said *Tiros VII*, which had tracked the major hurricanes of 1963, 1964, and 1965, was now commanded to transmit photos only occasionally. (AP, *NYT*, 6/18/67)

- NASA and USAF were expected to sign an agreement in fall 1967 for a joint lifting-body vehicle research program, *Technology Week* reported. NASA would probably have overall management responsibility of program, which would include test flights by USAF's SV-5P and NASA's HL-10 lifting-body vehicles. (*Tech Wk*, 6/19/67, 3)
- McDonnell Corp. was proposing an enlarged 9- to 12-man version of the

Gemini spacecraft to meet NASA's requirements for a manned advanced logistics space ferry in the 1970's, George Alexander reported in *Aviation Week*. He predicted that MSC would award a $150,000 extension to a $300,000 study contract already held by McDonnell to further explore the concept. (Alexander, *Av Wk*, 6/19/67, 20)

- NRL announced world's lowest recorded temperature—less than one-millionth of a degree Kelvin above absolute zero—had been achieved by Naval Research Laboratory scientists Dr. Arthur Spohr and Edwin Althouse through use of magnetic cooling technique, described as nuclear cooling. Previous lowest recorded temperature had been about 1.3 millionths of a degree above. The difference was of great significance because in this range most physical properties, including degree of orientation of atomic nuclei, varied markedly with even a slight change of temperature. At absolute zero, nuclei of most materials would have the highest state of order possible for the material. New record at Naval Research Laboratory would further advance development of an apparatus to study interaction forces between nuclei and to examine effects of oriented nuclei. (NRL Release)

June 20: NASA successfully conducted sixth of 10 planned rocket-launched flight experiments to investigate possible parachute landing systems for Voyager program. Two-stage Honest John-Nike rocket launched from WSMR ejected 40-ft cross-type parachute, one of three designs being studied, at 130,000-ft altitude where earth atmosphere resembles that of Mars. Parachute descended to earth carrying 200-lb payload. Test was part of LaRC's Planetary Entry Parachute Program. (NASA Release 67-162)

- USAF launched unidentified satellite with Titan III-B booster from WTR; satellite reentered Jun. 30. (*Tech Wk*, 6/26/67, 15; *Pres Rep 1967*)

- Apollo spacecraft 017 was mechanically mated to Saturn V booster at KSC in preparation for AS-501 unmanned mission scheduled for launch in third quarter of 1967. KSC officials said no major problems had been discovered in additional inspection of Saturn V's 2nd stage [see May 24]. (*Marshall Star*, 6/21/67, 1)

June 21: NASA Argo D-4 (Javelin) sounding rocket launched from NASA Wallops Station carried 120-lb payload containing an ionosphere experiment prepared by GSFC and Southwest Center for Advanced Studies to 493-mi (793-km) altitude. Flight was coordinated with the overpasses of *Alouette II*, the Canadian satellite, and *Explorer XXXI*, the U.S. satellite, launched pickaback by NASA Nov. 28, 1965. Primary purpose of mission was to measure properties of the ionosphere and to correlate results with simultaneous measurements made by the two Isis satellites. (WS Release 67-24)

- NASA Aerobee 150 sounding rocket launched from WSMR carried Univ. of Minnesota experiment to 131-mi (210-km) altitude to measure composition and determine temperature of atmosphere in altitude range 62-124 mi (100-200 km). Experiment used three mass spectrometers and a yoyo despin system. Rocket and instrumentation performed satisfactorily. (NASA Rpt SRL)

- All-weather landing capability would be a requirement for SST operations because long-range SST used fuel too quickly to permit prolonged holding in airport traffic patterns, Dr. Richard M. Head, ERC, told NASA Hq. press briefing on aeronautics research. He predicted that landings under zero visibility could be practical within a decade. Dr. Head also noted that, because of sun's influence on the ionosphere, SSTs with VHF radios might be subject to communications blackouts during the 1,200 mi of a transatlantic flight when they were too far from land for line-of-sight VHF reception. Use of satellites to relay communications might be a possible solution to the problem, he suggested. (Schmeck, *NYT*, 6/22/67, 6)

- MSFC had awarded three parallel contracts—$94,000 to Boeing Co.; $95,000 to TRW, Inc.; and $96,000 to General Electric Co.—for further project definition work on the Voyager program. Objectives of the contracts, effective through Oct. 15, were to furnish NASA with current data on Voyager spacecraft systems design and perform engineering study tasks. (MSFC Release 67-129)
- U.S.S.R., East Germany, Czechoslovakia, Hungary, and Poland withdrew from live international telecast scheduled for June 25 [see Feb. 7] for political reasons. U.S.S.R. charged that Western broadcasters were conducting "a slanderous campaign" against the Arab nations in the current Middle East crisis. (Gent, *NYT*, 6/22/67, 67M)
- Astronomers at Lick Observatory, Santa Cruz, Calif., were beginning a new study of star movement against the background of galaxies so distant they appeared to be fixed in space, Lick astronomer Dr. Stanislavs Vasilevskis told a Santa Cruz news conference. "The Lick proper motion study will open a new approach to research in astronomy using modern electronic equipment." A primary purpose of the study would be to gain further information on the rotation of the earth's galaxy, he said. Completely automated equipment installed at Lick for the study was financed by $367,120 NSF grant. (AP, *NYT*, 6/25/67, 12)

June 22: Senate Committee on Aeronautical and Space Sciences voted to report $4.851 billion NASA authorization bill (S. 1296)—$249 million less than $5.1 billion NASA had requested for FY 1968. Committee reductions included $120 million from Apollo Applications (AA) program and $71.5 million—total amount requested—from Voyager program. Responding to Committee report, NASA Deputy Administrator Dr. Robert C. Seamans, Jr., issued a statement saying that the affected programs were of "first-rank importance to mankind. The scientists of this country have expressed deep concern that the United States is not making enough progress in the exploration of the planets.

"We feel that many members of Congress will want to give further consideration to the President's recommendations" for the AA program. (Reistrup, *W Post*, 6/24/67)

- House completed general debate on NASA FY 1968 authorization bill (H.R. 10340) and was scheduled to resume discussion of bill June 27 when amendments would be considered and final vote taken. During debate Rep. Delbert L. Latta (R-Ohio) criticized the $444.7 million recommended by Committee on Science and Astronautics for the Apollo Applications (AA) program: ". . . we could spend [that money] toward research and possible cure of cancer, stroke, and heart disease, rather than some nebulous undefined program that will take place when—and I emphasize this point—and after we get to the moon. This money is to be spent on hardware after we get to the moon. No one knows when we are going to get there . . . [and] if we spend this money perfecting these vehicles and this hardware today, it might be out of date when we get to the moon. . . ." Rep. Olin Teague (D-Tex.), Chairman of House Committee on Science and Astronautics' Subcommittee on Manned Space Flight, defended the $444.7-million recommendation. "The manned space flight program—Apollo, Apollo applications and advanced manned missions—drives the Apollo effort into its final phases and begins the work of realizing benefits for man on earth. It supports national security, fosters national growth, expands knowledge, and retains options for the future at a cost the country can well afford. To delay this program is, in effect, to cancel it. Such a decision would put the space plum in the hands of the Soviet Union. . . ." (NASA LAR V/68; CR, 6/22/67, H7782, H7797)

- NASA test pilot William H. Dana flew X–15 No. 3 to 82,000-ft altitude and 3,682 mph (mach 5.52) in flight at Edwards AFB. Purposes of test: (1) cold-wall heat transfer; (2) step-panel heat transfer; (3) boost guidance checkout; (4) sonic boom study; and (5) PCM checkout. (NASA Proj Off)
- NASA test pilots Fitzhugh Fulton and Donald Mallick flew X–70 No. 1 to mach 1.8 and 55,000-ft altitude in test to check out stability and control of handling qualities, by-pass airflow calibration, and nose ramp flutter. (XB–70 Proj Off)
- MSFC awarded contract extensions—$149,914 to Bendix Corp. and $149,485 to Boeing Co.—to conduct specified design study tasks on the Local Scientific Survey Module (LSSM), a vehicle being proposed for transporting men on the lunar surface. Modifications extended through August 1967 two parallel six-month contracts awarded the firms June 29, 1966. (MSFC Release 67–130)

June 23: NASA and Dept. of Interior's Bureau of Sport Fisheries and Wildlife announced an agreement whereby KSC would add 11,436 acres to the 46,530 acres currently available for controlled public usage under Bureau-managed programs. (NASA Release 67–163)

June 24: Christopher C. Kraft, Jr., Director of Flight Operations at MSC, received Virginia Press Assn.'s (VPA) "Virginian of the Year" award at joint banquet of the VPA and National Newspaper Assn. (NNA) during NNA's annual meeting in Richmond. Kraft said U.S. might miss 1970 target date to land the first man on the moon. "If we are lucky, and if everything works well, we can probably do it. But if we have a normal share of problems, as happens with any large research effort, then I don't think we can." (AP, W Star, 6/26/67, B5)

June 24–25: Series of three NASA Nike-Cajun sounding rockets launched from Natal, Brazil, carried GSFC-instrumented experiments to obtain wind, temperature, pressure, and density data at intervals within 24-hr period near the fall equinox. These data would be used to investigate validity of the thermally driven diurnal tide theory of the upper atmosphere. Rocket and instrumentation performance was satisfactory, and good experimental results were expected after data analysis. (NASA Rpt SRL)

June 25: Thirteen nations participated in first live global telecast linking five continents. Program was transmitted to a potential viewing audience of nearly 500-million persons in more than 30 countries via NASA's *Ats I* and ComSatCorp's *Early Bird I* and *Intelsat II–B* comsats. Entitled "Our World," two-hour telecast featured the Beatles, Van Cliburn and Leonard Bernstein, Japanese shrimp fishermen, a Canadian mermaid, and a Wisconsin farmer. Five Communist nations—including U.S.S.R., whose *Molniya I* comsat was to have been used to transmit the telecast— had withdrawn from the program June 21 as a protest against Western involvement in current Middle East crisis. (NASA Ann, 6/20/67; NET Proj Off)

During week of June 25: Ceremonies at NASA's Rocket Engine Test Site at Edwards AFB marked delivery of the millionth ton of cryogenic rocket propellants and pressurants. Fluids were used by North American Aviation, Inc.'s Rocketdyne Div. to test fire F-1 engines prior to shipping them to MSFC for use in Saturn V 1st stages. (MSFC Release 67-139)

June 26: NASA's *Mariner V* spacecraft successfully completed two scheduled functions in preparation for Oct. 19 Venus flyby mission by switching its amplifiers to increase radio transmitting power and, 30 min later, turning off its battery charger on commands from Goldstone, Calif., Deep Space Network station. All systems onboard spacecraft, launched from ETR June 14, were continuing to operate satisfactorily. (NASA Release 67-165)

- KSC awarded Federal Electric Corp. a one-year, cost-plus-award-fee contract containing provisions for four additional one-year renewals to provide operational communications and instrumentation support services at KSC. Estimated value for five-year period beginning July 1 was $85 million. Federal Electric would operate and maintain instrumentation for telemetry, tracking, instrument calibration, and computation equipment and would provide computer programming services. (NASA Release 67-167)

June 27: Both House and Senate considered NASA FY 1968 authorization bills. Final votes would be taken June 28. (NASA LAR VI/68)

- NASA Deputy Administrator Dr. Robert C. Seamans, Jr., visited MSFC and MTF to discuss progress in AA and Saturn launch vehicle programs. He would visit MSC June 28. (MSFC Release 67-132; *Marshall Star,* 6/28/67, 1)
- The wife of Soviet Cosmonaut Aleksey Leonov, first man to walk in space, gave birth to their second daughter, Soviet press reported. (UPI, *W Post,* 6/28/67, A18)
- Survey by *New York Times* correspondents indicated that Western Europe had taken little action to close the "technology gap" with the U.S., Brendan Jones reported. Discussions and meetings had helped clarify the problems but had also caused discouragement about achieving any quick solutions. Survey, which covered 10 countries, also indicated that increases being made in research spending in Western Europe were still comparatively small. Plans for combining technical knowledge by Common Market countries or increasing exchange of technology with U.S. through NATO were in only the proposal stages. (Jones, *NYT,* 6/27/67, 51)
- France exploded a nuclear device of "low yield" near Mururoa Atoll in the Pacific. (AP, *NYT,* 6/28/67, 3)

June 28: House and Senate both passed NASA FY 1968 authorization bills which were less than the $5.1 billion originally requested by NASA. Senate bill (S. 1296) authorized $4.851 billion: $4.136 billion for R&D; $648 million for administrative operations; and $67 million for construction of facilities. House bill (H.R. 10340) authorized $4.791 billion: $4.076 billion for R&D; $648 million for administrative operations; and $67 million for construction of facilities.

Senate defeated two amendments by Sen. William Proxmire (D-Wis.) to reduce NASA authorization—one by $317 million; one by $98 million. An amendment by Sen. Charles Percy (R-Ill.) which would require NASA to keep Senate Committee on Aeronautical and Space Sciences "fully and currently informed" was also defeated. Committee Chairman Sen. Clinton P. Anderson (D-N.Mex.) said the proposal had been considered by the Committee and rejected as unnecessary: "It's not needed. What is needed is enforcement and compliance with existing legislation."

House bill authorized $200.7 million less than Committee on Science and Astronautics had recommended [see May 16]. An amendment by Rep. Richard L. Roudebush (R-Ind.) to an amendment by Rep. James G. Fulton (R-Pa.) cut $65 million from $455 million proposed by Committee for AA program. Fulton's original amendment had recommended cutting the program by $250 million. Remaining reductions, totaling $135.7 million, were contained in surprise recommittal motion offered by Representative Fulton at the end of the day. They included: (1) $6.2 million from advanced missions (funding eliminated); (2) $21.5 million from Voyager; (3) $20 million from nuclear rockets; (4) $78 million from launch vehicle procurement; and (5) $10 million from sustaining university program. Amendment also contained provision to establish an independent safety review board within NASA. Establishment of the board had been urged by Rep. Donald Rumsfeld (R-Ill.) as one means of avoiding another accident like the Jan. 27 Apollo fire. Differences between Senate and House authorization bills would be resolved by joint House/Senate Conference Committee. (*CR*, 6/28/67, S9078-94, H8150-200; UPI, *NYT*, 6/29/67, 11; Reistrup, *W Post*, 6/29/67, A6)

- AFCRL successfully launched 65-lb payload by Trailblazer II booster from Wallops Station in experiment to study reentry communications problems by carefully measuring plasma sheath properties and their effects. Booster's 3rd and 4th stage engines fired at 187-mi altitude to propel nosecone downward; at 60-mi altitude rocket achieved a velocity of 18,000 fps creating an envelope of plasma identical to the envelopes which block radio and microwave communications with reentering spacecraft. Since a pure plasma was essential to the success of the experiment, reentry vehicle had no ablative coating and burned up completely in the atmosphere. AFCRL would use results of measurements to develop techniques for predicting exact behavior of a radiating system during reentry. (OAR Release 67-9)

- FRC awarded RCA a $2.59 million, fixed-price contract to design, fabricate, and install an instrumentation radar system at NASA's Ely, Nev., tracking site. Scheduled for completion by Dec. 30, 1968, system would be used to provide highly accurate position and velocity data on such high-performance aircraft as the X-15 and the XB-70. (NASA Release 67-172)

- NAE announced establishment of an Aeronautics and Space Engineering Board to advise NASA and other Federal agencies. Board would operate under contract with NASA that estimated expenditures of $186,500 during first year of operation. Chairman was Dr. H. Guyford Stever, President of Carnegie Institute of Technology. Col. Robert J. Burger, Executive Secretary of Scientific Advisory Board, USAF Hq., would become Executive Director Aug. 1, following his retirement from USAF. Board's first meeting would be held in Washington, D.C., in mid-July. (NAE Release)

- On presentation of Distinguished Service Medal to L/G Leighton I. Davis

(USAF), Commandant of Industrial College of the Armed Forces, Air Force Chief of Staff Gen. John P. McConnell cited him for outstanding management of range support of U.S. ballistic missile and space vehicle programs and as DOD Manager for Manned Space Flight Support Operations—specifically, Projects Mercury and Gemini. Davis had been Commander of Air Force Eastern Test Range (AFETR) and National Range Div. from May 1960 through June 1967. (AFSC Release 140.67)

- On presentation of NASA Exceptional Service Medal to M/G David M. Jones (USAF), Commander of AFETR, NASA Associate Administrator for Manned Space Flight Dr. George E. Mueller stated: "As Deputy Associate Administrator for Manned Space Flight (programs) and as Acting Director, Apollo Applications, General Jones played an important role in guiding the definitions and development of future manned space missions." Jones had been assigned to NASA from December 1964 through April 1967. (AFSC Release 137.67)

June 29: USAF launched two satellites: *EGRS–9* and *Aurora I*. Thor-Burner II boosters were used in WTR launchings. *EGRS–9*, a Secor satellite, would provide an aid to geodetic survey of the earth sphere. *Aurora I* would study charged particles that precipitate in the upper atmosphere. (*Pres Rep 1967*)

- Maj. William J. Knight (USAF) made successful emergency landing of X–15 No. 1 at Edwards AFB 59 sec after takeoff because of an electrical power failure. (X–15 Proj Off)
- NASA awarded Philco-Ford Corp.'s Western Development Laboratories a $45,757,000, multiple-incentive contract modification for continued engineering and operational support at MSC's Mission Control Center (MCC). (NASA Release 67–173)
- Eight Apollo astronauts flew from New York to Iceland for eight-day expedition to gather geological data from an area believed to be similar to the lunar surface. Fifteen more astronauts were expected to join the expedition June 30. (*NYT*, 6/30/67, 13)
- NASA released "Opportunities for Participation in Space Flight Investigation" (NHB 8030.1A) describing opportunities available to scientists and technical groups for including their experiments and equipment on NASA missions. Handbook, which superseded earlier edition published in July 1965, described each of NASA's present flight programs and noted opportunities available in automated and manned flights and in flight-related programs so that scientists would have an up-to-date review of missions currently being sponsored by NASA and could develop compatible experiments. (NASA Release 67–169)

June 30: MSFC awarded two contracts: (1) a $2.4-million, two-month contract extension to Chrysler Corp. for procurement of long-lead time items for additional Uprated Saturn I 1st stages. Extension would enable Chrysler to continue procurements necessary to maintain capability to assemble four Uprated Saturn I boosters annually; (2) a $14,811,540 fixed-price-incentive-fee contract to NAA for 60 additional H–1 rocket engines for use on 1st stages of Uprated Saturn I boosters. Order increased total number of engines purchased to 322. Delivery would continue through September 1968. (MSFC Releases 67–136, 67–138)

- USAF named four USAF pilots to train for DOD's Manned Orbiting Laboratory (MOL) program: Maj. James A. Abrahamson; L/Col. Robert T. Herres; Maj. Donald H. Peterson; and Maj. Robert H. Lawrence, Jr., the first Negro selected for a mission in the Nation's space program. All were completing training course at Edwards AFB Aerospace Research Pilot School. They would join 12 previously selected pilots in September to begin project training. (USAF PIO; *NYT*, 7/1/67, 1, 8)

- French Army and Foreign Legion evacuated Hammaguir Range in accordance with treaty granting Algerian independence. Sahara Missile Proving Grounds had been evacuated March 31. (*NYT*, 2/11/67; *Av Wk*, 3/27/67, 50)
- Patent for a large circular vehicle which could roll across the rugged lunar terrain while its crew remained upright was granted to Martin Marietta Corp. executive Dr. Arthur A. Ezra. Open-sphere, 60-ft-dia vehicle had frame of circular metal rims and a closed cabin in the center which was suspended from gimbals and weighted so that floor would always remain level. Crew would steer vehicle with jet engines attached to cabin. (Jones, *NYT*, 7/1/67, 26)

During June: NASA had failed to locate *Surveyor III*'s transmitting frequency after repeated attempts to reactivate spacecraft since May 23. Efforts to reestablish contact would be resumed on a limited basis after end of third lunar night which would begin July 1. (NASA Proj Off; *SBD*, 6/5/67, 194)

- Contract to develop ESRO's TD-1 and TD-2, largest satellites to be designed in Europe, had been awarded to MESH consortium. European members of MESH—Engins Matra of France, E.R.N.O. of West Germany, S.A.A.B. of Sweden, Hawker Siddeley Dynamics Limited of Great Britain—would get £8-million contract to develop the two satellites; Engins Matra would be prime contractor. Hawker Siddeley Dynamics Limited and Engins Matra had jointly designed, developed, and manufactured *Esro II* satellite. TD-1 and TD-2 would be launched in 1970 and 1971 and would perform astrophysics missions. (*S/F*, 6/67, 198)
- Full size dynamic model of Canadian Defence Research Board's ISIS-A—third ionosphere satellite for Canada—had successfully completed series of vibration tests at GSFC. Joint Canadian/U.S. program of International Satellites for Ionospheric Studies would include ISIS-B and ISIS-C during the 1960s as solar activity would pass through its intensity, causing a variety of changes in ionosphere and outer space. Launch of ISIS-A was planned for early 1968 [see Mar. 23]. Like its predecessors (*Alouettes I* and *II*), ISIS-A would be spheroid and have 11,000 solar cells on its outer surface to power internal batteries. (*S/F*, 6/67, 197-8)
- Variety of ion engines could be used in satellite stationkeeping and attitude control, and solar-powered ion engines for primary propulsion would be feasible for Voyager-class missions of early 1970s, *Space/Aeronautics* reported. Nuclear-powered ion propulsion was still about a decade off. (*S/A*, 6/67, 92)
- Computer system which furnished selected scientists with biweekly listings of the new technical developments in their fields was put into operation at Ft. Monmouth, N.J., Hq. of USA Electronics Command. Designated Selective Dissemination of Information (SDI), system utilized abstracts of scientific and engineering reports prepared at Defense Documentation Center, Cameron, Va. Abstracts were classified into 7,144 subjects, inserted into computers, and then registered on reels of magnetic tape bearing document profiles which described specific subject or subjects covered by the documents. Tapes were then sent to Ft. Monmouth where document profiles were matched with "interest" profiles of subscribers.

Ft. Monmouth Commander M/G William B. Latta said the goal of the system was "to increase the productivity of our technical personnel by making sure that each scientist is kept abreast of the latest work in his specialty.

"Our researchers will thus be able to make maximum use of the scientific advances of their co-workers throughout the country and avoid costly duplication." (Lyons, *NYT*, 6/2/67, 22)

- Col. Maynard E. White (USAF, Ret.), former Deputy Chief of Staff for Personnel, USAF Hq. in Europe, was appointed Director of Manned Space Flight Program Control, OMSF. (NASA Ann)

July 1967

July 1: USAF Titan III–C booster launched from ETR successfully inserted six satellites—three Initial Defense Communications Satellite Program (IDCSP) comsats, one Lincoln Experimental Satellite *Les V*, one DOD Gravity Experiment (DODGE) satellite, and one Despun Antenna Test Satellite (DATS)—into circular, near-synchronous orbits.

Powered flight of Titan III–C was close to planned parameters. Transtage and payload were inserted into parking orbit with 103-mi (166-km) apogee and 89-mi (143-km) perigee. First transtage burn made necessary course corrections; second transtage burn moved stage and payload into transfer orbit with 20,920-mi (33,666-km) apogee and 111-mi (179-km) perigee. Third burn put satellite dispenser frame and six satellites into final orbit with 20,809-mi (33,488-km) apogee and 20,453-mi (32,915-km) perigee. Between 3:21 and 3.23 pm EST satellites were ejected separately into slightly different orbital paths.

Three IDCSP comsats, combined with 15 others launched June 16, 1966, and Jan. 18, 1967, completed Pacific link of DOD's IDCSP system designed to provide reliable, full-time radio communications between Washington, D.C., and U.S. troops in Vietnam and other distant outposts. *Les V*, first all solid-state UHF band comsat, would transmit radio signals and test communications to front-line troops. DODGE satellite would extend 10 metal booms to test satellite's stability, determine whether it could keep one side always facing earth, and transmit first color TV pictures of earth taken from orbit with a 22° camera. DATS, advanced version of IDCSP comsats, carried new antenna designed to transmit 75% of radio signal strength—rather than current 15%—back to earth. USAF officials reported all satellites were functioning properly. (DOD Proj Off; UPI, *NYT*, 7/2/67, 20; *W Post*, 7/2/67, A4; *Av Wk*, 7/10/67, 39; *Aero Tech*, 7/17/67, 20)

- AFSC's Ballistic Systems Div. (BSD) and Space Systems Div. (SSD) were combined to form new Space and Missile Systems Organization (SAMSO) with headquarters in Los Angeles. New organization would be commanded by M/G John W. O'Neill, former Commander of Electronic Systems Div. M/G John L. McCoy, former BSD Commander, was named Deputy Commander for Missiles; M/G Paul T. Cooper, former SSD Commander, was appointed Deputy Commander for Space. (AFSC Release 101.67; SSD Release 67–207)

- MSFC awarded McDonnell Douglas Corp. a $496,024 contract to study advanced versions of Saturn S–IVB orbital workshop. Study would explore concepts for a follow-on workshop for the Uprated Saturn I and several more sophisticated versions for the Saturn V, with initial emphasis on uses of the stage in Apollo Applications (AA) program. (MSFC Release 67–145)

July 1
- Princeton Univ. astronomers were studying photos of ultraviolet spectra of Venus and Jupiter taken by a rocket-borne telescope launched from WSMR May 5 in an effort to provide additional data on Venusian atmosphere, *New York Times* reported. Photos, taken at 118-mi altitude, provided "50 times more detail than ever obtained before," Dr. Donald C. Morton, director of Princeton's rocket astronomy program, explained. Princeton astronomers would continue analyzing spectrograms to determine composition of planets' gaseous atmospheres, and would compare findings with data transmitted by U.S.'s *Mariner V* and U.S.S.R.'s *Venus IV* spacecraft—both expected to reach Venus by late October. (*NYT*, 7/2/67, 21)
- M/G Vincent G. Huston, former Commander of Air Force Eastern Test Range (AFETR) and recipient of NASA Leadership Medal, became Commander of AFSC's National Range Div. He succeeded L/G Leighton I. Davis, who became Commandant of the Industrial College of the Armed Forces in Washington, D.C. (AFSC Release 103.67)

July 2: SSTs would solve sonic boom problem by flying at subsonic speeds where necessary over densely populated areas, Secretary of Transportation Alan S. Boyd said on ABC TV's "Issues and Answers." Since development of SST was "very worthwhile" to national prestige, balance of payments, U.S. leadership in aviation industry, and advances in related technology, it should be continued despite rising costs of Vietnam war, he said. Citing urban congestion as U.S.'s major transportation problem, Boyd said DOT was trying to develop a balanced system of transporting people to and through cities. Even with expanded airport facilities, relief by aviation would be limited by ground congestion, he said. (Smith, *W Post*, 7/3/67, A17)
- NASA personnel changes: (1) John S. Brown, Division Director in the Office of Legislative Affairs, NASA Hq., became Deputy Assistant Administrator for Legislative Affairs; (2) John P. Jewett, Assistant Chief for Administration, Nevada Extension of AEC/NASA Space Nuclear Propulsion Office (SNPO–N), became Chief of SNPO–N. (NASA Ann, 7/2/67, 7/25/67)
- France exploded nuclear device from a balloon over Mururoa Atoll in the Pacific. Test, originally scheduled for July 14, was last in summer 1967 series of three designed to perfect a trigger for hydrogen bomb France planned to explode in 1968. (UPI, *W Star*, 7/3/67, B4)

July 3: Minor fire broke out during an unmanned vacuum chamber test of Lunar Module (LM) descent stage mockup at WSMR. NASA officials said vehicle was under a vacuum soak condition when fuel escaped into the vacuum chamber; fire apparently developed from spontaneous combustion. It was extinguished immediately by the water deluge system, and damage was believed to be minor. (WSMR Release)
- Rep. Emilio Q. Daddario (D-Conn.), Chairman of House Committee on Science and Astronautics' Science, Research, and Development Subcommittee, emphasized that, "with the immensity of consequences and the irreversible nature of many technological changes, the propensity for risk taking must be coupled with a deeper assessment of both deficits and benefits." His statement, *Technology Assessment*, to subcommittee proposed a seven-step approach which would place technology within the total societal framework; "identify all impacts in the natural, social, economic, legal and political sectors;" and examine causes, effects, and alternatives. A three-phase Subcommittee program would include hear-

ings and seminars, a request for National Academies of Science and Engineering working group on pilot projects, and a number of other analyses. (Text)

- "Unless U.S. space program moves much faster than now seems likely, U.S. consumption of helium this year . . . may be down as much as . . . 25%," *Chemical and Engineering News* reported. Since Jan. 27 Apollo fire and resulting "slowdown" in space program, use of helium to pressurize rocket fuels—which accounted for 45% of helium consumption—had dropped to about half of its former level. Despite the 1967 recession, however, helium industry was optimistic that trend would be reversed when manned space flights were resumed in 1968. (*C&E News*, 7/3/67, 18–9)

- USN had selected six companies to compete for six-month study contract to design ship-based antimissile missile system which could support Nike-X, William Beecher reported in *New York Times*. Companies were Hughes Aircraft Co., Martin Marietta Corp., Boeing Co., Aerojet-General Corp., McDonnell Douglas Co., and Raytheon Co. Code-named Seaborne Anti-Ballistic Missile Intercept System (SABMIS), it would be deployed on submarines and warships stationed in international waters off Communist China and U.S.S.R. to intercept long-range missiles launched from those countries. A USA Nike-X official commented: "The Navy system is both feasible and attractive. It could intercept enemy ICBM's before they can deploy most of their penetration aids; it would reduce the number of missiles Nike-X would have to contend with; and it would tremendously increase the enemy's technology problems by forcing him to become very sophisticated indeed if he wants to stand any chance of getting his ICBM's past both the mid-course and terminal phase defenses." (Beecher, *NYT*, 7/4/67, 1, 2)

- ERC Director James C. Elms announced appointment of Dr. Francisc C. Schwarz as head of ERC's Power Conditioning and Distribution Laboratory. (ERC Release 67–22)

- Col. Robert A. Merchant (USMC, Ret.), former Assistant Chief of Staff, USMC Hq., became Director of the new International Technology Affairs Div., Office of International Affairs, NASA Hq. He would be responsible for coordinating dissemination of NASA scientific and technical information abroad, developing recommendations for the disposition of export control cases, and exploring broader use of space technology overseas. (NASA Ann, 7/3/67)

- Harold W. Yates, former chief engineer of Barnes Engineering Co.'s Field Research and Systems Dept., became Director of the National Environmental Satellite Center's Satellite Experiment Laboratory. (ESSA Release 67–17)

July 4: U.S.S.R. launched *Cosmos CLXVIII* into orbit with 268-km (167-mi) apogee, 199-km (124-mi) perigee, 89-min period, and 52° inclination. Equipment functioned normally. Satellite reentered July 12. (*SBD*, 7/6/67, 12; GSFC *SSR*, 7/15/67)

July 5: George H. Hage, former Engineering Manager of Boeing Co.'s Lunar Orbiter Program, became NASA Deputy Associate Administrator for Space Science and Applications (Engineering). His appointment had been announced May 18. (NASA Ann, 7/6/67)

July 6: NRL Aerobee sounding rocket launched May 17 had detected x-rays from quasar 3C–273 (believed to be 1.5-billion light years from earth) and Virgo A (a galaxy believed to be 30-million light years from earth),

July 6

Dr. Herbert Friedman, Superintendent of Naval Research Laboratory's Atmospheric and Astrophysics Div., announced at news conference in Washington, D.C. Rocket experiment, he said: (1) provided first evidence that x rays were produced as far away as the "rim" of the universe; (2) detected quasars, previously detectable only by their radio waves or by visible light seen through high-power telescopes; (3) lent support to theory that universe is saturated with background radiation from explosion of universe 10 billion yrs ago; and (4) detected three new x-ray sources, apparently outside our galaxy and not traceable to any known object. These findings, Dr. Friedman said, indicated that x-ray astronomy had "come of age" as a tool to explore the universe. To explore x-ray sources further, he suggested installing x-ray detection equipment on proposed satellites such as NASA's orbital workshop and launching larger equipment for longer periods of time to map the whole sky for x-ray sources. (Text; Reistrup, *W Post*, 7/7/67)

- USAF successfully launched an Advanced Ballistic Re-Entry System (ABRES) vehicle from Vandenberg AFB with Atlas-D booster. (*Aero Tech*, 7/17/67, 13)
- NASA's Lunar Orbiter Incentive Evaluation Board awarded Boeing Co. a $1,053,405 bonus for the *Lunar Orbiter III* mission. Board said pictures produced were "of high quality, properly exposed and with good positioning," although of unsatisfactory quantity, and noted that secondary site photography "had considerable value scientifically." Spacecraft, launched from ETR Feb. 4, had transmitted 154 of 211 photos taken and was gathering data on meteoroids, radiation, and lunar gravitational field. (NASA Release 67–177)

July 7: NASA Aerobee 150 sounding rocket launched from WSMR carried MIT x-ray experiment to 88-mi (142-km) altitude to search for undiscovered x-ray sources and to obtain data on celestial locations and energy spectra of "discrete celestial X-ray sources" in three regions. Rocket and instruments performed satisfactorily. (NASA Rpt SRL)

- NASA Associate Administrator for Space Science and Applications Dr. Homer E. Newell and OSSA Director of Space Applications Programs Leonard Jaffe discussed impact of space research on science and technology in *Science*. As a result of space research, they said, science and technology could contribute to the arts by furnishing new media for human expression; enrich the humanities by furnishing new media for broader basis for understanding himself and his place in nature; provide the historian with a better approach to history because of scientific methodology; and enable archeologists more accurately to date materials and identify their sources.

  As the entire world "absorbs the increased understanding of man and nature that science generates, and reflects that understanding in its literature, its social, political, and economic institutions, and its application to human daily living . . . it may be hoped that this common bond will . . . give increased motivation to solve peaceably the problems that beset the world . . . [and] provide increased means for doing so. . . . To use properly and effectively these powerful tools of our times is an inescapable challenge to men and governments the world over." (Newell, Jaffe, *Science*, 7/7/67, 29–39)

- ComSatCorp filed notice with FCC of three proposed contract awards totaling more than $13 million for antenna subsystems and related equipment to be installed at West Virginia, California, Puerto Rico, and

Hawaii earth stations. Fixed-price contracts proposed included $7,600,000 to Philco-Ford Corp. for four antenna subsystems; $4,500,000 to Raytheon Co. for four ground communications equipment subsystems; and $1,015,000 to Cutler-Hammer, Inc., for four low-noise receiver subsystems. Construction on three new earth stations and expansion of facilities at Hawaii station were expected to begin in late 1967. (ComSatCorp Release 67-36)

- H. Frank Hann became Director of the Financial Management Div., Office of Administration, NASA Hq. He had been Acting Director since April 3. (NASA Ann, 7/12/67)

July 8–9: The 1967 Soviet air show—first in six years—was held at Domodedovo Airport near Moscow. Program included aerial displays and ground exhibits of seven new types of supersonic jet fighters, including one VTOL and two variable-sweep-wing aircraft similar to U.S. F-111; Proton and Vostok spacecraft; World War II fighters; and four-engine IL-62 commercial aircraft. U.S.S.R. claimed that its new variable-sweep-wing models—a fighter and a missile launcher—were first and lightest in the world and could fly at mach 1 at sea level and mach 2 at higher altitudes. New VTOL aircraft was reportedly capable of rising vertically 130–160 ft before beginning forward flight. Western sources speculated that absence of any large new bombers suggested that U.S.S.R., like U.S., was concentrating on versatile missile-launching aircraft which had long-range striking power. (Reuters, *NYT*, 7/9/67, 17; *W Post*, 7/10/67, A10; *Av Wk*, 7/24/67, 38–44)

July 9: USN decision to study a ship-based antimissile missile system [see July 3] was praised by Hanson Baldwin in the *New York Times*. "Strategically and tactically the role of the Navy has been transformed by the nuclear age; the depths of the sea, the space above the atmosphere and every part of every land mass on earth are now accessible to modern naval power." Proposed SABMIS system would be especially effective because: (1) nuclear submarines' ability to remain completely submerged indefinitely gave them unequaled defense invulnerability; (2) submarine-launched ballistic missile would draw enemy attack toward seas rather than toward populated land as land-based missiles do; (3) sea-based system would be more effective against enemy missile containing several warheads, each capable of maneuvering along different trajectory; and (4) system would be able to intercept enemy missiles in midcourse, rather than terminal phase. (Baldwin, *NYT*, 7/9/67)

- Effect of sonic boom on SST flight routes and sales prospects was discussed by Secretary of Transportation Alan S. Boyd, FAA Administrator William F. McKee, and Director of SST Development B/G J. C. Maxwell (USAF, Ret.) in May and June testimony released by House Committee on Appropriations' Subcommittee on the Dept. of Transportation. Conditions under which supersonic flight would be permitted over populated land areas had still not been determined, Gen. Maxwell explained. "The available sonic boom data indicate that the present SST design may possibly be restricted from supersonic operations over populated land areas . . . [and we] have based our SST program decisions on the conservative assumption that this design will be operated primarily on water."

DOT did not have to force the American public to tolerate sonic booms to make SST program an economic success, Secretary Boyd assured the Committee. "We are satisfied that we can have a . . . highly successful program, assuming the sonic boom is intolerable over populated areas. . . . The only thing that is involved here is whether we sell more aircraft, not whether the program is a success. . . ." Even if SST were limited to subsonic flight over inhabited areas at least 300 aircraft could be sold initially, he said, and market studies indicated "that there is a possibility of going up to 1,200 aircraft by 1990." (Transcript, 41, 294, 942)

- NATO had begun to modernize its communications network by using comsats, Clyde H. Farnsworth reported in the *New York Times*. Initial $900,000 test phase of project had been inaugurated by Supreme Allied Commander in Europe Gen. Lyman L. Lemnitzer in a transmission from his headquarters in Casteau, Belgium, to Naples, Italy, via one of DOD's 15 Initial Defense Communications Satellite Project (IDCSP) satellites. U.S., which had proposed the NATO project, was permitting NATO to use the IDCSP satellites temporarily. Following approval of the second phase of the project, $45 million—of which U.S. would pay 25%—would be spent to link 12 alliance countries via two 100-lb comsats owned by NATO and launched by USAF into synchronous orbits over the Atlantic. Two ground stations would be in U.S. and one in each NATO country except France (who had not been invited to participate in the project), Luxembourg, and Iceland. (Farnsworth, *NYT*, 7/9/67, 17)

July 10: *Chicago Daily News* urged House/Senate Conference Committee on NASA FY 1968 authorization bill to accept Rep. Donald Rumsfeld's (R-Ill.) proposals to establish an independent safety review board and to require NASA to keep Congress fully informed on its operations [see June 28]: "There is ample justification for these requirements.

"Part of it can be found in the scandalous record of fumbling, bumbling and pure carelessness unearthed in the investigation of the blazing death of three [Apollo] astronauts . . . [and part from] the perils of an alliance between big government and the big industries that serve it and profit by its contracts.

"Congress . . . has a unique obligation to exert discipline over the executive department's workings. In exotic fields like those of the Central Intelligence Agency and NASA there is an understandable tendency to resent such surveillance. But those same agencies, for all the high average caliber of their work, have demonstrated that when supervision is lacking both arrogance and carelessness flourish." (*C Daily News*, 7/10/67)

- Capt. H. L. Anderton (USN, Ret.) became Chief of Communications and Tracking Branch, Electronics and Control Div., OART. Before his retirement from USN July 1, he had been Deputy Director, Aeronautics Div., Office of the Deputy Chief of Naval Operations (Development). (NASA Ann, 7/18/67)

July 9–15: Yugoslavia would build an earth station for use with communications satellites, the official press agency Tanyug announced. In addition, the Government had given its approval for "incorporating Yugoslavia into the world system of telecommunications via manmade satellites." U.S. experts speculated that statement indicated Yugoslavia was seeking to become the first Communist country to join INTELSAT consortium managed by ComSatCorp. Speculation was based in part on Yugoslavia's failure to attend April 3–15 Moscow meeting of European Communist-bloc nations which resulted in April 16 announcement by U.S.S.R. of plans to create a separate comsat network. (*NYT*, 7/21/67, 3)

July 10–13: Accelerated space effort by DOD was discussed in four-part *Long*

Island Newsday series by AP writer Howard Benedict. U.S., he said, was quietly developing a strong military space capability because: (1) U.S.S.R. "is vigorously pushing for a military space capability"; (2) Titan II and Minuteman missiles "are threatened by new warhead guidance system 10 times more accurate than any previous system"; (3) Polaris missiles deployed at sea "are threatened by new satellite reconnaissance devices" that may soon be able to locate submerged submarines; (4) U.S. missiles designed to penetrate Soviet defenses were threatened by potential Soviet capability to neutralize and destroy them before they reached enemy territory; and (5) "world strategic situation has been changed sharply by Red Chinese achievements in nuclear explosives and by steadily decreasing costs which will permit other nations to deploy missile forces."

DOD was developing "fantastic" weapons and military equipment, Benedict noted: "reconnaissance satellites that will spot a soldier hiding in underbrush or a missile buried underground; satellites to locate submerged submarines; communications satellites that will enable infantrymen . . . to converge on strategy; rockets to knock down enemy satellites; laser and radiation beams to pulverize space or ground targets; rocket-borne nuclear bombs to destroy missile warhead rockets with great power and versatility; and manned spaceships that will take off and land like conventional airplanes." USAF's Manned Orbiting Laboratory (MOL) program, he said, would determine whether man is more effective at performing tasks currently done by unmanned satellites and whether space itself could become a battlefield. (Benedict, *LI Newsday*, 7/10–13/67)

July 11: MSC engineer Caldwell C. Johnson was granted patent for a cocoon-like escape device for orbiting astronauts. Device, which would be stored in an external compartment on the spacecraft, consisted of an insulated zippered nylon bag with a retrorocket. Astronaut would don his pressure suit with oxygen supply, seal himself in the bag, and fire retrorocket to thrust toward a landmark on earth. He would then inflate a bladder to shape the bag into a sphere for stability during reentry. Sphere would deflate during reentry, notifying astronaut when it was time to deploy parachute for safe landing. (U.S. Patent Off; MSC *Roundup*, 7/7/67, 8; Jones, *NYT*, 7/15/67, 29)

- Second Saturn V booster's 1st and 2nd stages were mechanically mated at KSC in preparation for AS-502 unmanned mission to test Apollo spacecraft's reentry heat shield. Booster's 3rd stage was added July 13; instrument unit, July 14. (KSC Proj Off; *Marshall Star*, 7/12/67, 1)
- NASA contract awards: (1) Brown and Root-Northrop Corp. were selected for negotiations of a one-year, $10-million, cost-plus-award-fee contract to provide jointly operational support services to MSC's laboratory and test facilities; (2) Graham Engineering Corp. and LTV Range Systems Div. were selected for competitive negotiations of a one-year, $10-million, cost-plus-award-fee contract for facility support services at MSC. (NASA Releases 67-180, 67-181)
- First aerial drop in test series to develop and qualify Earth Landing System (ELS) for heavier Block II Apollo spacecraft was conducted by Northrop Ventura at El Centro, Calif. Tests were studying the use of a two-stage reefing system for three main ELS chutes and larger diameter drogue parachutes for the ELS. Block I Apollo spacecraft used only a single-stage reefing. (*NAA Skywriter*, 7/21/67, 1, 2)

July 11

- X–24A (formerly SV–5P), Nation's newest manned lifting-body vehicle, was publicly displayed and officially turned over to USAF in ceremonies at Martin Marietta Corp.'s Baltimore plant. The 5,000-lb wingless vehicle, constructed by Martin Marietta Corp. under AFSC contract, would be air-launched from B–52 aircraft at AFFTC and would be driven by Thiokol rocket engines to higher altitudes and supersonic speeds before gliding to controlled landing. Flight tests would be conducted by five USAF and NASA pilots in late 1967 under joint NASA/USAF lifting-body program. (AFSC Release 59.67; Clark, *NYT*, 7/12/67, 1)
- First regular data reports of activity on sun's "invisible hemisphere"—side visible from earth only once every 13.5 days—were being made to ESSA's Solar Disturbance Forecast Center in Boulder, Colo., by NASA's *Pioneer VI* and *VII* satellites, in orbit around sun. Satellites' observations on physical characteristics of solar storms would allow continuous scientific study of solar surface and provide Apollo astronauts with additional warning time to avoid solar radiation hazards. (NASA Release 67–174)

July 12: Soviet space officials hinted that they were conducting an advanced lunar exploration program and said that series of spacecraft from *Luna IX* through *Luna XIII* would not be used again, *Space Business Daily* reported. Officials did not elaborate on the "new material" that would be used, but indicated that new plans would include unmanned mobility on lunar surface. (*SBD*, 7/12/67, 42)

- American Security Council (ASC) released results of study prepared for House Armed Services Committee which concluded that U.S.S.R. would achieve a marked nuclear superiority over U.S. by 1971 unless steps were taken immediately to improve American strategic capabilities. Study, conducted by ASC's National Strategy Committee under chairmanship of former AFSC Commander Gen. Bernard A. Schriever (USAF, Ret.), found that U.S. had "exchanged its goal of a war-winning strategic superiority for a strategy of mutual deterrence" while U.S.S.R. "is driving hard toward a goal of overwhelming superiority. . . ." Report predicted that by 1971 U.S.S.R. would have a 30,000- to 50,000-mt nuclear delivery capability, contrasting with 6,000- to 15,000-mt for U.S., and recommended a rapid arms buildup or development of an antiballistic missile system. It warned that "by placing our sole reliance for deterrence on ICBMs, POLARIS and POSEIDON missiles, and the remaining manned bomber force . . . [U.S.] places itself in the dangerous position of having only one option left if it is faced with a Soviet ultimatum to surrender. It must either fire its offensive nuclear weapons or give up its sovereignty. An ABM [Antiballistic Missile System] would at least strengthen the hand of the President if he is confronted with such a fateful decision, and it might stay the hand of an enemy at a critical moment. . . ." (Text; ASC *News*)

July 12–19: "Cold flow" test series of Phoebus 2 nuclear reactor was successfully conducted at NRDS as part of NASA–AEC Project Rover program to develop a nuclear-propelled rocket. Purpose of tests was to check reactor design under gas-flow conditions and at liquid-hydrogen temperatures; measure variations in pressures, temperatures, and flow rates during simulated "start-up" operations; and obtain initial experience on the hook-up and test operations of Phoebus 2 reactors in test facility. Assembly used for experiments in "cold flow" tests was identical to that used in "hot" tests, except that it contained no fissionable material and did not produce a nuclear reaction. (NASA Proj Off; UPI, *NYT*, 7/22/67, 11)

July 13: Short circuit of electrical connector in Atlas-Centaur's upper stage caused 24-hr postponement of Surveyor IV launch. (UPI, W *Star*, 7/13/67, B11)

- Adm. David L. McDonald, Chief of Naval Operations, and Gen. John P. McConnell, Chief of Staff of the Air Force, presented Distinguished Flying Crosses to the military astronauts who had made space flights in special Pentagon ceremony. Included were posthumous awards to L/Col. Virgil I. Grissom (USAF) and L/Col. Edward H. White II (USAF), who died in Jan. 27 Apollo fire. Capt. Alan B. Shepard, Jr., speaking on behalf of his fellow astronauts, said they were "proud members of the military, but also . . . representatives of a great and significant national effort," who had grown "to enjoy the [space program's] successes and live with its failures." (DOD Release 642–67; AP, B *Sun*, 7/14/67, 6; *SBD*, 7/17/67, 68)
- House Appropriations Committee reported $1.53-billion DOT authorization bill (H.R. 11456) which included $142.3 million for development of two SST prototypes. Bill was approved by House July 18. (AP, *NYT*, 7/14/67, 5)
- Senate passed $2.6-billion AEC FY 1968 authorization bill (H.R. 10918) which included $7.3 million for a new proton accelerator in Weston, Ill. Amendment by Sen. John O. Pastore (D-R.I.) to delay approval of the Weston site until the community or the state passed open-housing laws which would enable Negro employees to find suitable housing near the plant was defeated. (*CR*, 7/13/67, S9484–521)

The *New York Times* later commented: ". . . credit goes to Senators . . . who exposed the real issues at stake, and who won a moral victory when the vote margin in Weston's favor proved much narrower than had been expected. . . .

"But there is an even more basic objection to any commitments or expenditures for this expensive research tool at this time. That objection is simply the irrelevance of a 200 billion electron volt accelerator to any real present national problem. . . . The budget cutters are now in full cry demanding reductions in already inadequate expenditures for human needs. It is a distortion of the national priorities to commit many millions now to this interesting but unnecessary scientific luxury." (*NYT*, 7/16/67, 12)

- NASA awarded Pratt & Whitney Div., United Aircraft Corp., a $458,000 contract to "study desirable characteristics of components for a quiet turbofan engine and define a development program" [see Feb. 1]. Contract was first awarded in NASA's Quiet Engine program to combine all known noise-control techniques to produce a 20,000-lb-thrust engine which would be 15–20 db quieter than present subsonic jet engines and could be used in current aircraft models. Program was being conducted by NASA as part of national noise abatement program. (NASA Release 67–184; Clark, *NYT*, 7/14/67, 27)
- NASA was studying feasibility of launching two navigation satellites to keep SST under constant surveillance and be alert to slightest malfunction, Mohammed Rauf, Jr., reported in the *News American*. NASA Chief of Navigation and Traffic Control Eugene Ehrlich, who conceived the plan, told Rauf in an interview that the system would provide maximum safety against accidents: "The program calls for the building of two

tracking stations on each side of the Atlantic. The wings of the plane and all its other important parts will be fitted with small transponders, which will continually relay information about the plane's condition to the satellites . . . [which would] relay the signals to the tracking stations, where giant computers will process them within minutes for appropriate action." Eventually, Ehrlich predicted, all aircraft and ships would be equipped with transponders "because the [proposed] navigation satellites . . . will be capable of handling the increasing amount of air and sea traffic predicted for this century.

"Four satellites can have the whole world under surveillance for air and sea traffic control, and will cut down both the cost of such control as exercised by conventional methods, and also the time required for it." (Rauf, *News American*, 7/13/67)

- Rep. Donald Rumsfeld (R-Ill.), speaking on the floor of the House, urged House to support Rep. Charles Mathias' (R-Md.) bill (H.R. 69) to establish a commission to review organization of Executive Branch: "Congressional review has often been piecemeal and after the fact. For example, the recent congressional hearing concerning the tragic Apollo spacecraft fire unearthed a shocking amount of information about NASA's operations which should have been reported to Congress long ago. . . . Legislation has been proposed to correct this, and hopefully, the provisions in the House version of the NASA authorization bill . . . will be accepted and thereby require NASA to keep Congress 'fully and currently informed'. . . . I am convinced such language is needed with respect to NASA. I am equally convinced that, if the operations of the executive branch are studied and reviewed for organizational deficiencies and economy of operation, there would be far less need to wage battles on the floor on a piecemeal, hit-or-miss basis." (NASA LAR VI/75)

July 14: Second anniversary of Mars flyby by NASA's *Mariner IV*, during which spacecraft flew within 6,118 mi of Mars, took 22 close-up photos of the planet, and transmitted scientific data on Martian atmosphere. Launched from ETR Nov. 28, 1964, *Mariner IV* had traveled 1.3 billion mi during its 959 days of flight and had operated continuously for more than 23,000 hrs—exceeding its 6,000-hr design lifetime by more than 400%. NASA announced that satellite, which was expected to continue operating satisfactorily until early 1968, had been assigned a new mission. It would be used in an experiment to conduct simultaneous comparison of solar radiation during August 1967 "solar system line-up" involving *Mariner IV*, *Mariner V* (launched June 14), the earth, and the sun. (NASA Release 67-182)

- U.S.S.R.'s *Venus IV* unmanned Venus probe, launched June 12, was about 5 million mi from earth and was performing satisfactorily, U.S.S.R. announced. Spacecraft was expected to reach Venus in mid-October. (AP, W *Star*, 7/15/67, 1)

- Bistatic radar detection—experimental technique for mapping celestial bodies by bouncing signals off them from an orbiting spacecraft—had been used successfully to map a portion of the lunar surface, Stanford Univ. Center for Radar Astronomy scientists Dr. G. L. Tyler, V. R. Eshleman, G. Fjeldbo, H. T. Howard, and A. M. Peterson reported in *Science*. Radar waves emitted Oct. 12, 1966, from an antenna on NASA's *Lunar Orbiter I* spacecraft had reached earth as two signals: one directly, and one after being reflected by the moon. Both underwent different shifts in frequency. From a measurement made on the ground scientists were able to tell the direction from the spacecraft to the site on the moon. Other measurements related to the same site enabled them to triangulate the position. (*Science*, 7/14/67, 193–5)

- U.S. space program was a sound investment which would pay "handsome dividends" to all Americans and to humanity in general, UCLA professor of chemistry Dr. Willard F. Libby told AAS meeting in Denver. Winner of the 1960 Nobel Prize in chemistry and discoverer of radio-carbon dating, Dr. Libby suggested that space program opponents "quit grumbling about going-to-the-moon extravagance with taxpayers' money" and realize that large expenditures were necessary to maintain U.S. world leadership in space.

  "And we can't stop at the moon. . . . Once we've ventured that far, we should establish a base there for manned takeoffs to Mars and Venus.

  "Now Congress is debating the merits of authorizing space exploration beyond the moon in the early '70s. They have duty to debate such a venture, costing more than $2 billion. But what . . . would Congress do with that money if it weren't spent on future space research?" Space research had already made great contributions to medicine, microelectronics, worldwide communications, and many other fields, he said, and could continue to achieve much more "for the good of humanity and the expansion of frontiers of human knowledge." (Lindbergh, *Denver Post*, 7/16/67)

- Harry N. Atwood, an associate of the Wright Brothers, died at age 83. On July 1, 1911, during 17-hr 12-min flight from Boston to Washington, D.C., Atwood became first man to fly over New York City. The following month he successfully flew his biplane, the "Baby Wright," 1,265 mi from St. Louis to New York City in 28 hrs 58 min, setting a new air distance record and winning a $10,000 prize. In 1927 Atwood announced the development of a motor that would "develop power without the use of a propeller, by exhausting exploding gases into the atmosphere or water, thus driving itself forward much as a rocket does." In 1935 he developed strong, lightweight plywood for aircraft bodies. (AP, *NYT*, 7/16/67, 65)

- U.S. tourist vacationing on beach at Harbour Island in the Bahamas had found a 10-ft-long piece of metal bearing words "United" and "destruct mechanism here," *San Diego Union* reported. Assuming metal to be part of a rocket, the tourist had notified KSC officials, who retrieved the part and determined that it was an insulation panel from the upper stage of an Atlas-Centaur used Oct. 26, 1966, to launch a Surveyor model on a test flight. (AP, *SD Union*, 7/14/67)

July 14–16: NASA's *Surveyor IV* (Surveyor D) was successfully launched from ETR by Atlas-Centaur booster (AC-11) on an excellent 65-hr lunar-intercept trajectory, but mission ended prematurely when communications with spacecraft were lost seconds before it was scheduled to softland on the moon. The abrupt termination of the spacecraft's signal made it impossible to determine whether spacecraft had landed safely or crashed out of control.

Launched by the last direct-ascent (single-burn) Centaur vehicle in Surveyor program, 2,290-lb *Surveyor IV* performed all required sequences on schedule, including a midcourse maneuver which narrowed estimated target miss from 124 mi to 2.4 mi. Landing sequence began as planned when spacecraft shifted its normal cruising attitude to position main retrorocket downward toward moon. Radio contact was lost two seconds before retrorocket ended its planned 42-sec burn to slow spacecraft's speed from 5,900 mph to 350 mph. Repeated attempts to reestablish radio contact with spacecraft failed.

Surveyor IV carried photographic equipment, a surface sampler, and a magnetic device on one of its footpads to test for magnetic materials in the lunar soil. Primary mission objectives: (1) to softland on the moon in the Sinus Medii and (2) to obtain TV photos of the lunar surface. Secondary objectives: (1) to conduct a vernier engine experiment; (2) to manipulate lunar surface with surface sampler in view of TV camera; (3) to obtain touchdown dynamics data; and (4) to obtain thermal and radar reflectivity data on lunar surface. Target zone in the Sinus Medii was almost in the exact center of moon's visible hemisphere and was rougher than the Ocean of Storms where *Surveyor I* and *Surveyor III* had landed. Benjamin Milwitzky, NASA Surveyor Program Manager, said the site was so rugged that spacecraft had only a 50–50 chance of landing intact. "But this region is extremely important to Apollo and that's why we're going there."

Surveyor IV was fourth in series of seven spacecraft in NASA's Surveyor program: *Surveyor I* (launched May 30, 1966) and *Surveyor III* (launched April 17, 1967) both softlanded successfully and transmitted photos to earth; *Surveyor II* (launched Sept. 20, 1966) did not softland because of ignition failure in one of its three vernier engines. Program was directed by OSSA Lunar and Planetary Programs Div.; program management was assigned to JPL. (NASA Release 67–172; NASA Proj Off; UPI, *NYT*, 7/15/67, 22; AP, *NYT*, 7/17/67, 16; Reistrup, *W Post*, 7/15/67, B4; 7/16/67, A5; 7/17/67, 1; UPI, *W Star*, 7/14/67, A1; 7/15/67, A2; 7/16/67, B5; 7/17/67, A1)

July 15: Dr. Athelstan F. Spilhaus, Univ. of Minnesota geophysicist, announced that he was resigning to become president of the Franklin Institute in Philadelphia, succeeding Wynn L. LePage who would become board chairman. (UPI, *NYT*, 7/16/67, 33)

July 16: *Lunar Orbiter II*, *Lunar Orbiter III*, and *Lunar Orbiter IV*, all circling the moon, were being used by NASA's Manned Space Flight Network personnel and stations for practice tracking exercises. Satellites served as "stand-ins" for the manned Apollo spacecraft which would eventually orbit the moon. (NASA Release 67–183)

- Jerald Kubat, formerly Special Assistant to the Apollo Program Director, OMSF, had been appointed Director, Apollo Program Control. He would have overall responsibility for program planning, configuration management, logistics, resources control, and procurement operations to ensure the attainment of Apollo goals. (NASA Ann, 7/17/67)

- A U–2 photographic aircraft was flown at Edwards AFB to test missile system that would intercept and then incapacitate enemy satellites, George Wilson reported in *Washington Post*. "The idea is to equip the United States with a missile or spacecraft that could blind reconnaissance satellites in time of war or national emergency.

"There is no intention now of ending the gentlemen's agreement under which both the U.S. and the Soviet Union spy on each other by satellite.

"But there may come a day, Air Force leaders say, when the U.S. will want to hide the movements of its bombers and missiles. . . ." (Wilson, *W Post*, 7/27/67, 17)

July 17: Cosmos CLXIX was launched by U.S.S.R. into orbit with 208-km (129-mi) apogee, 144-km (89-mi) perigee, 88-min period, and 50° in-

clination. It reentered later in the day. (*Aero Tech*, 7/31/67, 18)
- Congressional reaction to the FY 1968 space budget pointed up the need for the Johnson Administration to "take a fresh, hard look at shaping a truly national space program" in the 1970s, Michael Getler suggested in *Aerospace Technology*. "What we have now, on one hand, is the already big and expensive civilian space program which, though enormously successful and of great importance to national technological strength, is unfortunately vulnerable to on-again/off-again Congressional support and to the continuing challenge of finding dramatic new undertakings which can attract the necessary backing.

    "On the other hand, we now have a rapidly growing military space program which appears to be on its firmest footing and which has demonstrated an ability to request and get new funds even when dollars are very tight. . . ."

    There were many areas where the two programs should remain separate, Getler said, but there were also many areas where cooperation would be advantageous: "Rather than representing a loss of funding, we believe that joint development [in some areas] will lend stability to programs that might otherwise run into trouble as single agency undertakings, will lend more logic to the national commitment to space, and end some of the sniping about parallelism that has hurt in the past. . . ." (Getler, *Aero Tech*, 7/17/67, 70)
- Six airlines reserved delivery positions for 16 SSTs at a cost of $750,000 each [see June 5], bringing to 129 the total number of delivery positions reserved. (AP, *NYT*, 7/18/67)
- Rep. William F. Ryan (D-N.Y.), speaking on the House floor, repeated his request that Congress reopen its investigation of NASA [see June 12]: "The Apollo hearings should be reconvened—not for the purpose of continuing further a technical discussion of possible technical causes for the Apollo fire—but to undertake a serious and far-ranging management review of the sort that would force NASA to take stock and develop sound management practices for the protection of the public's investment in the space program. This is not to mention the safety of our astronauts. . . ." (*CR*, 7/17/67, H8750)
- RCA announced it had received $38-million NASA contract [see June 2] to maintain and operate portions of the Satellite Tracking and Data Acquisition Network (STADAN). (*WSJ*, 7/17/67, 3)
- Many valid conclusions could be drawn from the displays and exhibits at the Soviet Air Show [see July 8–9], Robert Hotz suggested in *Aviation Week*. Program indicated that U.S.S.R. had: (1) replaced "peaceful" Soviet aerial image projected at Paris International Air and Space Show with a "mighty brandishing of military might"; (2) begun to devote an increasingly large effort toward hardware and tactics for modern, non-nuclear war; and (3) exhausted her "creative design capability in the current generation of aerospace technology" and was relying more heavily on Western design concepts. (Hotz, *Av Wk*, 7/17/67, 21)
- Legal Subcommittee of U.N. Committee on the Peaceful Uses of Outer Space meeting in Geneva reached provisional agreements on aiding and returning astronauts who landed in foreign countries and on payment of damages caused by space launches. Agreements would be submitted to full U.N. Committee for approval and then placed before U.N. General Assembly for final action. (UPI, P *EB*, 7/18/67; UPI, *W Post*, 7/18/67, A3)

July 17–20: NASA personnel changes: (1) Albert P. Little, former Chief,

July 17-20, 1967

Program Plans, OMSF Plans and Analysis Directorate, had been appointed Assistant to the Assistant Administrator for Administration effective July 17; (2) R/Adm. Roderick O. Middleton (USN), former Deputy Director of OMSF Missions Operations Directorate, had been appointed Apollo Program Manager at KSC, replacing M/G John G. Shinkle (USA, Ret.), who resigned May 18; (3) Stanley de Jongh Osborne, a general partner of Lazard Freres & Co. investment banking firm in New York, had been sworn in as consultant to NASA Administrator James E. Webb on Government-industry relations and on the impact of space technology on the national economy; and (4) Erskine E. Harton, Jr., former Assistant Safety Director, had been appointed Acting Safety Director, replacing George D. McCauley who recently retired from Government service. (NASA Ann, 7/20/67; NASA Releases 67-189, 67-190)

July 17-21: Upper atmosphere, ionosphere, solar-terrestrial physics, sounding rockets, and probes were discussed in London Symposium on Results of the International Years of the Quiet Sun (IQSY 1964-65) and the concurrent annual plenary meeting of the Committee on Space Research (COSPAR). Nigel Calder, in *Science*, cited views of Univ. of Illinois physicist S. A. Bowhill: "The F-region of the ionosphere is now becoming fairly well understood, and the outstanding problems are yielding to theoretical treatment. The new explanations often involve large-scale movements of charged particles in the upper atmosphere. Attention will now turn to lower layers (the D-region and below) where recent studies indicate unexpected interactions between the ionosphere and the stratosphere." (Calder, *Science*, 8/11/67, 666-8)

July 18: JPL scientists abandoned attempts to contact *Surveyor IV* spacecraft resting on the moon [see July 14-16]. A JPL spokesman said both Canberra and Goldstone tracking stations had sent signals to the spacecraft July 17, but had received no response. A committee was attempting to determine whether *Surveyor IV* had softlanded on the moon as planned or crashed out of control. (AP, W Star, 7/18/67, A3; AP, NYT, 7/20/67, 11)

• A minimum of four manned space flights would precede the first U.S. attempt to land men on the moon, Astronauts Joe H. Engle and Alfred M. Worden, recently returned from a geology survey in Iceland, told NASA Hq. press conference. Three of the four missions would involve rendezvous with another spacecraft in earth orbit. Astronauts also said that they and their colleagues had even more confidence in the Apollo program now than they had before the Jan. 27 fire. (B Sun, 7/19/67; W Post, 7/19/67)

• House passed $1.53-billion DOT FY 1968 authorization bill (H.R. 11456) which included $142.3 million for SST prototype development. Amendment by Rep. Clark MacGregor (R-Minn.) to delete SST funds was defeated. (CR, 7/18/67, H8813-58)

• Christopher C. Kraft, Jr., Director of Flight Operations at MSC, received the Spirit of St. Louis medal from the American Society of Mechanical Engineers (ASME) meeting in Cocoa Beach, Fla. He was cited for his "meritorious service in the advancement of aeronautics." (NYT, 7/19/67, 32)

• LeRC awarded four contracts totaling over $5 million for major systems of the SERT II (Space Electric Rocket Test) spacecraft: (1) $2,462,678 to Fairchild Hiller Corp. for system definition selection and qualification of components for spacecraft support unit; (2) $1,308,065 to Hughes Aircraft Co. to develop prime power conditioning and control system for ion thrusters; $1,055,757 to Westinghouse Electric Corp. to develop backup power conditioning and control system; and (4) $290,987 to Cutler-Hammer Corp. to develop a radio frequency interference experiment to measure interference with radio communications generated by the ion engines. SERT II was an orbital mission scheduled for late 1968 to advance the development of ion engines as propulsion units for future long-duration space missions. (NASA Release 67-188)

- Astronauts on long-duration space flights might have to accept greater radiation exposure than the amount considered safe for industrial workers and, in some cases, this additional exposure could heighten the risk of cancer and shorten life expectancy, NRC's Space Science Board concluded in report entitled *Radiobiological Factors in Manned Space Flight*. Board recommended that criteria different from those used in industry be set up to gauge radiation risks in space flight and to decide what was acceptable. "Manned space flights are high risk endeavors. They take place in a hostile environment that cannot be permitted to encroach beyond the limits of human endurance if the missions are to succeed." Report noted, however, that radiation was "only one of many recognized and accepted potential risks" to space flight and shielding against it added weight to the spacecraft. Consequently, protection against radiation had to be balanced against other needs vital to the success of the mission. (NRC Release; Schmeck, *NYT*, 7/18/67, 38M; Reistrup, *W Post*, 7/18/67, A3; AP, *W Star*, 7/18/67, A3)

- AFSC had presented its ninth Craftsmanship Award—highest of three honors that could be won by a defense contractor under USAF's Industrial Zero Defects Program—to RCA's Aerospace Systems Div. To qualify for the award, a firm had to show performance records for at least 18 mos clearly reflecting achievements against predetermined goals. USAF personnel then validated performance data and determined adequacy and realism of the goals. (AFSC Release 71.67)

- A monument had been erected on the spot where Cosmonaut Vladimir M. Komarov was killed when his *Soyuz I* spacecraft crash-landed April 24, *Pravda* reported. It was located on the Orenburg Steppe, 1,207 km (750 mi) southeast of Moscow. (Reuters, *NYT*, 7/19/67; *W Post*, 7/19/67, C7)

July 19: First experimental tactical communications by satellite among USN, USAF, and USA units had been successful, DOD announced. Messages had been exchanged among airborne aircraft, a submarine, a ship, and fixed and mobile terminals via Lincoln Experimental Satellite (*Les V*), one of six satellites launched by USAF July 1. (DOD Release 661-67; AP, *NYT*, 7/21/67, 6)

- USAF announced that first unit to be equipped with F-111 variable-sweep-wing aircraft would be activated at Nellis AFB, Nev., in summer 1967. (DOD Release 664-67)

- Rep. John J. Rhodes (R-Ariz.), speaking in the House on behalf of the Republican Policy Committee, urged that House/Senate Conference Committee on NASA FY 1968 authorization bill accept key Republican amendments adopted by the House. Amendments would reduce NASA's FY 1968 budget request by $201.4 million, establish an independent safety panel in NASA, and require NASA to keep Congress fully and currently informed of problems. (*CR*, 7/19/67, H9045)

July 19

- President Johnson, in letter to Congress transmitting NSF's FY 1966 report on Federal activities in the area of weather modification, wrote: "This report provides clear evidence that progress is being made toward our goal of developing the capacity to modify the weather for the benefit of all mankind.

    "In the period covered by this Report, we found that: (1) Precipitation from some types of clouds may be increased by as much as ten per cent by seeding; (2) Seeding of thunderstorm clouds may reduce significantly the number of lightning strikes; (3) The incidence of hail may be reduced by heavy seeding; (4) Large bodies of cold ground fog may be dissipated through the use of dry ice or silver iodide; and (5) Mathematical models will be increasingly useful for experimentation with techniques for controlling hurricanes and tornadoes. . . ." (PD, 7/24/67, 1035)

July 19–22: NASA successfully launched *Explorer XXXV* Interplanetary Monitoring Platform (IMP–E) from ETR by three-stage Thrust-Augmented Delta booster on direct-ascent lunar-transfer trajectory. On July 21 retromotor burned 23 sec, decreasing spacecraft's velocity and permitting lunar capture; retromotor separated as planned two hours later and spacecraft entered elliptical orbit with 4,780-mi (7,692-km) apolune, 497-mi (800-km) perilune, 11.5-hr period, and 147° inclination.

Sixth of 10 Interplanetary Explorers planned by NASA and second designed to collect data at lunar distances, *Explorer XXXV* carried eight experiments, seven scientific and one engineering—a solar cell damage study. Preliminary data indicated that all experiments were performing satisfactorily. Primary mission objective was "to place the spacecraft into either a captured lunar orbit or a geocentric orbit with apogee near or beyond the lunar distance, to investigate out to and at lunar distances and to obtain scientific data on the characteristics of the interplanetary plasma and the interplanetary magnetic field." Secondary objective, if spacecraft achieved lunar orbit, was collection of data on dust distribution around the moon, the lunar gravitational field, the weak lunar ionosphere, and the radiation environment. IMP series was managed by GSFC under OSSA direction; two of the five previously orbited satellites—*Explorer XXXIII* (launched July 1, 1966) and *Explorer XXXIV* (launched May 24, 1967)—were still providing data. (NASA Releases 67–178, 67–193; NASA Proj Off; *NYT*, 7/20/67, 14; 7/22/67, 3; 7/23/67, 63; AP, W *Star*, 7/19/67, 1; 7/20/67, A9; 7/21/67, 1; *W Post*, 7/25/67, A8)

July 20: NASA test pilot William H. Dana flew X–15 No. 3 to 84,000-ft altitude and 3,682 mph (mach 5.51) at Edwards AFB. Purpose of test flight was to check (1) cold-wall heat transfer, (2) wavy panel heat transfer, (3) boost guidance, (4) PCM, (5) horizontal tail loads, (6) nose gear loads, and (7) tip-pod accelerometer. (NASA Proj Off)

- NASA's *Mariner V* Venus probe, launched June 14 from ETR, had traveled 55,327,194 mi of its 216-million-mi flight to Venus and was continuing to operate normally, JPL project officials announced. On July 23 central computer and sequencer would change amount of data transmitted from $33\frac{1}{3}$ bits per second (bps) to $8\frac{1}{3}$ bps to meet slower transmission requirements of rapidly increasing communications distance. New transmission rate would remain in effect for duration of flight. (NASA Release 67–194)

- NASA Aerobee 150 sounding rocket launched from WSMR carried Univ. of

Minnesota experiment to 131-mi (210-km) altitude to measure composition of atmosphere and determine temperature in altitude range 62–124 mi (100–200 km) using three mass spectrometers. Two of three spectrometers did not function properly; however, data obtained appeared satisfactory. (NASA Rpt SRL)

- ComSatCorp announced NASA would launch fourth satellite in the INTELSAT II series Sept. 20. Plans called for satellite to be launched from ETR into synchronous orbit over the Pacific to increase communications capability and serve as a backup against interruption of satellite communications service. (ComSatCorp Release)
- Recent cuts in NASA's FY 1968 budget were due in part to lack of Congressional understanding of the U.S. space program, NASC Executive Secretary Dr. Edward C. Welsh told AIAA's Joint Propulsion Conference in Washington, D.C. "Many members of the Congress understand" certain areas of the space program, but "not enough of our legislators seem to have sufficient comprehension of the vast benefits flowing from space activities. And if many members of Congress and very many of their constituents don't know about these benefits, then they may well feel justified in cutting back on the amount of money to be spent. . . ." To help increase understanding, Dr. Welsh recommended that the space industry avoid "confining their advertising to superlatives" about products and communicate the overall benefits of the space program. (Text)
- A. O. Tischler, director of OART's Chemical Propulsion Div., received AIAA's 1967 Wyld Propulsion Award for leadership and management of rocket programs. He was cited for his "outstanding contributions to the research and development of Liquid Rocket Propulsion Systems and overall management of the NASA Liquid Rocket Propulsion Program." (NASA Release 67-186)
- Rep. William F. Ryan (D-N.Y.), speaking on the House floor, suggested that SST program was progressing too rapidly. ". . . it is not clear why the United States is so anxious to rush into a costly program of SST development involving many unsolved problems and great economic risk on the strength of a nebulous national prestige. It may well be that our prestige will lose a great deal more if we show ourselves to be a nation whose commitments and investments are based largely on a hysteria about our 'image'—largely and simply because the British and French are building the Concorde—unable to wait and learn from their experience—and unable to take the advice of a multitude of technical and economic experts within our own country as to the inadvisability of this effort. . . ." (*CR*, 7/20/67, H9128)
- European leaders had realized in recent months that only they could close the "technology gap" between their countries and the U.S., Dr. Donald F. Hornig, Special Assistant to the President for Science and Technology, told *New York Times* reporter Evert Clark. Dr. Hornig reached this conclusion after a 16-day tour of six European capitals. The belief that U.S. had created the gap and should therefore assume the burden of closing it "has entirely disappeared," he said. "What we found [on the tour] was a realistic approach, a realization that most of these things are basic problems, ones they have to tackle themselves." This realization was "a most important advance" that represented an "enormous closing of the understanding gap" and meant that discussion of the many problems making up the larger technological gap "has now come down to earth." (Clark, *NYT*, 7/21/67, 15)

- Barry Goldwater, 1964 Republican presidential candidate, told National Retail Hardware Assn. in San Francisco that he believed U.S.S.R. had orbited military weapons: "We have every reason to believe that the Russians have a weapon in orbit that can be called down at any time against any target." Goldwater said he could not prove his statement, but that information from Soviet technical journals indicated it was accurate. (AP, B *Sun*, 7/21/67)

July 21: A new star just bright enough to be visible to the naked eye had been discovered in the constellation Delphinius by G. E. D. Alcock of Peterborough, U.K. Discovery was confirmed by M. P. Candy, British Astronomical Society, who estimated star's brightness as magnitude five, which is near the limit of visual observation. (SciServ, *NYT*, 7/21/67)

- Wind tunnel tests at USAF's Arnold Engineering Development Center (AEDC) in Tullahoma, Tenn., indicated "NASA has miscalculated the control spacemen would have over the Apollo spacecraft at the critical point it reenters the Earth's atmosphere," the *Nashville Tennessean* wrote in a copyrighted story. Loss of control "means that the Moon spacecraft's maneuvering ability would be cut, perhaps to a dangerous degree." Results of tests, which had been forwarded to MSC, "will cause NASA to alter its computer projections of the Apollo's reentry path." (AP, *H Chron*, 7/21/67)

- July 14 promotion of career administrator Herman Pollack from Acting Director to Director of State Dept.'s Office of International Scientific and Technological Affairs was accepted reluctantly by the scientific community, *Science* reported. "The role and potential of the office . . . [is a] fuzzily defined relationship between science, technology, and foreign policy," and the scientific community had tended to think of the office as its own. ". . . many elder statesmen of science . . . are not altogether pleased to find . . . [the directorship] in the hands of a nonscientist," but have admitted grudgingly "that Pollack has done an outstanding job [as Acting Director] and that the office is likely to improve still further now that he is free of the uncertainties of an acting appointment." (*Science*, 7/21/67, 292)

- Center for European Research (CERN), West Germany, and France signed an agreement in Geneva to share the estimated $19.5-million cost of constructing a bubble chamber in which particles of smashed atoms could be tracked photographically to provide further information on the properties of matter. Scheduled to be built at CERN Hq. near Geneva, the structure would be cylindrical, about 11.5 ft in dia and 10 ft high—the largest bubble chamber in existence. (*NYT*, 7/22/67, 11)

July 22: NASA successfully launched first of five Nike-Apache sounding rockets from NASA Wallops Station but canceled remaining four in the dusk-to-dawn series of launches because of unfavorable weather. Rocket ejected a trimethylaluminum (TMA) vapor trail of bluish color between 50- and 125-mi altitudes in experiment to measure wind velocities and directions at various altitudes in the upper atmosphere. Launchings were conducted for GCA Corp., under contract to GSFC. OSSA had overall program responsibility. (NASA Release 67–179; WS Release 67–25; *W Post*, 7/23/67, A3)

- NAA's incentive fees for the Apollo Command Module (CM) destroyed in Jan. 27 fire would be reduced by NASA to $41 million, Sen. Clinton P. Anderson (D-N. Mex.), Chairman of Senate Aeronautical and Space

Sciences Committee, told the press. He said information provided him showed "that amount is $15 million less than what North American's estimated potential incentive earnings would have been if the accident had not occurred." Senator Anderson said he believed the settlement was realistic and fair: "A stricter penalty would have served no purpose and could have adverse effects on the relationship between [NASA and NAA]." (Text)

- Twelve science experts met in Washington, D.C., to discuss implementing the April 13 promise made by 20 Latin American Presidents at Punta del Este, Uruguay, to "harness science and technology for the service of our [440 million] peoples." MIT Chancellor Dr. James R. Killian, U.S. representative, told the *New York Times* that "it would be of enormous value and importance if we were able to devise a quick transfer of what is already available, both within the other Latin countries and between the United States and the others." Strengthening of existing "centers of excellence" in science and engineering, and the creation of multinational centers that would overcome the problems of national boundaries could rapidly increase the diffusion of skills among Latin American industries and society, he said. (Clark, *NYT*, 7/23/67, 52)

- John Cobb Cooper, former president of International Institute of Space Law and founder of McGill Univ.'s Institute of International Air Law, died in Princeton, N.J. Cooper devised a legal code for outer space in 1961 when he was legal adviser to the International Air Transport Assn. Code included recommendations that: (1) the lower boundary of outer space be fixed at the point above the earth's surface where it is possible to launch a satellite into orbit at least once around the earth; (2) outer space have the same status as the high seas, and that no one be permitted to assert sovereignty over outer space or any celestial body in it; (3) satellites and other spacecraft have the nationality of the country that launched them unless otherwise agreed; and (4) a nation have the right to take action in outer space for its self-protection and self-defense. (*NYT*, 7/24/67, 27)

July 23: Dr. Roger W. Heyns, Chancellor of the Univ. of California at Berkeley, was named chairman of NASA's Ad Hoc Science Advisory Committee, replacing Dr. Norman F. Ramsey of Harvard Univ., who resigned to devote more time to his new duties as president of the Universities Research Assn. Established in March 1966, the Committee advised NASA on the conduct of future space programs, including manned programs. (NASA Release 67-191)

- NASA had awarded 12 inventors a total of $9,100 for the advancement of space technology: (1) Warren E. Armstrong, Donald S. La France, Carroll Z. Morgan, Lloyd B. Ryland, and Hervey H. Voge of Shell Development Corp., $3,000 for their contribution relating to a catalyst for monopropellant decomposition of hydrazine used in liquid propulsion technology; (2) Erwin Baker of Hughes Aircraft Co., $1,400 for a constant lift device used to simulate lunar landing conditions for spacecraft; (3) Joseph A. Goodrich and Kenneth T. Ingham, NAA, $1,200 for a turbine blade-locking device to retain rotor blades on rotor wheels for turbines employed in the J-2 engine program; (4) Richard G. Turner, RCA, $1,000 for a thermocouple assembly used in thermo-vacuum tests for TV systems in Ranger, Tiros, Lunar Orbiter, and other similar programs; (5) Richard A. McKay, JPL, $1,000 for temperature control system for circulating fluids used in mixing units for propellant development systems; and (6) John B. Schutt, GSFC, and Charles M. Shai, Electro Mechanical Research Corp., $1,500 for an alkali-metal silicate protective coating used for thermal control on spacecraft and for fire-resistant paint in both Government and commercial applications. (NASA Release 67-185)

July 24: Soviet launch rate for recoverable Cosmos reconnaissance satellites had leveled after six-year increase, *Aviation Week* reported. Number launched as of mid-July 1967 was same as number launched during same period in 1966, and annual 1967 total was expected to be similar to annual 1966 total of 21. "The Russians now are launching recoverable reconnaissance satellites on the average of one every 16–17 days, each of which remains in orbit for approximately eight days. When the program began in 1962, the launch rate averaged one every 73 days and the following year the pace accelerated to an average of one every 46 days. In 1964 the figure was every 28 days, in 1965 the frequency had increased to an average of one every 19 days and last year it began to taper to one every 17 days...." (*Av Wk*, 7/24/67, 26)

- France had taken two steps to strengthen the Franco-German Symphonie communications satellite program, Donald E. Fink reported in *Aviation Week*: (1) she had reorganized the industrial teams to ensure that French and German companies received equal shares of work; and (2) she had reassured representatives at the European Ministerial Space Conference (EMSC) in Rome that she was still interested in participating in planned European comsat system. "The moves demonstrate France's determination to develop a national regional comsat system—with financial and technical assistance from Germany—to strengthen its position in the International Communications Satellite (Intelsat) system. This drive is a direct challenge to U.S. domination of Intelsat...." Fink said French officials at the EMSC had justified development of the Symphonie system by saying it would be based on state-of-the-art technology: "This would permit the European effort . . . to be directed toward developing a second-generation advanced system." (Fink, *Av Wk*, 7/24/67, 27)

- Soviet aircraft designer Alexander Yakovlev had criticized U.S. exhibit at the Paris International Air and Space Show [see May 26–31], *Aviation Week* reported. U.S. space pavilion was too complicated for the general public, he said, whereas Soviet exhibit showed "simplicity and clarity." Although U.S. showed several of its commercial aircraft, it alienated visitors, as it had in 1965, by exhibiting military weapons. "The United States couldn't restrain itself, for example, from again showing the Phantom bomber, which is sadly known for its piratical attacks in Vietnam. But everything shown in the Soviet Union's salon had peaceful significance." (*Av Wk*, 7/24/67, 44)

- LaRC was developing a new airfoil that could improve performance of advanced variable-sweep-wing fighter aircraft, increase efficiency of subsonic jet transports, reduce helicopter blade tip stalling, and virtually eliminate local shock wave formation, B. K. Thomas, Jr., reported in *Aviation Week*. Airfoil, which was actually a "transonic wing," delayed boundary layer separation at high subsonic speeds, thereby delaying the rapid rise in aerodynamic drag in the transonic region. It was currently undergoing wind tunnel tests at LaRC, but no aircraft flight testing was planned before FY 1970. (Thomas, *Av Wk*, 7/24/67, 25)

- MSFC had awarded Astro Space Labs, Inc., a $93,136 contract for further development and testing of a remotely controlled device to handle tools, equipment, and men in space. Called a serpentuator, device was invented by Hans Wuenscher of MSFC's Manufacturing Engineering Laboratory. It consisted of links connected by powered hinges, which could be remotely controlled from either end of the device; one end would be connected to the spacecraft; one would be free. Small versions of the device could be powered by a hand pump or batteries; larger versions, by electric motors. (MSFC Release 67–151)

During week of July 24: West German Cabinet approved $456-million five-year plan which would substantially expand national space program. Commenting on the approval, Dr. Gerhard Stoltenberg, Minister for Scientific Research, told the press that enlarged national program was necessary to create a basis for improved cooperation with other technologically advanced nations. Plan would emphasize projects that provided a "rapid spin-off" of technical knowledge for industry and the economy, he said. Among the areas of special interest would be new materials, energy sources, electronic components, and new methods of planning and managing. (*NYT*, 7/30/67, 13)

July 25: USAF launched unidentified satellite from WTR using Thor-Agena D booster. (*Pres Rep 1967*)

- First color photos of the full earth (full disc) had been taken by DOD satellite *Dodge*, launched July 1 and designed to use 10 extendable booms for stabilizing satellite by earth's gravitational force. The Johns Hopkins Univ. Applied Physics Laboratory, under contract to Navy's Air Systems Command, used a gravity gradient stabilization method and near-synchronous orbit to point *Dodge* always in right direction and to keep it within line of sight of northeastern U.S. 5 days out of every 12. Slow-scan TV systems aboard *Dodge* obtained TV images which were transmitted to earth station; color images were obtained by successive exposure of vidicon tube through blue, green, and red filters. The two TV cameras also observed fluctuations of satellite from vertical and studied solar pressure effect on satellite's booms. (DOD Fact Sheet)

- NASA personnel changes: (1) Dr. Joseph F. Shea, Deputy Associate Administrator for Manned Space Flight, resigned effective Aug. 1 to become a vice president of Polaroid Corp.; (2) B/G Carroll H. Bolender (USAF), OMSF, was named manager for the Lunar Module (LM) at MSC's Apollo Spacecraft Program Office, replacing Dr. William A. Lee, who resigned for personal reasons. Bolender would be responsible for management of the LM program including design, development, and fabrication of the vehicle by Grumman Aircraft Engineering Corp., LM prime contractor; (3) Col. Maynard E. White (USAF, Ret.), former Deputy Chief of Staff for Personnel, USAF Hq. in Europe, became Director of Program Control, OMSF, replacing William E. Lilly, who was appointed Assistant Administrator for Administration March 8. White would be responsible for the total financial, administrative, and business management support services for OMSF and the administration and direction of approved facility projects in support of the manned flight program. (NASA Releases 67–196, 67–198; O'Toole, *W Post*, 7/26/67, A10; AP, *W Star*, 7/26/67)

- U.K., France, and West Germany would develop a short-to-medium-distance aircraft designed to reduce cost of air travel by the early 1970s, John Stonehouse of the British Ministry of Technology told a London news conference. Known as the European Airbus Project, plan called for a

300-passenger twin-engine aircraft with a 1,931-km (1,200-mi) to 2,253-km (1,400-mi) range to begin commercial service by 1973. Expected to cost $532 million, Airbus would be designed jointly by Sud-Aviation of France, Hawker Siddeley of U.K., and Arbeitsgemeinschaft Airbus of West Germany. (*NYT*, 7/26/67, 61M)

- France and U.K. did not believe Concorde supersonic aircraft, scheduled to enter commercial service in 1971, would be forced out of the world market by U.S. SST in 1974, Sir George Edwards, chairman of an Anglo-French management organization for the Concorde, told the Washington, D.C., Aero Club. One distinct selling advantage for the Concorde was its price—about $16,000,000, compared to an estimated $40,000,000 for SST. He predicted 150 Concordes would be sold by 1975 and about 40–50 per year after that date. (Sehlstedt, B *Sun*, 7/26/67)
- AFSC announced it would purchase eight production models of Titan III–C Standard Space Launch Vehicle. New boosters, along with four of the 17 original R&D vehicles which had not yet been flown, were expected to "support mission requirements through the 1969–1971 period." (AFSC Release 126.67)

July 26: NASA awarded Boeing Co. a $2.275-million cost-plus-fixed-fee contract for procurement of long-lead-time materials for two additional Saturn V launch vehicles. Contract, which would expire Jan. 1, 1968, was first Saturn V procurement in support of Apollo Applications (AA) program. (NASA Release 67–200)

- NASA selected Martin Marietta Corp. to negotiate 27-mo, $25-million, cost-plus-incentive-award-fee contract for payload integration of experiments and experiment support equipment on Apollo Applications (AA) spacecraft. Tasks would be performed at NASA's three manned spaceflight centers: (1) MSFC work would involve the orbital workshop and Apollo Telescope Mount (ATM); (2) MSC work, the meteorological and earth resources payloads; and (3) KSC work, the test integration planning and support for launch operations. Selection of contractor followed competitive definition phase in which Martin Marietta Corp. and Lockheed Missiles and Space Co. studied AA payload integration under parallel, $2-million, fixed-price contracts. (NASA Release 67–199)
- NASA Administrator James E. Webb testified on NASA FY 1968 authorization bill before Senate Committee on Appropriations' Subcommittee on Independent Offices. Asked by Sen. Spessard Holland (D-Fla.) to make a choice between a substantial cut in funding for the Apollo Applications (AA) program and the Voyager program, Webb replied that both were vital to the U.S. space effort. "The Apollo application is a small investment to expend on something you have already spent $15 billion to get and it seems to me that this is important.

  "On the other hand, the United States, if it retires from the exploration of the planetary field, in my view . . . [will face] the most serious consequences because the Russians are going to be moving out there and our knowledge of the forces that exist in the Solar System can affect the Earth and can be used for many purposes to serve mankind or for military power. . . ." Criticized by Sen. Holland for refusing to make a choice, Webb said he did not want "to give aid and comfort to anyone to cut out a program. I think it is essential that we do them both." (Transcript, 2382–3)
- 20th Anniversary of USAF, established as a separate military service under 1947 Armed Forces Unification Act. (EH)

July 27: USAF launched two unidentified satellites from WTR on one Atlas booster. Satellites were later identified as *OV 1-12* and *OV 1-86*. (Pres Rep 1967)

- Sen. Joseph S. Clark (D-Pa.), speaking on the Senate floor, urged President Johnson to appoint a commission to study the need for construction and deployment of an antiballistic missile (ABM) system. ". . . the ABM question is of such a magnitude that it is essential to have a careful and objective evaluation of the course the United States should follow. I do not believe . . . that the military-industrial complex is objective enough to advise the United States Congress or the President on how we should proceed. This being the case, I strongly suggest that a temporary blue ribbon commission drawn from all sectors of national life is the best way to bring a thorough inquiry into the issues.

  "Our very national survival may be at issue in the ABM controversy. It is time we put the best minds in the country to work." (CR, 7/27/67, S10364-6)

- NASA's Apollo program was "back on the track" six months after the Jan. 27 fire at KSC, Washington *Evening Star* columnist William Hines concluded after a tour of NASA centers. "Half a year after that unforgettable night of Jan. 27 the process of trauma, shock, and recuperation has run its course. Much remains to be done, of course—and it will be another nine or 10 months before astronauts fly again—but things are moving once more.

  "At Downey . . . where Apollo spacecraft are built, at Houston, where flights will be controlled, and at Cape Kennedy, where they will begin, there is a noticeable air of confidence (but not over-confidence) mingled with caution (but not over-caution).

  "A sadder but wiser Apollo team has emerged from the disaster at pad 34 with the realization that 'pretty good' simply is not good enough. . . ." (Hines, W *Star*, 7/27/67, A1)

- DOD officials were concerned about serious mechanical problems that had "undermined the readiness" of the Minuteman II ICBM, William Beecher reported in the *New York Times*. "As recently as this spring, reliable sources say, nearly 100 missiles—about 40 per cent of all the advanced Minuteman 2 missiles and 10 per cent of the entire 1,000-missile Minuteman force—were out of action because of trouble with the weapon's guidance and control system." Although DOD's "more or less official view" was that the problem was "worrisome but well in hand," many officials "while observing that mechanical problems will always group up in any complex electronics system and can be corrected, cite the Minuteman 2 trouble in questioning the wisdom of the near total reliance that Mr. McNamara places on intercontinental ballistic missiles in the strategic field." [See July 28.] (Beecher, *NYT*, 7/28/67, 1, 2)

- LaRC would negotiate with Ling-Temco-Vought, Inc. (LTV), for 24-mo, $10-million, firm-fixed-price contract to provide 15 Scout launch vehicles, NASA announced. Deliveries would begin about the 10th month at a rate of one vehicle a month. Scout was a multistage booster which could launch a 320-lb satellite into 300-mi orbit or a 50-lb satellite into an 11,000-mi orbit. (NASA Release 67-201)

- NASA and Centre National d'Études Spatiales (CNES) were conducting series of laser tests with five orbiting satellites—three U.S. and two French—to make precise geodetic measurements, NASA reported. In experiments,

July 27

a laser beam was fired at a satellite equipped with a reflector designed to return beam to the transmitting station. Time elapsed from transmission to reception was measure of the distance between station and satellite. If satellite's position was known by independent means or if several stations fired lasers simultaneously, it was then possible to determine distance between observing ground stations. NASA said initial tests, which began in April from stations in U.S., Europe, and North Africa, had been so successful that it might soon be possible to prove or disprove the theory of continental drift by determining the movements of continents relative to one another. (NASA Release 67–197)

- A fir tree had been tested in the Royal Aircraft Establishment's wind tunnel at Farnborough, U.K., to study the effects of high storm winds on forests. (Can Press, *NYT*, 7/27/67, 3)

July 28: NASA's *Ogo IV* (OGO–D) Orbiting Geophysical Observatory was successfully launched by a Thrust-Augmented Thor-Agena booster from WTR into polar orbit with 564-mi (908-km) apogee, 256-mi (412-km) perigee, 97.9-min period, and 86° inclination. Primary mission objective was "to acquire data for studies of latitude-dependent atmospheric phenomena, from an attitude stabilized platform, during the period of perigee rotation from the Northern Hemisphere across the Arctic pole into the Southern Hemisphere." Secondary objective was to conduct "detailed correlative investigations in atmospheric physics, energetic particle physics and polar region physics with known spacecraft attitude for the greater part of one diurnal cycle." Carrying 20 experiments, 1,240-lb satellite was fourth of six spacecraft in NASA's Orbiting Geophysical Observatory (OGO) program. *Ogo I* was launched Sept. 4, 1964; *Ogo II*, Oct. 14, 1965; and *Ogo III*, June 7, 1966. To date these missions had provided over 450,000 experiment hours of data on earth's environment and had resulted in over 100 scientific papers and published reports. OGO program was managed by GSFC under OSSA direction. (NASA Release 67–187; NASA Proj Off; AP, *NYT*, 7/29/67, 7)

- NASA successfully conducted at WSMR second of five balloon-launched parachute tests to determine effectiveness of parachutes as decelerators in soft-landing unmanned instrumented capsules on Mars. On ground command, 815-ft-long balloon released disc-shaped flight unit containing parachute, eight acceleration rockets, and 500-lb payload at 129,000-ft altitude, where earth atmosphere is as thin as that of Mars. Rockets then propelled unit to 140,000-ft altitude; parachute deployed and descended with payload into planned recovery area. Test was part of LaRC's Planetary Entry Parachute Program to investigate parachute landing systems for Voyager program. (NASA Release 67–170; *Roswell Daily Record*, 7/28/67; AP, *NYT*, 7/29/67, 7)

- ComSatCorp informed FCC that it was prepared to finance, build, and operate a pilot program for domestic satellite services. Explaining program ComSatCorp proposed March 31, ComSatCorp Chairman James McCormack said pilot system would include two synchronous satellites serving 34 domestic ground stations. Estimated total cost was $104.7 million: $35.7 million for two satellites, $22 million for 34 earth stations, and $47 million for operating costs during two years. Charges for each type of service could not yet be established accurately, but ComSatCorp would provide, without fee, the capacity equivalent to two television channels for noncommercial TV service. McCormack concluded: "We believe that the pilot program offers the American people

July 28: NASA's *Ogo IV* folded for vacuum chamber tests before launch into polar orbit to study the sun's influence on the earth's environment during high solar activity. In orbit, with booms and solar panels extended, *Ogo IV* is 49 ft long and 20 ft wide.

a beginning to their realization of the benefits of satellite communications, a beginning to their dividend from space technology in which the Federal Government and American industry have played such an important role. . . ." (ComSatCorp Release 67-38)

- Secretary of the Air Force Dr. Harold Brown released a statement on the Minuteman II IBCM [see July 27]: "The effectiveness of a missile depends not only on its alert rate but on the reliability and accuracy with which it will fly to its destination and place its payload on a predetermined target. The overall performance of Minuteman II when fired from alert status has been outstanding by any standard. . . .

"Therefore, while the Minuteman II guidance and control system is proving to require more maintenance than predicted earlier, the alert Minuteman II missile force is indeed reliable. . . ." (DOD Release 696–67)

- MSC Director of Flight Operations Christopher C. Kraft, Jr., said that new technological developments might enable NASA to make up some of the time lost when three astronauts died in Apollo capsule. In commencement address at St. Louis Univ.'s Parks College of Aeronautical Technology, Cahokia, Mo., he said, "If things go very right for us, we may still put a man on the moon by 1970." He predicted space exploration would contribute solutions to problems posed by world's growing population, specifically in communications, transportation, food production, and weather forecasting. (*St. Louis P–D*, 7/29/67)

- Increased appropriations for riot-torn cities should take precedence over NASA's request for funds to finance unmanned exploration of the planets, *Washington Daily News* suggested in an editorial. "James E. Webb . . . is concerned over cuts in the agency's $5 billion budget including reduction in money for unmanned exploration of Venus and Mars. Russia, he says, is about to send up the biggest . . . rocket you ever saw. In the present state of domestic affairs, we'd say let 'em!

"From all appearances our scientists already have more and more exact information about conditions on Venus and Mars than our police and war-on-poverty experts have on Los Angeles and Cleveland. . . ." (*W News*, 7/28/67, 16)

July 29: Seventh attempt to launch ELDO's Europa I rocket from Woomera, Australia, failed when mechanical fault developed in release system of the French Coralie 2nd stage. (Reuters, *C Trib*, 7/31/67)

- President Johnson had approved USN recommendation that Navy Navigation Satellite System be released for use by U.S. civilian ships and for commercial manufacture of shipboard receivers, Vice President Hubert H. Humphrey told a group of marine scientists at Bowdoin College, Brunswick, Me. "Our all-weather satellite system has been in use since 1964 by the Navy and has enabled fleet units to pinpoint their positions anywhere on the earth. The same degree of navigational accuracy will now be available to our non-military ships." Humphrey noted that close allies would probably request permission to purchase U.S. receivers and said "policy and procedures for responding to these requests are currently under consideration." (Text; Wilford, *NYT*, 7/30/67, 1, 28)

- Supersonic aircraft fares should be 25% higher than subsonic fares, ATA president Stuart G. Tipton, touring facilities where Anglo-French Concorde supersonic aircraft was being built, told a Paris press conference. He predicted that $18-million price of the Concorde would make it necessary for airlines to impose a premium fare at first. Tipton believed French prototype would be flown on schedule in February 1968. (*NYT*, 7/29/67, 38)

July 30: NASA's *Nimbus II* meteorological satellite had made the first well-substantiated observation from space of a volcanic eruption when it recorded the eruption of Iceland's Surtsey volcano in 1966, U.S. Geological Survey reported. "During August 1966, the *Nimbus II* satellite was in a near-polar orbit and passed over the island the day after the eruption began with its infrared sensing devices working.

"The Surtsey eruption in its entirety was recorded as a minute black spot on infrared images during more than eight orbital sweeps of the *Nimbus II* satellite—first on August 20 and definitely identified as late as October 3." *Nimbus II* was participating in a three-dimensional study of Surtsey by scientists from U.S. Geological Survey and AFCRL using infrared survey techniques. Experiment marked first time volcanic heat had been monitored almost simultaneously from ground, aircraft, and space. Scientists believed the ability to measure heat radiation could be applied to studies of volcanoes on the earth, moon, and planets. (USGS Release, 7/30/67; Clark, *NYT*, 7/30/67, 41)

July 31: U.S.S.R. launched *Cosmos CLXX* into orbit with 208-km (129-mi) apogee, 145-km (90-mi) perigee, and 50° inclination. All equipment functioned normally, but satellite reentered later in the day. (*SBD*, 8/10/67, 202)

- ComSatCorp reported net income of $2,088,146 for first six months of 1967: a net operating loss of $755,603 was offset by a net interest gain of $2,843,749. ComSatCorp entered full commercial operations May 1. (ComSatCorp Release 67–39)

- Latest Harris poll indicated American people, by 54%-to-34% margin, did not believe U.S. space program was worth an annual $4-billion expenditure. By 60%-to-30%, public would oppose continuing program at present rate of expenditure if U.S.S.R. were not active in space. People were divided sharply by income and age: (1) low-income persons opposed Apollo manned lunar landing program by almost three-to-one, while upper income favored it by nearly two-to-one; (2) 62% of persons over 50 opposed program, while 58% under 35 favored it. A 1965 poll had showed public supporting space program by 45%-to-42% margin. (*W Post*, 7/31/67, A2)

- The aerospace industry had a special responsibility to try to respond to the critical challenges posed by the Vietnam war and the urban riots, Robert Hotz suggested in *Aviation Week*. "Perhaps the aerospace industry has been so much absorbed in its traditional task of providing defense against external enemies that it has never given much thought to the contributions it can make to strengthen this nation internally. We submit that the events of this summer should stimulate the industry to take a long hard look in re-evaluating its role on the American scene and to make a bolder and more imaginative bid to offer its services to meet the new challenges of our times...." (Hotz, *Av Wk*, 7/31/67, 11)

- House/Senate Conference Committee agreed on $4.86-billion NASA authorization bill—$234.2 million less than NASA had requested for FY 1968. It included $42 million for Voyager program, $347.7 million for AA program, $73 million for nuclear rocket program, and $157.7 million for launch vehicle procurement. Bill, which required Senate and House approval, was passed by the Senate Aug. 2. (NASA LAR VI/83; Committee Release; Clark, *NYT*, 8/1/67, 13)

- NASA investigative board concluded that immediate cause of the M2–F2 lifting body vehicle crash at Edwards AFB May 10 was "an unusually low landing flare maneuver and premature ground contact" resulting from an unusual set of circumstances "that individually would not have ended in an accident." These circumstances were: (1) "Pilot was overburdened in his normally exacting task by a combination of events that disoriented and distracted him and denied him normal height information; (2) The large amplitude roll oscillation during final approach . . . caused a temporary loss of lateral control . . . and changed the landing heading"; (3) presence of rescue helicopter near path of landing heading represented a collision potential; (4) lack of visual height cues in landing area; and (5) unavoidable absence of radioed attitude callouts from chase aircraft. Board recommended: (1) easing pilot workloads in landing lifting-body vehicles; (2) maintaining an enlarged cleared area where inadvertent landings might occur; and (3) improving the information flow on flight planning, briefing, and monitoring procedures. (FRC Release 20–67; NASA Release 67–205)

- NASA would negotiate with Bendix Corp. for one-year, $25-million contract extension to provide KSC launch support services. (NASA Release 67–204)
- A Polaris A–3 missile fired from nuclear submarine U.S.S. *Will Rogers*, submerged off the coast of Cape Kennedy, traveled 1,500 mi down Eastern Test Range (ETR) and impacted close to target area near Antigua Island. Scheduled to begin her first patrol in September, U.S.S. *Will Rogers* was 41st and last ballistic weapon submarine in U.S. fleet. (Baldwin, *NYT*, 8/1/67, 13)
- William C. Schneider, formerly Mission Director for NASA's Apollo Applications (AA) program, was named Apollo Mission Director and Apollo Program Deputy Director for Missions, OMSF. Schneider would be responsible for management, direction, and coordination of mission and flight plans, schedules, and associated activities. (NASA Release 67–206)

During July: Editorial comment on proposed NASA FY 1968 authorization bills [see June 28]:

Science: "The political atmosphere surrounding the U.S. space program is today murkier and less hospitable than at any time since 1961 when President Kennedy decided to send men to the moon. This does not necessarily mean that NASA is in any danger of falling off its $5-billion-a-year budgetary plateau . . . [but] it is clear that, where Congress is concerned, the agency has lost much of its innocence and therefore its plans are to be reviewed with caution and skepticism.

"Some members of Congress are remarkably frank in indicating that their principal interest in the space program lies in the economic benefits it brings their districts. . . . Indeed, one might be justified in predicting that the major budgetary struggles of the future will be less concerned with how much money NASA gets than with how NASA spends the money it does get." (Carter, *Science*, 7/14/67, 170–3)

Washington *Evening Star:* "If continuity and momentum are to be maintained in the space program . . . the [congressional] conferees must act with special wisdom this year.

"Two . . . items in the House bill should be embodied in one final version. . . . One would establish an independent safety panel in NASA, and the other would require . . . [NASA] to keep Congress 'fully and currently informed' of problem areas in the space effort.

"Both proposals were strongly opposed by administration spokesmen on grounds that seem shallow and legalistic. It would be well to embody both in the final 1968 authorization bill, if only to put NASA on firm notice that it is time to shape up." (W *Star*, 7/7/67, A16)

New York Times: "To read the record of the Congressional debate on NASA's budget is a melancholy experience. . . . Behind the rhetoric it was easy to see the pressure of the many corporations and communities now sharing in NASA's largeness and fearful of the impact of any major cut in this flow of golden manna. . . .

"A rational 1968 space budget would have been substantially lower than either the House or Senate version, and would have cut particularly severely at the Apollo project. . . . It would have changed the present emphasis . . . to lay greater stress on the exploration of the cosmos with relatively cheap, unmanned rockets. Instead, the enormously expensive manned lunar landing effort is left virtually intact, while the highly promising program for sending unmanned Voyagers to Mars and other planets could be entirely eliminated.

"The lobbyists can rejoice, but the nation's real interests have been dealt a severe setback." (*NYT*, 7/3/67, 10)

- AIAA President Harold T. Luskin, in *Astronautics and Aeronautics* editorial, proposed a joint Soviet-American safety program to reduce hazards of manned space flight. He urged that representatives of U.S. and U.S.S.R. meet to exchange technical data on spacecraft safety features, to set up a common communications network for use in emergencies, and to plan joint rescue operations. "Cooperation for safety could be a step forward in promoting the greatest possible success of manned space exploration. The world would regard it as a meritorious indication that nations can find ways to compete and cooperate at one and the same time. . . .

 "Let both countries work toward the day when a flight crew is saved through cooperative action. That event would be the finest space spectacular, the best space first!" (*A&A*, 7/67, 33–4)

- AFCRL Space Physics Laboratory scientists Drs. Graham R. Hunt and John W. Salisbury said, "Major 'hot spot' on the moon is not volcanic," a conclusion based on measurements made by AFCRL imaging device to produce thermal photographs of infrared emissions of lunar surface during dark of moon. Images had been obtained by using infrared imager in association with AFCRL Strawberry Hill Observatory's optical telescope at Concord, Mass. The hot spot, associated with crater Tycho, had heat distribution pattern very similar to solar illumination pattern just before lunar sunset. Because of clear relationship between heat release and solar illumination, Salisbury and Hunt had concluded that crater was emitting stored solar heat rather than volcanic heat. (*OAR Res Rev*, 7/67, 12–3)

- International Council of Scientific Unions' Inter-Union Commission of Solar-Terrestrial Physics (IUCSTP) had planned coordination of observations and data exchange during and beyond period of solar maximum activity expected to arrive before end of 1968. Members had been preparing list of projects in six major areas: solar activity; particles and waves in interplanetary space; particles, fields, and waves in magnetosphere; external geomagnetic field; ionosphere; and aeronomy. Much if not most of this work would be carried out from ground-based observatories and geophysical stations, but spaceborne instruments would be essential to supply key information not obtainable from ground. IUCSTP recommended that "ground-based, rocket and balloon programs should . . . take advantage of the fact that certain space vehicles will be making related measurements at particular times." (NAS–NRC–NAE *News Report*, 8/9/67, 6–7)

- *Federal Support for Academic Science and Other Educational Activities in Universities and Colleges*, NSF's report to OST for fiscal years 1963–66, showed these Federal obligations were $3.0 billion in 1966, more than double 1963's $1.4 billion. Although R&D support increased from $813.2 million to $1,257.7 million, its percentage share of the total dropped from 58 to 42. Federal agencies expected a drop in annual rate of growth in R&D support from 16% in 1963–66 to 8% in 1966–68. Support for nonscience activities, however, had increased ninefold, from $85 million in 1963 to $847 million in 1966. (Text)

- Joint Committee on Atomic Energy reported that internal Chinese Communist strife had had little effect on nuclear weapon program, that Chinese Communists could possibly launch a "low order" missile attack on U.S. by early 1970s, and that their progress in developing thermonuclear weapons had been "excellent." Joint Committee report, *Impact of Chinese Communist Nuclear Weapons Progress on United States National Security*, was based on testimony by officials from State Dept., DOD, AEC, CIA, and nuclear weapon laboratories. (Text)
- Low-orbit space operations from a manned orbital base "has the potential to reduce operating costs by a factor of two" for the mission traffic anticipated for post-1970, G. A. Sears of the RAND Corp. had concluded. His paper, *Orbital Basing: Key to Low Cost Space Operations?*, compared present earth-based manned spaceflight program with concept of basing spacecraft at manned space station. Spacecraft would be injected into mission orbits as required and later rejuvenated for reuse by rendezvous with station. Spacecraft would have to operate nearly coplanar with base, a constraint acceptable to many but not all low-orbit missions; however, as method of reducing costs, orbital basing "deserves attention on a par with recovery-to-earth and reuse and extended-life concepts," Sears stated. (Text)

August 1967

August 1: NASA's *Lunar Orbiter V* (Lunar Orbiter E) unmanned spacecraft was successfully launched by Atlas-Agena D booster from ETR on mission to photograph the lunar surface [see Aug. 5–27].

Agena 2nd stage fired to boost 860-lb spacecraft into parking orbit of 100-mi (161-km) altitude, reignited after 30-min coast period to inject spacecraft on 89-hr translunar trajectory, and separated. On schedule, *Lunar Orbiter V* deployed its four solar panels and two antennas and locked its five sensors on the sun. On Aug. 2 the star tracker encountered slight difficulty in locating Canopus because of "glint" from the sun and "earthshine," but obtained a fix at 18:34 GET, before midcourse maneuver. At 06:00 on Aug. 3 midcourse maneuver with 26-sec engine burn was successfully conducted to slow spacecraft's speed and alter its aiming point slightly.

Primary objectives of NASA's *Lunar Orbiter V* mission were: (1) to place three-axis-stabilized spacecraft into high-inclination lunar orbit; and (2) to obtain photography of selected scientifically interesting areas on lunar surface and supplemental photography of candidate Apollo lunar landing sites. Photos would cover five Apollo sites, 36 scientific sites, and areas of the moon's hidden side previously unphotographed. Spacecraft would also provide precision trajectory information to refine definition of lunar gravitational field, monitor micrometeoroids and radiation intensity in lunar environment, and serve as a target for tracking operations by Manned Space Flight Network stations.

Last in series of five spacecraft designed to improve knowledge of the moon, *Lunar Orbiter V* differed from four earlier missions because it would concentrate on targets of primary interest to science. *Lunar Orbiter I* (launched Aug. 10, 1966), *Lunar Orbiter II* (launched Nov. 6, 1966), and *Lunar Orbiter III* (launched Feb. 4, 1967) were in direct support of Apollo and Surveyor programs, identifying at least eight areas suitable for manned landings. *Lunar Orbiter IV* (launched May 4, 1967) completed broad photographic survey of 99% of moon's front face with 10 times finer resolution than best existing telescopic views. Lunar Orbiter program was managed by LaRC under OSSA direction. Tracking and communications were the responsibility of JPL-operated Deep Space Network. (NASA Proj Off; NASA Release 67–192)

- NASA announced that *Mariner V* Venus probe—8,225,000 mi from earth, 86,830,000 mi from sun, on trajectory toward Venus—had detected two solar flares of medium intensity. Data transmitted to JPL from spacecraft via Goldstone, Calif., tracking station showed large increases in measurements of solar radiation at 1:30 pm and 8:47 pm EDT. Occurrence of flares was confirmed by visual observations at Solar Geomagnetic Monitoring Service, Ft. Belvoir, Va. (NASA Release 67–209)

- House Committee on Science and Astronautics' Science, Research, and Development Subcommittee released report, *Science, Technology, and Public Policy during the Eighty-Ninth Congress, January 1965 through December 1966.* Rep. Emilio Q. Daddario (D-Conn.), subcommittee chairman, emphasized: "Congress does indeed recognize the impact of science and technology on modern society and is moving expeditiously to meet its challenge and shape its potential on behalf of human welfare [and] . . . is becoming aware of the difficulties and dangers which technologically applied science may carry in its genes—and is searching for effective means to counter them."

    NSF Director Dr. Leland J. Haworth, in statement before Subcommittee, said: "In all honesty, it must be admitted that thus far there have been no broad, systematic, and intensive studies designed to develop an understanding of alternative policy structures and options open to the Nation as a whole and to assess the associated opportunities and risks, as well as the impact, which such policies and choices might have on the individual agencies, disciplines and the Nation's goals." Haworth referred to the major lack in the studies concerning long-range national needs. (Text)

- KSC held dedication ceremonies for its $1.1-million Visitor Information Center. One hour later a visitor purchased the 500,000th ticket for the center's guided tour. (LaMont, *Cocoa Trib*, 8/1/67)

- B/G Harold C. Teubner (USAF), formerly with Hq. USAF comptroller's office, became AFSC Deputy Chief of Staff for Comptroller, succeeding M/G Wendell E. Carter, who was being reassigned to the Pentagon as Deputy Assistant Secretary of Defense (Comptroller/Information). (AFSC Release 152.67)

- Miami's Opa Locka Airport was the Nation's busiest airport in FY 1967, with a record 596,949 takeoff and landing operations, FAA reported. Chicago's O'Hare International was second with 588,527 operations. (FAA Release 67–61)

- Scientists at Royal Aircraft Establishment, near Farnborough, U.K., were firing bullets at raindrops because they believed damage inflicted on a bullet impacting with a raindrop was same as damage that would be inflicted on an aircraft flying at high speed through a squall. Scientists feared raindrops, relatively minor problem to subsonic aircraft, might cause serious fuselage and wing erosion to the Concorde SST. (Reuters, *NYT*, 8/1/67, 29)

- Cost estimates for developing Anglo-French Concorde supersonic transport had risen to $1.47 billion, House of Commons Committee of Public Accounts revealed in a report. Additional $75.4 million would probably be incurred by Ministry of Technology in R&D costs directly connected with the aircraft. Estimates for development costs were $476 million in November 1962 when Concorde project began. (*NYT*, 8/3/67, 44)

August 2: Senate passed, by routine voice vote, and sent to the House the $4.86-billion NASA FY 1968 authorization bill (S. 1296). House/Senate Conference Committee had agreed on a $4.86 billion bill July 31. (*CR*, 8/2/67, S10578–80)

- Five of the 11 scientist-astronauts selected for NASA's Apollo program were announced in the Nation's press. NASA released complete and official list Aug. 4. (AP, *NYT*, 8/3/67, 8; UPI, *W Post*, 8/3/67, A1)

- NASA Arcas sounding rocket launched from Point Mugu, Calif., carried Naval Ordnance Test Station experiment to 33-mi (52-km) altitude to flight-test internally modified version of standard ROCOZ payload, designed to measure ozone distribution for support data on *Ogo IV* satellite. No experimental results were obtained because of loss of telemetry. There was evidence of payload malfunction near time of experiment ejection. (NASA Rpt SRL)

- MSC selected NAA's Rocketdyne Div. to negotiate $5-million, cost-plus-fixed-fee contract for design, development, and qualification of a backup injector for Apollo Lunar Module (LM) ascent engine. Contract would provide initially for design feasibility and development testing with provision for delivery of four production injectors if required. Two of the injectors would be used in ground testing at WSMR and remaining two would be held for possible use in flight engines. NASA said present injector, which was being built by Bell Aerospace Corp., was causing unstable combustion in engine chamber and excessive erosion of ablative thrust chamber. Bell was attempting to correct the deficiencies, but "it was decided to develop an injector of a different design in the event the problems cannot be corrected." (NASA Release 67-207)

- The number of scientists, engineers, and physicians immigrating to the U.S. had increased 77% between 1956 and 1966, according to report released by House Committee on Government Operations' Research and Technical Programs Subcommittee. Report said this influx from poor nations was an "involuntary gift of valuable resources.

  "Such a loss by the developing countries is of direct concern to the United States. To the extent that it undermines development, it also defeats a major United States foreign policy objective for the sake of which this country is currently spending about $3.7 billion per year...." Reduction of this influx would probably have to be the joint responsibility of the U.S. and the nations that were losing their skilled professionals, report said. (Text; *Science,* 8/18/67, 783; *NYT,* 8/3/67, 13)

- Office of Science and Technology released an interim report on USAF test flights at Edwards AFB to study the effects of the sonic boom June 3–23, 1966, and Oct. 31, 1966–Jan. 17, 1967, 11 types of aircraft made 367 supersonic and 261 subsonic flights. Report concluded that flights caused little or no physical harm to test structures, 393 test subjects, or 220,570 test animals, and noted relative annoyance of booms of differing intensity and of booms compared to jet aircraft engine noise. Between 33% and 98% of the test subjects objected to booms in the 2.0- to 3.5-lb-psf overpressure range that SSTs were expected to produce under the worst possible conditions. Other preliminary conclusions: (1) sonic booms sounded louder outside than inside; (2) walls of houses were more effective in blocking out jet engine noise than sonic booms; and (3) annoyance increased faster with increasingly bigger booms than with increasingly louder engine noise. (Text; Clark, *NYT,* 8/3/67, 1, 24)

- A farmhouse in the village of Mauron, France, collapsed seconds after a loud bang was heard, killing three workmen and seriously injuring a fourth. French military and civil officials immediately began an investigation to determine whether a sonic boom had been responsible for the accident. The *New York Times* later commented: "The latest discoveries about sonic boom . . . provide scant comfort for enthusiasts of supersonic commercial air transports.

  "Further research will undoubtedly turn up additional useful information, but by now the results of several years' investigation of this phenomenon are beginning to fall into a pattern. In a world where everyone was healthy and vigorous, where all buildings were relatively new and well constructed, sonic booms would be just one more source of annoyance and frayed nerves among the many such 'blessings' modern civilization and technology have produced. But in the real world—where any large community has many sick and infirm people and many old and poorly constructed buildings—sonic booms, especially if repeated frequently, pose appreciable hazards to the more fragile human beings and structures. . . ." (Hess, *NYT*, 8/3/67, 24; *NYT*, 8/3/67, 30)

August 3: NASA Aerobee 150 sounding rocket launched from WSMR reached 93-mi (130-km) altitude in flight to measure the radiation from celestial objects in the spectral region, 2,800–1,200 Å, 40–50 Å, and 2–8 Å. Experiments performed satisfactorily, but parachute did not deploy and payload crashed because of faulty wiring. Recovery was not required to obtain scientific data. (NASA Rpt SRL)

- Astronaut-Aquanaut M. Scott Carpenter (Cdr., USN) was detached from NASA at USN's request for assignment in Navy's Deep Submergence Systems Project (DSSP). One of the Nation's seven original astronauts, Carpenter became the second American to orbit the earth when he piloted *Aurora 7* spacecraft on three-orbit mission May 24, 1962. Active in USN's Man-in-the-Sea program since 1965, he set a world record in underwater work during Aug. 28–Oct. 10, 1965, Sealab II experiment when he lived 205 ft down in the Pacific for 30 consecutive days. Under DSSP he was tentatively assigned as an aquanaut in Sealab III, a 60-day experiment 600 ft down in the Pacific scheduled to take place in 1968. As USN's senior aquanaut, Carpenter's duties would include responsibility for Sealab III ocean floor operations, experiments, and equipment and for coordination of aquanaut team training.

  NASA Administrator James E. Webb commented on Carpenter's departure. ". . . we are grateful to him and the United States Navy for his services over the past eight years. During the past two years, he has demonstrated that many of the technologies, techniques, and psychophysiological factors related to space flight have direct applications to the Man-in-the-Sea program. Thus, Cdr. Carpenter becomes the first astronaut to return to his parent service in order to apply the skills, knowledge, and experience acquired during his assignment to NASA. . . ." (NASA Release 67–208)

- MSFC had successfully completed 11-mo dynamic test program of Saturn V booster and Apollo spacecraft, which qualified the Saturn V configuration as dynamically and structurally sound. Conducted by Boeing Co., program included dynamic tests of Saturn V's 1st, 2nd, and 3rd stages, Instrument Unit (IU), and Apollo spacecraft. (MSFC Release 67–161)

- Dr. Smith J. De France, Director Emeritus of ARC, was presented the National Aeronautic Assn.'s (NAA) Elder Statesmen of Aviation Award at ARC. Award honored outstanding Americans for contributions to aeronautics. Dr. De France was cited "in recognition of his significant and enduring contributions over the years to the progress of aeronautics, and his demonstrated qualities of patriotism, integrity, and moral courage worthy of emulation." (ARC *Astrogram*, 8/17/67, 2)

- ComSatCorp had issued three separate RFPs for construction of buildings and facilities at three new earth stations, at Cayey, Puerto Rico; Green Valley, W. Va.; and Jamesburg, Calif. Construction was scheduled to begin in 1967 and end in late 1968. Also issued was an RFP for multiplex subsystems at the three new stations and at existing stations in Andover,

Me., and Paumalu, Hawaii. All RFPs were filed with FCC. (ComSatCorp Release 67-40)

August 4: ELDO's Europa I rocket, launched from Woomera Rocket Range after 10 postponements, crash-landed when French Coralie 2nd stage failed to ignite. Malfunction was traced to equipment associated with the release system designed to separate the Coralie from U.K.'s Blue Streak 1st stage. Rocket, which should have flown 4,828 km (3,000 mi) into the Pacific, landed 965 km (600 mi) north of Woomera in the Simpson desert. (Reuters, *NYT*, 8/5/67, 7)

- NASA announced selection of 11 civilian scientist-astronauts: Dr. Joseph P. Allen, 30, a physicist research associate at the Univ. of Washington; Dr. Philip K. Chapman, 32, a naturalized citizen born in Australia and staff physicist at MIT's Experimental Astronomy Laboratory; Dr. Anthony W. England, 25, a graduate fellow in geophysics at MIT and the youngest man ever to be named an astronaut; Dr. Karl G. Henize, 40, an astronomy professor at Northwestern Univ. and experimenter in the Gemini program; Dr. Donald L. Holmquest, 28, who would report for duty in one year after completing his medical internship at Baylor College of Medicine; Dr. William B. Lenoir, 28, an assistant professor of electrical engineering at MIT; Dr. John A. Llewellyn, 34, a naturalized citizen born in Wales and associate professor in chemistry at Florida State Univ.; Dr. Franklin S. Musgrave, 31, a post-doctoral fellow at Univ. of Kentucky with a doctorate in medicine from Columbia Univ., a Ph. D. in physiology from Univ. of Kentucky, and four other college degrees; Dr. Brian T. O'Leary, 27, who held a Ph. D. in astronomy and was in the NASA trainee program at Univ. of California Dept. of Astronomy's Space Sciences Laboratory; Dr. Robert A. Parker, 30, assistant professor of astronomy at Univ. of Wisconsin; and Dr. William E. Thornton, 38, who recently completed two-year tour of duty with Brooks AFB Aerospace Medical Div.

  The new astronauts, except Dr. Holmquest, would report for duty at MCS Sept. 18. After two weeks orientation, they would begin "ground school" training, which would include orbital mechanics, astronomy, computers, spacecraft orientation, general mathematics and physics refresher courses, and field trips for contractor facility orientation. In March they would start Air Force flight training to become qualified jet pilots. The new group, which increased number of NASA astronauts to 56, was sixth class to be selected and second to be chosen specifically for scientific education. They were selected from a group of 69 nominees submitted to NASA in March by NAS after evaluating 923 applications. (NASA Release 67-211; MSC *Roundup*, 8/18/67, 1-3)

- Conclusion in Mar. 24 *Science* article by Bruce C. Murray of Cal Tech and his associates about probable contamination of Mars and Venus by U.S.S.R.'s *Zond II* and *Venus III* received comment from Dr. Richard W. Porter, member of International Relations Committee of NRC's Space Science Board, and from Britain's leading radio astronomer, Sir Bernard Lovell. Murray and associates had presented information: tracking data from Jodrell Bank Experimental Station had indicated that *Zond II*, launched Nov. 30, 1964, had been on collision course with Mars, terminating Aug. 6, 1965; Mstislav Keldysh, President of Soviet Academy of Sciences, had said *Zond II* would pass within 1,448 km (900 mi) of Mars; *Venus III*, launched Nov. 12, 1965, may have crashlanded on Venus' surface Mar. 1, 1966; radio communications had been maintained with *Venus III* throughout the flight but were lost as the probe approached Venus; no telemetric data were received in final moments before impact.

In a letter to editor of *Science*, Porter wrote: "Their conclusion with respect to Venus is based largely on what was *not* said by the Soviets about sterilization of the flyby bus and on the authors' assumption that the capsule separation and bus deflection maneuvers were not made automatically, even after loss of radio contact." Porter said authors' conclusion about Mars was also based on insufficient evidence and noted that in discussions with "highly placed" Soviet scientists during spring 1965, he had received impression that *Zond II*'s miss distance was likely to be much greater than 900 mi.

In his letter to editor of *Science* regarding Murray and associates' statements on probable contamination of Mars and Venus, Sir Bernard Lovell noted that *Zond II*'s signals were so strong that there should have been no difficulty in tracking the probe at Mars encounter; also, miss distance of 1,500 km (932 mi) was within the accuracy of Jodrell Bank calculations. Sir Bernard Lovell would "prefer to accept the statement of the President of the [Soviet] Academy [of Sciences (Keldysh)] about the miss distance. . . ." (*Science*, 3/24/67, 1505-11; 8/4/67, 487-8)

- After analysis of *Surveyor IV* failure and *Surveyor I* and *III* successes, NASA *Surveyor IV* Technical Review Board recommended that significant changes in spacecraft hardware be avoided. Board was unable to identify cause for failure of *Surveyor IV* mission, but indicated that spacecraft's performance had been virtually flawless from launch until communications signal disappeared abruptly less than three minutes before touchdown. (NASA Proj Off)
- President Johnson appointed Paul R. Ignatius, Assistant Secretary of Defense (Installations and Logistics), as Secretary of the Navy, succeeding Paul H. Nitze, who became Deputy Secretary of Defense July 1. John T. McNaughton, who had been nominated and confirmed as Nitze's successor in June, was killed in an aircraft crash July 19. The President approved reassignment of Thomas D. Morris, Assistant Secretary of Defense (Manpower), to replace Ignatius. (*PD*, 8/7/67, 1100)

August 5: NASA Nike-Apache sounding rocket launched from WSMR carried Dudley Observatory instrumented payload to 93-mi (150-km) altitude to test modified parachute recovery system and evaluate performance of Dudley Observatory micrometeoroid collection payload before flight during meteor shower. Malfunction of recovery system resulted in hard impact of scientific payload and loss of collection material. Flights using modified recovery system were discontinued until cause of malfunction was determined. (NASA Rpt SRL)

- NASA's selection of two foreign-born astronauts [see Aug. 4] was praised by Erwin D. Canham in the *Christian Science Monitor:* "The Australian, Dr. Philip K. Chapman . . . and the Welshman, Dr. John A. Llewellyn, . . . are both naturalized Americans. They are, one may assume, part of the brain drain which has brought so many able persons to the United States in the last quarter of a century. Nothing like this flow of talent has been seen on such a scale in history before.

"The outreach of science also has become profoundly international. It is a deeper bond than we have yet recognized it to be. Someday, perhaps, it will transcend politics. That is the challenge: to bring political institutions into line with the basic unities which are evidenced in men's constantly changing relationship to the physical universe. . . ." (Canham, *CSM*, 8/5/67)

August 5–27: NASA's *Lunar Orbiter V* became fifth U.S. spacecraft to circle the moon. It entered lunar orbit following successful deboost maneuver which reduced its speed by 1,440 mph and permitted lunar capture. Initial orbital parameters: apolune, 3,734 mi (6,023 km); perilune, 121 mi (194 km); period, 8 hrs 30 min; and inclination, 85°. Launched from ETR Aug. 1, spacecraft performed 513 attitude changes, responded to 4,524 commands, and recorded one micrometeoroid hit. Spacecraft systems were functioning normally. Photographs taken beginning Aug. 6 included 23 previously unphotographed areas of the moon's far side, first picture showing the "full earth"; 36 sites of primary interest to science; and five additional candidate Apollo landing sites. Transmitted were 212 medium- and 212 high-resolution photos. Final readout was completed Aug. 27. Mission substantially filled in blanks in previous LO photography so that entire lunar surface, front and back, had now been photographed at resolutions about 10 times better than obtainable from earth-based observations.

Harold Masursky, U.S. Geological Survey, told news conference at JPL that several of the pictures indicated there might be frozen liquid on the floor of perpetually shaded craters near the lunar poles. The nature of the "fluidal materials" was open to speculation, he said, but "meandering rills appeared to have been caused by some form of material flowing from a volcanic eruption—similar to terrestrial streams found in desert areas on earth." (NASA Proj Off; AP, *W Post*, 8/9/67, A3; UPI, *C Trib*, 8/15/67; *NYT*, 8/15/67, 1)

August 6: New York State Univ. had acquired two of 12 Atlas F ICBM sites near Plattsburgh, N.Y., for use as earth and space research laboratories. Sites had been turned over to GSA for disposal by USAF after deactivating and removing Atlas ICBM's. (*NYT*, 8/6/67, 2)

August 7: NASA issued three modifications totaling $30.9 million to NAA's Saturn V 2nd-stage contract: first, valued at $5.5 million, provided for equitable adjustment of 2nd-stage changes previously ordered; second, valued at $5.8 million, covered seven changes involving KSC ground support equipment; and third, valued at $19.4 million, was for 12 changes covering alterations of selector switches and insulation of the entire 2nd stage. (NASA Release 67–212)

- Sen. William Proxmire (D-Wis.), releasing testimony earlier submitted to Senate Appropriations Committee's Independent Offices Subcommittee, urged that $427 million be cut from NASA's FY 1968 authorization bill (S. 1296). "We have to establish our priorities, and when it comes to parceling out money the space program stands far below defense, education and many other vital programs." Specifically, Proxmire recommended: (1) temporarily delaying initiation of programs that would commit U.S. to a manned Mars mission and other deep-space probes; (2) cutting AA funds; and (3) cutting advanced missions funds. (Text)
- USAF launched unidentified satellite from WTR using Thor-Agena D booster; satellite reentered Sept. 1. (*Pres Rep 1967*)
- Congressional cuts in NASA's FY 1968 authorization bill (S. 1296) indicated that a new era was beginning for NASA and the U.S. space program, Robert Hotz suggested in *Aviation Week* editorial. "We do not think it was the intent of this Congress to either repudiate or emasculate the

U.S. national space program. We think it was the aim of these legislators to mark the end of the era when NASA got everything it wanted simply because it asked for it in a strident and persistent voice. We think it was also the goal of Congress to put the national space program in a better perspective in relation to such other high-priority problems as the Southeast Asia war and the urban riots. . . ." Hotz predicted that the future of the space program as a valid, continuing national goal "will depend in the long run on the successful development of its new hardware and the successful operation of this equipment on missions of increasing complexity and greater value. The most significant contribution that NASA and its industrial contractor complex can make to the future of the space program now is to bend their every effort to assure that the Apollo hardware will function flawlessly, not only for its missions to the moon but also for long-duration earth-orbital operations. . . ." (Hotz, *Av Wk*, 8/7/67, 17)

- U.S. and Soviet delegates attending COSPAR's International Years of the Quiet Sun (IQSY) symposium in London [July 17–21] had recommended that cooperative solar studies begun during IQSY be extended into the period of maximum solar activity, Herbert J. Coleman reported in *Aviation Week* [see Jul. 17–21]. NRL scientist Dr. Herbert Friedman, President of the Inter-Union Commission on Solar Terrestrial Physics (IUCSTP) which initiated the proposal, said the application of new technologies, including space techniques, "had opened up new fields of investigation such as the relationship between the solar wind and the magnetosphere. A period of increased solar activity could provide valuable data." Soviet physicist R. V. Pushkov pledged full Soviet cooperation in extending the program. (Coleman, *Av Wk*, 8/7/67, 67)
- ComSatCorp's plans to launch a satellite by 1970 for relaying VHF communications from transoceanic airliners might be delayed by high costs, *Electronics* reported. "As far as the FAA is concerned, the only issue is cost. The agency is still reeling from a Comsat proposal earlier this year pricing such communications services at $5 million per channel per year; the FAA told Comsat it was not interested at that price. To make matters worse, one FAA official says current estimates from hardware makers are even higher." (*Electronics*, 8/7/67)
- Second phase of NATO's proposed comsat project (NATO/Satcom) [see July 9] was being delayed by dispute over which countries would develop satellite systems, *Aviation Week* reported. Several member countries reportedly opposed use of U.S.-supplied, Philco-Ford Co. Skynet satellites because it would eliminate participation by European avionics industry and give U.S. dominant role. Satcom officials were considering alternate plans in the event that agreement on original Phase II proposal could not be reached: (1) delaying satellite development until Phase III to allow sufficient time to reach agreement on the work-sharing formula; or (2) dropping satellite development from Phase II and constructing a ground station network compatible with DOD's IDCSP satellites which were being used temporarily during initial test phase. (*Av Wk*, 8/7/67, 28)

August 8: NASA Aerobee 150 sounding rocket launched from WSMR achieved 180-mi (289-km) altitude in flight to measure: (1) vertical profiles of neutral and ion composition and electron densities; and (2) dissociation of molecular oxygen and atomic nitrogen density using mass spectrometers designed for the POGO satellite. Preliminary data indicated that the

rocket and all experiments had performed satisfactorily. (NASA Rpt SRL)
- NASA Nike-Apache sounding rocket launched from Wallops Station carried Univ. of Illinois-GCA Corp. experiment to 88-mi (141-km) altitude to determine electron density in lower ionosphere (during period of non-blanketing Sporadic-E), electron collision frequency, electron temperature, and positive ion densities of species in the 8- to 60-a.m.u. range. Rocket and instrumentation performed satisfactorily, and excellent results were anticipated. (NASA Rpt SRL)
- U.S.S.R. launched *Cosmos CLXXI* into orbit with 220-km (137-mi) apogee, 145-km (90-mi) perigee, and 50° inclination. All equipment was functioning normally. Spacecraft reentered the same day it was launched. (SSR; *Aero Tech*, 8/14/67, 13)
- House passed $4.86-billion NASA FY 1968 authorization bill (S. 1296), completing congressional action on the authorization. (*CR*, 8/8/67, H10054–60)
- NASA had modified TRW Systems' $34.7-million contract for Orbiting Geophysical Observatory (OGO) satellites to a $53.7-million fixed-price incentive contract. Revised agreement provided for assembly, test, and launch of OGO–D, OGO–E, and OGO–F, in addition to incorporating previously authorized work. (NASA Release 67–213)
- Report by House of Commons Estimates Committee urged U.K. to appoint a space minister and initiate a five-year national space program at a cost of $84–$98 million annually. Program should be directed toward the 1971 launch of a comsat which could be used for both commercial and military operations. Report criticized U.K.'s current space effort for lack of purpose and organization, noting that money "had been poured into international projects without a properly conceived national program." (*W Post*, 8/9/67, A16; *SBD*, 8/10/67, 205)
- Eastern Airlines named A. Scott Crossfield, NACA test pilot from 1953 to 1961, Systems Director of Research and Development for Flight to help prepare airline for its first Concorde and Boeing 2707 supersonic aircraft. Crossfield in 1953 became the first man to fly at twice the speed of sound when he piloted the rocket-propelled D–558–II aircraft. He was also the first pilot to fly the X–15 to mach 2.97 (Nov. 15, 1960). In 1961, he became division director of test and quality assurance for NAA's Hound Dog Missile, Apollo, Saturn, and Paraglider projects. For his contributions to aviation he was awarded the Harmon International Aviator's Trophy in 1961 and the Robert J. Collier Trophy in 1962. (*NYT*, 8/9/67, 61)

August 8–9: NASA successfully launched five Nike-Apache sounding rockets with chemical cloud experiments and one Arcas sounding rocket with a small live animal payload from NASA Wallops Station between 10.43 p.m. and 7:27 a.m. Five Nike-Apaches ejected vapor trails between 50- and 125-mi altitudes to measure wind velocities and directions at various altitudes in the upper atmosphere. Nike-Apache launched at dawn was sodium experiment which created reddish-orange trail; remaining four were trimethylaluminum (TMA) experiments which created bluish-green trails. Data were obtained by photographing continuously the motion of the trails from five ground-based camera sites within 100-mi radius of Wallops. Arcas sounding rocket carried payload with small live white rat to 25-mi altitude; payload descended by parachute and was recovered from the Atlantic by helicopter. Subsequent examination by Wallops scientists indicated rat was in excellent condition.

Nike-Apache launches were conducted for GCA Corp., under contract to GSFC. Arcas launch was part of Wallops Station's Bio-Space Technology Training Program to assist biological experimenters in evaluating the engineering and operational aspects of spaceflight research. (WS Release 67–27)

August 9: U.S.S.R. launched *Cosmos CLXXII* into orbit with 301-km (187-mi) apogee, 202-km (125-mi) perigee, 89-min period, and 52° inclination. Equipment functioned normally. Satellite reentered Aug. 17. (*Aero Tech,* 8/14/67, 13; GSFC *SSR,* 8/31/67)

- *Explorer XXXV*, launched July 19, had completed 33 orbits of the moon. Results from GSFC's magnetic field experiment aboard spacecraft indicated moon was not magnetized. Capture of interplanetary magnetic field lines by the moon and formation of lunar magnetosphere (theorized by T. Gold in 1966) was not substantiated. A lunar bow shock wave had not yet been observed when moon was located in the interplanetary medium or the magnetosheath of the earth. (NASA Proj Off)

- A balloon carrying 9,000-lb instrumented payload for NASA was successfully launched from Chico, Calif., by Univ. of California Space Sciences Laboratory. Approximately 10 hrs later, however, parachutes failed to deploy after ground stations had commanded payload to separate and payload fell into the Pacific. Equipped with flotation gear, it landed within 20 mi of the recovery barge, but had not yet been located. Believed the heaviest ever carried by a balloon, payload contained instruments to: (1) measure cosmic radiation, nuclear interactions of primary particles, and interference background of secondary particles; and (2) gain information on mass and momentum of incoming particles. Scientists were uncertain whether instruments had survived the impact, but said they had obtained excellent telemetry data during 9½-hr flight. (NASA Release 67–218)

- NASA had selected RCA Defense Electronics Products Div. for negotiation of a $5-million, cost-plus-fixed-fee contract for 16 pairs of flight-qualified dual spacesuit communications (SSC) systems to be used by Apollo astronauts on the moon. SSC systems—each consisting of AM and FM transmitters and receivers, warning system, and related equipment—would (1) permit two Apollo astronauts on lunar surface to maintain constant voice communications with each other and with flight controllers on earth and (2) simultaneously transmit telemetry data from each astronaut back to the lunar module (LM) for relay to earth. Contract would be managed by MSC. Delivery of first two flight units was scheduled for October 1968. (NASA Release 67–214)

- Rep. George P. Miller (D-Calif.), Chairman of the House Committee on Science and Astronautics, praised NASA's Lunar Orbiter program on the House floor: "In the short span of just under 1 year, all five Lunar Orbiters have been launched and have provided the world with more information about the moon than has been obtained in all previous history. Lunar Orbiter photography has certified four potential landing sites for our Apollo astronauts, and has provided detailed photographs for 99 percent of the front side of the moon and 60 percent of the back side.

  "Successful completion of the present mission [*Lunar Orbiter V*] will provide supplementary data for the certification of five more Apollo landing sites, detailed photography of virtually the entire moon's surface, front and back, and very close view of 36 areas of great scientific interest on the front face of the moon.

"I wish to compliment the NASA team that has made the Lunar Orbiter program one of the most successful in the 9-year history of that agency. . . ." (CR, 8/9/67, H10175)

- Informed sources said U.S.S.R. had launched *Cosmos CLXX* (July 31) and *Cosmos CLXXI* (Aug. 8) to test new parachute system designed to avoid the malfunction blamed for the death of Cosmonaut Vladimir Komarov. U.S.S.R. attributed the April 24 crash-landing of *Soyuz I* to a tangled parachute that failed to decrease spacecraft's speed during reentry. Both satellites had gone into orbit but had reentered the same day as launched. (AP, B *Sun*, 8/10/67; AP, C *Sun-Times*, 8/11/67)
- French cabinet authorized France to sign U.N. space law treaty [see Dec. 8, 1966]. Signed by representatives of 62 nations Jan. 27, treaty now had about 80 signatories. (*NYT*, 8/10/67)

August 9–11: Officials from NASA Hq. and field centers and MSFC prime and local support contractors participated in MSFC conference on organizational communication. Purposes of conference were to: (1) review the state-of-the-art and current academic pursuits in organizational communication in government and industry; (2) exchange experiences and knowledge among NASA personnel; and (3) identify areas of common interest with MSFC contractors. Representatives from U.S. Civil Service Commission and selected universities attended as observers. (MSFC Release 67–154)

August 10: NASA's *Mariner IV* (launched Nov. 28, 1964) and *Mariner V* (launched June 14) satellites drew into a unique "solar system lineup" which would enable JPL scientists to obtain their first three-point measurements of solar plasma and space magnetism. Orbiting about 70 million mi apart on a direct line from the sun, with earth approximately in the middle, satellites were expected to remain in about the same position relative to each other until Oct. 10. Positions would enable JPL to compare the densities and velocities of solar rays simultaneously from three points in space—two satellites and earth. (JPL Release; Wilford, *NYT*, 8/13/67, 54)

- NASA had converted $13.5-million systems-integration portion of Chrysler Corp.'s Uprated Saturn I contract to cost-plus-incentive-fee agreement and extended contract through April 1969. Under new arrangement, which raised total value of systems-integration portion to $35.5 million, Chrysler's fee would be judged according to the performance, quality, and timeliness of the work, which would be performed at MSFC and Michoud Assembly Facility. (NASA Release 67–216; MSFC Release 67–168)
- Military, congressional, and industrial sources agreed that F–111A variable-sweep-wing aircraft had developed new technical difficulties and that old problems had not yet been completely solved, Hanson Baldwin reported in the *New York Times*. Original plans had called for some F–111A's to reach Vietnam for combat tests in fall 1967; "now December or January appear to be the target dates if 'fixes' to the present difficulties are successful." Sources said that excessive vibration—a problem believed to have been completely solved—was recurring in tests. Aircraft was also experiencing difficulty with its speed brakes, and the problem of partial engine stall during flight persisted. Col. Charles W. Reed, project officer for the F–111A at Nellis AFB, Nev.,

August 10

told Baldwin there were no "real serious problems" with the aircraft and said that accelerated testing and training project which began at Nellis in July was "ahead of schedule." He conceded, however, that aircraft at Nellis had not yet carried bomb loads. Flight and combat testing of the F–111A was continuing at Edwards AFB and Eglin AFB. (Baldwin, *NYT*, 8/11/67, 1, 2)

- The universe was 70 billion yrs old—about seven times older than generally believed, Soviet astrophysicists Iosif Shklovsky and Dr. Nikolay Kardashov told a Moscow news conference. When it was formed it first expanded, then remained static for about 50 billion yrs, and now was expanding again. Scientists said they made their conclusions after analyzing the red light shift in spectrograms made of quasars—a method of measuring the distance and velocity of celestial objects. (AP, *NYT*, 8/13/67, 69)

August 11: NASA Aerobee 150 sounding rocket launched from WSMR carried ARC-instrumented payload to 88-mi (141-km) altitude to collect for analysis meteoritic debris during the peak of meteor shower. Three deployable module support arms held four module trays each, to entrap meteoritic debris. Rocket and instrumentation performed satisfactorily. (NASA Rpt SRL)

- OGO "handlers" were hoping that *Ogo IV* (launched July 28) might be economical enough with its gas supply to permit it to continue for one year at the peak of its investigative capabilities. Achievement of one-year performance would enable satellite to gather data on earth, its upper atmosphere, and its near-space environment during period of peak solar activity, which occurs every 11 yrs. *Ogo IV*'s gas system, used to supply thrust to series of small attitude-stabilized jets, had been improved by installation of gas bottle filled with krypton instead of argon. This increased gas pressure from 3,000 psi to 4,000 psi and doubled available impulse from 900 to 1,800 pound-seconds. NASA reported that if present gas expenditure rate were maintained they were assured the minimum 50 days of three-axis stabilized operation that would permit mission success. (NASA Release 67–210)

- Universities should not rely so heavily on NASA and DOD to advance and support basic science, Cal Tech president Lee A. DuBridge suggested in *Science*. A moderate fraction of these agencies' budgets was necessarily used to advance basic research, he said, and in the space program "the resulting technologies are providing a valuable tool for carrying on scientific investigations which would otherwise be impossible.

  "However, it must . . . be stressed that neither NASA nor any other agency charged with implementing a national-policy goal is intended to be a philanthropic agency authorized to provide benefactions to university science departments. They are agencies seeking to get a job done, and they turn to universities only when the universities can render a service. . . ." DuBridge urged that Federal Government develop a more adequate and balanced program for strengthening basic science by charging "suitable agencies (principally the National Science Foundation) with this particular task" and providing adequate funds for carrying it out. (DuBridge, *Science*, 8/11/67, 648–50)

- Commercial applications of a miniaturized TV aerial developed by USAF after four years of research were being explored by at least 100 electronic companies, Jack Gould reported in the *New York Times*. Firms believed device, originally developed for military communications, might accommodate public preference for small antennas without sacrificing picture quality. Inventor Edwin M. Turner, attached to the Air Force at Wright-Patterson AFB, told the press that the "heart" of the miniaturized antenna was the use of transistors. (Gould, *NYT*, 8/12/67, 27)

August 12: GSFC scientists were predicting that NASA's *Echo I* comsat, launched Aug. 12, 1960, would decay from orbit during the next year, AP reported. Once a perfect sphere 100 ft in dia, balloon had deflated and its speed in orbit was steadily decreasing. It had been launched to demonstrate that large inflatable spheres could be used as passive communications reflectors in space. (AP, *H Post*, 8/13/67; Volker, *M News*, 8/12/67)

- AIAA Board of Directors endorsed hobbyists who built model rockets from paper, plastic, and balsa wood, but warned against the hazards of amateurs mixing their own fuels and experimenting with homemade metal rockets. Action preceded the National Model Rocket Championship contest scheduled to open in Mankato, Minn., Aug. 15. AIAA, which had for many years withheld endorsement of any type of nonprofessional rocket experiments, recommended that model rocketry be conducted only with adult supervision. (Wilford, *NYT*, 8/14/67, 37)

- U.S. and Australia were engaged in a joint program to build a highly technical installation in the Australian desert to intercept information transmitted by Soviet reconnaissance satellites, UPI reported. On Aug. 11, DOD statement said only that U.S. was "cooperating with the Australian government in a space defense research project." U.S. officials admitted privately, however, that project was a "ferret" installation to detect radar, electromagnetic radiation, and other electronic impulses emanating from Soviet satellites. (UPI, *W Star*, 8/12/67)

August 14: President Johnson, in a message to Congress on communications policy, announced that he was appointing a 15-man task force headed by Under Secretary of State for Political Affairs Eugene V. Rostow to formulate "a national communications policy." Among questions the group would examine were: (1) how soon would a domestic satellite system be economically feasible? (2) should a domestic system be general purpose or specialized? and (3) should there be more than one system? Task force's report was due in one year. The President also: (1) reaffirmed U.S. commitments made in 1962 and 1964 in support of the development of a global system of comsats "to make modern communications available to all nations"; (2) pledged U.S. support to the continuation of the INTELSAT consortium, saying, "We seek no domination of satellite communications to the exclusion of any other nation—or group of nations"; (3) stated that U.S. "should take no action in the establishment of a domestic [comsat] system which is incompatible with our support for a global system"; and (4) urged U.S.S.R. and the nations of Eastern Europe to join INTELSAT. "Here is a rare opportunity to join in an activity to bring benefits to all nations and loss to none." He suggested there might be an eventual linkage between Soviet Molniya system and INTELSAT system.

The President concluded: "Historians may write that the human race survived or faltered because of how well it mastered the technology of this age.

"Communications satellites now permit man's greatest gifts—sight, expression, human thoughts and ideas—to travel unfettered to any portion of our globe. The opportunity is within our grasp. We must be prepared to act." (*PD*, 8/21/67, 1146–54; *WSJ*, 8/15/67)

August 14

- ComSatCorp Chairman James McCormack, welcoming President Johnson's message to Congress on communications policy, offered full cooperation "in forwarding the President's objective of bringing high quality, dependable satellite communications to users throughout the world." He reaffirmed support of a "global system, and the continuation and expansion of [INTELSAT] as the foundation on which to build a permanent structure for this system." (ComSatCorp Release)
- William S. Aiken, Jr., former Chief of Operations Research Branch, OART, was appointed Deputy Director, Aeronautical Vehicles Div., OART. (NASA Ann, 8/15/67)
- NASA's Manned Space Flight Network would be tested for the first time with the unmanned Apollo 4 mission, tentatively scheduled for early October, Barry Kalb reported in the Washington *Evening Star*. Dr. Friedrich O. Vonbun, chief of GSFC's Mission and Trajectory Analysis Div., told Kalb in an interview that the "reliability of the system is 99.5 percent." Network included a new system, he said, which combined functions of tracking, signaling, and voice communication. Previously a separate system was required for each. Only certain portions of the network would be used for the Apollo 4 mission, Von Bun said. The entire system consisted of 13 earth orbit tracking stations, three deep-space stations (each with a backup unit), five tracking ships, and eight specially fitted jet aircraft. Three deep-space stations, located in Goldstone, Calif., Madrid, Spain, and Canberra, Australia, were designed specifically for "communicating with vehicles at lunar distance." (Kalb, *W Star*, 8/14/67, A6)
- National Aeronautics and Space Council was accumulating data on various approaches to earth resources satellites with the goal of forcing competing agencies to agree on a single coordinated program for FY 1969, *Aerospace Technology* reported. A decision was expected by mid-September. The Depts. of Interior and Agriculture favored an operational program, but NASA believed current technology justified only an R&D approach at this time. (*Aero Tech*, 8/14/67, 3)
- Electronics Research Center was studying a navigation satellite system capable of providing location information to ships and aircraft accurate to within 0.1 mi, *Aerospace Technology* reported. System would be competitive with recently declassified Navy Navigation Satellite System [see July 29], but would be aimed primarily at small users such as private aircraft and pleasure craft. User cost for receiving equipment would range from $1,000 to $5,000. Current phase of program involved three contractors: (1) Philco-Ford Corp., which had completed study on fan beam navigation satellites; (2) RCA, which was studying a hyperbolic L-band ranging system providing continuous radiation of the signal; and (3) TRW Systems which was designing a synchronous-orbit satellite system, also operating in the L-band. The next phase, pending NASA Hq. approval, was laboratory testing followed, possibly, by flight experiments in Applications Technology Satellite (ATS) or Apollo Applications (AA) program. (*Aero Tech*, 8/14/67, 10)
- The time had come to "fix" the responsibility for the total results of U.S. technology, Michael Getler suggested in *Aerospace Technology*. There was no single group or agency to "link the broad spectrum of American technological development with social development. There is no single group with the expertise available to probe all forms of technology and its effects, to pump funds into a truly important new applied sci-

ence not able to capture headlines, to speed the transfer of defense and space technology into other areas, and to put the special interest groups on the line for blockage of improvements within the state-of-the-art. There is no group directly responsible to the citizenry to develop and carry through legislation to try and save us from tomorrow's smog and urban clutter; for we have not, by a long shot, run out of self-inflicted ills. . . ." Getler urged that the overall responsibility be given to Congress. "It is the Legislative branch of Government, properly augmented with a permanent staff of specialists, that is best equipped to serve as a responsive arbiter between the good life of our people, the promise of technology, and the pressure of the marketplace." (Getler, *Aero Tech*, 8/14/67, 62)

- Soviet mathematician Dr. Alex M. Letov, visiting Huntsville for the Guidance, Control, and Flight Dynamics Conference sponsored by Alabama chapter of the AIAA, said he supported the exchange of technical information between U.S. and U.S.S.R. After delivering an important technical paper, Letov remarked: "I am hopeful that I may have initiated some steps in this area of exchange. I am completely satisfied with my presentation and the response of the American scientist. In my report I will reflect this." One of the world's leading authorities on spacecraft guidance and control, Letov outlined a complex theory for stabilizing a large manned spacecraft, which he considered a major problem. "You can't consider the other problems of engineering if your system is not stable. It must be stable." (*KC Star*, 8/15/67)

August 14–18: U.S. scientists "still cannot provide definite protection from the biological hazards" of 1,000-day duration planetary mission, according to experts at Blacksburg, Va., Bioastronautics Conference, sponsored by VPI and NASA. Discussions had indicated: "Extremely long periods of ground simulation and experiments in space with animals will be necessary before man can undertake a 1,000-day journey to a near planet like Mars." Scientists had discussed several problem areas: ability of man to function in weightlessness for more than 90 days (midcourse goal of Apollo Applications program); weight, volume, and content of food, water and atmosphere requirements; and biological reactions of crew to spacecraft contaminants. Toxicologists considered reaction to be "most insidious potential problem," respiratory system taking on contaminants in "subtle, long-term reactions." (Normyle, *Av Wk*, 8/21/67, 17; 9/11/67, 75–8)

August 15: NASA Administrator James E. Webb, testifying at a closed hearing before House Committee on Appropriations' Subcommittee on Independent Offices and Department of Housing and Urban Development, said NASA hoped to keep an approximate $5.5-billion annual budget following a manned lunar landing. ". . . we believe it would be wise to keep the budget approximately level, approximately at where it is now or even going up a half billion dollars a year, and use this capability for some further landings on the moon, for ability to move around, for synchronous orbital work with large stations using men for a multiplicity of other purposes related to the earth, and further studies in space. . . ."

"This is very much what the Russians are doing. They are developing a very large number of options and maintaining the ability to select those that are most useful to them at a particular time and move rapidly with it and creating the image that they are out in front with respect to all these modern technologies." Webb said he believed four recent launchings of large unmanned spacecraft by U.S.S.R. were "preparations for manned flights shortly to come. I believe they are flights to work out the difficulties that resulted in the death of their cosmonaut." (Transcript, pp. 8, 34)

- Selection of NASA's sixth astronaut class [see Aug. 4] indicated that astronaut "glamour" grew with each new peril, Neal Stanford reported in *Christian Science Monitor.* "Scientists . . . compete more intensively for the few scientist-astronaut jobs open than do jet pilots in the military services for the pilot jobs available.

  "The glamour has not entirely gone out of space flying. What has happened is that its perils have become more apparent, its scientific opportunities more exciting, and its financial rewards less considerable.

  "There have been 6 selections of astronauts, 4 of flier astronauts, and 2 of science astronauts. In this sixth and latest selection 939 [sic] scientists applied for the 11 positions open. In the previous selection of pilot astronauts only 403 applied for 19 positions open. . . ." (Stanford, *CSM*, 8/15/67)

- Proposal to operate a 55-passenger STOL airliner in the greater New York area as a three-year demonstration project had been advanced to Federal, state, and city transportation officials, Robert L. Cummings, president of New York Airways, told *New York Times*. Developed jointly by New York Airways, McDonnell Douglas Corp., and Cutler-Hammer, Inc., proposal called for four-engine turboprop that could cruise at speeds ranging from 50–250 mph and land on strips 1,200–1,400 ft long. Financing the estimated $12–$15-million demonstration project would probably require "some sort of contract operation, presumably with a Government agency," Cummings said. (*NYT*, 8/16/67, 66)

August 15–17: Divers Arthur Pachette and Glen Taylor set records for longest and deepest underwater excursions outside a protective capsule in series of three saturation dives in Gulf of Mexico. Sponsored by Ocean Systems Inc., and Esso Production Research, tests studied new deep-diving system designed to enable men eventually to walk and work on the ocean floor at depths of 1,000 ft or more. During tests divers descended 636 ft in a steel hull sphere, donned rubberized "wetsuits" and, breathing a helium-oxygen gas mixture, worked on a dummy oil well outside the capsule for nearly seven hours. Ocean Systems spokesman said they experienced no decompression problems and were able to manipulate nuts and bolts and replace valves. In Westinghouse Electric Corp. tests in June, divers descended to 600 ft in similar operation, but did not remain at the depth long enough to achieve saturation—condition in which the body tissues become completely saturated with gases contained in breathing mixture. (Wilford, *NYT*, 8/25/67, 19; *Aero Tech*, 8/28/67, 22)

August 16: USAF launched unidentified satellite from WTR using Titan III–B booster; satellite reentered Aug. 28. (*Pres Rep 1967*)

- Astronauts Walter M. Schirra, Donn F. Eisele, and Walter Cunningham toured McDonnell Douglas Test Facility near Sacramento to inspect 3rd stage of the Saturn V booster which would be used to launch them on first manned Apollo mission in 1968. Schirra told newsmen that NASA would not be "pressed by the press" into flying the mission on a specific date. Asked if he believed pressure from the press was in any way responsible for the Jan. 27 fire, Schirra replied: ". . . I found that whenever we . . . get together with the members of the press that you're right away asking us, 'Well, now when are you going to fly?' That's almost the first question, and it's not a very unique question. . . . It seems you want that answer that may be hard news, but when we give you that answer then if we don't make that date, for some reason or another it appears that we're failing our goal. While the goal is to have a successful flight at a reasonable time rather than an early failure. . . ." (Transcript)

- NASA's *Mariner V* spacecraft had reached the halfway point in elapsed time in its 128-day, 216-million-mi flight to Venus. Scheduled to fly by the planet Oct. 19, spacecraft had traveled 98,667,000 mi in 64 days. It was scheduled to reach midway mark in distance Aug. 22. (NASA Release 67–223)
- The Federal Government had involved itself in the SST program because failure to do so would have resulted in a loss of jobs and progress in U.S. aviation industry, M/G J. C. Maxwell (USAF), Director of FAA SST Development, told the National Space Club. "The aircraft industry had been challenged by a powerful consortium [Concorde] of two large aircraft companies fully subsidized by two major governments. Our aircraft companies did not have the resources to meet this challenge. The government acted." Federal participation had resulted in some major side effects, Maxwell said. "We were able to make available to all potential SST contractors a vast amount of technological information that the government had acquired in its own huge aircraft endeavors. The flight test experiences of the B–70, the SR–71, and the F–111 were and are being provided to our SST contractors. . . . We were able . . . to contribute some 49,000 hours of SST wind tunnel tests conducted by NASA. The military . . . has logged some 300,000 hours of supersonic experience. . . . The SST program is founded on the experience and know-how from other programs sponsored and paid for by the government—and that includes the space effort.

  "Another benefit . . . [was] the fact that the resources of the whole aviation industry were brought into the project. Our search for the best design was, in every sense of the word, a competition. The best talent and genius of American aeronautical engineering was put to work." (Text)

August 17: Photo of the earth in the near full phase transmitted by *Lunar Orbiter V* [see Aug. 5–27] was a "fraud, a fake, a piece of trickery or deceit," Samuel Shenton, Secretary of the International Flat Earth Research Society, told newsmen in London. "Look at the photograph the satellite supposedly took. Notice those lines across the bottom of the picture. They show it's a composite, a mock-up." Asked why U.S. should seek to perpetrate such a deception, Shenton replied: "It's all part of the great global earth conspiracy, and it's a shame." Members of the International Flat Earth Research Society believed that the earth is not only flat, but static in space. They also maintained that the theory of gravity was advanced only to support the belief that the earth spins. (AP, B *Sun*, 8/18/67)

- St. Louis Univ. scientists Drs. F. C. Bates and Albert Pallman were studying structure and circulation of the lower Martian atmosphere to aid NASA in deciding whether to use a hard- or soft-landing spacecraft on first Mars mission. (UPI, *NYT*, 8/17/67, 24)
- North American Air Defense Command (NORAD) announced officially that no flying saucers, no extraterrestrial spacecraft, and no "unfriendly

objects" had ever been identified by air forces over U.S. or Canada: "There have been thousands of reports about official air force stands on UFOs from the widest variety of sources over the years—but so far as the command is aware, it never has been shown anything but natural atmospheric or astronomical phenomena or airborne objects—birds, insects, hardware—seen under unusual circumstances, or misinterpreted." (Thomis, *C Trib*, 8/18/67, 1)

August 18: House Committee on Appropriations reported $4.6-billion NASA FY 1968 appropriation bill—$516.6 million less than NASA had requested and $282 million less than NASA FY 1968 authorization bill (S. 1296). Committee said the bill was "less than would be recommended under less stringent fiscal circumstances," but it would support "a viable space program, and one that is consistent with available resources.

"The reductions . . . should in no respect be interpreted as an expression of a lack of confidence in our space endeavors. . . ." Cuts included: (1) $50.5 million from Apollo program; (2) $154.7 million from AA program; (3) $27.5 million from Nuclear Rocket Program; (4) $37.7 million from Tracking and Data Acquisition; and (5) $71.5 million from Voyager (funding eliminated). (NASA LAR VI/91; *WSJ*, 8/21/67, 4; O'Toole, *W Post*, 8/19/67, A4)

- First two Apollo spacesuits incorporating changes recommended by NASA's Apollo 204 Review Board were scheduled to arrive at MSC in early September for thermo-vacuum and compatibility testing. New suit, designated A–7L, was an improved version of the original A–6L suit. Changes, primarily replacing flammable materials with nonflammable or low-flammability materials, included substitution of: (1) Beta fabric, a nonflammable fiberglass cloth, for Nomex, a high-temperature nylon, in outside layer; (2) Nomex for a more flammable nylon inner layer; and (3) fire-resistant Kapton/Beta fiberglass insulation for aluminized Mylar-Dacron insulation. In addition to its greater fire-resistant properties, spacesuit, which was white rather than blue, was more comfortable and mobile than A–6L. It would be used in all manned Apollo missions during prelaunch and launch phases, and reentry. First production model of spacesuit, manufactured by International Latex Corp.'s Government and Industrial Div. under MSC contract, was scheduled for mid-October delivery. (NASA Release 67–222)

- Rocket engine for Apollo service module's propulsion system had been undergoing tests at Arnold Engineering Development Center during past $3\frac{1}{2}$ years. Initially, engine performance was unknown; it would use an ablatively cooled thrust chamber and a high expansion ratio nozzle of thin metal and would operate at relatively low combustion chamber pressures. Initial tests were on configuration of engine's fragile nozzles. Follow-on tests, for prequalification of engine combined with heavy duty replica of propellant tankage, had checked ballistic performance. Simulated high-altitude testing would lead to eventual man-rating of propulsion system. (AEDC Release 67–212)

- NASA had awarded NAA's Rocketdyne Div. a $1.4-million contract for continued technological investigation of the advanced aerospike rocket engine. Studied by OART for $1\frac{1}{2}$ yrs, the engine utilizes a doughnut-shaped combustion chamber which discharges exhaust gases against the surface of a short center cone. In conventional rocket engines, gases are expanded inside long bell-shaped nozzles. Current concept specified an engine 8 ft in diameter and $4\frac{1}{2}$ ft high—about 50% shorter than conventional bell nozzle engines. A new "aerodynamically compensating nozzle" would permit use of the engine at sea level or high altitude, making it potentially suitable for both lower and upper stages of launch vehicles. Contract completion date was May 31, 1968. (NASA Release 67-221)

- A completely fireproof version of the spacecraft fabric Velcro had been perfected and would be used on NASA's first manned Apollo mission in 1968, William Hines reported in the Washington *Evening Star*. In tests conducted at MSC early in August, the all-metallic fabric did not ignite, even in pure oxygen atmosphere at 16.5 psia. Velcro Corp. vice president Jack Mates told Hines in an interview that NASA had not yet placed an order for the improved Velcro, but that his company was beginning production of a substantial quantity. In the Jan. 27 Apollo fire, flames in the spacecraft had quickly spread along the surface of ordinary nylon Velcro adhesive pads which were used to hold frequently used equipment. (Hines, *W Star*, 8/18/67)
- Inexpensive and easily acquired property was not necessarily the best site for a new airport, according to airport site selection guide published by FAA. Other factors such as accessibility to users, construction costs, future expansion possibilities, obstructions, anticipated community response to aircraft noise, and proximity to other airports could be even more important factors. Airport site selection was a local responsibility, guide noted, but all sponsors of civil airport projects were required to notify FAA of proposed construction so agency could review project's effect on surrounding airspace. (FAA Release 67-62)

August 19: Negotiations were being conducted in Washington, D.C., to move Canada's High Altitude Research Project (HARP) from McGill Univ. to Univ. of Vermont, Jay Walz reported in the *New York Times*. Move was a result of Canada's decision to cut off funding because it believed program should be self-supporting. HARP, which maintained a launch site in Highwater, Quebec, near the Vermont border, used 16-in naval guns to launch payloads inexpensively into the ionosphere for scientific and meteorological experiments. Project, which received about $1.2 million annually from U.S. Army, employed 50 persons at McGill Univ., 70 at Highwater, and 60 at a launch site in Barbados, West Indies. (Walz, *NYT*, 8/20/67, 20)

- An unusual sequence of Soviet launches had led some U.S. officials to conclude that U.S.S.R. was testing techniques for reentering warheads from space, Evert Clark reported in the *New York Times*. "This could mean that the Russians are developing weapons to be stationed in orbit. Weapon re-entry techniques can be tested with or without the use of weapons, and the same re-entry techniques can be used for either nuclear or conventional weapons." Clark said all of the flights (1) were very short, (2) had been launched from Tyuratam range, (3) had entered orbits with 49° inclination, and (4) had been given a Cosmos number. A resolution adopted by the U.N. General Assembly in 1963 asked all nations to refrain from orbiting weapons of mass destruction. The question of whether resolution prohibited the development of such weapons had never been clarified, Clark said. (Clark, *NYT*, 8/20/67, 17)
- Hugo Gernsback, an inventor, author, editor, and publisher who was often called the father of modern science fiction, died in New York at age 83. Gernsback described radar in 1911—35 yrs before communications experts bounced a radar signal off the moon. In 1927 he began pub-

lishing *Amazing Stories*, one of the earliest magazines devoted entirely to science fiction. In 1928 he sponsored New York's first television broadcasts: images only slightly larger than postage stamps were received on crude scanners owned by 2,000 amateurs. (*NYT*, 8/21/67, 29)

August 20: NASA published the first extensive chart of the far side of the moon. Compiled by USAF Aeronautical Chart and Information Center using NASA-supplied photos, map covered more than 75% of the moon's hidden side. Approximately 85% of the photos used were provided by NASA *Lunar Orbiters I, II, III,* and *IV*; remaining 15% were provided by Soviet *Zond III*. Map was released to assist astronomers at the International Astronomical Union, convening Aug. 22 in Prague, in naming features on moon's far side. (NASA Release 67–220)

- USN successfully conducted first underwater launch of the Poseidon missile off the coast of San Clemente, Calif. Missile was expected to be operational by 1970 aboard 30 of the 41 Polaris submarines. (*W Post*, 8/21/67)

August 21: President Johnson signed $4.86-billion NASA FY 1968 authorization bill (S. 1296) into law. Noting that on Aug. 18 House Committee on Appropriations reported $4.6-billion NASA FY 1968 appropriation bill, $516.6 million less than NASA had requested, the President said: "Under other circumstances I would have opposed such a cut. However, conditions have greatly changed since I submitted my January budget request. . . .

"I recognize—as also must the Congress—that the reduction in funds recommended by the House Appropriations Committee will require the deferral and reduction of some desirable space projects. Yet, in the face of present circumstances, I join with the Congress and accept this reduction." (*PD*, 8/28/67, 1192–3)

- Two-stage Nike-Cajun sounding rocket launched from Natal, Brazil, failed when rocket strayed from planned trajectory. Failure was first in Brazil's space program which opened Dec. 15, 1965, with the launch of a Nike-Apache sounding rocket. (Reuters, *W Post*, 8/23/67)
- NASA Arcas sounding rocket launched from Point Mugu, Calif., carried Naval Ordnance Test Station experiment to 33-mi (53-km) altitude to flight-test internally modified version of standard ROCOZ payload, designed to measure ozone distribution for support data on *Ogo IV* satellite. Rocket and instrumentation performed satisfactorily. Scientific data were usable. (NASA Rpt SRL)
- Charles W. McGuire, formerly an aerospace engineer in the test directorate of NASA's Gemini program, became Director of Safety for the Apollo Program. He would be on the headquarters staff of and report directly to the Apollo Program Director while being administratively assigned to the Safety Office, OMSF. (NASA Ann, 8/7/67)
- Maj. William J. Knight (USAF) flew X–15 No. 2 to 3,409 mph (mach 5.10) and 90,000-ft altitude in first flight with a new ablative coating and sealer designed to permit aircraft to fly at mach 7.4 without basic airframe modification. Flight also tested: (1) stability and control with dummy ramjet and ablative; (2) Hycom phase-II camera (KA–51A); (3) ramjet local flow; (4) ramjet separation characteristics; and (5) wing tip accelerometer. (NASA Proj Off; *Av Wk*, 8/28/67, 33)
- Special NSF panel to screen radio astronomy proposals advised NSF Director Leland J. Haworth to accept immediately only two of the six recommendations presented to the panel during July hearings: (1) Cornell

August 21: The X-15 research aircraft stands with the XB-70 research plane outside the hangar at NASA's Flight Research Center. The X-15's new white ablative coating is designed to protect its metal skin from overheating during speeds of more than mach 6.

Univ.'s proposal to improve accuracy of the surface of 300-m antenna at Arecibo, Puerto Rico; and (2) Cal Tech's proposal to add seven dish antennas to one already in existence at Owens Valley Observatory. Panel rejected or deferred: (1) National Radio Astronomy Observatory's (NRAO) proposal for a Very Large Array (VLA) consisting of 36 25-m dish antennas distributed along three arms of a Y; (2) Northeast Radio Observatory Corp.'s (NEROC) proposal for 135-m fully steerable dish antenna under plastic steel-ribbed radome; (3) Associates for Radio Astronomy's (ARA) proposal for 100-m fully steerable dish antenna without radome; and (4) Committee on Institutional Cooperation's proposal for a device to study upper atmosphere and ionosphere. Eight-member panel was chaired by Princeton Univ. physicist Dr. Robert H. Dicke. (McElheny, *Science*, 8/18/67, 782–4; 8/25/67, 907–10)

- Rep. William L. Hungate (D-Mo.) expressed on the House floor concern about NASA's FY 1968 budget request "at a time when the Nation's economy is threatened by inflation and when we are spending $24 billion annually on the war in Vietnam. . . .

  "As an example of untimely NASA spending, $21,100,000 of the funds requested are to be devoted to research directly applicable to the supersonic transport project. . . . Before more millions are poured into supersonic transport, some consideration should be given to solving the problems created thus far by this costly innovation. . . ." (*CR*, 8/21/67, H10888)

- ComSatCorp would issue RFP's in 1968 to build a 10,000-channel satellite—Intelsat IV, *Electronics* reported. Procurement of four or more of the satellites would have to be approved by INTELSAT, but "chances of a go-ahead look good." ComSatCorp's plan was based on design studies recently completed by Hughes Aircraft Co. and Lockheed Aircraft Corp.

August 21

for a high-capacity, multipurpose satellite [see March 23]. (*Electronics,* 8/21/67)

- A space rescue blanket for campers and sportsmen was being manufactured by National Research Corp., using a superinsulation material it had developed to store cryogenic materials in spacecraft, Douglas W. Cray reported in the *New York Times.* Made of a plastic base one half of one thousandth of one inch thick with an aluminum coating one millionth inch thick, blanket weighed two ounces, measured 56 by 84 in, and could be folded into a cigarette-pack size. It was waterproof and windproof, remained flexible at temperatures as low as $-60°$ F, and could reflect 80% of user's natural body heat. (Cray, *NYT,* 8/21/67, 45, 48)
- Lewis Research Center's Quiet Engine Program should produce a significant reduction in jet-engine noise and reverse the rising noise levels which had created serious problems at major airports, Michael L. Yaffee reported in *Aviation Week.* Expected to cost $50 million, program was seeking to develop a new turbofan demonstrator engine that would show a reduction of 15 PNdb (perceived noise in decibels) on takeoff and 20 PNdb on landing, below noise of current transport engines. Definition phase of Quiet Engine Program began under $458,000 contract LeRC awarded United Aircraft Corp.'s Pratt & Whitney Div., July 13. Remaining $750,000 allotted to program for FY 1967 was being used to purchase supporting services and hardware. (Yaffee, *Av Wk,* 8/21/67, 38–9)

August 22: House voted 312–92 in favor of $4.6-billion NASA FY 1968 appropriation bill (H.R. 12474)—NASA's lowest in five years. This was the sum recommended by the House Appropriations Committee [see Aug. 18]. Two amendments by Rep. William F. Ryan (D-N.Y.)—one to reduce R&D funds by $100 million, one to reduce administrative operations' funds by $37 million—were defeated by voice vote. (NASA LAR VI/93; *CR,* 8/22/67, H10909–30)

- Leonard Rawicz, former patent counsel for GSFC, had been appointed NASA Assistant General Counsel for Patent Matters, succeeding Robert F. Allnutt, who was named NASA Assistant Administrator for Legislative Affairs June 15. (NASA Release 67–226)
- Tass announced that U.S.S.R. would build more than 20 relay stations during 1967 so that Moscow telecasts could be transmitted via Molniya comsats to Siberia, the Soviet Far East, and the "extreme north." (Reuters, *W Post,* 8/24/67; Clark, *NYT,* 8/25/67)
- USAF announced it was sending teletype messages between two aircraft flying 10,000 mi apart via a satellite in 20,000-mi-altitude earth orbit. Scheduled over six-month period, communications tests were being conducted over Pacific and Atlantic Oceans, North and South America, the polar regions, and Europe. (AFSC Release 144.67)
- A college which would train 2,000 pilots and technicians annually for the expanding aviation industry was proposed in report by a study group sponsored by the Economic Development Administration. Report recommended that various segments of the aviation industry join to form a nonprofit organization to operate flight training portion of the college. Academic side would be handled by Arizona State Univ. Report estimated college would cost $13-million and could be opened by 1969. (UPI, *NYT,* 8/23/67, 73)

August 22–31: International Astronomical Union (IAU) met in Prague.
Dr. Robert B. Leighton of Cal Tech reported that there were about

three times as many craters on Mars as preliminary photos transmitted in 1965 by *Mariner IV* spacecraft indicated. Citing data from two-year study of photos using new computer-enhancement technique, Dr. Leighton said pictures showed 300 clearly defined craters and more than 300 additional possible craters. Original estimate from unprocessed pictures was less than 100. Mars appeared to be about as densely pitted as the moon but its craters were smoother because of more effective erosion processes. Several photos showed a number of linear features similar to lines formed by rifts and faults on the moon. These features were not the "much-discussed" canals of Mars because they were too narrow to be visible from earth, Dr. Leighton said, but they might be related to some of the markings identified earlier as canals. (NASA Release 67–225)

Dr. and Mrs. Pierre Connes of Haute Provence Observatory and Meudon Observatory, France, respectively, reported evidence that might indicate Venusian clouds are not formed by water. Observation technique which greatly increased precision with which planets' spectra could be analyzed showed spectral lines of hydrogen chloride and hydrogen fluoride in the light reflected from Venus. When hydrogen fluoride combines with water it creates hydrochloric acid which absorbs radar waves; yet radar waves are known to penetrate Venutian clouds, returning echoes to the surface of the earth. (Sullivan, *NYT*, 8/24/67)

The "mysterium phenomenon"—peculiar radio emissions from the Milky Way which display almost all the characteristics of artificial interstellar communication—was the subject of a report by Dr. T. K. Menon, National Astronomy Observatory, Green Bank, W. Va., and a later panel discussion. Dr. Menon said most recent observations of the emissions in the nearby part of the galaxy had been made simultaneously from California, Massachusetts, and U.K., using almost the entire width of earth as a baseline for measurement. They had identified sources of emissions which were smaller than the smallest star as seen by the human eye and discovered some of these emissions were fluctuating on time scales measured in hours or even less. Panel raised the possibility that emissions might be from stars in the process of formation. They were being transmitted at four closely spaced wavelengths associated with hydroxyl, a molecule formed of a single oxygen and hydrogen atom, and were being given off when such molecules shed energy. Emissions also appeared to have various properties associated with a maser—man-made device in which a radio or light wave is beamed through certain atoms whose energy has been raised, enormously amplifying original wave and, if wave is confined to a single axis, producing a beam of extreme intensity. (Sullivan, *NYT*, 8/26/67, 1, 10; 8/28/67, 22)

Lunar Committee postponed naming of surface features on moon's far side pending the results of *Lunar Orbiter V* mission. A list of 228 names for approximately 500 major features had been submitted by Soviet scientists, but U.S. delegates cited errors in their designations and asked that features be assigned numbers until a more definitive map could be prepared using *Lunar Orbiter V* photos. Commission agreed to interim number designations and accepted U.S. map temporarily, assigning study group to report on specific names at next assembly. (Sullivan, *NYT*, 8/25/67, 13; NAS–NRC–NAE *News Report*, 10/67, 6–7)

IAU president Dr. Pol Swings of Belgium confirmed an observation made by two obscure Irish astronomers in 1882. Astronomers Lohse and

Copeland had observed a comet during the day, apparently with a handheld spectroscope, and reported that it glowed strongly with spectral lines of iron as it passed the sun. All earlier observations of light emitted from comets as they neared the sun indicated they were composed of very light materials, and the idea that they were rich in iron was viewed skeptically. However, photos and electric recordings taken of the comet Ikeya-Seki as it passed the sun in 1966 showed unmistakable spectral lines of iron, Dr. Swings said. (Sullivan, *NYT*, 8/27/67, 51)

Dr. Gordon H. Pettengill, MIT's Lincoln Laboratory, said that results from an experiment which studied Martian terrain by bouncing radar signals off the planet indicated: (1) Mars was roughly as mountainous as earth; (2) there were no large bodies of water on Mars; and (3) there was no clear correlation between height of land and presence of dark areas on Mars. (Sullivan, *NYT*, 8/30/67, 15)

Dr. Bradford A. Smith, New Mexico State Univ., confirmed report by French astronomer Pierre Guerin that atmosphere of Venus was moving around the planet in a clockwise direction at a remarkable rate of speed. Conclusion was based on photos of Venusian atmosphere taken successively both at night and during the day, which recorded changes in wavelengths of ultraviolet light. (Sullivan, *NYT*, 8/30/67, 15)

August 23: USAF launched unidentified satellite from WTR using Thor-Burner II booster. (*Pres Rep 1967*)

- The *New York Times* commented on the task force appointed by President Johnson to formulate a national communications policy [see Aug. 19]: "The . . . task force . . . has an unprecedented opportunity to break internal governmental deadlocks that stand in the way of a domestic satellite system and other needed forward steps in communications. It can do much to advance public television and to give all TV, both commercial and noncommercial, a firmer root in public service. Perhaps most important of all, it can point a path for the interconnection of international satellites so that all countries can be encouraged to see and speak to one another across the heavens. . . .

    "The major challenge is to clear channels through which revolutionary new developments in communications can be used with maximum effectiveness for total human service and enlightenment. The last thing that should happen is for Congress to postpone action on President Johnson's proposal for a Corporation of Public Broadcasting. Such a planning body should be in existence to move speedily on any sound recommendations the task force makes." (*NYT*, 8/23/67, 40M)

August 24: Cosmos *CLXXIII* was launched by U.S.S.R. into orbit with 528-km (328-mi) apogee, 280-km (174-mi) perigee, 92-min period, and 71° inclination. All equipment functioned normally. Satellite reentered Dec. 17. (UPI, *NYT*, 8/25/67, 5; GSFC *SSR*, 12/31/67)

- U.S. and U.S.S.R. submitted to disarmament conference in Geneva identical texts of a draft treaty to prevent the further spread of nuclear weapons. Treaty would bar the five nuclear powers from transferring nuclear weapons or nuclear explosive devices to nations that did not have them or from assisting them in the production of nuclear weapons. After a draft was agreed to by 17-nation disarmament conference, it would be submitted to U.N. General Assembly for discussion.

    President Johnson commented: "If we now go forward to completion of a worldwide agreement, we will pass on a great gift to those who follow us.

"We shall demonstrate that—despite all his problems, quarrels, and distractions—man still retains a capacity to design his fate, rather than be engulfed by it.

"Failure to complete our work will be interpreted by our children and grandchildren as a betrayal of conscience, in a world that needs all of its resources and talents to serve life, not death. . . ." (PD, 8/28/67, 1204–5; Hamilton, NYT, 8/25/67, 1)

- NASA test pilots Fitzhugh Fulton and Donald Mallick flew XB–70 No. 1 to mach 2.27 and 58,000-ft altitude in flight at Edwards AFB to check out: (1) three pacer airspeed calibration points; (2) photos of the tufted right upper canard; (3) boundary layer noise data; (4) stability, control, and handling qualities; (5) bypass close unstart of left inlet; (6) nose ramp flutter data; and (7) radar airspeed calibration. (NASA Proj Off)
- Ten cosmonauts were practicing parachute jumps over the water, indicating that future Soviet space missions might attempt American-style splashdowns, *Krasnaya Zvezda* reported. All previous Soviet manned space missions had come down on dry land. (UPI, NYT, 8/25/67, 29)
- Soviet aircraft would drop bombs into two large areas of the Pacific from Sept. 2–10 in series of seismic tests which would aid Soviet study of world oceans and research into forecasting tidal waves and earthquakes. Tass asked foreign governments to warn their ships not to enter first area from 10:00 a.m. to 6:00 p.m. local time, Sept. 2–5; or second area during same hours Sept. 5–10. (Reuters, NYT, 8/25/67, 12)
- Republic Aviation Corp. offered $10,000 each to 12 well-trained men willing to spend 270 consecutive days in 28-ft-long simulated space chamber. Applicants had to be qualified psychologists, microbiologists, or systems engineers. Tests, at Wright-Patterson AFB under $646,434 USAF grant, would try to determine how men would react during a long space voyage. (AP, NYT, 8/25/67, 16)

August 25: Dr. Homer E. Newell, NASA Associate Administrator for Space Science and Applications, would become NASA Associate Administrator Oct. 1, NASA Administrator James E. Webb announced. He succeeded Dr. Robert C. Seamans, Jr., who became Deputy Administrator Dec. 21, 1965. In his new position, Dr. Newell would work closely with Webb and Dr. Seamans on the planning, development, and conduct of the space program. D. D. Wyatt, Assistant Administrator for Program Plans and Analysis, would report directly to Dr. Newell. (NASA Release 67–228)

- NASA Aerobee 150 sounding rocket launched from WSMR carried Lockheed Missiles & Space Co. experiment to 76-mi (125-km) altitude to determine x-ray spectra in the 0.3- to 3-kev energy range of several previously discovered cosmic sources and to search for variation with galactic latitude of diffuse x-ray background. Rocket and instrumentation performed satisfactorily. (NASA Rpt SRL)
- *Mariner V* spacecraft had traveled more than 114 million mi of its 216-million-mi flight to Venus and was now closer to Venus than to earth, NASA announced. JPL officials said spacecraft's star tracker had been successfully updated for the first time Aug. 24. This electronic change of the tracker's "look-angle," commanded by the spacecraft's on-board computer, would occur three more times during the flight. (NASA Release 67–229)
- Prof. Jacques-Emile Blamont of the University of Paris would receive the International Academy of Astronautics' (IAA) 1967 Daniel and Florence

Guggenheim Award, IAA President Dr. C. Stark Draper announced. Given in recognition of an outstanding contribution to the progress of astronautics through work done during the last five years, award would be presented during International Astronautical Congress to be held in Belgrade Sept. 24–30. (IAA Press Release 32)

- Maj. Michael J. Adams (USAF) flew X–15 No. 3 to 82,000-ft altitude and 3,136 mph (mach 4.71) at Edwards AFB. Purposes of test flight were to check: (1) phase II tail loads; (2) boost guidance checkout; (3) traversing probe; (4) tip pod accelerometer; and (5) boundary layer noise. (NASA Proj Off)

August 26: First Saturn V booster was moved to its launch pad on a crawler-transporter from KSC's Vehicle Assembly Building. Three and one half mile trip took about 10 hrs. The 364-ft tall booster would undergo at least six weeks of additional testing at the pad before being used to launch unmanned Apollo 4 mission, tentatively scheduled for October. Apollo Program Director M/G Samuel C. Phillips (USAF) told newsmen at KSC that the movement of the Saturn V was an important milestone in the Apollo program. He predicted that, barring any major problems, U.S. could still achieve its goal of landing a man on the moon in 1969. He warned, however, that "it's a long way to the moon and we have a major job ahead. Dangers go with this business. We've done everything we can to improve the equipment. But, remember, we're treading on new ground and it's a risky business. We feel we can overcome the risk and make it work." (NASA Release 67–219; AP, W Star, 8/27/67, A3; NYT, 8/27/67, 1, 29; W Post, 8/27/67, 1)

- FAA announced allocations of $70.2 million in Federal matching funds to construct and improve 386 public civil airports under FY 1968 Federal-aid Airport Program (FAAP). Program provided $64.1 million to improve 348 existing airports and $6.1 million to construct 38 new public airports. (FAA Release 67–64)

August 28: National Academy of Sciences had concurred in House Committee on Government Operations recommendation that outside experts as well as those in Government be used to advise decisionmakers of R&D deficiencies and remedies. Committee released report of NAS and Bureau of the Budget comments on recommendations to improve Federal R&D decisionmaking process, which also showed NAS had concurred that Executive Office should examine program-level recommendations but had warned of danger of attempting to determine gaps or establish levels of support on too centralized a basis. NAS also had voiced strong reservations on a cost-benefit approach to R&D evaluation. (Text)

- Ralph W. Tyner, former Industrial Relations Officer at GSFC, became Deputy Director for Labor Relations at NASA Hq., replacing C. Stuart Broad, who transferred to DOD July 17. (NASA Ann, 8/25/67)

- NASA Associate Administrator for Space Science and Applications Dr. Homer E. Newell wrote to the Editor of *Aerospace Technology:* "In 1973, NASA proposes to conduct the first of the Voyager series of missions to Mars, using the Saturn launch vehicle. We wish to call the attention of the scientific community to the opportunity of proposing experiments for this initial flight. . . .

". . . Regardless of the level of funding support obtained for Voyager this fiscal year, it is our plan to begin evaluating scientific proposals on Nov. 1, 1967, and to select scientists for participation in the planning and development steps by Feb. 1968. . . ." (*Aero Tech*, 8/28/67, 8)

- ARC was expected to award a one-year, $100,000–$150,000 study contract within a few months to define the characteristics of an economical general aviation aircraft engine for 1985, *Aviation Week* reported. Contract would be part of NASA's renewed emphasis on general aviation and would be designed to support in-house work that ARC had been conducting for the last year. Basic objective was to find areas where new technology could be applied to present or future propulsion systems that would lead to a more economical general aviation engine. (*Av Wk*, 8/28/67, 113)
- Report released by House Committee on Government Operations' Subcommittee on Military Operations said that DOD's Initial Defense Communications Satellite Program (IDCSP) had many inadequacies and urged that steps be taken "forthwith" to reinforce it by adding synchronous satellites. Report argued that undue reliance on present system would stretch it "skintight," impairing its capabilities if unexpected failures developed. The subcommittee expressed concern over DOD's apparent intent to use IDCSP "for a longer period than originally planned" and its tendency to lean "rather heavily on the future promise of technological breakthroughs and improvements in the state-of-the-art." It conceded it would be wrong to settle for a limited system when new technological advances could be made, but also felt it would be an error to inflate the possibilities of present system as an excuse to defer needed improvements. (Jones, *M/S Daily*, 8/29/67)
- USAF had awarded letter contract to Lockheed Missiles & Space Co. to begin Program 612, a new reconnaissance satellite system, *Aerospace Technology* reported. It was estimated that program would cost $350 million. (*Aero Tech*, 8/28/67, 3)
- European aerospace industry assn., Eurospace, was urging formation of an optional regional communications satellite system to provide TV and telephone coverage for Europe, Donald E. Fink reported in *Aviation Week*. An expanded version was also being studied to link the continent with the entire Mediterranean basin, the Middle East, and major portions of Africa. Designated Eurosat, system would consist of one 375-lb comsat launched into stationary orbit by an ELDO Europa II booster and at least 18 ground stations in Europe and North Africa. Fabrication of satellite and ground stations would be restricted to European aerospace companies. Initial cost estimate for satellite system was $100 million. (Fink, *Av Wk*, 8/28/67, 29)
- ComSatCorp had requested proposals for construction of its new laboratories near Clarksburg, Montgomery County, Md., to house initial staff of about 300 [see May 5]. Mission of laboratories would include research in radio-frequency transmission, communications processing, spacecraft systems, and physics. (ComSatCorp Release 67-41)

August 28–30: AIAA meeting on "Space Program Issues of the '70's" was held in Seattle.

Rep. Joseph E. Karth (D-Minn.), Chairman of House Committee on Science and Astronautics' Space Science and Applications Subcommittee, said in panel discussion that the civilian part of the Nation's space program "no longer has a very high priority," partly because of the Vietnam war and partly because NASA had not convinced the public and Congress of the economic and technological benefits of space research. Spending on space exploration was not likely to be increased significantly for several years and accomplishments in the 1970s could be "severely limited." He said: "Recent Congressional action on the 1968 NASA budget was a direct reflection of these changing [economic, political, and social] values; it was also a reflection of how the mood of a majority of those in the Congress has changed [and] there is little doubt that the worth of the space program, in terms of economic, political and social values, has been judged and found wanting by large numbers of the public and their representatives in government." He added that, "in spite of great advances, scientific and technological programs are among the least understood of our national efforts [but, on the other hand] every nation in the world that is unable or fails to carry out an aggressive R&D program also fails to develop an economy that educates, feeds, houses and clothes its people." Karth urged that NASA consider "an evolution to a functional organizational arrangement . . . on such a breakdown as: earth orbital, lunar, solar, planetary and deep space." He hoped to see "the emergence of a better system" for long-range planning which was not simply a "pasting together of manned and unmanned plans," but said that "long-range prospects for progress are enormous [and] these prospects must be sold to the public and top policy-makers in terms of what makes sense for the overall benefit of society." (Text; Wilford, *NYT*, 8/30/67, 17)

NASC Executive Secretary Dr. Edward C. Welsh described national space program planning as having "characteristics of complexity, flexibility, and continuity." He recalled that NASC had recommended to the President that the Nation undertake the Apollo program, with its "within this decade schedule" and added: "Apollo project was planned at a time when we were in a secondary position to the Soviets in space competence [and] when we needed a policy decision and specific plans to accomplish that decision in order to energize our entire space effort. Along with a number of other space projects, this major project has given substantial spirit and substance to our program and we are no longer behind any country in over-all space performance." Welsh concluded: "'this country does not now need such a special source of impetus [like Apollo] to keep it going in the right direction in space [and] to maintain our competence or sense of purpose." (Text)

Dr. Albert J. Kelley, formerly Deputy Director of ERC and currently Dean of Boston College's Graduate School and School of Business Administration, argued that clearly defined post-Apollo goals had to be formulated if full potential of the Nation's space program were to be realized. "If we set forward well-defined goals and plans we have a good possibility of continuing to maintain a strong national space program; if not, then we stand a chance of losing by default much of our gains to date, but even more important of not realizing our potential payoffs in the 70's." (*SBD*, 8/30/67, 312)

August 28–31: Aviation Subcommittee of Senate Committee on Commerce held hearings to explore needs, problems, and means of maintaining an adequate airport system for a forecast 400% air passenger growth by 1980. Proposals heard on financing included passenger head tax, fuel tax, landing fees, Federal loan program, interest subsidy, and expanded FAAP funds. Among other proposals were noise standards, engine and airframe redesign, CAB diversion of traffic, reliever airports, air taxis, helicopter service, and high-speed ground transportation. Also recommended to the committee were increased state participation, tightened congressional control, and a national air transport system. (Text)

August 29: Sen. Margaret Chase Smith (R-Me.) expressed concern on Senate floor that the Nation's space program suffered from a "serious lack of centralized leadership within the administration. Leadership is needed to weigh and decide on the merits of proposed space projects; to coordinate program requirements and development efforts of civil and defense agencies; and to establish long-range national space goals and to seek the resources necessary to achieve them.

"I believe that this failure to assert administrative responsibility has resulted in the costly and needless development of two families of launch vehicles—one for NASA and one for the Department of Defense. Currently, NASA and the Department of Defense are each separately undertaking manned space programs having similar, if not identical, basic objectives—to determine whether man can effectively operate in space. . . ."

Criticizing President Johnson's support of NASA FY 1968 budget cuts recommended by House Appropriations Committee [see Aug. 21], Sen. Smith said: "I am mindful that some significant budgetary decisions must be made because of our commitments in Vietnam and social strife at home. But is the Congress to supply the initial decisionmaking power for the executive branch? Is so little consideration and concern given to determining the required resource level for space technology programs that across-the-board reduction recommended by a legislative committee can be accepted so lightly?

"There are many of us who feel that science and technology may well be the key to our future existence—to the very survival of freedom in the world of tomorrow. In the past, I believe that the administration also held this to be true. However, the present lack of responsible action in regard to our space program leads me to wonder whether the administration plans and manages the future course for our country with any foresight or merely reacts to events with fear and trepidation." (*CR*, 8/29/67, S12399)

- NASA Deputy Administrator Robert C. Seamans, Jr., reporting on gravity of budget situation in the light of pending congressional action, said that study of program alternatives would continue while hiring of additional personnel would be suspended. He gave elements of NASA Hq. guidance on restricting activities, to provide maximum flexibility for future decisions when budget and expenditure limits were clarified. (Text)
- A 180-man crew working a 10-hr day six days a week had completed 14-week construction job of 15-building installation in the Australian desert, which would be used to intercept information transmitted by Soviet reconnaissance satellites [see Aug. 12]. (AP, *NYT*, 8/29/67, 24)

August 30: NASA engineers maneuvered *Lunar Orbiter III*, launched Feb. 4, by firing spacecraft's velocity control engine 125.5 sec, with intent to simulate planned orbit for manned Apollo. Planned 100-mi circular orbit for Apollo was approximated in 196-mi apolune and 89-mi perilune achieved by maneuver from *Lunar Orbiter III*'s prior 1,133-mi apolune and 87-mi perilune. Data on lunar gravity fields would be obtained for Apollo program. Lunar Orbiter series of four spacecraft orbiting around moon was providing valuable tracking target activity for NASA's Manned Space Flight Network (MSFN). (NASA Release 67–233)

- NATO intelligence had learned that French President Charles de Gaulle would begin extensive cooperation with U.S.S.R. in the field of space exploration, including the orbiting of nuclear weapons, Henry J. Taylor reported in the Philadelphia *Evening Bulletin.* Taylor claimed U.S.S.R. was devoting a "substantial percentage" of its total military budget to the orbiting of nuclear weapons. (Taylor, P *EB,* 8/30/67)

August 31: Two NASA Nike-Tomahawk sounding rockets, launched from Andoeya, Norway, carried GSFC payloads containing chemicals to produce barium clouds. Peak altitude attained by rockets would be determined by triangulation and by magnetometer measurements; no radar track was made. Missions of the cooperative Norway-U.S. project were (1) to compare electric field measurements made by two techniques, double probe and barium release; and (2) to analyze electric fields from observed motions of neutral and ionized barium clouds during an aurora condition. In this first of three pairs of launchings, good data were obtained and excellent results were expected. (NASA Rpt SRL; NASA Release 67-234)

- U.S.S.R. launched *Cosmos CLXXIV* into orbit with 39,750-km (24,699-mi) apogee, 500-km (311-mi) perigee, 715-min period, and 64.5° inclination. (GSFC *SSR,* 8/31/67)

During August: "Despite the rancor and doubt of the moment [about U.S. space program], the evidence is persuasive that the nation has been well served indeed," *Fortune* magazine said in an article reviewing NASA problems, shortcomings, and accomplishments. The article, "Jim Webb's Earthy Management of Space," said: "Webb . . . has developed a managerial doctrine as extraordinary as the space mission itself. Essentially it is a bold extension of the systems-management principle first applied in the Pentagon for weapon development. This approach means that the Pentagon tries to relate design and procurement plans to the weapon system's maximum potential, to other weapon systems, to cost effectiveness, to foreseeable strategic situations, and to other special considerations. In NASA's case, the 'system' is usually an exotic research objective, like landing a package of instruments on Mars. Such an undertaking, of course, requires management consideration of costs and engineering feasibility, but NASA also tries to relate it to the social, economic, and even political impact the new knowledge may have on some of the most basic mysteries of human life. In one sense, the system becomes an exercise in applied philosophy. To master such massively complex and extensive problems, the agency has mobilized some 20,000 individual firms, more than 400,000 workers, and 200 colleges and universities in a combine of the most advanced resources of American civilization." (Mecklin, *Fortune,* 8/67)

- The Nation's press commented on photos transmitted by *Lunar Orbiter V.* *New York Times:* "Orbiter 5 is currently adding some enduring contributions to mankind's cultural heritage as photographic by-products of the space effort.

"Some of the close-up pictures of the moon have been among the wildest, most disordered and most surrealistic scenes ever viewed by human eyes. Yet its distant views of the far side of the moon convey a sense of desolate tranquillity, of peace achieved after many storms.

"This new art form has been produced by machines hundreds of thousands and millions of miles away, in realms where men have never been. Yet many of these pictures have far more power to move than much of the output of contemporary artists. . . ." (*NYT,* 8/13/67, 10E)

Washington *Evening Star:* "The pictures of the moon we have been getting back from our assorted orbiting cameras have been fine and fascinating, but the most interesting of all has turned out to be the picture of the earth from the distance of the moon.

"This is natural enough: self-portraits are always the best portraits, especially from the point of view of the sitter.

"But the thing about this self-portrait is the inescapable conclusion that the earth is really a rather lovely place, when seen from 214,806 miles away...." (W *Star*, 8/18/67)

- Choice of spacecraft atmosphere, according to *Space/Aeronautics,* "won't be pure oxygen, but whether the diluent will be nitrogen, helium or latecomer neon is far from clear." It emphasized oxygen regeneration would require difficult hardware tradeoffs to satisfy variety of missions. (*S/A*, 8/67, 71)
- Federal civilian employment of professional scientific and technical personnel rose by 9% from October 1962 to December 1964, NSF stated in report on occupational characteristics of scientific and technical employees of Federal Government in December 1964. Engineers, largest of three major groups—scientists, engineers, and health professionals— made up 40% of the total 189,500 in 1964. Scientists made up about 36%. DOD continued to be the major Government employer. (Text)
- Soviet northern cosmodrome, whose existence was first publicly announced at British Interplanetary Society's November 1966 meeting following radio-tracking activities at Kettering Grammar School [see Dec. 12, 1966], had maintained steady rate of satellite launches since beginning with Cosmos satellite in March 1966. New launch site, about 100 mi south of Archangel and between towns of Yarnema and Tarasova, still had not been acknowledged by Soviet authorities. (Perry, *S/F*, 8/67, 274)
- First issue of Soviet magazine *Space Biology and Medicine* had been published in Moscow. Contributors included medical experts, biologists, physiologists, and psychologists. Subjects included space psychophysiology, and life support systems. (*S/F*, 8/67, 271)

September 1967

September 1: Need for design modifications had required shipment of a manned orbital workshop to the McDonnell Douglas Space Systems Center, Huntington Beach, Calif., said Marshall Space Flight Center. Workshop, built by McDonnell Douglas, was model for the Saturn S–IVB stage and when modified would represent Saturn stage as it was to be used in Apollo Applications program as a manned space laboratory. Some 50 ft long and 21.5 ft in diameter, workshop was used at MSFC as an engineering tool in designing structures, equipment, and experiments for the initial mission of 28 days. (MSFC Release 67–181)

- FAA proposed maximum speed of 250 knots per hr (288 mph) for all aircraft operating below 10,000-ft mean sea level (MSL) in move designed to give pilots more time to "see and avoid" other air traffic in Nation's most heavily traveled airspace. Proposed rule would mean a change only for en route aircraft below 10,000 ft MSL. Aircraft arriving at an airport and flying below 10,000 ft MSL had previously been placed under speed limitations: an indicated air speed of 250 knots for all such aircraft within 30 mi of their destination; within immediate terminal area, 200 knots for turbine-powered aircraft and 156 knots for piston-engine aircraft. FAA cited growing numbers of high-performance aircraft using airspace below 10,000 ft MSL where virtually all VFR (visual flight rules) flying is done, as well as about half of all IFR (instrument flight rules) flying. FAA said proposed rule would promote safer and more efficient utilization of airspace, since pilots of many high-speed aircraft probably would choose to operate above 10,000 ft MSL for optimum performance and economy. (FAA Release 67–66)

September 2: The second of three pairs of NASA Nike-Tomahawk sounding rockets in cooperative Norway-U.S. project [see Aug. 31] was launched from Andoeya, Norway, carrying GSFC payloads, one of which contained chemicals to produce four barium clouds. Investigating electric fields during an aurora condition, experimenters obtained good data. (NASA Rpt SRL)

- Soviet Union announced a new series of rocket tests believed designed to perfect a sea-landing system for Russian astronauts in preparation for a resumption of manned space flights. Soviet news agency, Tass, said the test of "carrier rockets" would begin Sept. 3. It warned all ships and planes to stay clear of two 80-mile-wide "bulls' eyes" in the Pacific Ocean, one about 900 miles northeast and one 300 miles northeast of Midway Island, from noon until midnight every day through Oct. 30. (Reuters, *W Post,* 9/3/67, A16)

September 3: Lunar Orbiter V's photographs had revealed flow patterns characteristic of hardened fluid material on the floor of the moon's Tycho Crater, NASA announced [see Aug. 5–27]. Preliminary study of new photographs tended to confirm prevailing theory held by scientists that the big crater had been formed by a great meteoroid crashing into the visible side of the moon. Some scientists believed much of the fluid material ejected when the three-mile-deep crater was formed might have resulted from the intense heat generated by the meteoroid's impact. They also suggested that the appearance of the dark halo resulted from the manner in which pulverized ejected material had been deposited. Material might have traveled outward from the impact point in a "base surge" close to the moon's surface, giving a windswept appearance to the terrain. NASA photo afforded scientists first close-up view of the crater floor, providing new clues to its origin. *Orbiter V*'s high resolution telephoto picture showed fractures, flow markings, and protruding dome-like hills with exposed layers. The Tycho region was one of the 36 areas of high scientific interest on the moon photographed by *Lunar Orbiter V*. (NASA Release 67-231)

- Soviet Academician Leonid I. Sedov hinted at cooperation between the Soviet Union and the United States in manned exploration of space. The vice president of the International Aeronautical Federation dropped the hint in an interview with Tass: "Flight to Mars and the creation of large orbital scientific bases . . . will certainly demand joint efforts of all nations." The Soviet Union previously had declined all suggestions of joining its space efforts with those of the United States, although it had a limited agreement with France. (Reuters, *NYT*, 9/5/67)

September 4: NASA-sponsored research on equipment used for surviving hard landing on the moon and planets had been conducted by Cal Tech-JPL engineers in simulated operations. Split-second JPL impact tests, which used powerful compressed air guns and a giant slingshot, hurtled equipment against solid backstops at speeds up to 500 feet per second (340 mph). JPL's engineer, Stan Taylor, explained: "During the brief instant that the test specimen is experiencing the 10,000 g shock, every part of it behaves as though it weighed 10,000 times its normal weight. The bolts supporting a one pound test item have to be good for about five tons." The experiments had proved that intricate, delicate components required in sophisticated unmanned spacecraft of today and tomorrow could be protected to withstand shocks of 10,000 g. (JPL Release)

- NASA's work on advanced air-breathing engines was aimed at development of significantly improved propulsion systems, primarily for second-generation supersonic transports, *Aviation Week and Space Technology* reported. Based mainly at LeRC, advanced engine component research program was designed to improve overall performance by improving efficiencies of individual components such as the compressor, combustor, turbine, and nozzle. Although program was directed primarily to supersonic powerplants, most anticipated improvements were expected to be applicable to future subsonic jet engines. (*Av Wk*, 9/4/67, 64–71)

- The House Committee on Government Operations in H.R. 613 urged DOD to proceed with plans for an advanced communications satellite system that would eventually replace the present Initial Defense Communications Satellite Program (IDCSP). The Committee said it "would be a mistake" to lean too heavily on the IDCSP because its limited channel capacity was already strained. Several possible system configurations were under study by DOD. (*Av Wk*, 9/4/67, 18)

- Current attacks on the SST project appeared to stem from ignorance of the

September 4

plan and a failure to appreciate what it would contribute to the economy and security of this country, reported columnist L/G Ira C. Eaker (USAF, Ret.). He had presented estimates of the sales volume and financial aspect of the SST programs which had been developed, checked, and rechecked by civil advisory groups, the airlines, and individual experts in transportation, industry, and finance. He said no Federal program since World War II had had the thorough analysis, careful examination, and close scrutiny accorded the present SST plan. He emphasized that the SST was not competitive in any way with social reform programs. On the other hand, it would stimulate continued growth of the economy, create employment, and provide a higher standard of living for thousands. (Eaker, *Detroit News*, 9/4/67)

- Michael E. Gluhareff, Russian-born engineer and associate of helicopter pioneer Igor I. Sikorsky, died in Bridgeport, Conn., at 74. Gluhareff was engineering manager of Sikorsky Aircraft, a division of United Aircraft. His many patents had included a dart-shaped plane, sailplane (glider), tail-less airplane with rear-mounted propeller, and device to use exhaust gases of engines to increase the speed and load of the plane (co-inventor). (AP, *W Post*, 9/6/67, B8)

September 5: ATS Hydrolic Communications Experiment system, designed to report river height and rainfall measurements automatically via satellite was the object of a cooperative program by the Weather Bureau's Office of Hydrology and NASA, ESSA announced. The synchronous satellite, *Ats I*, would be the communications link between automatic measurement devices at three locations in the U.S. and the Office of Hydrology in Washington, D.C. Purpose was the development of a fast, reliable, operational system by which potential flood conditions would be relayed instantly so public warnings could be issued. Sites chosen were Benton, Ark., which started transmitting data to the satellite on Aug. 1; Salem, Ore., which began operating Aug. 4; and Sacramento, Calif. (ESSA Release 67–68)

- Trained monkeys completed a two-month simulated space flight which indicated that astronauts could return their spacecraft to earth even after a near-fatal dose of solar radiation. Sixteen rhesus monkeys, trained to operate an instrument panel, were used in a test at Los Alamos Scientific Laboratory. Some of the animals were exposed over a 10-day period to from 500 to 1,000 roentgens of gamma radiation. Three of the monkeys in the high-dose group died but the others survived. This evidence of monkey vitality was of great importance to man because the U.S. target date for moon landings would come at a peak of the sun's 11-year-cycle for emitting deadly radiation. (*C Trib*, 9/5/67)

September 6: Early Bird I communications satellite passed its test as a device for expediting clearance of passengers and cargo through airports when jet aircraft would be put in service in 1970. Moments after an airline flight took off from Frankfurt, Germany, for Dulles International Airport, data on passengers and cargo were sent via satellite to Government officials at Dulles so they could process the data in advance of the plane's arrival. Officials in Frankfurt relayed facsimiles of the cargo manifest, health certificates, and Customs declaration forms of the 120 passengers. (*NYT*, 9/7/67)

- NASA Nike-Apache sounding rocket, launched from Puerto Rico in Univ. of Illinois experiment to probe lower regions of ionosphere up to about 125 mi (200 km), achieved satisfactory results. In first of three such rocket flights, measurements were made simultaneously with those of Arecibo Ionospheric Observatory's giant radar-radio telescope, 30 mi from Vega Baja Airport launch site. World's largest radar-radio telescope took ionospheric readings, using its 1,000-ft-dia wire mesh reflector. (NASA Proj Off; NASA Release 67-230; WS Release 67-28)

- Army contract awards to the Western Electric Company, totaling $231.4 million, for Nike-X work had been announced. Specific totals and contract purposes: $215.2 million for continued R&D; $13.2 million for deployment planning activities; and $3.0 million for support facilities. (DOD Release 837-67)

September 7: Reorganization of the Bureau of the Budget had created the Economic, Science and Technology Division, headed by John D. Young, reported Jerry Kluttz, in *Washington Post*. Young, at one time NASA's Assistant Administrator for Administration, had headed the task force that recommended the shakeup. Reporting to his new division would be the Departments of Commerce and Transportation, AEC, NASA, NSF, Small Business Administration, and the regulatory agencies. Both NASA and AEC were transferred from the Military Division. (Kluttz, *W Post*, 9/7/67)

- Navy contract award had been made to Radio Corporation of America in letter contract for $2.0 million, for six navigation satellites. The Special Projects Office, USN would be contracting activity. (DOD Release 846-67)

September 7-9: NASA's *Biosatellite II* (Biosatellite-B) was successfully launched from ETR by a two-stage Thrust-Augmented Improved Delta launch vehicle into orbit with 202-mi (326-km) apogee, 187-mi (302-km) perigee, 90.8-min period, and 33.5° inclination. The 940-lb research spacecraft consisted of three main sections: an adapter section which would remain in orbit; the reentry vehicle carrying the retrorocket and heat shield for reentry into the earth's atmosphere; and, within the reentry vehicle, the experiment capsule containing the scientific experiments, life support equipment, parachutes, and radio beacon to aid in recovery. *Biosatellite II* carried 13 experiments to determine effects of the space environment on various life processes.

Performance of the spacecraft was satisfactory except for some difficulty with completely reliable response to command transmission from the ground stations and out-of-spec accelerations during the first few orbits. The spacecraft's temperature, attitude control, and atmospheric pressure readings were normal, and few additional anomalies were experienced. Mid-air recovery of NASA's *Biosatellite II* experiment capsule, by aircraft at 3:15 pm EDT Sept. 9, within 15 mi of predicted impact location of 7°15'N and 162°1'W, marked a new first. *Biosatellite I*, which had remained in orbit for two months after the retrorocket failed, landed in the ocean near Australia following reentry Feb. 15, but was not recovered.

The most important single question for *Biosatellite II* was whether the changes produced in organisms by radiation are slowed or hastened under weightless conditions. Biosatellite scientists had noted that early recovery had resulted in less risk to the experiments than if recovery had been delayed beyond the planned three days. Despite the delayed launch and early recovery, scientists reported only a small decrease in the expected data return from the experiments. Indications were that all programmed events in orbit, such as fixation of frog eggs, fixation and

Before launch.

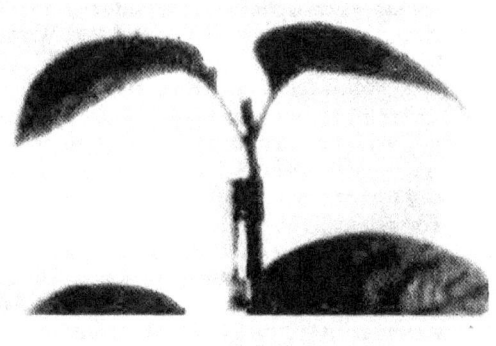

4 hours and 40 minutes of weightlessness.

12 hours and 29 minutes of weightlessness.

17 hours and 40 minutes of weightlessness.

September 7–9: Pepper plant, subjected to 45 hrs in orbit aboard NASA's *Biosatellite II*, shows progressive effects of weightlessness on leaf orientation. Scientists believe loss of gravity causes growth hormone to move from bottom to top of leaves, in turn causing cells on top to grow and induce downward curling.

feeding of amoeba, fixation of wheat seeds, and pepper plant camera shutter opening, had worked as planned. Successful recovery of plant and animal life carried on a 45-hr journey through space by *Biosatellite II* was expected to provide important clues to how life forms would develop in a weightless environment or how radiation might affect future generations.

Biosatellite II was under management of ARC and under direction of OSSA. The Delta launch vehicle was managed by GSFC. Communications and tracking were handled by NASA's Satellite Tracking and Data Acquisition Network, operated by GSFC. *Biosatellite II* was built by the General Electric Reentry Systems Dept. The Delta launch vehicle was built by Douglas Aircraft Co. (NASA Proj Off; NASA Release 67-239; *Av Wk*, 9/18/67; AP, *W Post*, 9/11/67; Myler, UPI, *W Post*, 9/15/67, 9/18/67; B *Sun*, 9/8/67; Yuenger, *C Trib*, 9/15/67)

September 8: Explorer XXXIV, launched May 24, demonstrated that all its spacecraft systems and 10 of 11 experiments were operational and performing satisfactorily. No scientific data had been received from TRW spherical electrostatic analyzer experiment since launch because of electrical system malfunction. (NASA Proj Off)

- Col. Joseph F. Cotton (USAF), pilot, and L/Col. Emil Sturmthal (USAF), copilot, flew XB-70 No. 1 to 59,700-ft altitude and 1,510 mph (mach 2.3) in flight at Edwards AFB. Purposes of test: longitudinal stability and control and handling qualities checkout; throat unstart checkout; variable nose ramp functional test; radar airspeed calibration; and handling qualities approach sidestep maneuvers. Flight was successful except for lost radar contact. (XB-70 Proj Off)

- NASA's *Mariner IV*, launched Nov. 28, 1964, after 1,014 days of flight, reached its closest approach to earth at 29.1 million mi (46.9 million km). (NASA Proj Off)

September 8-10: Engineers and scientists on three continents secured aerial recovery of *Biosatellite II*, approximately 15 mi from the recomputed impact point over the mid-Pacific, and softlanded *Surveyor V* on the moon at a speed just over eight miles an hour and within 18 mi of its original target site. The work had involved hundreds of men in tracking stations (certain of which could track simultaneously up to three separate objects: deep space, manned capsule, or near-earth satellite) and control centers in the U.S., South America, and Australia. *Surveyor V* and *Biosatellite II* projects were carried out by two different NASA teams. Participants had included scientists and engineers from JPL and Hughes Aircraft Co. for *Surveyor V*, ARC and GSFC for *Biosatellite II*, and tracking-station and control-center personnel from around the world. (NASA Release 67-241; NASA Proj Off)

September 8-30: NASA's *Surveyor V* (Surveyor-E) unmanned spacecraft was successfully launched by Atlas-Centaur launch vehicle from ETR on a two-burn ascent mission to softland in the Apollo area of interest on the moon and to take television pictures of the lunar surface around its landing area. *Surveyor V* was the first of the Surveyor series to carry an instrument to study the chemical characteristics of the lunar soil and the first to attempt a landing in the eastern portion of the Apollo zone. The coast period between burns was 6 min 44 sec. Spacecraft insertion into a lunar trajectory occurred at 04:15:12.9 EDT Sept. 8. The trajectory appeared to be satisfactory. After separation from the launch vehicle, *Surveyor V* automatically acquired the sun and responded to

September 8–30: First chemistry set on the moon, *Surveyor V*'s alpha-backscattering instrument slid several inches downhill Sept. 12 when it was gently nudged by firing of the spacecraft's three vernier rockets. The picture at left shows the instrument as deployed Sept. 11. The picture at right, after half-second engine firing, shows edge of instrument's flange covered with lunar soil. Fragments against uphill edge of instrument range from a few tenths to one inch in size.

commands to turn its transmitter to low power. Spacecraft telemetry indicated that all systems were operational. The *Surveyor V* midcourse correction was accomplished at 21:45 EDT Sept. 8. Immediately after the correction, the pressure in the helium supply tank was determined to be decreasing, indicating that the regulating valve did not make a positive closure. Correction of leak was not possible; a descent strategy was developed to overcome the resultant degradation of the vernier engine retrograde capability.

Surveyor V softlanded on the moon at 20:46:38 EDT on Sept. 10. The landing site was in the Sea of Tranquility at approximate coordinates 23.19°E, 1.52°N, some 18 mi from the target point. The spacecraft took 200- and 600-line pictures of the lunar surface and conducted an alpha backscattering experiment. At lunar sunset, 6:40 am EDT, Sept. 24, 18,006 high-quality lunar photographs had been obtained (more than the combined totals from *Surveyors I* and *III*). The alpha-backscattering instrument had operated for 83 hours on the lunar surface, providing excellent data on the relative abundance of chemical elements in lunar material. To optimize the conditions for lunar-night survival, the spacecraft was operated for short periods at 2½-hr intervals, using compartment heaters, until the battery charge was down to 30 ampere hours, about 4–6 days after lunar sunset. The spacecraft then would be placed in hibernation for the remainder of the lunar night.

The vernier engine erosion experiment was conducted at 01.38 EDT on Sept. 13, to obtain information on effects of impingement of a high velocity jet-gas stream on the lunar surface. TV pictures were taken of the area under study before and after vernier engine firing. Initially no significant erosion was noted, but follow-on operations with low sun angle photographs and shadow progression studies revealed a crater-like depression under vernier engine number 3 which had not been visible at the higher sun elevations. *Surveyor V* accomplished all mission objectives. Before touchdown, radar reflectivity data were obtained; touchdown dynamics data were determined at landing time (bearing strength of lunar surface as the ultimate objective). Thermal data were obtained (accumulation of data throughout the mission was the ultimate objective).

The Surveyor program was managed by JPL under OSSA direction. Hughes Aircraft Co., under contract to JPL, designed and built the Surveyor spacecraft. Tracking and communications were the responsibility of JPL-operated Deep Space Network. (NASA Proj Off; NASA Release 67–227; B *Sun*, 9/11/67; AP, W *Star*, 9/25/67; Hines, W *Star*, 9/29/67; W *Post*, 9/14/67; O'Toole, W *Post*, 9/30/67; Clark, *NYT*, 9/30/67)

September 9: Sen. John O. Pastore (D-R.I.), Chairman of Joint Committee on Atomic Energy, speaking at launching of Navy's latest nuclear submarine, the *Narwhal*, in Groton, Conn., called for "full speed ahead" on building an ABM system. (*Av Wk*, 9/18/67; *CR*, 9/19/67, S13207–9)

- To astronauts, Maj. Edwin E. (Buzz) Aldrin, Jr. (USAF), and Capt. James A. Lovell, Jr. (USN), had won the 1967 Harmon International Aviation Trophy, the Clifford B. Harmon Trust announced in New York. Trophies also would go to Miss Sheila Scott of Britain, who had established speed records during a 28,633-mi solo flight around the world in a single-engine plane, and to Alvin S. White. White, now manager of supersonic flight research and development for Trans World Airlines, was named the outstanding airplane pilot for test flights of the XB–70 at mach 3 as chief test pilot for North American Aviation Company. The awards are made annually for exceptional feats of individual piloting skill. (*NYT*, 9/10/67)

- Stanford Linear Accelerator Center was dedicated. Two-mile-long building would contain the world's longest research instrument, one of the most complex and precise machines ever built by man. Director of the Stanford Linear Accelerator Center would be Professor Wolfgang K. H. Panofsky. The energy of the linear accelerator would be three times that of any other electron accelerator in the world, producing more than 20 billion electron volts. Dr. Glenn T. Seaborg, Chairman of AEC, in his speech of dedication, hoped that "these forces of science, these common quests that men can undertake and work together on, . . . will in large part be responsible for the fulfillment of one of man's most basic needs— that of being able to live together in peace and understanding." Dr. Seaborg emphasized that the center "is open . . . to qualified scientists from all parts of the United States and the world." (AEC Release S–35–67)

September 10: Soviet Union reportedly had been giving top priority to efforts to develop multiple warheads for its big missiles, William Beecher wrote in *New York Times*. If the report was correct and if the Soviet effort succeeded, Moscow could increase several-fold its ability to hit a number of targets in the U.S. simultaneously. In Britain, Air Commodore Neil Cameron described the new weapon in *Royal Air Force Quarterly*. "It is widely believed that the Russians have developed the multiple individually guided reentry vehicle, known as MIRV, which could revolutionize nuclear strategy. . . . It would mean that each Soviet missile dispatched could carry ten or more individually guided warheads in the megaton class, which could neutralize the Americans' numerical lead in this field and make a counter-force strategy as far as the Soviets are concerned a realistic possibility." (Beecher, *NYT*, 9/10/67; AP, *B Sun*, 9/15/67; Cameron, *RAF Quart*, Autumn 67, 175–83)

September 11: An all-metal payload shroud which separated in three sections would be used when an Air Force Titan III launch vehicle placed a DOD communications satellite into space in the fall of 1968. Developed and produced by Douglas Aircraft Co., Inc., under a $2.6-million contract, the shroud would be made up of fairings to protect the sensitive instrumentation of scientific payloads during passage through the earth's atmosphere. (AFSC Release 171.67)

- *Cosmos CLXXV* was launched into orbit with 386-km (240-mi) apogee, 210-km (130-mi) perigee, 92.0-min period, and 72.9° inclination. Satellite reentered Sept. 19. (*W Post*, 9/12/67; GSFC *SSR*, 9/30/67)

September 12: Two NASA Nike-Tomahawk sounding rockets, launched from Andoeya, Norway, carried GSFC electric field experiments. Four barium clouds released by the first were photographed from all sites for analysis of electric fields from motions of neutral and ionized barium clouds during aurora condition. The second rocket was launched to correlate electric field measurements with measurements of magnetic fields, flux, and energy spectra of low-energy particles, but failed on loss of radio-frequency signals. (NASA Rpt SRL)

- NASA Aerobee 150 sounding rocket, launched from WSMR to verify launch compatibility of rocket and modified Nike-Ajax military rail launcher, failed on premature staging at 150 ft; launch verification was successful. (NASA Proj Off)

- The world's largest aircraft would begin to take shape in October when three sections of the Air Force's C–5A Galaxy would be joined to form a nearly 230-foot-long structure, AFSC announced. The aircraft, scheduled for completion early in 1968, had been under construction by Lockheed-Georgia Co. at Marietta, Ga., under AFSC contract. With engines and operating equipment installed—minus fuel and payload—the C–5A would weigh 323,900 pounds. It would be capable of airlifting all types of combat and support forces anywhere in the world—complete with tanks, equipment, and supplies—within 24 hours. Its total length would be 246 feet, with a wingspan of 223 feet and a tail tip more than 65 feet off the ground. (AFSC Release 149.67)

- An Army Pershing ballistic missile fired from Utah by U.S. troops overshot the White Sands Missile Range and smashed into the Mexican countryside. The Mexican government gave permission for a U.S. helicopter to cross the Rio Grande to look for it. No injuries or property damage were reported. The test firings had been part of an annual training for West German Air Force personnel and U.S. troops based in Europe. The Utah-to-New Mexico range had been used because there was no range large enough in Europe. (*W Post*, 9/13/67)

- *Cosmos CLXXVI* was launched into orbit with 1,581-km (982-mi) apogee, 206-km (128-mi) perigee, 102.5-min period, and 81.9° inclination. Soviet Union had successfully launched its second unmanned satellite in three days, according to Tass. (*NYT*, 9/14/67; GSFC *SSR*, 9/15/67)

September 12–15: The Soviet supersonic transport, the Tu–144, designed to fly at 1,559 mph, should be ready for flight tests in the first quarter of 1968, according to AP. It was understood that the Soviets had hoped to get the plane in the air before the celebration of the 50th anniversary of the Bolshevik Revolution on Nov. 7. A visiting British aviation delegation saw the incomplete Tu-144 prototype on Sept. 12 and was told by its builders that it would fly in the first quarter of next year at the earliest. Soviet Deputy Minister of Civil Aviation, Leonid Zheludev, said "we need to test it very well, and we cannot say when it will be put into operation on civilian passenger flights." The British-French supersonic Concorde, similar in delta-wing shape and in projected performance to the Tu-144, was scheduled to make its first flight by February 28, 1968, and go into passenger service in 1971. (AP, W Star, 9/13/67; AP, W Post, 9/16/67)

September 13: ComSatCorp announced the selection of B. F. Parrott & Company, Roanoke, Va., as contractor for general construction work ($1.1 million) on the new comsat earth station in Preston County, about 35 miles southeast of Morgantown, W. Va. The high-capacity earth station would serve the Atlantic area and would be capable of sending and receiving all forms of international communications via commercial satellites—television, telephone, telegraph, and data. The station would be in operation by the fall of 1968. Earth stations also had been started near Cayey, Puerto Rico, and near Jamesburg, Calif. Three existing stations were located at Andover, Maine; Brewster Flat, Washington; and Paumalu, Oahu, Hawaii. (ComSatCorp Release 67-43)

- Safety of combat air crews in Southeast Asia would be greatly improved by application of a new material (polyurethane foam) first used on the Indianapolis speedway. AFSC's engineers had successfully adapted the foam, which would be installed in the fuel tanks of combat aircraft in Vietnam. Under an Air Force contract, the Monsanto Research Corp., Dayton, Ohio, confirmed compatibility of the foam with military aircraft fuels. It would virtually prevent explosion in a direct hit on the fuel tank by machine gun tracer bullets or other incendiaries and would suppress slosh in the fuel tanks during flight. (AFSC Release 158.67)

- TV coverage of space launches and press coverage of rising costs of space program had been almost the only source of public exposure of U.S. space effort, said Aerospace Systems Group Vice President J. Lynn Helms at Rotary International meeting during Michigan Univ.'s Sesquicentennial Celebration. Speaking on "A Citizen's Return on Investment," he outlined many accomplishments of U.S. space program that affect daily life, stating that "money spent in the quest for new information keeps coming back to us again and again . . . a society that continually invests part of its annual profits into the future is a progressive society. Progressive societies serve their people better, in living standards, medicine, and health, and consideration for and by other nations of the world." (Text)

- Dept. of State announced appointment of Office of Outer Space Affairs Director Robert F. Packard as director of new Office of Space and Environmental Science Affairs. This office would be a part of the International Scientific and Technological Affairs offices, whose responsibility is to advise and assist the Secretary on scientific and technological factors affecting foreign policy. Packard would be responsible for the international aspects of U.S. programs in space, atmospheric science, marine science, and earth sciences. Packard had served from 1961 to 1962 in the Office of the Special Assistant to the Secretary for Atomic Energy and Outer Space. (State Dept. Release, 9/13/67)

September 14: Richard J. Keegan, Chief of the Procurement Division, GSFC, was appointed to new position of NASA Assistant Director of Procurement (Policy and Review). (NASA Ann)
- The Institute for Strategic Studies of London, England, had said that the U.S. was way ahead of the Soviet Union in missiles, but that the Russians were closing the gap, reported *New York Times.* The Institute's annual report, "The Military Balance," said deployment of a limited ballistic defense system around Moscow had increased protection of Soviet land-based ICBMS. The new Soviet defense system now being installed along the eastern Baltic coast—the so-called Tallin Line—might be primarily an extension of radar coverage. The report also noted a "probable reduction in the combat effectiveness" of the Chinese armed forces because of the domestic upheaval. (*NYT*, 9/15/67)
- NASA Arcas sounding rocket was launched from Barking Sands, Hawaii, to 38-mi (60-km) altitude in NOTS experiment to measure incident solar ultraviolet irradiance in support of *Ogo IV*'s mission. First of a series of seven, rocket failed to attain experimental results because of parachute malfunction. (NASA Rpt SRL)
- France delayed the signing of an agreement with Britain and West Germany for development of a European "airbus" to carry 250–300 passengers in the 1970s, reported Reuters. Officials insisted that there were only "minor differences" and that the agreement, reached last July, would be approved. Britain's Rolls-Royce Company would supply the engine, the RB–207, to be constructed in cooperation with the French firm of SNECMA and MAN of West Germany. France would build the aircraft. (Reuters, *W Post*, 9/15/67)

September 15: *Ats I* (Advanced Technical Satellite) was used experimentally to relay weather photographs of current storms to the U.S. Weather Bureau's National Hurricane Center in Miami and other hurricane watchers in the Caribbean. The satellite was launched by NASA into its synchronous orbit on Dec. 7, 1966, and was positioned over Christmas Island in the Pacific at an altitude of 22,000 miles. The satellite had been used as a testbed for advanced meteorological, communications, and satellite control systems, and also had made scientific measurements of its orbital environment. Photographs were recorded and radioed to earth by the *Essa V* weather satellite while passing over the storm area. The collected photographic data were then compiled as a mosaic and gridded by computers at ESSA before delivery to the Goddard Space Flight Center ground station at Mojave, Calif., for radio transmission to the *Ats I*. The satellite then retransmitted the gridded data to earth by means of its weather fascimile (WEAFAX) experiment. Any ground station equipped with the relatively inexpensive Automatic Picture Transmission receivers could read out the pictures. Mosaics would be transmitted daily on an experimental basis in this manner between 9 and 10 pm, EDT. WEAFAX was a joint NASA–ESSA experiment. (NASA Release 67–243)
- *Mariner IV*, launched Nov. 28, 1964, survived apparent micrometeoroid shower. Spacecraft's cosmic dust detector registered 17 hits within 15 min, while *Mariner IV* traveled between orbits of earth and Mars. Spacecraft was 29.6 million miles from earth and had traveled about 1.4 billion miles in its 1,020 days of flight. Micrometeoroid shower had caused temporary change in spacecraft's attitude but no loss of power. Scientists concluded, from one-degree temperature drop inside

spacecraft, that only the thermal shield was damaged. Within about a week, spacecraft was operating without any apparent effect from incident. Since photographing Mars in July 1965, *Mariner IV* had been used for engineering experiments and acquisition of scientific data. (O'Toole, *W Post*, 9/16/67, A7; NASA Proj Off)
- USAF launched unidentified satellite from WTR using Thor-Agena D booster; satellite reentered Oct 4. (*Pres Rep 1967*)
- Findings of 14-mo study by Denver Research Institute (DRI), under NASA contract, of methods governing acquisition of new technology in 62 firms in four industries—electric batteries, printing and reproduction, industrial controls, and medical electronics—were summarized in *Science*. Report, "The Channels of Technology Acquisition in Commercial Firms and the NASA Dissemination Program," concluded that (1) few organizations were vigorously seeking to acquire new technological know-how from space research; (2) little had been done to increase the technical awareness of the manufacturer outside the space and military fields; and (3) the technologist faced with an immediate problem was inclined to use information in standard manuals two to five years behind the state of the art. The report noted that "most individuals felt it too difficult to retrieve relevant material from the mass of government publications and indicated that they expected to learn of important government-developed technology through trade and professional channels. In several firms, those interviewed felt it wasn't really practical to keep up with and use government technology unless one's firm had government R&D contracts." (Greenberg, *Science*, 9/15/67, 1016–18)
- William Markowitz, physics professor at Marquette Univ., summarized his study of the reported flight dynamics of UFOs (unidentified flying objects) in *Science*. Study was made because of letter in Oct. 21, 1966, *Science* by Dr. J. Allen Hynek, consultant to USAF's Project Blue Book, urging investigation of residue of puzzling UFO cases by physical and social scientists. Markowitz's findings: (1) the control of reported UFOs by extraterrestrial beings is contrary to the known laws of physics and (2) the data published do not justify the holding of investigations. He added that he would not believe that earth had ever been visited by any extraterrestrial until he was shown such a visitor. (Markowitz, *Science*, 9/15/67, 1274–79)
- A Soviet astronomer, V. Makarov, had said UFO reports were not all hoaxes, according to the magazine *Technica Molodezhi*. The astronomer added that he did not believe they were full of little men from far out in space either. Previously, UFOs had been written off by Soviet commentators as inventions of the foreign press or the hallucinations of disturbed persons. Mr. Makarov wrote that "the files of some private persons and organizations dealing with these phenomena either as their official job or out of curiosity are filled with considerable information. Whatever they may be, we cannot say that some phenomena which are still hard to explain exist in reality." Mr. Makarov said data had come from radar stations, pilots, and scientists "about whose common sense and honesty we have no right to be doubtful." (*NYT*, 9/14/67)

September 16: The likelihood of aircraft collisions and ship collisions would be reduced considerably by an advanced navigation satellite, NASA's Eugene Ehrlich, chief of NASA's Navigation and Traffic Control program,

September 16

explained. By providing accurate position fixes, a system of such satellites also could reduce the separation distances between aircraft; automatically steer ships across oceans; warn high-flying supersonic jets of radiation dangers; guide craft around bad weather; monitor airplane systems for possible trouble; and pinpoint ditched planes or ships in distress. "The navigation satellite of 1970–1980 will be an entirely different spacecraft from the presently operational navigation satellites of the United States Navy." The Navy orbited its first Transit navigation satellite in 1960. Attention had focused on the Navy system recently when the U.S. announced that the previously classified network would be made available to America's civilian ships. The Navy satellites orbit at 700 miles. NASA envisioned a network of evenly spaced synchronous satellites 22,300 miles above the earth. A ground craft could be in touch with as many as six of them simultaneously for extremely accurate positioning. Mr. Ehrlich said advanced satellites "will contain a position determination capability, communications service, weather relay service and scientific sensors." He said the Maritime Administration and advanced thinkers in the shipping field had been looking to the day of the automated ship. "The satellite-derived position information would be fed directly into the ship's computer to keep the ship on a preset path." Mr. Ehrlich also had said that the reduction of the present widely spaced separation standards for aircraft was estimated to be capable of saving some $46.5 million annually for the North Atlantic region alone. (B *Sun*, 9/18/67)

- The 20,000 concerns that work for NASA under contract have laid off employees at the rate of 4,000 a month, largely because of cutbacks in funds, James E. Webb, NASA Administrator, told an audience at the dedication of the Olin E. Teague Research Center, Texas A&M Univ., College Station, Texas. He said about 100,000 persons had been laid off and the figure might reach 200,000. Rep. Olin E. Teague (D-Tex.), in a short conference following the dedication, said lack of money would be the biggest detriment to U.S. space programs. (UPI, *NYT*, 9/18/67; *Bryan Daily Eagle*, 9/17/67)
- *Cosmos CLXXVII* was launched into orbit with 267-km (166-mi) apogee, 201-km (125-mi) perigee, 89.1-min period, and 51.7° inclination. Satellite, announced as part of a program of space research, reentered Sept. 24. (GSFC *SSR*, 9/30/67; *Krasnaya Zvezda*, 9/19/67, 1)

September 17: NASA Arcas sounding rocket was launched from Barking Sands, Hawaii, to 35-mi (56-km) altitude as second in NOTS series to measure incident solar ultraviolet irradiance in support of *Ogo IV*'s mission. Rocket and payload performed satisfactorily, and achieved all experimental objectives. (NASA Rpt SRL)

September 18: NASA Nike-Apache sounding rocket launched from NASA Wallops Station carried GSFC payload containing pitot-static probe to measure atmospheric density, pressure, and temperature in observing diurnal variation in the fine structure of the region from 15 to 110 km. Probe was launched in conjunction with thermosphere probe experiments. The 2nd stage of the sounding rocket did not ignite because of low current application. (NASA Rpt SRL)

- NASA Nike-Apache sounding rocket launched from NASA Wallops Station carried Univ. of Michigan payload to 178-mi (286-km) altitude on flight to measure simultaneously density and temperature of neutral N2 and of electron temperature and density; to compare day and night measurement of these parameters; to compare a new design for the Omegation Gauge System with the flight proven system; and to compare the 120-km region of the pitot-static probe and thermosphere probe density measurement. All experimental data appeared good. Excellent results were expected (NASA Rpt SRL)

- Secretary of Defense Robert S. McNamara announced his approval for production and deployment of a "thin" Nike-X antiballistic missile (ABM) system at a meeting of UPI editors and publishers in San Francisco. He explained: "Were we to deploy a heavy ABM system throughout the United States, the Soviets would clearly be strongly motivated to so increase their offensive capability as to cancel out our defensive advantage." He opposed a more advanced, heavy, anti-Soviet system and approved the "thin" ABM system because "the system would be relatively inexpensive—preliminary estimates place the cost at about $5 billion—and we would have a much higher degree of reliability against a Chinese attack than the much more massive and complicated system that some have recommended against a possible Soviet attack." McNamara stated: "An ABM deployment designed against a possible Chinese attack . . . would provide an additional indication to Asians that we intend to deter China from nuclear blackmail, and thus would contribute toward our goal of discouraging nuclear weapon proliferation among the present nonnuclear countries."

  The Secretary of Defense said the defensive move was being made with China in mind, not the Soviet Union. He warned the Soviet Union, however, that if it decided to expand its limited antimissile system, the U.S. would respond by increasing its offensive forces to maintain its overwhelming retaliatory capacity.

  The decision was generally applauded by congressional leaders but also unleashed an offensive—expected to be focused in the Joint Committee on Atomic Energy—for building a "heavy" defense system against Soviet missiles, reported John W. Finney in *New York Times*.

  The proposed "thin" ABM system would provide each region of the U.S. with at least one battery of warhead interceptors, reported AP. The exact locations of the batteries, each with several missiles, remained a secret. (DOD Release 868–67; *Av Wk*, 9/25/67, 11; Beecher, *NYT*, 9/19/67, 1; Finney, *NYT*, 9/19/67, 1; AP, *SF Exam*, 9/19/67, 5)

- Interchange of ideas, technology, hardware, and personnel had helped to make progress in space possible, said B/G Edmund F. O'Connor (USAF), MSFC's Industrial Operations director, in a talk before the Tennessee Valley Chapter of the Air Force Association, at Huntsville, Ala. "Today's space program is a joint effort of both military and civilian agencies, and there exists a cooperative give-and-take relationship in the finest sense," he stated. (MSFC Release 67–191)

- Britain's leading radio astronomer, Sir Bernard Lovell, was convinced the Soviet Union would attempt one or more space spectaculars during 1967, possibly during October, reported Richard Lewis of *Chicago Sun-Times*. October would mark the 10th anniversary of the launching of the first satellite, *Sputnik I*, and the 50th anniversary of the Bolshevik Revolution. The possibilities included a flight around the moon or a lunar landing with animals as passengers, a manned space station in earth orbit, and the first photographs of the planet Venus, the astronomer said. Sir Bernard Lovell said: "I have always believed that 1967 was the year they would try the circumlunar (around the moon) flight, but

I should think they would do it unmanned because of the difficulties of the return at lunar flight velocity." (Lewis, *C S/T*, 9/18/67, 16)

September 19: Secretary of Defense Robert S. McNamara's decision to go ahead with "thin" Nike-X antiballistic missile (ABM) system [see Sept. 18] was praised by Sen. Clinton P. Anderson (D-N.Mex.), Chairman of Senate Committee on Aeronautical and Space Sciences, on floor of Senate: "I indicate my support of this decision [and] under present world circumstances, we have no choice but to go ahead with [ABM] system." (*CR*, 9/19/67, S13207-9)

- Orbiting Geophysical Observatory (OGO) satellites, *Ogo I, II, III*, and *IV*, transmitted data simultaneously for the first time. Launching dates for four satellites had been: *Ogo I*, Sept. 4, 1964; *Ogo II*, Oct. 14, 1965; *Ogo III*, June 6, 1966; and *Ogo IV*, July 28, 1967. (NASA Release 67-252)

- USAF launched unidentified satellite from WTR using Titan III-B booster; satellite reentered Sept. 30. (*Pres Rep 1967*)

- Fabrication of five S-IIs (Saturn V's 2nd stage) would be accomplished under a contract modification awarded by NASA to North American Aviation, Inc. Purchase of the five stages would complete the S-II requirement for the 15 Saturn V launch vehicles currently approved for development as a part of the Apollo program. The first of the stages would be delivered to NASA in Feb. 1969. The Marshall Space Flight Center, Huntsville, Ala., managed the Saturn program. (NASA Release 67-244)

- The three Pegasus spacecraft launched by NASA in 1965 had far exceeded the most optimistic expectations and were still operating as the end of a one-year lifetime extension drew near. The spacecraft were launched aboard the last three Saturn I launch vehicles on Feb. 16, May 25, and July 30, 1965. They had completed their assigned tasks—to measure the meteoroid environment of near-earth space—and had telemetered back to earth much information on other subjects of interest to space scientists. Scientific results of Project Pegasus had been covered in an interim report prepared by MSFC's Space Sciences Laboratory. (MSFC Release 67-192)

- Examples of air, sea, and space technology "working together as partners for the better understanding of our total environment" were cited by Dr. Robert C. Seamans, Jr., NASA Deputy Administrator, in a speech to the 2nd International Buoy Technology Symposium, Marine Technology Society, Washington, D.C. The Interrogation, Recording and Location System (IRLS), developed by Goddard Space Flight Center, would demonstrate the use of a satellite to locate and determine the position of sensors, receive data from the sensor, record that data on the spacecraft, and later relay the data to ground stations. Possible terrestrial sensors are meteorological stations or buoys, oceanographic buoys, gauges for measuring the strains in the earth's crust, drifting balloons, ice islands, or any of a wide variety of data platforms located on the surface of the earth or in its atmosphere. Seamans said, "It is quite easy to conceive how such a system could begin to solve many of the problems hindering world-wide fixed or free buoy systems for the study of currents and ocean-depth profile data." He also noted that "photographs and images returned by such satellites as Nimbus, Tiros, and Gemini are being analyzed to assess their oceanographic value." (Text)

- Three Lewis Research Center engineers received "I-R 100" awards at National Industrial Research Week ceremonies in New York City. The "I-R 100" awards were sponsored by *Industrial Research Magazine* and presented in observance of National Industrial Research Week. LeRC recipients William D. Klopp, Peter L. Raffo, and Walter R. Witzke were cited for the development of tungsten RHC, the highest strength metal known at temperatures above 3,500°F. The alloy would have eight times the tensile strength of pure tungsten at the same temperature with no sacrifice in ductility at low temperatures. At temperatures below 250°F, pure tungsten would be very brittle and alloying it to increase strength usually made it more brittle. (LeRC Release 67–60)
- Addressing the House on means for a terminal-phase missile interception system, Rep. William R. Anderson (D-Tenn.) called for a sea-based antiballistic missile intercept system (SABMIS). Pointing to the popular estimate that Red China would have a ballistic missile delivery capability in the early 1970s, the Congressman stated: ". . . deployment of a SABMIS unit would place in the seas close to an adversary's homeland and across his 'launch trajectory window,' a mobile, partly submerged screen of antimissile forces." (*CR*, 9/19/67, H12137)
- Mayor Henry Maier of Milwaukee, addressing a conference of UPI editors and publishers in San Francisco, said: "The central city cannot be saved without a reallocation of our national resources to spend as much to build decent cities for man on earth as we spend on sending a man to the moon." (CTNS, *C Trib*, 9/19/67)
- U.S.S.R. launched *Cosmos CLXXVIII* into orbit with 310-km (193-mi) apogee, 137-km (85-mi) perigee, 88.6-min period, and 50° inclination. Satellite, announced as part of a program of space research, reentered on same day it was launched. (GSFC *SSR*, 9/30/67)

September 20: In informal remarks at Kennedy Space Center, NASA Administrator James E. Webb predicted that congressional cuts in the space budget would slow up the launch rate and lead to the phasing out of Apollo equipment after man landed on the moon. Webb said that only nine of the 15 Saturn V launch vehicles could be launched before 1970, instead of 13 planned launchings. "I don't think there's any doubt that to absorb this cut for $500 million, after already having absorbed a $600 million cut back in 1964, means that we will have to slow up the launch rate," he stated. The cut had also knocked out any new planetary missions before 1975. Space budget cuts would likely mean the elimination of the Saturn IB from the middle-term space program, Webb explained in a brief press conference. NASA was committed to the purchase of 16 Saturn IB launch vehicles. Three had been flown; the next one would propel the first manned Apollo into earth orbit in mid-1968. NASA planned to switch manned flights to the three-stage Saturn V as soon as possible. The first Saturn V flight, an unmanned test, was scheduled for Oct. 19. The Saturn IB had been planned for continued use on flights not requiring the huge power of the Saturn V, but Webb said that emphasis instead would switch to the Air Force's Titan IIIM launch vehicle, being developed for the military Manned Orbiting Laboratory. Webb listed several future checkpoints: (1) fewer orders for new Apollo spacecraft equipment after the moon landing, in such programs as Apollo Applications; (2) definition of the next system after Apollo, which could be a 100,000-lb space station, or something bigger, launched with Saturn V, or a new launch vehicle; (3) development of new and better concepts since planetary missions have been knocked out until 1975; and (4) massive reorganizational changes a possibility for NASA.

Webb and West Germany Minister for Scientific Research Dr. Gerhard Stoltenberg, at the press conference, stated jointly: five West German scientists would participate in studying Apollo program results, and two research satellites and a solar probe would be launched in West German-U.S. cooperative program. Stoltenberg also said that West German industry would be main contractor for an ESRO satellite for 1969 launching, the "first German satellite." Webb said Associate Administrator Dr. Homer E. Newell had been asked to examine question "How do we . . . decide whether a mission should use man or not use man?" (Text; UPI, W Star, 9/21/67; Hill, H Chron, 9/22/67)

- NASA Arcas sounding rocket, launched from Barking Sands, Hawaii, to 33-mi (53-km) altitude, third in NOTS series to measure incident solar ultraviolet irradiance in support of *Ogo IV*'s mission, achieved excellent results. Rocket and payload performed properly. (NASA Rpt SRL)
- U.S. public and communications users would be deprived of "early attainable benefits" if ComSatCorp's proposed pilot program for domestic satellite services were delayed, ComSatCorp stated. "In the end, a pilot-scale demonstration is essential, and nothing is gained by postponing for over a year the planning and building of the demonstration model," James McCormack, ComSatCorp's chairman said. Comments had been filed before the FCC in the pending domestic satellite case, in reply to the Ford Foundation, which had asked FCC on Sept. 18 to withhold approval of ComSatCorp's pilot program. ComSatCorp argued that valuable experience could be gained by implementation of the pilot program, that this would supplement, not hinder, the work of the Task Force appointed recently by President Johnson to develop national telecommunications policy. Mr. McCormack had said the pilot program would provide vital experience for voice, broadcast, data, and the press in the economical utilization of high-capacity satellite and earth station facilities. (ComSatCorp Release 67-47)
- ComSatCorp announced the selection of Vanguard Construction Corporation, NYC, as contractor for general construction work ($864,990) on the new earth station near Cayey, about 30 miles south of San Juan, Puerto Rico. The earth station would serve the Atlantic area, including communications between the U.S. and Puerto Rico, and would be capable of sending and receiving all forms of international communications via commercial satellites—television, telephone, telegraph, and data. The station would be in operation in the fall of 1968. (ComSatCorp Release 67-44)
- New rules designed to improve substantially the passengers' chances for surviving airplane crashes were announced by the FAA. For the first time, airplane manufacturers would have to demonstrate a 90-sec emergency evacuation (instead of present two-minute evacuation) using a full and representative passenger load before they would be issued type certificates. The new rules ("Crashworthiness and Passenger Evacuation Standards for Transport Category Airplanes") would require extensive safety improvements in both airline equipment now in service and in designs of new planes not yet type-certificated. In general, the new rules would take effect October 24, 1967. Aircraft now in service or already type certificated would have to meet most of the retrofit changes

required under the new rules by October 1, 1969. (FAA Release 67–67)
- The Soviet Union had urged the U.N. Committee on the Peaceful Uses of Outer Space to take action on the Soviet draft treaty on rescuing cosmonauts in emergencies, reported *Space Business Daily*. The Soviets had expressed regret that the Legal Subcommittee had not completed action on the proposal. Soviet representative P. D. Morozov said the Soviets would not object to amendments. He praised the recent U.N. space treaty as "a great achievement of the progressive forces on the road of using outer space in the interests of mankind. It is a serious step forward in the development of cooperation and mutual understanding between states and peoples." Morozov also noted that the April meeting of "experts of the socialist countries outlined new steps for broader cooperation in space meteorology, biology and medicine," as well as communications. (*SBD*, 9/20/67, 100)
- Japanese defense specialists had expected the first Red Chinese test of an intercontinental missile by the end of this year, reported *Washington Star*. Before they would test a full-scale missile, Chinese would test-fire the first stage of missile. This could come at any time. However, Prof. Tetsushi Okamoto of Tokyo Institute of Technology believed that Peking regime would first launch a satellite, probably on Oct. 1. Fear of China's progress toward a system that could deliver a nuclear bomb lay behind Secretary of Defense Robert S. McNamara's Sept. 18 announcement that U.S. would begin production of a limited antiballistic missile system designed to guard against a Chinese attack. China's ICBM was expected to be fired from a site at Shwangchengtse, north of Chiuchuan in Lanchow, central China. (Axelbank, *W Star*, 9/20/67)
- France and Quebec would launch a joint communications satellite for relaying French television programs to Canada, UPI reported. The satellite would permit programs to be relayed without depending on existing American Early Bird satellites, informed sources said. (UPI, *C Trib*, 9/21/67)

September 21: Defense of the U.S. supersonic transport program against charges the Administration had bankrolled an industrial monopoly was made by Secretary of Transportation Alan S. Boyd in a speech before New York's University Club, reported David Hoffman of *Washington Post*. Critics had alleged that Boeing and General Electric could reap huge profits because of a loosely worded Federal contract. Boyd had said "there are very good economic reasons to believe" that at least 500 planes would be sold, and if so, the Federal treasury would get back its $1.25 billion and another $1.1 billion in royalties. He had said the SST program would involve no monopoly for Boeing and General Electric. Far from being a monopoly, the SST program would provide economic activity for: two prime contractors, 20,000 subcontractors, 12 U.S. airlines, 14 foreign airlines, and 65 R&D firms working with seven Government agencies. (Hoffman, *W Post*, 9/22/67)
- Hawker Siddeley P–1127 Kestrel VTOL strike fighter crashed on ferry mission over Great Britain with heavy damage. Pilot, Sqdn. Ldr. Hugh Rigg, ejected and sustained cuts and bruises.

  In U.S., LARC had decided against repairing one of two P–1127s obtained for flight testing; NASA's aircraft had crashed at WS Aug. 21 with heavy damage. Parts would be used as spares for remaining aircraft. (*Av Wk*, 10/2/67, 17)

September 21–22: Rep. Emilio Q. Daddario (D-Conn.), Chairman, House

Committee on Science and Astronautics' Science, Research, and Development Subcommittee, addressed House Seminar on Technology Assessment: "The past few years have brought a change in attitude toward science and technology, both in the public which is more technically literate, and in the Congress with its enhanced understanding. Faith in science, and awe of technology, have been supplanted by a recognition of a grave responsibility for decision—that is, what should we do with what we know? Technology assessment is a major key to discharging that responsibility. We are now turning to the natural sciences and asking them to move further in achieving a collective wisdom with politics, law, economics, and social interests for the management of technology. . . . The technical community must introduce a concern for public policy impacts and methods of operation—at an early point and in considerable detail. We believe that technology assessment will involve the scientific method and will be largely accomplished by scientists and engineers. But the purpose of assessment is to enable decisions for the public good." (Text)

September 22: *Cosmos CLXXIX* was launched into orbit with 157-km (98-mi) apogee, 141-km (88-mi) perigee, 87.3-min period, and 49° inclination. The satellite was announced as part of a program of space research; like *CLXXVIII*, it reentered on the same day it was orbited. (GSFC *SSR*, 9/30/67; *Pravda*, 9/23/67)

- More than 2,000 major construction and supply firms in the U.S. received invitations to bid on the second phase of NASA's Electronics Research Center at Kendall Square, Cambridge, Mass., announced the New England Division Engineer, Corps of Engineers. The construction package would include three brick-faced buildings: one at least 10 stories high to house laboratories, offices, and cafeteria; 350-seat auditorium; and a center support building for service utilities. The Electronics Research Center had been in operation since Sept. 1, 1964, in temporary quarters at Technology Square near the permanent site in Kendall Square. (USA Release, 9/22/67)

- Expanded memory units for tracking facility computers of the Apollo lunar landing program had been added by NASA as a modification to its fixed price contract with the Sperry-Rand Corp. The work would be performed at Sperry-Rand's Univac Defense Systems Div., St. Paul, Minn., which would expand existing memory bank units for digital data processing systems from a storage capacity of 32,000 to 48,000 words. The systems would be installed in ground stations and aboard ships in the worldwide tracking network for Apollo flights. Work would be performed under the direction of GSFC. (NASA Release 67-246)

- North American Aviation, Inc., merged with Rockwell-Standard Corp. to create North American Rockwell Corp. (North American Rockwell Corp. Release 092267)

September 23: Rep. George P. Miller (D-Calif.), chairman of the House Committee on Science and Astronautics, described correction of "serious procedural deficiencies in the flow of timely information" from NASA to the Committee, J. V. Reistrup reported. A letter from Rep. Miller to Rep. William F. Ryan (D-N.Y.) had explained the corrective measures, worked out with NASA Administrator James E. Webb. NASA would notify the Committee staff immediately of any accident involving death or serious property damage; NASA Oversight Committee would be briefed every three months on Committee-defined "critical problem areas";

NASA would report significant program changes and problems as soon as they occurred; and NASA would furnish budget data in October, updating the data monthly. (Reistrup, *W Post*, 9/23/67)

September 25: The scientific data required for the Orbiting Geophysical Observatory (*Ogo IV*) primary mission had been obtained, and extended data acquisition with three-axis stabilization was continued. With the exception of two experiments which had ceased operating after several weeks in orbit and an oscillatory perturbation introduced by an experimental antenna, all observatory systems were functioning and operating well. From July 31 to Sept. 17, *Ogo IV* had acquired over 19,000 hours of data on latitude-dependent atmospheric phenomena, energetic particle characteristics, VLF noise and magnetic fields, and incoming solar radiation. Initial reports from experimenters were impressive and underscored the importance of interdisciplinary measurements and global coverage for studies of the near-earth environment. *Ogo I, II,* and *III* were functioning and had obtained useful data. On Sept. 19, all four OGOs were transmitting data, in over 50 experiments providing high-data-rate measurements at widely separated spatial locations within the magnetosphere. The period of coordinated operation of four OGOs would continue for the next several weeks. (NASA Proj Off)

- USAF launched unidentified satellite from WTR on Scout launch vehicle. (*Pres Rep 1967*)
- Plumbing leaks in the first model of LEM had delayed flight testing at Cape Kennedy. Portions of the plumbing system had been sent back to the plant of the main contractor, Grumman Aircraft Engineering Corp., Bethpage, N.Y. The plumbing leaks had virtually eliminated any chance of launching this year the first model of the craft that would land American astronauts on the moon. (AP, *W Post*, 9/26/67, A9)
- Congressional cuts in the NASA budget had forced an agency-wide hiring freeze, *Aerospace Technology* reported. No job vacancies might be filled or new positions created unless there were exceptionally mitigating circumstances. JPL expected to lose about 200 engineers during the freeze, but no estimates had been released by NASA field centers. There was no indication of how long the freeze would last. (*Aero Tech*, 9/25/67, 3)
- Airport development sparked the economy of five communities in Texas, South Carolina, California, Maryland, and Minnesota, the FAA had reported. "The Airport—Its Influence on the Community Economy," a study report related economic growth to general aviation (non-airline) airport development at Hereford, Tex.; Sumter, S.C.; Hayward, Calif.; Frederick, Md.; and Fairmont, Minn. Public airports had been developed or improved in each of the communities with Federal assistance under the Federal-Aid Airport Program administered by FAA. (FAA Release 67–70)
- USAF Space and Missiles Systems Div. had awarded a contract to Lockheed Missiles & Space Co., Sunnyvale, Calif., for $1.2 million for Agena launch services at Vandenberg AFB Oct. 1, 1967, through Sept. 30, 1968. (DOD Release 896–67)

September 25–30: International Astronautical Federation 18th Congress was held in Belgrade, Yugoslavia.

Detailing systems engineering activities of U.S. manned space flight program, NASA Saturn Apollo Applications Director Charles W. Mathews said that efforts had encompassed three flight systems—Mercury, Gemini, and Apollo—and included many other aspects, worldwide instrumentation and communications networks, control centers, recovery support elements, and scientific and technical experimental equipment. Thus, systems engineering had had to be all inclusive in providing for logical development of flight hardware, successful conduct of flight operations, and achievement of maximum benefits from each mission. (Text)

T. J. Gordon, L. M. Dicke, and J. S. Nieroski of McDonnell Douglas Corp.'s Douglas Missile & Space Systems Div. (MSSD) presented "Economics of Commercial Space Stations." Space station would pay its own way by taking a share of increased customer profits. According to MSSD representatives, customers would be "those enterprises who, by virtue of their use of the orbital services, can offer new or higher quality profits or services, or lower operating costs for existing products or services." Possible space station uses were described as locating fish, reducing ocean shipping transit time and hazards, locating petroleum and mineral reserves, controlling and routing air traffic, reducing air and water pollution by detecting sources, improving fresh water reserves by surveying sources, and improving agricultural and forest productivity by use of detectors of disease, damage, and other factors. (Haggerty, *J/Armed Forces*, 11/25/67, 9)

Soviet Academician L. I. Sedov described "Ten Years of Space Exploration in the Soviet Union" and said: "Cosmic systems have become necessary not only for the study of fundamental problems of physics and the Universe, but also for various practical purposes of importance to national economies [and] there is no doubt that cooperation will favourably affect many sides of international life. We hope that this cooperation and its benefits will grow deeper in the future." (Text)

Worldwide cooperation in space was urged by MSFC's former advanced mission planner, Dr. H. H. Koelle, now professor of space technology at Technical Univ. of Berlin. To get around tremendous costs and avoid reduction in space activity in all countries, he urged "international organizations to plan and coordinate advanced manned missions in the 1980s." He said space powers should begin by increasing coordination of mission objectives in their national space programs. For a cooperative concept, he called for annual investment of $4 billion and work force of 350,000 persons, with 200,000 persons from U.S., 60,000 from U.S.S.R., 60,000 from Western Europe, and 30,000 from rest of world. (Fink, *Av Wk*, 10/2/67, 21–2; Van Osten, *Aero Tech*, 10/9/67, 43–7)

September 26: The first manned Apollo space flight might be delayed from spring to mid-summer of 1968 by modification of the Apollo spacecraft by its prime contractor, North American Rockwell Corp., according to NASA spokesman Julian Scheer. Scheer, NASA Assistant Administrator for Public Affairs, told UPI the anticipated delay, among other things, could be because of considerable changes in the capsule after last January's fire. (UPI, *W Post*, 9/27/67, A6)

- An ordinance outlawing sonic booms was adopted by Santa Barbara's city council, Reuters reported. This was believed to be the first such municipal ordinance in the U.S. The mayor had voted against it. He explained, "I do not believe the law is enforceable." Councilman Klaus Kemp, introducing the ordinance, said booms caused damage to homes and physical injury to individuals. At Hamilton AFB a spokesman said "presumably nothing" would be done about the new law. The presumption was based on court decisions ruling the Federal Goverment (FAA) had sole jurisdiction over aircraft movements. "We are not attempting to prohibit supersonic flights . . . [but] proposing that unnecessary flights over or immediately adjacent to this city at unreasonable and unnecessarily low altitude be prohibited," said Kemp. (Reuters, *NYT*, 9/28/67)

- *Cosmos CLXXX* was launched into orbit with 350-km (218-mi) apogee, 206-km (128-mi) perigee, 90.1-min period, and 72.9° inclination. Satellite reentered Oct. 4. (*Krasnaya Zvezda*, 9/28/67, 3; GSFC *SSR*, 10/15/67)

September 27: President Johnson proclaimed Oct. 9 Leif Erikson Day to give "national recognition to Leif Erikson today, when men of similar courage and imagination are confronting equally formidable challenges in the heavens and under the seas." (*PD*, 10/2/67, 1350)

- Development of an advanced direct sensing system for the Nimbus-D Meteorological Satellite, scheduled for launching in 1970, was subject of NASA contract awarded to Radiation Inc., Melbourne, Fla. Interrogation Recording and Location System (IRLS) would involve development of a satellite-borne device and remote platform electronics to obtain measurements of atmospheric and other data from fixed and free floating sensory platforms, including balloons and buoys. Work would be performed under the direction of GSFC. (NASA Release 67–247)

- NASA appointed an interim working group composed of Dr. Alfred J. Eggers, Dr. Floyd L. Thompson, and Gen. Jacob E. Smart (USAF, Ret.) to review NASA safety procedures and prepare a plan under which the Administrator could establish an Aerospace Safety Advisory Panel. Gen. Mark Bradley (USAF, Ret.), Garrett Corp., would serve as a consultant to the chairman, Dr. Eggers. The action would implement Section 6 of Public Law 90–67 (the NASA Authorization Act for FY 68). Eggers was Special Assistant to the Administrator and Deputy Associate Administrator for Advanced Research and Technology. Thompson, Director of NASA's Langley Research Center, served as chairman of the Apollo 204 Accident Review Board. Gen. Smart was NASA's Assistant Administrator for Policy. (NASA Release 67–249)

- Space Nuclear Propulsion Office, Germantown, Md., had extended its contract with Aerojet-General Corp. for development of nuclear propulsion. Aerojet, which had been the prime contractor since 1961 on the NERVA project, would receive an estimated $14.7 million on the interim contract for work to be performed through Nov. 30. The Space Nuclear Propulsion Office, a joint operation of AEC and NASA, would receive $6.8 million of the funds on the contract extension from NASA, the balance from AEC. (NASA Release 67–250)

- Rep. Theodore R. Kupferman (R-N.Y.), speaking on the House floor, expressed concern about the "noise pollution problem" [see Feb. 1]. He quoted from a speech which he had presented on Sept. 12 to a luncheon meeting of the U.S. Chamber of Commerce, held in Washington, D.C.: "It is possible to have noise control and industrial development without penalty to each other. . . . The idea that noise is a necessary price of industrial and economic progress is as antiquated as is the belief that contaminated waters and a polluted atmosphere must also accompany civilization's material advances. . . . Noise pollution, however, unlike water and air pollution, is only now beginning to receive a proper share of public attention." (*CR*, 9/27/67, H12606–13)

September 27

- Dr. Edward U. Condon, Director, University of Colorado's study of Unidentified Flying Objects (UFOS), said that "such studies ought to be discontinued unless someone comes up with a new idea on how to approach the problem." Condon said "it's been extremely difficult to gather information on this subject" and asked anyone with any "hard information to speak now—or forever hold your peace." He stated: "We have approached the problem in an unprejudiced way, seeking to find whatever there is to find by objective scientific means with no scientific results so far. Some of the difficulties associated with belief in visitors from outer space are well set forth in an article by Prof. William Markowitz in *Science* for Sept. 15 and should be studied by all who wish to be informed on the subject," he added. (Harkins, Boulder *Camera*, 9/27/67; AP, *LA Times*, 9/28/67)

September 27–October 2: NASA launched for ComSatCorp the fourth of INTELSAT II series of commercial communications satellites (*Intelsat II–D*)—designated *Pacific II*—and second of them in the Pacific. Launch was from ETR by Thrust-Augmented Delta (DSV-3E) launch vehicle. Satellite went into transfer orbit over Pacific at approximately 176° east with 29,056-mi (37,105-km) apogee, 186-mi (299-km) perigee, 658.8-min period, and 26.4° inclination. Satellite would supplement services of and replace current Pacific satellite, *Intelsat II–B* (174°E), in event of its failure. NASA support nominally would terminate after about three days, subsequent to providing backup transfer orbit data and calculations. Final orbit and spin axis orientation would be made by ComSatCorp with onboard hydrogen peroxide system. Approximately 15 days after launch, satellite would be available for commercial use. Satellites could handle TV data transmissions, or up to 240 voice channels (480 circuits) between "first class" ground stations such as Andover. Part of satellite capacity would support NCS/NASCOM for Apollo program. It would serve as backup against interruption of satellite communications service and would add communications capability across the Pacific.

On Sept. 30, apogee motor was ignited at 7:45 pm EDT, on command from ComSatCorp-operated earth station at Paumalu, Hawaii, and kicked new satellite from its elliptical transfer orbit into a planned, near-synchronous orbit over Pacific. Reorientation and velocity correction maneuvers were carried out Oct. 1 and 2 to place spin axis in orbit plane and to improve circularity with predicted orbit (resulting from these maneuvers) of apogee, 22,531 mi (36,260 km), perigee, 21,805 mi (35,091 km), period, 1,430.5 min. drift rate, 1° east per day. As *Intelsat II–D* approached *Intelsat II–B*, launched Jan. 11, now located between 174° and 175°, drift rate would be reduced and intersatellite interference tests would be conducted to determine minimum angular separation which could be employed without mutual interference (NASA cooperating with ComSatCorp in planning tests). The INTELSAT satellite system, in addition to this new *Pacific II* satellite, included the *Pacific I* satellite and two satellites serving the Atlantic area, *Early Bird I*, launched Apr. 6, 1965, and *Atlantic II*, launched Mar. 22, 1967. (NASA Proj Off; ComSatCorp Releases 67–45 and 67–48)

September 28: NASA Administrator James E. Webb testified before the Subcommittee on Independent Offices, Senate Committee on Appropriations. His prepared statement made these principal points: NASA would not be able to achieve for this Nation "a position of leadership in space" for less money than an annual budget of between $5.5–$6.0 billion a year.

At the proposed level of funding, after launch vehicle #216, production of the Saturn IB and the earth-orbital portion of the Apollo Applications Program would be terminated. Rather than continue to use the Apollo system for work in orbit at conclusion of Apollo flights, NASA would accept a hiatus in the continuous development of manned space flight, deciding on a fresh start for the years ahead. A new study would be begun in FY 1968 and FY 1969 on purposes and form of future development of spacecraft and "the national family of large launch vehicles that should be available in the 1970's." This "basic change in the United States space outlook" would make it "clearly prudent to preserve the physical facilities which have been built and paid for and which would be required if a decision is made to construct a large solid rocket vehicle." Because the budget level approved by the House would prevent beginning work on NERVA II, development of a flight weight engine of the NERVA I class, "with probably from 60,000 to 75,000 lbs of thrust" would be initiated in FY 1968. Because we would have "no choice" but to accept the fact that Voyager cannot be initiated in FY 1968, and, in order to preserve "a capability for future planetary programs," NASA would propose about $10 million for the "lunar and planetary program for studies and supporting research in the technology of planetary exploration." NASA would propose no changes in the budget for the Apollo program but would "accept the legislative history as a clear signal that we should proceed with a hard-hitting carefully managed effort to achieve the objectives of the Apollo program." NASA would be able to launch "a total of nine Saturn V unmanned and manned missions before the end of Calendar Year 1969." A further reduction in funding in FY 1968 could only cause "complete havoc" and would be a blow to the national aeronautical and space effort from which it would be extremely difficult to recover.

Webb told the subcommittee: "In this situation, we cannot ask, as we otherwise would have done, that the Senate restore in full the amount reduced by the House. . . . We are here to ask the committee that the Senate approve the amounts passed by the House and accepted by the President, and to make a clear record in the legislative history that will give us the flexibility we will need to permit us to carry out the strongest possible program within the amounts made available to us. . . . Long leadtimes requiring timely initiation of new programs and implying commitment to proceed over a period of years would . . . make it perfectly clear to all that we cannot somehow find a way to carry forward the same program and give the Nation the same assurance of a strong position in aeronautics and space with a $517 million cut in our budget."

He said he had been in touch with the President in the last 24 hours and "he still believes the recommendations he made in January represent" the best budget for the country. But the President also believed his proposed 10 percent income tax surcharge "is more important to press for than extra money for the space bill." Webb said he had drastically reshaped the space program to comply with the $4.5 billion fund for this year voted by the House and approved by the President. But Subcommittee Chairman Warren G. Magnuson (D-Wash.) and several senators, with major space installations in their states, gave strong indications they would plan to restore some of the millions of dollars cut by the House. Webb promised to revise his multi-billion-dollar space plans

September 28

again "and bring you the best program possible for the money." (NASA LAR VI/101; Text; Testimony; AP, *NYT,* 9/29/67, A3; 9/30/67, 52)

- Senate Appropriations Committee activities were listed by *Wall Street Journal:* voted $142 million for this year's development of the SST; doubled the House-approved $20 million for highway safety grants; added $50 million to FAA's budget for improved air safety (versus the $7 million requested by the Administration); and, in all, voted nearly $1.7 billion for the Department of Transportation's FY 1968 budget. Sponsor for the improved air safety figure of $50 million had been Sen. John Stennis (D-Miss.), chairman of the Appropriations Subcommittee that handled FAA's budget, who proposed $20 million for operation of air control towers and overtime payment for their crews, plus another $30 million for such new safety equipment as airport ground lighting and improved radar. Sen. Stennis had stated his hope that someday air safety projects could be financed partly with special user charges. (*WSJ,* 9/28/67, 3)
- The LTV Range Systems Division, LTV Aerospace Corporation, Dallas, was awarded a contract valued at about $10 million to provide facilities support services to the Manned Spacecraft Center, Houston. The services would cover 12 mo, beginning Dec. 1, 1967, contained provisions for four one-year renewal options. (NASA Release 67–253)
- USAF Manned Orbiting Laboratory Systems Program Office, Los Angeles, Calif., issued a $3.5-million initial increment to an $8.7-million fixed-price contract to Hamilton Standard, Windsor Locks, Conn., for development, procurement, and support of astronaut pressure suit systems for the MOL program. USAF Space and Missile Systems Organization issued a $4-million increment to an $8.1-million definitive contract for engineering services in support of the Agena space vehicle program. (DOD Release 912–67)

September 29: The Senate had unanimously passed S.J. Res. 109, which would mark the 50th anniversary of the Langley Research Center in October 1967. (NASA LAR VI/102)

- Dr. Donald H. Menzel of the Harvard College Observatory said in a letter to *Time* that he planned to propose naming features of moon's far side for the three American astronauts and one Russian cosmonaut who lost their lives in accidents connected with space research. Features were to be named at 1970 meeting of International Astronomical Union. (*Time,* 9/29/67, 11)
- A contract for operation of NASA's Scientific and Technical Information Facility at College Park, Md., had been renewed by NASA with Documentation, Inc., Bethesda, Md., at an estimated $5.6 million for one year. The company had been operating the facility since its inception in 1962. The facility used modern computer techniques to store and retrieve a large volume of scientific and technical reports for the benefit of Government, university, and industrial users. (NASA Release 67–255)
- Western Electric received a $43.4 million addition to an existing Nike-X research and development contract. Approximately $18.0 million of this was initial implementation of the decision announced Sept. 18 by Secretary of Defense Robert S. McNamara to deploy a Communist Chinese-oriented ABM system. The remainder of the announced addition of $43.4 million would be for continuing research and development on the Nike-X system, supplementing the FY 1968 contract signed early in September. Under that contract Western Electric had received initially

$215.3 million for continued research and development effort on Nike-X. The Communist Chinese-oriented ABM system would consist of two types of radar and two interceptor missiles. (DOD Release 992–67)

September 30: NASA Aerobee 150 sounding rocket, launched from NASA Wallops Station to 134-mi (216-km) altitude, in AFCRL experiment to measure incident solar radiation in upper atmosphere from 1,206 to 284 Å with monochromator while similar instrument was in operation in *Oso III* satellite. Monochromator operated successfully; data were somewhat degraded by partial instrument failure. (NASA Rpt SRL)

- ComSatCorp Report to Shareholders for third quarter of 1967 indicated operating revenues were at record high of $5,224,219. For second quarter in a row, operations showed over $0.5-million loss, but income from interest more than offset loss, to provide net income of over $0.8 million. (Text)

During September: Suggestions for restoring public confidence in and acceptance of the U.S. space program were advanced by William Leavitt in *Air Force and Space Digest:* "We can husband our resources by looking carefully at the question of whether we need two expensive and competitive manned orbital laboratories, one run by the National Aeronautics and Space Administration and one run for the Defense Department by the Air Force. . . .

"We can increase the funding and sharpen the planning of the unmanned working satellites. . . . We can begin to look seriously at the potential of aerospace systems analysis and engineering in the solution of nonspace and nonmilitary public problems, with an emphasis on building into these techniques . . . social, economic and political factors. . . . And we can begin to put to work in our schools, suburban and urban, many of the space-age training techniques that have been a beneficial by-product of the technological age we live in." (*AF/SD*, 9/67, 158–159, 162, 165)

- M/G Alvin R. Luedecke (USAF, Ret.), Deputy Director of JPL, resigned He had come to JPL on Aug. 1, 1964, after five years as general manager of the Atomic Energy Commission and 25 years as an Air Force officer. (JPL *Lab-Oratory*, Sept. 67)
- NAA announced that Igor I. Sikorsky, popularly known as the father of the helicopter, had been selected to receive the 1967 Wright Brothers Memorial Trophy. Presentation would be made at the Dec. 14, 1967, annual memorial dinner in Washington, D.C. (NAA Release)
- NSF's Office of Economic and Manpower Studies published report showing independent nonprofit institutions employed more than three times as many scientists and engineers in January 1965 as they did in January 1954. *Scientific Activities of Nonprofit Institutions, 1964* said expenditures for R&D performance in these institutions were a little more than 5½ times as much as in 1953, increasing the nonprofit sector's share of the Nation's total R&D outlay from two percent in 1953 to three percent in 1964. (Text)
- More than two thirds of 1966's public intercity travel had been by air, *Space/Aeronautics* reported. While long-haul air traffic demand would be shifted to aircraft like the SST, the airlines would need to come up with short-haul systems of equal caliber. Past studies, sponsored by FAA, had revealed that V/STOLs, complex and costly to operate as they are, would capture a substantial share of the short-haul market with

their time savings and convenience and would, in fact, even increase the market by inducing more travel. (S/A, 9/67, 102–15)
- General aviation fatal accidents for the first time fell below 3 per 100 million hours, the CAB estimating 2.5 for 1966. (National Pilots Assn News, 9/67)
- The overall national security significance of space, in both tactical and strategic terms, were discussed by Dr. Harold Brown, Secretary of the Air Force, in *General Electric Forum*. "The Department of Defense does not view space as a mission, but rather as a place," he said, explaining broad Air Force responsibilities and goals in space. He cited representative features of the military unmanned space programs: (1) ballistic missile warning; (2) antisatellite defense; and (3) nuclear detonation detection. (Brown, *General Electric Forum*, Autumn 1967)
- U.S.S.R. and Great Britain planned to sign a joint technological agreement shortly as result of British government and industry team visit to Moscow during September, reported *Aviation Week*. Agreement would cover automatic landing, aircraft engine technology and reliability, engine noise, and sonic boom research, Minister of State John Stonehouse, team leader, stated. Britain had substantial lead over U.S.S.R. in these fields, but Stonehouse contended sharing of information would lead to new export markets for British hardware in Soviet and Sovietbloc countries. (*Av Wk*, 10/2/67, 16)
- Earth-orbiting satellites, according to Soviet cosmonaut Pavel Popovich's report in *Aviation and Cosmonautics*, would render practical help to agriculture, assessing crop and forest resources and discovering areas of vegetation affected by disease, fires, locusts, and underground springs. His report specified regularly photographing areas under cultivation from sowing to harvesting and processing data in computers to obtain yield forecasts. (S/F, 9/67, 310)

Soviet radio astronomer Professor Vsevolod Troitski, in statement to Tass, said new information had provided convincing proof that moon's surface was sufficiently strong to support spacecraft landings. He stated that substance composing top layer of moon's surface consisted of loosely connected particles averaging 0.01 in and that contacts between particles were very weak. On an average, porous substance covered moon to a depth of about 13 ft, according to Troitski. (S/F, 9/67, 305)

October 1967

October 1: Ninth anniversary of NASA, established by the National Aeronautics and Space Act of 1958.
- NASA Associate Administrator for Space Science and Applications Dr. Homer E. Newell became NASA Associate Administrator. His appointment had been announced Aug. 25. Dr. Newell would be replaced as Associate Administrator for Space Science and Applications (Sciences) by Dr. John E. Naugle. (NASA Release 67-228; NASA Ann, 10/20/67; Hines, W *Star*, 10/3/67, A4)
- NASA announced that solar radiation and magnetic properties around planet Venus as well as in interplanetary space would be studied as *Mariner V* (launched June 14) approached closer to Venus. Flight plan would take spacecraft within 2,500 mi of surface of Venus Oct. 19. NASA hoped to acquire valuable information to supplement that learned from *Mariner II*, which in 1962 had detected no radiation belts and no magnetic field at miss distance of 21,600 mi. Added investigations included: properties of solar wind; electron count by radio beam passing through Venus' atmosphere; study of so-called "solar plasma cavity" on dark side of Venus; and investigation of unexplored region 54 million mi from sun. (NASA Release 67-248)
- Introduction to current U.S. research on use of spacecraft to study oceans, "United States Activities in Spacecraft Oceanography," was published by National Council on Marine Resources and Engineering Development and jointly prepared by NASA, Naval Oceanographic Office, Bureau of Commercial Fisheries, and ESSA. Scientists reported on progress in photographic and spectral scanning; when done from orbital locations, scanning yielded data equivalent to that obtained by thousands of globally distributed ocean-surface sensors. "Although data acquired from space will initially appear coarse compared with measurements from surface platforms, a sensing system with global coverage reporting as often as desired, and with increasing scope and accuracy will offer unique opportunities for broad unified synoptic analysis," study reported. (Text)
- *New York Times* editorial praised success of NASA's *Surveyor V*, launched Sept. 8: "If present indications are confirmed, Surveyor 5 has made one the fundamental scientific discoveries of the century . . . that earth's natural satellite is made up of the same kind of material as is the earth.

"Surveyor 5's historic finding, if backed by later experiments, must force new consideration of the possibility that the moon was torn from the earth. . . . Certainly much more evidence is required . . . [but] Surveyor 5 has performed a notable feat. Moreover it has demonstrated again how much can be learned from instruments delivered by relatively inexpensive unmanned rockets without the great dangers and huge costs required to send a man to the moon and return him safely to earth." (*NYT*, 10/1/67, 12E)

October 2: NASA Administrator James E. Webb announced resignation of Deputy Administrator Dr. Robert C. Seamans, Jr., effective Jan. 5, 1968. "Dr. Seamans joined the government with a commitment to serve two years, but . . . served seven years . . . [and] has now decided to return to private life," Webb said. "His departure will leave a wide gap [in NASA leadership]." A former MIT aeronautical engineering professor, RCA engineer, and member of NACA technical committees, Dr. Seamans joined NASA in 1960 as Associate Administrator. In 1965 he was appointed Deputy Administrator and was presented NASA's highest award, the Distinguished Service Medal, for distinguished service to U.S., outstanding leadership of NASA, and "ability to bring together into a framework of effective action thousands of industrial organizations, tens of thousands of scientists and engineers, and hundreds of thousands of aerospace workers." He had also received Naval Ordnance Development Award (1945); AIAA Lawrence Sperry Award (1961); and Godfrey L. Cabot Aviation Award (1965). (NASA Release 67–257; O'Toole, *W Post*, 10/3/67, A3)

- NASA awarded supplementary or new research grants and contracts totaling $2,059,104 to 23 universities, colleges, and private institutions. (NASA Release 67–254)

- Texas' role in U.S. space program was summarized by NASA Associate Administrator for Manned Space Flight Dr. George E. Mueller for the Texas State Society. "Many thousands of Texans working for dozens of Texas firms are creating through their . . . [R&D] effort the new technology required for space exploration. One measure of Texas contributions may be made in terms of the $771,000,000 of work that . . . [its] firms have performed on NASA's programs. In addition, Texas colleges and universities are not only conducting high level research for NASA, but are training future space scientists and engineers under NASA grants." He noted also that Texas was birthplace of seven astronauts and five astronaut wives and "home town" of all 56 astronauts, based at MSC. (Text)

- Soviet satellites were studying solar x-ray sources in an effort to protect cosmonauts from deadly radiation storms, according to *Pravda. Cosmos CLXVI* (launched July 17) and *Electron II* (launched Jan. 30, 1964) had gathered "extensive information" about x-ray sources and had discovered that "in principle, by recording an x-ray flare, it is possible to warn cosmonauts of the approach of radiation danger . . . [and] enable spaceship crews to take necessary protective measures." (UPI, *NYT*, 10/3/67, 2)

- Under reduced gravity conditions on moon, most natural and comfortable gait for an astronaut would be "a lope at about ten feet per second," advised Amos A. Spady, Jr., LaRC scientist. He and fellow scientist, Donald Hewes, had tested subjects on specially constructed device—simulating effects of astronauts moving in moon's lower gravity. Simulated lunar gravity had not guaranteed that all motion would be higher, faster, or farther. Scientists had discovered that, at most speeds, lunar stride would be longer—sometimes twice as long—but number of steps per minute would be reduced by as much as half. (Weil, *W Post*, 10/2/67, B4)

- In response to charge by Harry Schwartz in the *New York Times* that "the costs to both countries [U.S. and U.S.S.R.] of . . . ten years of space rivalry have been enormous" and prediction that "the economic and technical problems ahead will force the space race to stop at the moon with cooperation replacing rivalry," AIAA Executive Secretary James J. Harford defended space program in letter to *New York Times:* "The modest scientific dividends which he [Schwartz] credits the space program with producing are puny, indeed, compared to the effects space technology will certainly have on civilization in eons to come. . . . Technologically, both the U.S. and the U.S.S.R. are enormously stimulated. . . . Politically, the space program has produced the first international treaty in history. . . . Philosophically, the space program has made the world vividly aware that it is, itself, a spacecraft with three billion passengers, traveling in a solar orbit, in a vast universe. . . .

"Everyone is for international cooperation, but two nations with ICBM's aimed at each other have understandable problems working out cooperative programs. . . . Perhaps the time for serious joint effort is coming. . . . In the meantime, a peaceful program costing less than 1 per cent of the G.N.P., which has produced new technology at a rate that only wars have equaled in the past, . . . should not be diminished because of serious domestic problems." (Schwartz, *NYT*, 10/2/67, 47; *NYT*, 10/14/67, 26)

- Ralph K. Bennett, writing in the *National Observer,* described F-111 aircraft [formerly TFX] as "an Airplane Lemon" and questioned its capabilities: "On paper it is the most sophisticated, flexible, yet economic air-weapons system yet devised—a single airplane capable of performing the divergent missions of both Navy and Air Force . . . [but] in reality it may be the most costly bungle in the . . . history of military procurement. . . ."

Sen. John L. McClellan (D-Ark.) later told Senate he agreed with Bennett: ". . . I am not implying that commonality [of one aircraft for USAF and USN] cannot be a valid and economical concept. There is no way, however, to build with identical parts an effective single weapons system which will carry out separate and widely diverse military missions. Mr. Bennett's article points out that the TFX program was an attempt to combine . . . drastically different design and performance characteristics. . . . The result is . . . [a] hybrid craft that cannot capably perform to the requirements of either service" (Bennett, *Natl Obs,* 10/2/67; *CR,* 10/4/67, S14225–7)

October 3: NASA Nike-Tomahawk sounding rocket launched from NASA Wallops Station ejected barium vapor at 75-, 115-, and 140-mi altitudes, forming three distinct green and purple clouds visible for hundreds of miles. Experiment was conducted for AFCRL to (1) measure electric fields and wind motion in upper atmosphere by photographing and tracking movement of ionized barium clouds, and (2) measure attenuation on passage of radio signals through ionized barium plasmas. (WS Release 67–32; UPI, *W Star,* 10/4/67, D2)

- NASA Aerobee 150 sounding rocket was launched from WSMR to 114-mi (184-km) altitude in GSFC experiment to obtain x-ray and EUV photos of the sun and to flight test SCAT solar pointing system. Flight, first for SCAT, proved to hold pointing to within a few arc sec. Rocket and other instrumentation performed satisfactorily. (NASA Rpt SRL)

- Maj. William J. Knight (USAF) set new winged aircraft speed record when he flew X–15 No. 2 to 4,534 mph (mach 6.72) and 99,000-ft altitude in flight at Edwards AFB to: (1) test Martin ablative coating and ramjet

October 3: Among those attending the 50th Anniversary Celebration and Inspection of NASA's Langley Research Center, Hampton, Virginia, Oct. 2-6, were (left to right): Dr. Floyd L. Thompson, Director of LaRC; James C. Elms, Director of NASA's Electronics Research Center; Dr. Robert C. Seamans, Jr., Deputy Administrator of NASA; and Dr. Abe Silverstein, Director of NASA's Lewis Research Center.

local flow; (2) check out stability and control with dummy ramjets and characteristics of external tank separation and wing tip accelerometer; and (3) conduct fluidic temperature probe. Previous speed record of 4,250 mph (mach 6.33) had been set by Major Knight Nov. 18, 1966. (X-15 Proj Off; FRC *X-Press,* 10/20/67, 3)

- U.S.S.R. successfully launched *Molniya I-6* comsat to relay television signals from the Soviet Far East to Moscow and on to Paris. Orbital parameters: apogee, 39,870 km (24,774 mi); perigee, 499 km (308 mi); period, 11 hr 55 min; and inclination, 64.96°. Tass said comsat would relay television, radiotelephone, and telegraph messages and form part of new television network scheduled to be in operation by 50th anniversary of Bolshevik Revolution Nov. 7. (GSFC *SSR,* 10/15/67; *W Post,* 10/4/67, A17; UPI, *NYT,* 10/4/67, 3)

- In accordance with congressional resolution, President Johnson issued Proclamation 3811 designating first week in October for commemoration of LaRC's 50th anniversary. He noted some of LaRC's numerous contributions to astronautics and aeronautics since its establishment in 1917 as the first national laboratory to conduct basic aeronautical research: (1) cowling for radial air-cooled engines in late 1920's; (2) scientific information and technical skills to increase performance and utility of World War II aircraft; (3) research facilities for advancement of SST; (4) investigations into concepts of advanced supersonic and hypersonic vehicles, helicopters, and V/STOL aircraft; (5) research which contributed to development of Scout launch vehicle and to management of Project Mercury, Lunar Orbiter Program, and Project Fire; and (6) support services for Gemini and Apollo programs through use of simulators and special laboratories.

Messages of congratulations from President Johnson, Vice President Humphrey, and Director of U.K.'s Royal Aircraft Establishment Sir Robert Cockburn were inserted in *Congressional Record* by Rep. Thomas N. Downing (D-Va.), who said: "... this Nation can be proud today and always of the system that maintains the Langley Research Center." At LaRC the week was marked by a daily inspection, with hundreds of industry, university, and Government visitors. (NASA Release 67–251; *CR*, 9/29/67, S13885; 10/2/67, H12787–9; 10/3/67, H12913–4; *PD*, 10/9/67, 1387–8)

- Senate Appropriations Committee ordered favorably reported, with amendments, NASA FY 1968 appropriation bill (H.R. 12474). Action restored $96 million of $516 million cut by House [see Aug. 22]. Two projects—Voyager Mars and NERVA II—were restored by increased funding in Senate move; as approved, bill totaled $4.7 billion. Bill moved to floor of Senate. (NASA LAR VI/104; O'Toole, *W Post*, 10/4/67, A2)
- Mass meeting in U.S.S.R.'s Palace of Congresses inside the Kremlin Wall heard L/C Aleksei A. Leonov recall heritage of fellow pioneers; he had been an associate of Col. Komarov who had been killed on Apr. 24 in the first flight of third-generation *Soyuz I* spacecraft. Cosmonaut Leonov said: "We cannot help recall our friend, test pilot Vladimir Mikhailovic Komarov. Grateful mankind will always preserve the memory of these wonderful people in its heart." The mass meeting was the highlight of Soviet celebration of the 10th anniversary of *Sputnik I*'s launching [see Oct. 4].

In the Soviet press on the same day, scientist Konstantin P. Feoktisov, veteran of three-orbit flight in 1964, listed a manned laboratory in orbit as first among coming space developments, then a moon landing, then a "starting to Venus and Mars." (UPI, AP, *NYT*, 10/5/67, C20)

October 4: 10th anniversary of the space age, inaugurated with the launching of first man-made satellite, U.S.S.R.'s 184-lb *Sputnik I*. During first decade of space age 803 spacecraft were successfuly orbited: U.S. orbited 529; U.S.S.R., 269; France, 4; and Italy, 1. U.S. total included 14 manned orbital space flights with 677 orbits and 1,993.4 man hours; U.S.S.R. had 9 flights with 310 orbits and 532.5 man hours.

Of 45 worldwide flights related to the moon, 20 were successful—including 5 softlandings—providing high-quality photos and on-site analyses of lunar surface. Some 24 planetary flights provided close-up photos of Mars and atmospheric data on Mars and Venus. Some 17 bioscience payloads, including mice, dogs, mold spores, plants, insects, a chimpanzee, and an anthropomorphic dummy, were successfully launched and recovered, providing data on effects of space flight on living systems. Space applications satellites totaled 292; advanced technology satellites, 26; and military support and observation satellites, 158. (EH; Sheldon, "Review of the Soviet Space Program," Report of the House Committee on Science and Astronautics)

- To commemorate first decade of space age, Deputy Assistant Secretary of Interior S. Fred Singer, first Director of U.S. Weather Satellite program, wrote in the *Washington Post* under dateline, "Oct. 4, 2007": "It is now a half century since the first man-made satellite went into orbit around the earth, and an opportune time to examine how space technology and exploration have contributed to the economic development of the United States and to human welfare generally. The effects have been subtle—none of the early spectacular predictions have come to pass: the planets have been explored but not colonized; rockets have become more powerful but rocket transportation has not replaced aircraft and ramjet vehicles." He cited three major technological changes as a result of space program: increased reliance on electronic computer and data processor; a managerial revolution which created highly trained technologist-managers; and increased understanding of natural environment. (Singer, *W Post*, 10/1/67, F1)

- U.S.S.R.'s Academician Dr. Anatoli A. Blagonravov, chairman of Soviet Commission for Exploration and Use of Outer Space, reviewed first decade in space in AP article: "The Space Era is not a chance phenomenon in the development of natural sciences. It is a legitimate and indispensable stage in the history of development of human society." (Blagonravov, AP, *W Star*, 10/1/67, C3)

- NASA test pilot William H. Dana flew X-15 No. 3 to 250,000-ft altitude and 3,886 mph (mach 5.35) at Edwards AFB. Purpose of test flight was to check: ultraviolet exhaust plume; solar spectrum measurement; micrometeoroid collection; air density; x-ray air density; ARC boost guidance; and tip-pod camera. (NASA Proj Off)

- Secretary of Defense Robert S. McNamara's decision to go ahead with "thin" Nike-X antiballistic missile (ABM) system was criticized by Rep. W. J. Bryan Dorn (D-S.C.) on floor of House: ". . . the best defense is an offense. Command of the skies, space, and the seas is the way to keep war and destruction from the shores of our homeland." He recommended use of "sea-based anti-ballistic missile intercept system." (*CR*, 10/4/67, A4923)

- MSFC used AEDC facilities to investigate preparation of spent rocket stage for human occupancy in space by dumping 840 gal of super-cold liquid nitrogen through "dead" or inactive J-2 rocket engine. Simulating 100,000-ft altitude, dumping experiment would pave way for astronauts to move inside orbiting stage and use tanks as living quarters for space station. (MSFC Release 67-203)

- Static launch from motionless rocket sled of F-111 aircraft's crew escape module at Air Force Missile Development Center (AFMDC) was successful. Under severe conditions pilot might encounter in emergency before takeoff, module would allow crew to escape. (*Av Wk*, 10/16/67, 19)

- For the Saturn V program NASA purchased nine S-IVB stages from McDonnell Douglas Corp. for $146.5 million, fulfilling requirements for currently approved 15 Saturn V and 12 Uprated Saturn I launch vehicles. Delivery would begin in April 1968 and end in May 1970. Purchase brought total S-IVB contract for both vehicles to $957,182,093.

    McDonnell Douglas also received a $24-million, cost-plus-fixed-fee contract extension for S-IVB launch services at KSC, which increased total value of contract to $34 million. Included were stage receipt, checkout, launch, and launch evaluation. (NASA Releases 67-258, 67-259)

- U.N. Secretary General U Thant stated that "the very important treaty governing the activities of states in outer space, on the moon and other celestial bodies, forged in the conference room of the United Nations, has reaffirmed the principle first stated in the General Assembly that outer space is the province of all mankind." [Space law treaty signed Jan. 27 would go into effect Oct. 10.] (*NYT*, 10/5/67)
- NASC Executive Secretary Dr. Edward C. Welsh was interviewed by *Christian Science Monitor*'s Neal Sanford. Comparing American and Soviet space budgets, he expressed belief that U.S.S.R. was spending more than twice the U.S. percentage of gross national product on space. "Their [GNP] is less than half of ours. While they are spending about the same amount of money that we are on space it is about twice the percentage of GNP. They are putting a greater priority on space than we are." Soviet space program for 1967 had been "the most active in . . . [their] history, so far as launches go," he said. "I don't know if they are cutting [their budget] back. They are going along at a good rate . . . [and] from the progress they are making, we assume they are going along at about the same funding." (Sanford, *CSM*, 10/4/67, 5)
- News on failures of Soviet manned space flights was "at least in part, 'managed' by Washington," Julius Epstein, research associate at Hoover Institution on War, Revolution, and Peace, wrote in *Los Angeles Times*: "According to reliable information, the CIA submitted a confidential document to the White House early this year reporting the deaths of at least 11 cosmonauts in addition to that of Komarov. The Washington decision not to disclose information on . . . failures was made during the last days of the Eisenhower Administration, presumably to protect intelligence sources." This decision, he said, was recorded in DOD–NASA agreement, Jan. 13, 1961. Mr. Epstein asked that the U.S. give an honest accounting of what it knew about Soviet space losses, so that U.S. losses could be assessed with proper perspective. (Epstein, *LA Times*, 10/4/67)
- FCC's concern for future communications facilities in Atlantic basin area was expressed in letters to ComSatCorp, AT&T, and international telegraph carriers. Data were requested on the various requirements, to permit evaluation of proposals for launching Intelsat IV in 1970 time frame and laying TAT V cable between U.S. and Spain about same time. Response and exchange of comments were requested by end of October. (Text)

October 5: NASA Aerobee 150 sounding rocket launched from WSMR carried NRL payload to 107-mi (173-km) altitude on flight to measure, with EUV heliograph and coronagraph, micrometeoritic particles close to rocket and to take pictures of solar disc. Excellent results were obtained; coronagraph package yielded more than 50 photographs of white-light solar corona from three to 10 solar radii. (NASA Rpt SRL)
- Astronaut Clifton C. Williams, Jr. (Maj., USMC), was killed during routine flight when his T–38 jet aircraft crashed and disintegrated near Tallahassee, Fla., apparently after he had blacked out from an oxygen deficiency. A Lunar Module specialist and *Gemini X* backup pilot, Williams was eighth astronaut to be killed in an accident and third to die in a T–38 crash: one other astronaut had died in an air crash; three in Jan. 27 Apollo fire; and one in an automobile accident. He would be buried in Arlington National Cemetery Oct. 9 with full military honors. (*W Post*, 10/6/67, A1; 10/10/67, B3)
- Senate, approving (71–1) DOT's FY 1968 appropriations bill (H.R. 11456), accepted figures of its Appropriations Committee, where special attention

had been given to air and highway safety programs [see Sept. 28]. Despite persistent efforts of Sen. William Proxmire (D-Wisc.), who would have allowed only $1 million for SST development, amount was fixed at $142 million. Senate-approved bill was sent to conference committee. (*CR*, 10/5/67, S14299-333)

- Displeasure with NASA's administrative officials was voiced by Rep. William F. Ryan (D-N.Y.), member, House Committee on Science and Astronautics, on floor of the House. "NASA continues to evade the serious responsibility of candor," said Rep. Ryan while inserting a series of documents in *Congressional Record*, including Feb. 15 letter from MSFC Director Wernher von Braun to S. K. Hoffman, NAA Rocketdyne Div. president, on ". . . hardware failure, most . . . traceable to human error, inadequate manufacturing, quality control, or test procedures."

  NASA Hq. news conference, called at 5:00 pm same day by NASA Administrator James E. Webb, accompanied by key NASA and industry officials, responded to Rep. Ryan's charges and explained that letter to Hoffman had been a case of normal Government-contractor business. Webb pointed out that subcommittees of the House Committee on Science and Astronautics were kept informed each month so that emergency matters could be brought to their attention.

  Dr. von Braun explained how inspection and quality control procedures were continued as rocket engines were processed at contractor's plant and at Government test sites. He stated that NASA-Rocketdyne coordination on inspection continuity and quality control were applied to both the rocket engine as an entity and the rocket engine/launch-vehicle-stage complex.

  As trend of news conference discussion turned from quality control to Saturn/Apollo weight problems, M/G Samuel C. Phillips, Apollo Program Director, said the heavier Lunar Module resulted from fire-hazard-type design changes. He stated that LM was now 31,500 lb against a 32,000-lb control weight. Referring to weight-lifting capability of Saturn V, NASA Associate Administrator for Manned Space Flight, Dr. George E. Mueller, said lunar payload had increased from 90,000 to 100,000 lb by increasing engine efficiency. Similar improvement was expected in LM's propulsion, which would ease the weight problem. (*CR*, 10/5/67, H13046-8; Transcript)

- Boeing Co. won seven USAF Industrial Zero Defects Program awards, including one for the largest value engineering change submitted to and approved by USAF—an improved cable connection for Minuteman missile launch facilities which saved USAF $7,749,258. Other awards: Award of Achievement to Boeing's Aerospace Group; two additional awards for value engineering to Boeing's Minuteman organization; and three awards to individual Boeing employees for cost-reduction ideas. (Boeing Releases S-9429, S-9434)

- Brian Todd O'Leary, one of NASA's 12 new scientist-astronauts, was appointed assistant professor of Astronomy Dept., Univ. of Texas. Although flight training portion of astronaut's program would continue for next two years, 30% of O'Leary's time would be available for scientific activities. (AP, B *Sun*, 10/6/67)

- F-111 aircraft, under development by USAF and USN, was defended by Secretary of the Air Force Dr. Harold Brown in letter to Sen. Stuart Symington (D-Mo.), ranking member of Senate Armed Services Committee's Preparedness Investigating Subcommittee: "There has never been and will never be a major development program that does not provide grist for the critic's mill, and I say this without questioning either the sincerity of the critics or the utility of the function they perform." He professed confidence in "comprehensive testimony" before congressional committees and asserted "the F-111 is the aircraft we want and need." (CR, 10/9/67, S14445)

- According to National Aeronautic Assn., Soviet authorities had filed with Fédération Aéronautique International two new world speed records and one new world altitude record for their Mikoyan E-266 twin-tailed fighter aircraft. Pilot Peter Ostapenko had completed 1,000-km (621-mi) closed course with 4,410-lb load at average speed of 2,910 km/hr (1,808 mph). With same load, pilot Mikhail Komarov had completed 500-km (311-mi) closed course at 2,928 km/hr (1,820 mph). Alexandr Fedotov reached an altitude of 98,462 ft. (Av Wk, 12/18/67, 54–5; NAA/PIO)

October 5–6: JPL engineers conducted series of exercises with NASA's *Mariner IV* spacecraft (launched Nov. 28, 1964) to demonstrate techniques for controlling *Mariner V* (launched June 14, 1967) if its startracker were to fail during Venus flyby Oct. 19. *Mariner IV*, after traveling more than 1.3-billion mi in solar orbit, successfully responded to commands by breaking lock on its unidentified guide star and locking instead on Canopus; breaking lock on Canopus; maintaining stability by turning on inertial control gyros for first time since December 1964; and pointing high-gain antenna toward earth and using it—with a 20-db signal strength increase at DSN ground stations—for first time since October 1965. Practice of procedures gave JPL engineers confidence in ability to stabilize *Mariner V* by gyros and to keep it oriented by ground command. Spacecraft would remain on inertial control with high-gain antenna oriented toward earth until completion of planned second midcourse maneuver motor firing and other engineering tests. (NASA Proj Off; NASA Release 67-264)

October 6: NASA announced that *Ogo IV* Orbiting Geophysical Observatory, launched from WTR July 28, had passed its primary test objective of collecting scientific data while remaining stable in all three axes for more than 50 days and was continuing to radio data from 18 of its 20 experiments. Unexpected oscillation in 60-ft-long experiment boom, apparently caused by heat from the sun, had been controlled by manually operating satellite's control system and thus expending less control-jet fuel than with automatic control. (NASA Release 67-252)

- Senate, in approving (60–5) NASA FY 1968 appropriations bill (H.R. 12474) of $4.7 billion, accepted figures of Senate Committee on Appropriations. Measure would go to conference to reconcile differences between House and Senate versions. NASA FY 1968 funding in three categories tallied: R&D Senate appropriation of $3.955 billion was increase of $96.0 million over House appropriation, decrease of $152.1 million from authorization, and decrease of $356.5 million from budget. Construction of Facilities Senate appropriation of $55.4 million was increase of $19.5 million over House appropriation, decrease of $14.6 million from authorization, and decrease of $21.3 million from budget. Administrative Operations Senate appropriation of $628.0 million was decrease of $20.0 million from House appropriation, decrease of $20.2 million from authorization, and decrease of $43.3 million from budget.

Sen. John C. Stennis (D-Miss.), Chairman, Senate Committee on

Armed Services' Preparedness Investigating Subcommittee, said: "Ten years ago we did not have the knowledge or the equipment to meet the Russian challenge.... But today ... we are at the crossroads of decision as to what our space program is going to be in 1970 and beyond...." Sen. Margaret C. Smith (R-Me.), ranking minority member, Senate Committee on Aeronautical and Space Sciences, stated that President's signing of FY 1968 authorization bill "clearly signaled a new policy of relegating the space program to a secondary position." (Testimony; *CR*, 10/6/67, S14367–82)

- Secretary of Defense Robert S. McNamara's decision to go ahead with "thin" Nike-X antiballistic missile (ABM) system was criticized by Sen. Gale W. McGee (D-Wyo.), member of Senate Appropriations Committee, on floor of Senate: "... there is a touch of tragedy involved. It lies in the fact that such a system, designed to guard us against possible attack by the unstable Chinese nation, can be effective in neutralizing the danger of nuclear holocaust but not the danger of continued guerrilla warfare as practiced by Mao-Tse-tung and his followers." (*CR*, 10/6/67, S14359)

- Vexing problem of the direction and period of Venus' rotation had acquired a new dimension in recent ultraviolet photographs, according to an article in *Science* by Bradford A. Smith of the New Mexico State Univ. Observatory. Smith recounted the history of 60 yrs of visual and spectroscopic observation of Venus that produced inconclusive evidence. Radar observation begun in 1961 by 1967 had produced what was considered firm evidence that the solid globe of Venus in 244 days rotated once in a direction retrograde to that of its planetary neighbors earth and Mars. Recent ultraviolet photos taken at the New Mexico State Univ. Observatory with a new 61-cm reflector indicated that ultraviolet clouds of the Venusian atmosphere rotated retrograde in five days. This large discrepancy between movement of the globe and atmosphere of Venus could be explained only by the existence in the Venusian atmosphere of a persistent and widespread planetary wind system blowing at more than 180 mph, Smith said. Although earth's jet stream reached such speeds, it was narrow and zonal, in contrast to the massive system that would have to exist on Venus if these observations were correct. (*Science*, 10/6/67, 114–6)

October 7: *New York Times* commented on congressional cuts in NASA's FY 1968 budget: "The Senate, ignoring the most obvious of all opportunities to cut nonessential spending, has not only matched the $4.5 billion the House voted for space excursion but tacked on nearly $100 million more to start new programs after the first Americans land on the moon. The certainty that another huge space appropriation will come out of conference on Capitol Hill adds special interest to the fact that twice recently the leading space scientist of the Soviet Union, Prof. Leonid Sedov, has publicly stressed the importance of large-scale international cooperation for manned flights beyond the moon.... Now that Professor Sedov has opened the door, the Administration has an opportunity to begin diplomatic explorations to see if the Russians really are serious about replacing senseless rivalry in space with rational international cooperation." (*NYT*, 10/7/67, 28)

Launches of military espionage spacecraft dominated U.S. space program, Soviet scientists Dr. V. Denisov and F. Soluyanov charged in *Krasnaya Zvezda*. According to the scientists, of 455 U.S. spacecraft launched by mid-1967, 346 were military satellites which the Pentagon "proposed to use . . . for reconnaissance on a so-called real-time scale, followed by an immediate transmission of information." Although U.S. "continues to persist that its aspirations are the peaceful utilization of space," they said, officials "do not hide the fact that NASA builds the scientific-technological foundation for the development of military-space facilities . . . [while DOD] finances and controls the planning . . . of space armament." (Miles, *P Inq*, 10/8/67, 13C)

October 8: President Johnson addressed International Conference on World Crisis in Education, Williamsburg, Va.: ". . . I would like to suggest . . . these challenges: How can we use what we already know about educational television to accelerate the pace of basic education for all the children of the world? How can we use modern technology to economize on that most essential and that most needed educational resource: the good teacher?" He added: "If future historians . . . seek a name for this period in America, I hope they will give consideration to calling it the age of education. If our children's children want to measure what we tried to achieve, I hope they will remember one thing: The American Government in only 3 years multiplied its commitment to education and to health four times over. Congress passed more laws and committed more funds to education and health in the last 3 years than in all previous history." (*PD*, 10/16/67, 1419–23)

October 9: Sen. Joseph S. Clark (D-Pa.), member of Senate Foreign Relations Committee, on Senate floor called proposed Nike-X ABM system "a very expensive flying 'erector set'" and said that Secretary of Defense Robert S. McNamara was following an unwise course, "militarily, economically, and diplomatically. . . . Even if we were to spend forty billion dollars or more on a so-called massive system, the resulting increase in our security would be zero. . . . The result would be a waste of a great deal of money . . ." at a critical time. He cited three flaws in arguments supporting ABM system: (1) justification based on approaching threat of Communist Chinese nuclear tipped missiles ("We have the capacity to devastate China many times over . . . and the Chinese know that," he said); (2) assumption that a thin ABM system would be effective against Communist China for any appreciable period; and (3) assumption that during so-called safe period in Chinese nuclear development, U.S., behind an ABM defense, would be genuinely safe from nuclear attack. "The inevitable conclusion," Senator Clark noted, "is that the proposed . . . system simply will not do the job which its proponents say it will do." (Text; *CR*, 10/9/67, S14413–27)

- Senate action in voting $4.6 billion for NASA for FY 1968 and $142 million for SST development was criticized by Baltimore *Sun:* "Approval of these sums by the Senate underscores, once again, the need for a set of Federal priorities—for establishing firm control over 'the course of fiscal developments,' as Chairman Mills of the House Ways and Means Committee put it." (*B Sun*, 10/9/67)

- Secretary of Defense Robert S. McNamara's decision to go ahead with "thin" Nike-X antiballistic missile (ABM) system was criticized by Sen. Joseph S. Clark (D-Pa.), member of Senate Foreign Relations Committee, on floor of Senate: "I urge the administration to reconsider its decision, which I believe to be wrong on three counts—militarily, economically, and diplomatically." He anticipated "resulting increase in tensions between the United States and the Soviet Union which this new escalation of the arms race is bound to provoke." (*CR*, 10/9/67, S14413–27)

October 9–11: Two Lunar Orbiters were intentionally crashed on lunar surface to free their radio frequencies. *Lunar Orbiter III*, launched Feb. 4, 1967, crashed Oct. 9, 1967. *Lunar Orbiter II*, launched Nov. 6, 1966, crashed Oct. 11, 1967.

In preparation for Oct. 18 lunar eclipse, *Lunar Orbiter V*'s orbit had been adjusted by burning spacecraft's velocity-control engine for 41 sec to increase spacecraft's period by 37 min and shorten length of time it would be in darkness during eclipse. Scientists wished to ensure that spacecraft's power system would survive expected five-hour dark period. (NASA Release 67–265; NASA Proj Off)

October 10: President Johnson, in White House ceremony marking entry into force of space law treaty, greeted ambassadors from 13 countries and key U.S. officials and renewed America's offer to cooperate fully with any nation in exploring planets and solar system, using tracking facilities, mapping the earth, exchanging bioscientific information, and communicating via comsats.

Responding, Soviet Ambassador Anatoli F. Dobrynin said: "These [international legal] principles . . . are aimed at insuring peaceful activities of states in outer space for the benefit of all mankind. We hope that [the treaty] will contribute to the settlement of major international problems still facing the mankind on our planet."

British Ambassador Sir Patrick Dean said: "We welcome [the treaty] all the more because the treaty removes outer space from the effect of the rivalry and dissension between nations. . . . It is a treaty about . . . rule of law and the extension of that field within which international law is to regulate the conflicts of interests, which arise between men and nations."

Secretary of State Dean Rusk said: "The treaty also takes steps to limit and reduce the competition in armaments, a terrible burden on peoples everywhere." He added: "It is evident that men and nations can, in fact, achieve the maturity necessary to embody in binding form their points of agreement, despite political differences in other areas." (*PD*, 10/16/67, 1425–8)

- Planned launch of Apollo 4/Saturn V for Oct. 17 was postponed until early November, according to NASA spokesmen at ETR, reported UPI. Troubles with ground support equipment had stalled countdown rehearsal. Once rehearsal was finished, engineers would evaluate results and set date for unmanned launch. (UPI, *W Post*, 10/11/67, A18)

- R/A Albert C. Read (USN, Ret.), commander of first heavier-than-air aircraft to cross the Atlantic (USN NC–4 seaplane), died in Miami, Fla. He had received the Distinguished Service Medal for the 4,500-mi, 23-day flight he made in May 1919, from Rockaway, N.Y., to Plymouth, U.K. and in 1965 was elected to the Aviation Hall of Fame in Dayton. (*W Post*, 10/12/67, B6)

October 11: NASA test pilots Fitzhugh Fulton and Donald L. Mallick flew XB–70 No. 1 to 58,000-ft altitude and 1,581 mph (mach 2.43) in flight at Edwards AFB. Purpose of test was to check lateral-directional stability, control, and handling qualities at mach 2.35 and 55,000 ft; inlet unstart at mach 2.4 and 57,500 ft; variable nose ramp functional test at mach 1.6 and 38,000 ft, and at mach 1.4 and 32,000 ft. Flight was successful. (XB–70 Proj Off)

October 11

- U.S.S.R. successfully launched *Cosmos CLXXXI* into orbit with 863-km (536-mi) apogee, 668-km (415-mi) perigee, 101.1-min period, and 99.16° inclination. Satellite reentered Oct. 18. (GSFC *SSR,* 10/31/67)
- USAF launched unidentified satellite from Vandenberg AFB employing a Thor-Burner II booster. (UPI, P *EB,* 10/11/67, 2; *Pres Rep 1967*)
- NASA announced organizational changes in preparation for changes that would occur when Deputy Administrator Dr. Robert C. Seamans, Jr., left NASA Jan. 1: (1) General Counsel Paul G. Dembling and Executive Secretary John R. Biggs would report directly to NASA Administrator James E. Webb; (2) Assistant General Counsel for Procurement Matters S. Neil Hosenball would fill new post of Deputy General Counsel; and (3) OMSF Secretariat representative Frank J. Magliato would serve as Special Assistant to the Administrator and direct organization of a Headquarters Communications Center. (NASA Release 67-263)
- Executive Order 11374, abolishing Missile Sites Labor Commission, and transferring its functions and responsibilities to Federal Mediation and Conciliation Service, was signed by President. (*PD,* 10/16/67, 1430-1)
- Sen. Henry M. Jackson (D-Wash.) expressed concern about U.S. defense policy in speech to the Hoover Institution on War, Revolution, and Peace meeting at Stanford Univ. He cited three specific ideas with which he disagreed: (1) exaggerated picture of U.S. as catalyst in Soviet desire to "possess a first-rate military establishment," (2) theory that military technology had reached a plateau and stabilized, and (3) seeming trend toward nuclear parity. Some scientists and defense planners, he said, had "a mirror image interpretation of Moscow's decisions . . . seeing them as reflex actions" and believed that if U.S. did not act U.S.S.R. would not act. He also argued, "Ordinary economic technology is always finding better ways to do things and there is no reason to suppose that military technology will cease in this effort." Sen. Jackson believed that "international peace and security depend not on a parity of power but on a preponderance of power in the peace-keepers over the peace-upsetters. Our aim is not . . . an unlimited accumulation of weapons . . . [but] to create and maintain . . . a relationship of forces favorable to the deterrent of adventurism and aggression." (Text)
- "The ceremonies in Washington, Moscow and London putting the space treaty into effect mark not only the first major East-West agreement since the nuclear test ban treaty of 1963 but a potential starting point for an even more momentous accord," *New York Times* commented on space law treaty which went into effect Oct. 10. It added that "recent statements by the Soviet Union's leading space scientist, Professor Leonid Sedov, indicate that the Russians too may be coming around to the conclusion that the costs of going it alone to the moon and beyond outweigh the dimming propaganda magic." (*NYT,* 10/11/67)
- Jonathan Spivak in *Wall Street Journal* commented on congressional cuts in NASA's FY 1968 budget: "Plans to launch unmanned satellites to Mars or Venus in 1971 and 1973 and conduct extended exploration of the moon by astronauts face probable abandonment because of a lack of funds. There will also be sharp curtailment of proposals for long-duration manned flights in orbit about the earth to enhance the nation's space technology and exploit practical and scientific applications of space. Thus, say space officials, the present outlook is that only a limited national space program will continue in the 1970s after the lunar landing." (Spivak, *WSJ,* 10/11/67, 24)

October 12: Vertical probe launched by U.S.S.R. carried instrumented payload to 4,400-km (2,734-mi) altitude to study "characteristics of ionosphere . . . general intensity of cosmic rays . . . doses of radiation for various protections at the time of the flight in radiation belts . . . [and] density of neutral hydrogen." Equipment and instruments—including a telemetry system—functioned normally. (UPI, *P Inq,* 10/14/67, 1; *SBD,* 10/16/67, 239)

- NASA Administrator James E. Webb held press conference on management changes: (1) George S. Trimble, Jr., Director of Advanced Manned Missions Program, OMSF, was appointed Deputy Director of MSC. (2) NASA Deputy Associate Administrator for Space Science and Applications Edgar M. Cortright was appointed Deputy Associate Administrator for Manned Space Flight. He would be replaced by Donald P. Hearth, Voyager Program Manager. (3) MSFC Deputy Director Eberhard F. M. Rees was named to serve as Special Assistant in Manufacturing Problems to MSC Apollo Spacecraft Program Manager George M. Low.

    Discussing space law treaty [effective Oct. 10] after making announcements, Webb said it was a valuable further advance of the rule of law into space but noted that laws could be broken. "We have an agreement in the Antarctic . . . not to introduce military operations or equipment and this has worked well. . . . We now have this agreement in space, and . . . the most important enforcer of these that relate to space technology is momentum and continued development of technology. The forward capability is what in essence has produced cooperation when we have had it. . . . [U.S.S.R.] and other nations must recognize that they're not going to have a monopoly in this field. [Our being] . . . there is the most important way to make sure we get the benefits of the treaty. Without continued development of this kind I am not sure what the treaty means." (Transcript; NASA Release 67–268)

- All-weather landing system (AWLS) was certified by FAA for use on USAF C–141 StarLifter aircraft under minimums of 1,200-ft visibility and 100-ft decision altitude. Capable of bringing the 316,000-lb fanjet transport to a landing within 12 ft of either side of runway center line and within 300 ft of either side of a touchdown point, AWLS would considerably improve all-weather landing capability. Developed by AFSC and FAA and produced by Lockheed-Georgia Co., AWLS would program aircraft's flight path, speed, angle of approach, and attitude; generate and provide data to pilot or autopilot; and be nearly self-sufficient. Ultimate goal of AWLS was to enable aircraft to make totally blind landings when ceiling and visibility both were zero. (AFSC Release 185.67)

- Satellites should be used to enable land-based air traffic control centers to track constantly commercial airliners flying over oceans, urged Pan American World Airways' chief electronic engineer Ben F. McLeod at NATO's Advisory Group for Aerospace Research and Development (AGARD) meeting at New York City. McLeod said that within five to 10 years traffic increases and new aircraft would demand improved air traffic control system with positions of hundreds of aircraft being sent automatically by satellites to ground stations for plotting. (*NYT,* 10/12/67, 50)

October 13: President Johnson presented Harmon International Aviation Trophies for 1967 to Astronauts James A. Lovell, Jr. (Capt., USN), and Edwin E. Aldrin, Jr. (L/C, USAF), and Alvin S. White, former NAA test pilot. Lovell and Aldrin were cited for successful *Gemini XII* mission

October 13: President Lyndon B. Johnson presents the 1967 Harmon International Aviation Trophy to two NASA astronauts and a veteran test pilot in the White House Rose Garden. From the left are North American Aviation XB-70 test pilot Alvin S. White, Astronaut Edwin E. Aldrin, Jr., Astronaut James A. Lovell, Jr., and the President.

(Nov. 11–15, 1966) during which Aldrin spent 5 hrs 28 min outside spacecraft. White was cited for mach 3 XB-70 flights. British aviatrix Sheila Scott, scheduled to receive award for her record-setting solo flight around the world in a single engine aircraft, was unable to attend ceremony. (*PD*, 10/16/67, 1436-7; AP, *NYT*, 10/14/67, 7)

- NASA Arcas sounding rocket was launched from Barking Sands, Hawaii, to 37-mi (59-km) altitude, fourth in NOTS series to measure incident polar UV irradiance in direct support of *Ogo IV's* mission. Excellent data were obtained. Rocket and payload performed satisfactorily. (NASA Rpt SRL)
- NASA announced it had asked Grumman Aircraft Engineering Corp. to deliver second Lunar Module (LM-2) in unmanned configuration, changing earlier plan for manned configuration. With the change, NASA could provide a backup for the Apollo V launch scheduled for early 1968 without a wait for modifications. If backup for LM-2 was not needed, manned launch would make use of LM-2 at later date. (NASA Release 67-266; Transcript)

- NASA Deputy Associate Administrator for Space Science and Applications (Engineering) George Hage was named Deputy Director (Engineering) of the Apollo Program, OMSF. (NASA Ann, 10/13/67)

  Inquiry into economic impact of building SST was made by Dael Wolfle in *Science* editorial. Congressional arguments on whether to build SST "illustrate why members of Congress want to increase congressional competence to assess the consequences of technological developments and proposals. . . . What Congress needs in deciding about the SST or other technological matters is essentially what it needs in deciding about taxes, military affairs, education, or other matters with which it deals: ability to ask the right questions and ability to evaluate critically the information it receives from advocates and opponents of proposed actions." (Wolfle, *Science*, 10/13/67)
- In its fourth year of full operation ESRO was entering a more expensive and ambitious phase and consequently was encountering financial, political, and organizational difficulties, John Walsh reported in *Science*. ESRO was currently appraising its eight-year program and seeking ways to coordinate its activities with ELDO, European Conference on Communications Satellites (CETS), and new European Space Conference. According to 1962 convention which became effective in 1964, ESRO was to have launched 300 sounding rockets and 10–12 satellites within eight years under $300-million budget. Although ESRO council was empowered to adjust budget authorization in response to major scientific and technical developments, it lacked unanimity necessary for change. As a result, Walsh noted, scheduled number of satellites—including $170-million Large Astronomical Satellite (LAS)—would probably not be launched. Expenditures for buildings and equipment for ESRO's new four-station satellite tracking, telemetry, and telecommand network, and its five main research installations had been particularly heavy. European Space Technology Center (ESTEC), for example, responsible for applied research and payload development, had a 456-man staff—nearly half of total ESRO staff—which was expected to reach 1,000 by the end of 1967. In addition to financial problems caused by expansion, Walsh said, ESRO had to overcome a sensitivity to what national aerospace industries would get back in relation to national contributions, differences over selection of projects, and lack of unity and coordinated effort. Differences of opinion between larger countries such as U.K., Germany, and France, which contributed 23.13%, 24.31%, and 20.17%, respectively, to ESRO budget, and smaller countries such as Denmark, which contributed 2.15%, had to be resolved before a real sense of purpose could be acquired, Walsh said. "ESRO until now has been primarily an agency providing technical services and support for university scientists who prepare actual experiments. As experiments grow more complex, more work will be done in ESRO labs and by industry," and new dimensions would be added to its tasks. (Walsh, *Science*, 10/13/67, 242–4)

October 14: USAF launched experimental reentry vehicle on Atlas-F booster from Vandenberg AFB in Advanced Ballistic Re-Entry Systems (ABRES) program. (*P Inq*, 10/15/67, 7)
- Saturn V's countdown rehearsal was completed; final flight preparations would be completed for scheduled launch in about three weeks, reported UPI. (UPI, W *Star*, 10/14/67, A3)
- Apollo spacecraft would carry two fire extinguishers, designed and

built for NASA by Southwest Research Institute, the *New York Times* reported. Foam from new extinguishers would provide fire protection for astronauts. Extinguishers would weigh about eight pounds, measure nine inches long with five-inch diameter, and would have removable safety pin and pistol-type grip with thumb button for release of foam. (*NYT*, 10/15/67, 29)

- *Christian Science Monitor*, commenting on space law treaty which went into effect Oct. 10, noted that space exploration had evolved this treaty which banned military installations and nuclear weapons from moon and planets and which could evolve into space cooperation between U.S. and U.S.S.R. Treaty did not halt use of reconnaissance satellites but it provided for international inspection of vehicles and installations on moon and other bodies. Although there would be no right of inspection to see whether satellites were carrying nuclear weapons, and vigilance would always be vital in this regard, treaty was one more visible proof that Moscow and Washington could make useful progress. Space law treaty was agreement, following Antarctic Treaty, which could remove installations of war from one more section of mankind's expanding environment. (*CSM*, 10/14/67)

October 14–15: Three NASA Nike-Cajun sounding rockets—first in series of five three-rocket launches—were successfully launched from Natal, Brazil, in GSFC program to isolate effect of changing sun angle on mesospheric temperature structure over six-month time period. Altitude would be computed by sound arrivals in track-through-grenade-sounding method. Of 19 grenades, 18 ejected and exploded. (NASA Rpt SRL)

October 15: JPL announced development of double vacuum chamber, designed by JPL's James B. Stephens, to simulate extreme effects of interplanetary space on spacecraft. Called Molsink (molecular sink of outer space), chamber was expected to determine what happened to paint, components, and even microorganisms on spacecraft. The eight-foot-dia chamber would be capable of capturing 99.97% of all condensable molecules emanating from test item by freezing the molecules to the walls at −400°F with super-cold helium gas. Conventional vacuum chambers had not been able to pump, or capture, more than half of gas molecules flying off test item. Space scientists believed Molsink chamber would provide reliable tool to measure and counteract problem of spacecraft out-gassing. In testing, long periods would simulate actual flight time of spacecraft to planets—four months to Venus, seven months to Mars, for example. The chamber was expected to be fully operational early in 1968, when testing would be stepped up for NASA's Mariner Mars 1969 exploration program. (JPL Release 459)

October 15–31: Surveyor V responded to commands on Oct. 15. Spacecraft had been in hibernation for its first lunar night that had begun Sept. 24, and had survived in fair condition. TV camera had weak video level; alpha-backscattering instrument was noisier; and thermal control was reduced in Compartment A. Thermal data, which exhibited effects of lunar eclipse on Oct. 18 (03:58–08:59 EDT) would be analyzed for both engineering and scientific implications.

Local sunset for second lunar day occurred on Oct. 23. Spacecraft operating time from turn-on command until second sunset was 232 hrs. Total of 1,043 TV pictures were obtained. By sunset *Surveyor V* had been configured for hibernating through second lunar night. Estimates were that engineering interrogations and compartment heating would be discontinued Oct. 31. (NASA Proj Off; UPI, *W Post*, 10/16/67, A3; AP, *NYT*, 10/17/67, 9)

October 16: *Cosmos CLXXXII* was successfully launched by U.S.S.R. Orbital parameters: apogee, 355 km (221 mi); perigee, 210 km (131 mi); period, 89.9 min; and inclination, 65°. Satellite reentered Oct. 24. (*NYT*, 10/17/67, 12; GSFC *SSR*, 10/31/67)

- FAA issued forecast of aviation activity at 22 large air transportation centers and selected airport facilities required to service increases in air activity. By 1980, takeoffs and landings were expected to increase by 270%; air passengers by 433%; air cargo on scheduled airlines by 1,400%; and general aviation (non-airline) aircraft by 146%. (FAA Release 67–77)

- U.S.S.R. had invited U.K. to help track its Venus probe, *Venus IV*, in telegram from Mstislav V. Keldysh, President of U.S.S.R. Academy of Sciences, to Sir Bernard Lovell, Director of Jodrell Bank Observatory, reported *Washington Post*. U.S.S.R. did not furnish code to signals; Sir Bernard had said that tapes of signals would be sent to Moscow. Keldysh's telegram referred to "this experiment's extraordinary importance and significance for mankind." Sir Bernard speculated that U.S.S.R. might attempt softlanding on planet Venus. Not since launch of *Venus I* in 1961 had the Soviets requested British working assistance. (Schmidt, *NYT*, 10/17/67, 13; *W Post*, 10/17/67, A17)

- Three possible military reasons why U.S.S.R. used one-orbit spacecraft in Cosmos series during period from Sept. 17, 1966, through Aug. 8, 1967, when seven were launched, were suggested by Phillip J. Klass in *Aviation Week:* launches evaluated (1) missile defense radars and ICBM penetration aids, (2) very-long-range ICBM attacking from south, or (3) orbital weapon. Other reasons had been noted—including test of recovery systems for Soyuz manned spacecraft. In all cases, a principal object reentered after one orbit and other objects after a day or two. Soviets had never mentioned orbital period nor had they explained low perigee of 90 mi—an indication spacecraft would remain in space for only one orbit or less.

    Orbital weapon theory was also suggested by Evert Clark, in *New York Times:* "Most analysts . . . now believe the flights are exploring the techniques of bringing down a nuclear warhead from an orbiting platform." [Such tests would not violate space law treaty which became effective Oct. 10, he said, because treaty did not prohibit engineering tests necessary to station weapons in orbit. Nor would it prohibit nuclear armed flights that were less than one full orbit, as these were.]

    Schedule for series of seven launches had been: two in 1966, Sept. 17 and Nov. 2, both unannounced, and five in 1967. *Cosmos CXXXIX* Jan. 25, *Cosmos CLX* May 17, *Cosmos CLXIX* July 17, *Cosmos CLXX* July 31, and *Cosmos CLXXI* Aug. 8. First two in series had exploded into many fragments during day of launch. (Klass, *Av Wk*, 10/16/67, 26–7; Clark, *NYT*, 10/17/67, 1)

October 17: Maj. William J. Knight (USAF) flew X-15 No. 3 to 3,818 mph (mach 5.18) and 277,000-ft altitude, becoming fourth USAF X-15 pilot to meet 264,000-ft-minimum-altitude requirement for USAF astronaut rating. Purposes of flight: (1) ARC boost guidance system evaluation; (2) solar spectrum measurement; (3) micrometeoroid collection; and (4) ultravoilet plume and tip-pod camera evaluation. (X-15 Proj Off; FRC *X-Press*, 10/20/67, 1)

- The five-out-of-five successful missions of $200-million Lunar Orbiter project to photograph moon to aid in selecting Surveyor and Apollo landing sites and to contribute to scientific knowledge of moon were reviewed at NASA Hq. press conference. Photographic data provided by *Lunar Orbiter I, II,* and *III* and *Surveyor I* resulted in selection of eight candidate sites. *Lunar Orbiter IV* provided basis for extending cartographic grid system for front-face mapping around to moon's hidden side by photographing eastern limb areas. *Lunar Orbiter V* photographed scientifically interesting areas on moon's front side, completed farside coverage, and provided supplementary photography of potential Apollo landing sites.

  Apollo Program Director, M/G Samuel C. Phillips, said that, when "approaching on a pretty flat trajectory at about 30 miles from the landing site . . . details of the topography and the radar return which is coupled into the automatic control and landing system [must be known]." In planning details of approach and landing maneuvers, Phillips said that "a very great deal" had been learned from Orbiter "about the rather unusual light conditions and the effect of light reflections from the lunar surface under various angles."

  Harold Masursky, USGS, explained Lunar Orbiter photographs showed that "apparently the earth and the Moon are very similar in their distribution of high and low ground." He hoped that "by examination and comparison of . . . geologic plot with the selenodetic information . . . [from] the Orbiter series, we will be able to plot the geopotential field . . . and tie it in with geology. . . . By studying the Moon we know that it has a long, complex geologic history, that many of the same kinds of things that happen on the earth happen on the Moon. We can now play these two against each other, compare their simularities, contrast their differences, and perhaps be able to arrive at a greater understanding of the fundamental processes that affect the earth."

  Strange moon features looking like watercourses might be dry-gulch arroyos, Harold Spradley said. Because watercourses had originated in volcanic craters and carried "fluidal materials," volcano's output—possibly slurry of ash and water—would exist long enough to cut its way through dry lunar surface material. Scientists agreed that orthodox existence of water on "the airless moon" was improbable.

  Clifford Nelson, LaRC Lunar Orbiter Project Manager, described results of experiments: ". . . density distribution [very nearly homogeneous] in the moon now seems to be resolved . . . center of mass of the moon is displaced toward the earth from the center of its geometrical figure, a situation which would have interesting scientific implications as to the stability of the moon and its origin and history." Meteoroid penetration rate "in the region covered by lunar orbiter spacecraft is less than one-half the rate in the near earth environment. . . ." He said all 13 MSFN stations had successfully tracked and communicated with lunar orbiter, and "orbit determination programs . . . in the initial stages of checkout . . . will continue with *Lunar Orbiter V*."

  NASA Associate Administrator Dr. Homer E. Newell stressed overall significance of Lunar Orbiter program: "Before the first Lunar Orbiter mission this country had never placed spacecraft into orbit about any other celestial body other than the earth. In all five . . . missions, the spacecraft were not only placed into initial orbits as desired, but then the orbits were changed in accordance with a plan . . . set up beforehand, in . . . a precision type of operation. . . . Lunar Orbiter and its companion, Surveyor, have shown that we can use sophisticated technology in complicated spacecraft for automated missions to other bodies of the solar system. And it is this kind of capability that we will be using again and again as we continue the exploration of our solar system." (Transcript)

- NASA's *Mariner V* would fly within 2,500 mi of planet Venus at 1.34 pm EDT, Oct. 19, at communications (direct) distance of about 49 million mi from earth, NASA announced. Launched from ETR June 14, spacecraft would have traveled about 217 million mi in its arching trajectory. Scientists and engineers hoped their instruments would record precise data on planet's atmosphere, ionosphere, temperatures, and perhaps even surface qualities. As *Mariner V* approached Venus, its scientific instruments would begin measuring planet's magnetic field, charged particles, gases present in upper atmosphere, and radiation levels. Spacecraft's flight path would curve behind Venus and would vanish from view of tracking stations on earth. Just before spacecraft went behind planet (as seen from earth) its radio signals would pass through atmosphere of Venus. Effect of Venusian atmosphere on *Mariner V*'s radio signals would be measured, thereby providing measure of density of planet's atmosphere. Current scientific theory on density ranged from five times earth's atmosphere to several hundred times. Density measurement by occultation was one of prime objectives of spacecraft's flight.

    Since launch of spacecraft, engineering and science subsystems had performed as planned. Master timer had commanded Canopus cone angle update as scheduled on Aug. 24, Sept. 10, Sept. 26, and Oct. 10; timer had also commanded spacecraft transmitter to switch to high-gain antenna on Oct. 2. Encounter sequence would be initiated by command from DSN station in Australia, at 10:49 pm EDT, Oct. 18. (NASA Proj Off; NASA Releases 67-260, 67-267)

- NASA Administrator James E. Webb reviewed NASA-university relations at dedication of Univ. of Illinois' Coordinated Science Laboratory building in Urbana, Ill.: "During NASA's first year of operation . . . [$3 million] was invested in research at universities. . . . The second year, approximately $6 million went to universities, and the third year about $14 million . . . [primarily for] solution of immediate problems. . . ." Appraisal of program in 1961 concluded the universities were already making significant contribution to the space program, but that with the right kind of help they could do more for themselves and for NASA. Expansion included support of research, construction of facilities, and establishment of graduate fellowships, until program reached peak in 1966 with 10,600 campus participants. Today, however, need was "to reduce expenditures for research, to preserve at minimum cost the essential strength needed for the future, and to increase the multidisciplinary flux that will magnify the values the nation can derive from the activities we can support." To find the best ways of reducing the program, he said, NASA had formed a "special task force to evaluate all our relationships with the academic community and to come up with specific recommendations as to least harmful methods we can use. There is no longer any doubt that NASA's university support programs will have to be redefined quite drastically in some areas." Since much of the program was funded in three one-year steps, this would help spread the impact over a longer time period. Ph. D. candidates now holding NASA grants would be able to obtain their degrees. Looking to the future, Webb urged the universities to think, reorganize, and work in interdisciplinary groupings. "The development of a few university teams working together across the traditional disciplines, combining the best efforts of technology, the social sciences and management capability which they have created and which make them the best available source of trusted information in our society can write a new chapter of history of problem solving in this nation." (Text)

- FAA adopted new rule banning unauthorized aircraft from designated manned and unmanned spaceflight recovery areas to prevent interference with aircraft and pararescue personnel participating in recovery operations. Areas selected by NASA and DOD and effective dates would be published before launch in a Notice to Airmen. (FAA Release 67-78)
- Progress in F-111B aircraft development was discussed by Assistant Secretary of Navy (R&D) Robert A. Frosch, at 1967 Electronics and Aerospace Systems Technical Convention, Washington, D.C.: ". . . we are convinced that in its primary air defense interceptor role the F-111B, equipped with the PHOENIX airborne missile control system, and firing multiple shots of the long-range PHOENIX air-to-air missile, represents the finest fleet air defense system available in the immediate future." He further stated: ". . . aircraft will probably not meet all of the initial specifications and the contractor will have to accept some responsibility for this lack. It is, of course, not unusual for a military aircraft that uses advanced state-of-the-art to fail to meet some of the specifications, the real question is whether it meets military needs." He added: ". . . we find that it meets our fleet air defense requirements better than any competing system available for study." (Text)
- F-5 supersonic fighter aircraft manufactured by Northrop Co. could be sold to Latin American governments, Covey T. Oliver, Assistant Secretary of State for Inter-American Affairs, was reported to have told key members of Congress. Because of shortage of pilots, facilities, and financial resources, only five nations would be interested—Peru, Brazil, Argentina, Chile, and Venezuela. Initial congressional reaction was reported favorable, the decision being displayed as no change in U.S. policy. State Dept. informants, however, conceded that recent reports of Peru's negotiations with France on purchase of 12 Mirage V jet fighters had a major effect in hastening DOD's and State Dept.'s action. (Welles, *NYT*, 10/18/67, 1)

October 18: NASA's *Oso IV* (OSO-D) Orbiting Solar Observatory was successfully launched from ETR with three-stage Thor-Delta booster to study the sun and its influence on earth's atmosphere. Orbital parameters: apogee, 358.8 mi (577 km); perigee, 336.7 mi (542 km); period, 95.8 min; and inclination, 32.9°. Fourth of eight spacecraft in NASA's OSO program to provide direct observation of the sun during most of an 11-yr solar cycle, *Oso IV* weighed 599 lbs, carried nine experiments, was designed for a six-month lifetime, and had two main sections: the wheel (lower) section provided stability by gyroscopic spinning and housed the telemetry-command equipment, batteries, electronic controls, gas spin-control arms, and six experiment packages; the sail (upper) section was oriented toward the sun and contained solar cells and three solar pointing experiments.

Primary purpose of *Oso IV* was to obtain high-resolution spectral data within the range 1Å–1350Å from pointed solar experiments in the spacecraft during major portion of one solar rotation and adequate operational support of spacecraft subsystems including raster scan of solar disk to carry out acquisition of these scientific data. Secondary objective was to obtain useful data from nonpointed experiments and from pointed experiments during more than one solar rotation. Experiments, designed to continue and extend work of *Oso I* (launched March 7, 1962), *Oso II* (launched Feb. 3, 1965), and *Oso III* (launched March 8, 1967) by collecting data on solar x-rays, gamma rays, ultraviolet radiation, and other solar activity, were provided by two American and two British universities, one U.S. Government agency, and one private company. *Oso IV* was first in series to carry foreign experiments and second (*Oso II* was the first) capable of scanning entire solar surface by means of a Solar Ultraviolet Scanning Spectrometer which could record a "picture" (digital number) of the sun and transmit it to receiving stations. Spacecraft also had an improved ground-control system which permitted it to receive up to 140 different commands—compared to 10 for *Oso I*, 70 for *Oso II*, and 94 for *Oso III*. Project officials at GSFC reported that satellite spin rate, power level, charge rate, and temperatures were normal and that both tape recorders were operating; all experiments were fully operational by Oct. 25. OSO program was managed by GSFC under OSSA direction. (NASA Proj Off; NASA Release 67–262)

- U.S.S.R.'s *Venus IV* planetary probe (also designated *Venera IV*) entered atmosphere of planet Venus at 12:34 am EDT and ejected instrumented capsule intended for softlanding on Venus. After aerodynamic braking, automatic parachute system provided gradual descent through 15-mi dense, hot, windy atmosphere, and capsule made [purported] softlanding at 2:08 am EDT. Mstislav Keldysh, President of Soviet Academy of Sciences, later presented scientific proof for softlanding [see Oct. 30]. Launched June 12, the 2,438-lb *Venus IV* had traveled trajectory of more than 210 million mi; 49 million mi would be covered by transmitted signals from capsule to earth antennas.

According to Soviet scientists, atmosphere as recorded by instrumented capsule had almost whole carbon dioxide composition with about 1.5% oxygen and water vapors; no significant traces of nitrogen; and a temperature range of 104–536° F. Reports indicated no noticeable magnetic field and no radiation belt, but detectable weak hydrogen corona. "Because Venus rotates so slowly around its axis [purported to be once in 244 earth days] half the planet is cold," said Soviet scientist Josif Shklovski. Because the other half heated to "monstrous temperature," he said surface of Venus would appear to be "a dry hot desert." He noted that temperature extremes would cause constant winds on Venus "of about 450 mph" [accounting for large amount of drift of capsule during descent].

Moscow news reports from *Izvestia* and Tass said: *Venus IV*'s mission had been accomplished and instruments had ceased functioning after radioing data from softlanding site on heat, pressure, and carbon dioxide hostile to most forms of life; mother spacecraft burned in Venus' atmosphere. Some reports did not claim softlanding. Soviet astronomer Vitali Bronshtein's claim that radio contact was being maintained with instrumented capsule was unconfirmed. Also, cessation of signals from instrumented capsule received varied explanations from international sources. Tass reported antenna of package had failed to point in the right direction and that package's bulk had blocked direct contact with earth. Jodrell Bank Observatory Director Sir Bernard Lovell expressed doubt that any electronic device could withstand high temperatures on Venusian surface. (Friendly, *W Post*, 10/18/67, 1; Kamm, *NYT*, 10/19/67, 1; Bishop, *WSJ*, 10/19/67, 4; AP, B *Sun*, 10/20/67; AP, *W Star*, 10/22/67, A3; *SBD*, 10/24/67, 283–5)

- Communist Party and Soviet government leaders hailed *Venus IV* instrument package landing on planet Venus as honor to Bolshevik Revolution's 50th anniversary, Nov. 7. (Shub, *W Post*, 10/19/67, A9)
- U.S.S.R. launched *Cosmos CLXXXIII* into orbit with 212-km (132-mi) apogee, 145-km (90-mi) perigee, 89.9-min period, and 49° inclination. Equipment and instruments functioned satisfactorily. Satellite reentered same day. (UPI, *NYT*, 10/20/67, 40; GSFC *SSR*, 10/31/67)
- A total lunar eclipse began at 3:10 am EDT; the earth cast an almost 900,000-mi-long moving shadow on its journey around the sun. A meteor shower—the Orionids—added to the meteorological event. Weather permitting, these events would have been visible over most of North America. Schedule for meteor shower was Oct. 16–21.

  During eclipse, infrared studies of lunar soil were made by MSFC Space Sciences Laboratory and Univ. of Georgia research team. Perfect weather at Athens, Ga., permitted 20 traverses of moon during the eclipse. Using university observatory equipment—infrared radiometer attached to large telescope—to look for volcanic hot spots, scientists saw crater Tycho against cooler background during eclipse. The more general radiation pattern would be heat being radiated from lunar surface, including anomalies. Results were highly successful, reported Space Sciences Laboratory's deputy director Gerhard B. Heller. (*W Star*, 10/15/67, B6; McCandlish, *NYT*, 10/19/67, 49; MSFC Release 67–214)
- JPL scientists Robert J. Parks and Jack N. James were awarded Franklin Institute's Stuart Ballantine Medal for 1967 by Institute President Wynn L. LePage. Parks and James were cited for excellence in systems engineering and extraordinary skill in applying techniques of electromagnetic communications to first successful reconnaissance on Mars by *Mariner IV*. The Ballantine Medal was awarded for outstanding achievement in the fields of communication and reconnaissance which employed electromagnetic radiation. (JPL *Lab-Oratory*, 11/67, 3)
- California study of education in urban poverty communities and accompanying social problems was made by Lockheed Missiles & Space Co. and California State Dept. of Education's Office of Compensatory Education. Systematic Effort to Analyze Results (Project SEAR) used systems approach and aerospace techniques in studying impact of compensatory education on students, parents, and communities. (Lockheed Release)
- Government authority to ground private planes during poor weather would lead to a substantial reduction in accidents, Flight Safety Foundation President M/G J. D. Caldara (USAF, Ret.) told fourth biennial convention of National Assn. of Air Traffic Specialists in Atlantic City. Study of 5,300 general aviation accidents in 1965 showed 250 charged to pilots who took off, lost control, or continued visual flight in poor weather. (AP, *NYT*, 10/19/67, 81)
- U.S.S.R. and North Vietnam signed agreement to establish a satellite tracking station in North Vietnam and cooperate in synchronic photog-

raphy of earth satellites. (Reuters, *NYT*, 10/19/67, 9)

October 18–19: National Security Industrial Association Symposium was held in Washington, D.C.

Rep. George P. Miller (D-Calif.), chairman of House Committee on Science and Astronautics, spoke on role of Congress in R&D in 1970s. Citing increasing "Government concern for the involvement of science and technology in the welfare and security of the American people," he said it was increasingly obvious that science and technology, a prime source of many environmental problems, must be means by which they are solved. He also called for realism in minimizing influence of competition between political philosophies on implementation of scientific knowledge for human problems. "We must direct more and more attention . . . to the effects of urbanization, such as the pollution of air, water and the land, the practical control of population, to offsetting the constantly decreasing supply of natural resources such as fresh water, arable land, and fuels." (Text)

Gen. Bernard A. Schriever (USAF, Ret.), speaking on techniques for forecasting technological advances, emphasized need for enlightened and comprehensive planning and bold, imaginative management. He described technological forecasting as one of most valuable new tools for planners and managers. Indicating that forecasting must go beyond advanced planning, he called for its use in determining long-range implications of technological progress. If a manager, keeping a program open to possible advances in technology, chose improved performance, then he could be assured that when his truly advanced item entered the market or the military inventory, it would be effective for the maximum number of years. (*Av Wk*, 12/4/67, 21)

October 18–22: NAS President Frederick Seitz cabled congratulations on *Venus IV* to President of U.S.S.R. Academy of Sciences M. V. Keldysh, adding, "As our two nations enter the era of direct planetary exploration, we affirm the increasing need for full and prompt exchange of scientific data and our willingness to further such exchange through all appropriate international media." Keldysh replied, "On behalf of the USSR Academy of Sciences, I thank you for your congratulatory message and in my turn congratulate American scientists and technicians on . . . *Mariner 5*." (NAS–NRC–NAE *News Report*, 11/67, 6)

October 19: NASA's *Mariner V* made successful flyby of planet Venus to obtain scientific data on nature and origin of planet. Distance from surface of Venus at time of closest approach—estimated at 1:39:21 pm, EDT—was 2,544 mi (4,094 km). Launched June 14, *Mariner V* was 49 million mi from earth at time of flyby.

All scientific instruments and spacecraft subsystems performed properly during encounter sequence, as they had throughout flight. Sequence of 15 significant events during encounter, occultation, closest approach, exit occultation, and return to cruise mode, had begun at 10:49 pm EDT Oct. 18 and had ended at 3:55 am EDT Oct. 23. Full recorded account of experiments performed by *Mariner V*—first and second playbacks—exhibited clear and steady signals. JPL scientists obtained meaningful scientific data from all of spacecraft's experiments.

Results of experiments included: solar plasma probe and magnetometer showed existence of solar wind/Venus interaction; trapped radiation found no energetic particles or radiation belts, and found magnetic moment less than 1% of earth's; S-band occultation estimated carbon dioxide and found ionospheric observations on daytime side indicated electron density; dual-frequency radio propagation provided data on Venus mass as 81.50% of earth's mass, and astronomical unit determination agreed with radar computations; and ultraviolet photometer found hydrogen corona comparable with that of earth, detected no oxygen, and found night airglow to be very faint ultraviolet emission detected at dark limb of planet—probably result of chemical reactions, particle bombardment of electrical discharge. Spacecraft's mission also was to accumulate data on interplanetary environment during period of increased solar activity. After return to cruise mode, *Mariner V* would resume report on interplanetary space weather. By end of November 1967, when earth was no longer in direction of high-gain antenna pattern, communications would be terminated; however, three-month-duration communications would be possible during next five years. (NASA Proj Off; Wilford, *NYT*, 10/20/67, 1; Sullivan, *NYT*, 10/22/67, 12)

- Cone-shaped RAM C-1 spacecraft was successfully launched from NASA Wallops Station by four-stage Scout booster at 1:33 pm, EDT. On termination of eight-minute ballistic flight, impact occurred 725 mi downrange about 150 nm northeast of Bermuda. Spacecraft, 15 in long and 26 in in dia with 12-in-dia hemisphere nose, plunged back into earth's atmosphere at 17,000 mph in flight test to study methods for preventing loss of radio signals from reentering spacecraft. No recovery was attempted.

  NASA's Project RAM (Radio Attenuation Measurement) was to measure effectiveness of water addition concept in maintaining communications during reentry. Additional experiments were also included to evaluate X-band telemetry system as a means of overcoming blackout: study plasma and ablation effects on antenna performance; measure attenuation as a function of signal frequency; and obtain measurements of ion densities in the flow field of the spacecraft in the presence of ablation and water injection.

  Water was injected at three points on spacecraft—on nose cap and on each side—with nitrogen gas as pressurizing agent. Ion density changes, due to varying amounts of water injection, were measured. X-band frequency experiment showed blackout period of communications was reduced by 6 sec, using water-injection method.

  Series of Arcasonde meteorological rockets were launched from Bermuda before and after RAM C-1 experiment to collect upper atmosphere density, temperature, and wind data. (NASA Proj Off; NASA Release 67–261; WS Releases 67–33, 67–34)

- NASA Aerobee 150 sounding rocket launched from WSMR reached 118-mi (182-km) altitude in flight to observe profile of solar hydrogen Lyman-alpha line and to measure radiation in relation to earth's planetary hydrogen. Experiments performed satisfactorily, but range schedule delay caused loss of coordination with *Mariner V* experiment. (NASA Rpt SRL)

- NASA Arcas sounding rocket was launched from Barking Sands, Hawaii, to 36-mi (58-km) altitude, fifth in NOTS series of seven to measure incident solar UV irradiance in support of *Ogo IV*'s mission. Excellent data were obtained. Rocket and payload performed satisfactorily. (NASA Rpt SRL)

- USAF F–111A aircraft crashed near Bowie, Texas; two crew members ejected safely in detachable cockpit. According to General Dynamics Corp. pilots David Thigpen and Max Gordon, aircraft had faulty hydraulic system which made maneuvering impossible. Crash was second for USAF test model; first had crashed Jan. 19, killing one crew member. USN test model (F-111B) had crashed April 21, killing both pilots. (AP, *W Star*, 10/20/67, A9; UPI, *W Post*, 10/20/67, A7)

- Fast growth of air travel market could be adversely affected by travel costs, lack of hotel space, unreasonable airport charges, and unnecessary restrictions on operating rights of U.S. carriers abroad, Air Transport Assn. of America's President Stuart G. Tipton told Dublin, Ireland, audience. Addressing Trans-Atlantic Travel Congress of European Travel Commission, he suggested that "serious consideration be given to the effect upon airlines of high airport costs," while he pointed out that, since 1946, average fare per passenger mile had dropped 38 percent. Referring to "the [Boeing] 747 era [as being] a little over two years away," he warned that "the lead time for building a new hotel, however, runs from three to five years." Air passengers from the U.S. alone had spent $773 million in the Europe-Mediterranean area and had contributed to the economic well-being of the area; however, he said, "U.S. flag carriers are paying $500 . . . for every landing at some European airports [while] their foreign flag counterparts flying into the U.S. pay only $90 per landing for the same aircraft." (Text)

- Injectors for Apollo Lunar Module's ascent engine had performed in over 100 test firings at Rocketdyne Div.'s laboratory near Los Angeles, the company reported. Injectors had returned automatically to stable combustion after explosive charge was set off to deliberately create rough combustion in several tests. They were developed as backup to the original injectors after rough combustion was noted in test firings. Three designs would be continued through feasibility testing; then the best design would be chosen for development. (Rocketdyne Release)

- Great Britain's Ministry of Technology had released funds to build 400-ft, fully-steerable, dish radiotelescope, said Dr. H. C. Husband, British delegate to international conference on large antennas for radioastronomy held in Cambridge, Mass. New telescope would be world's largest—almost twice as large as 210-ft Goldstone model. Dr. Husband's announcement had come just one day after announcement by West German delegate, Prof. O. Hachenberg of Univ. of Bonn, that West Germany planned to complete a 330-ft dish in 1969. (*W Post*, 10/20/67, A7)

- Czechoslovakian magazine *Vencerni Praha* reported that unidentified flying objects (UFOs) had been sighted over Prague in the years 1130 and 1142. Early sightings, described by Prague historians as "fiery flying dragons," were later followed by frequent sightings in 18th and 19th centuries identified by one contemporary scientist as "a mass of damp particles, ignited by friction and static electricity." (AP, *NYT*, 10/19/67, 83)

October 20: NASA awarded contracts: (1) IT&T Federal Electric Corp. received a $4.6-million, one-year, cost-plus-award-fee contract extension for logistical and technical information support services at MSC; (2) IBM Federal Systems Div. was selected for negotiations on a three-month, $998,000 contract extension for computer programming and engineering support services at GSFC's Real Time Computer Complex (RTCC); and (3) Bendix Corp. was selected for negotiation of a $4,500,000, 18-mo, cost-plus-incentive-fee contract for "design, development, qualification, and delivery of long-duration cryogenic gas storage tanks" for first 56-day manned Apollo Applications (AA) flight. Bendix contract award would not be made until NASA's FY 1968 appropriation and expenditure levels were established. (NASA Releases 67-269, 67-270, 67-271)

October 22: *Molniya 1-7* comsat was launched by U.S.S.R. into orbit with 39,740-km (24,693-mi) apogee; 456-km (283-mi) perigee; 11-hr 54-min period; and 64.7° inclination. Tass said comsat would form part of new system to relay telephone, television, and telephone signals. (GSFC *SSR*, 10/30/67; AP, *NYT*, 10/24/67, 31; *SBD*, 10/24/67, 287)

- NASA Arcas sounding rocket was launched from Barking Sands, Hawaii, to 36-mi (59-km) altitude, sixth in NOTS series of seven to measure incident solar UV irradiance in support of *Ogo IV*'s mission. Excellent data were obtained. Rocket and payload performed satisfactorily. (NASA Rpt SRL)

- Informed U.S. military sources said U.S.S.R. had now been convinced the aircraft carrier would be vital to international conflicts of today, and was even now building its first aircraft carrier, a "baby flattop," reported UPI. Adm. Ephraim P. Holmes, Supreme Allied Commander, Atlantic, and his chief of staff, V/A William E. Ellis, said little was known of specifications but he believed it would have long-range capabilities. (Oestreicher, UPI, *W Post*, 10/23/67, A3)

October 23: NASA announced that primary mission objectives of *Biosatellite II* (launched Sept. 7) had been accomplished. Biological experiments had been conducted to investigate synergistic and antagonistic effects of weightlessness and controlled gamma radiation aboard attitude-controlled spacecraft. Spacecraft recovery—essential to success—had been completed as planned. Environmental control of capsule and experiment packages had kept organisms alive while radiation source was exposed for flight duration. Dr. Rudolph H. T. Mattoni, principal investigator for bacteria experiment, said bacteria had increased 20 to 30% faster in weightless condition. He concluded "cellular turnover may be higher in man" during weightlessness and could be harmful to astronauts during long-duration space flights. He suggested that if further research proved growth of human cells to be faster in weightless state, future spacecraft might have to be equipped with some form of artificial gravity or astronauts might have to carry drugs to prevent too-rapid turnover of their body cells. Since the 45-hr experiment with bacteria equaled many years of human life, this would not be significant on flights of a few weeks or even a few months. The pepper plant experiment showed abnormal growth, indicating that plant leaves also depend on gravity. Other experiments did not show significant change and were still being analyzed. (NASA Proj Off; AP, *NYT*, 10/15/67, 46)

- Rep. Emilio Q. Daddario (D-Conn.), chairman of House Committee on Science and Astronautics' Science, Research and Development Subcommittee, spoke on "A Challenge to the Scientific Community," at autumn meeting of NAS, at Univ. of Michigan, Ann Arbor. He urged scientists to become more actively involved in helping Government leaders find effective ways of dealing with national problems, and he believed that Congress should take an active interest in the "state of science" because scientific activity "is critical to the deliberations and the decisions of our political representatives." In turn, he urged an understanding by scientists and engineers of "pressures and functioning of a government which must be responsive to every shade of thought which our nation harbors." (Text)

- Passage of P.L. 90–112, DOT's FY 1968 funding, provided general funds of $1.5819 billion, including $142 million for SST development and $70 million for advanced-funding airport grants [see Sept. 28 and Oct. 5]. (P.L. 90–112)
- House adopted conference report on amended NASA FY 1968 appropriations bill (H.R. 12474) and sent report to Senate, which adopted report, leaving two amendments in disagreement. Senate had added $96 million to original House bill for R&D on Voyager and Nerva programs. House approved only $10 million of Senate's increase on R&D item; insisted on its original approval of $35.9 million for construction facilities for Voyager and Nerva programs, turning down Senate boost of $19.5 million; and approved Senate-voted $20 million cut for NASA's operations. (CR, 10/23/67, H13785–93; S15142; NASA LAR VI/115)
- Rep. James G. Fulton (R-Pa.), ranking minority member of House Committee on Science and Astronautics, during debate on NASA appropriations bill on floor of House, said: "China is now proceeding with the development of a rocket with nuclear propulsion. The Chinese nuclear rocket program is based upon a nuclear reactor through which passes liquid hydrogen that is stored in the rocket. This same fuel . . . is also used as a coolant for the nozzle of the rocket." He also said Communist Chinese "are proceeding with the development of a sounding rocket called Caditi . . . [and] telecommunications satellites."

    He explained that Communist Chinese believe, "to operate successfully in space, big booster power, high energy fuels, and nuclear propulsion are necessary. The University of Nanking . . . is conducting training courses in various applications of rocket programs. The Chinese space research center of Sinkiang now has programs in the study of materials necessary for space experiments. In addition, it should be seriously considered that the Chinese rocket institute—Balon Roditi—is interested in the development of rockets 'for all purposes.' This, of course, includes military as well as peaceful uses. Peking is not a signatory to the recent peace treaty for the peaceful uses of outer space. . . .

    "The nations of the West, including Russia, should not underestimate Chinese competence in these matters and the pace of their research and development programs. The U.S. Government and the American people must be alerted to the swift progress of the Red Chinese nation in space research and development, high energy propulsion, and large boosters . . . [and they] must realize the tremendous advances being made in the Red Chinese nuclear research field. . . . It is my belief and prediction that the [Communist] Chinese will orbit . . . [their] first . . . satellite by January 1968. . . ." (CR, 10/23/67, H13785–93)
- U.S.S.R.'s intensified study of near-earth environment indicated their scientists planned long-duration manned orbital flights, reported *Aviation Week*. Two investigations were cited: *Cosmos CLXVI*, launched June 16, in near-earth radiation study for protection of future cosmonauts in orbit during high sunspot activity; and "vertical cosmic probe," reaching altitude of 2,734 mi on study of upper atmospheric layers, ionosphere, and near-earth space. The "vertical cosmic probe," launched Oct. 12, provided Soviets with information on electron and positive ion concentrations, general intensity of cosmic rays, and radiation doses through various types of protection during flight within radiation belts. *Cosmos CLXVI* satellite provided solar flare data for continuous three months.

Similar to these Soviet studies were those conducted using NASA's *Oso IV* (launched Oct. 18), which would monitor sun so scientists could better determine how phenomena such as radio "blackout" and weather on earth were affected by solar activity. (*Av Wk*, 10/23/67, 27–8; NASA Release 67–273)

- Engine noise survey designed to study noise prediction during ground runups was postponed by NASA, reported *Aviation Week*. Tests, which would have been combined wth USAF studies of propulsion system calibration for AMSA, were given lower priority because these experiments would have caused two-week stand-down between flights. (*Av Wk*, 10/23/67, 23)

October 23–27: AIAA's Fourth Annual Meeting and Technical Display was held in Anaheim, Calif.

Reviewing history of NACA and NASA, GISS Executive Officer Arthur L. Levine noted that forces which accounted for creation of NACA in 1915 were both different from and far less intense that those behind creation of NASA in 1958. NASA played a much more significant role in the Nation and in Federal activity in science and technology than NACA, he said, and NASA's leadership was subjected to a wider variety and greater intensity of pressures from within Government and from outside interests.

NASA Historian Eugene M. Emme reviewed historical evolution of U.S. aeronautics and astronautics in context of major national policies and decisions. Citing effects of World Wars I and II, the "cold war" of the 1950's, and launching of Sputnik, he described how they helped determine the pace and utilization of technological advances and reviewed major U.S. turning points and relevant factors contributing to the Apollo decision. (Program)

M/G J. D. Caldara (USAF, Ret.), President of Flight Safety Foundation, Inc., suggested commonality between safety problems and solutions for underwater, aircraft, and space systems based on recognition, reasoning, and responsibility: "This process [of problem solving] spells out the basic building blocks in the structure of the man/machine relationships . . . involving safety. . . . If every scientist-engineer accepts his rightful responsibility and applies his imagination and experience to his task, the end result will be . . . [operations that are] 100% efficient and . . . 100% safe." (Text)

GSFC engineers F. O. Vonbun and J. T. Mengel proposed development of a worldwide tracking and communications system for manned planetary missions similar to that used for Apollo. Planetary Manned Space Flight Network (PMSFN)—consisting of three 30- to 50-ft-dia ground stations, three 210-ft-dia dish antennas, and three synchronous orbiting tracking stations (SOTS)—would obviate need for other radar, ships, and aircraft. SOTS, they said, while providing continuous communications, tracking, and telemetry functions, would be more cost-efficient by removing requirement for numerous new tracking stations. (Text; Program)

Addressing Honors Banquet, USC President Dr. Norman Topping stressed need for expanded program to search and disseminate data produced in space research. Although such data did reach social problems occasionally, "the arrival in the right place at the right time is largely a coincidence," he noted. To realize the total potential of modern discovery, scientists had to communicate outside their separate, narrow fields, find a common language with the layman, and work together to "find places for their discoveries within the social, ethical, and moral framework of our civilization." (Text)

Princeton Univ. economics professor Oskar Morgenstern, participating in President's Forum, "The Worth of the Space Program," contended that the space effort, as a public investment of natural resources, had to be evaluated by comparing its benefits with other ways the resources could have been used. Although no basic discoveries as important as atomic fission, the transistor, or the laser could be attributed to space exploration, he said, U.S. could still afford a high degree of space exploration and could expect discoveries "of a magnitude that will astonish even this generation which has seen so much and is prepared to expect that our knowledge will increase without limit."

Dr. T. Keith Glennan, President of Associated Universities, Inc., and first NASA Administrator, submitted that the most important enduring contribution of the space endeavor to U.S. future was education. Progress in higher education, he said, had occurred as a result of the compelling challenge of space exploration along with NASA's "clear and evident desire to assure an adequate flow of highly trained manpower. . . ." Current statistics on NASA investments in higher education were very impressive, he noted, but if NASA's budget were trimmed further its support to education would have to be diminished.

Dr. Willard F. Libby of UCLA predicted that space program would "come out to be the best bargain the American people have ever made— both for its practical value in technological developments, and for its contributions to pure knowledge through science." Space program had already contributed to and answered questions about geosciences, interplanetary sciences, astronomy, and chemistry but most consuming question was that of extraterrestrial life in solar system. (Texts)

Major AIAA awards presented at Honors Convocation included: Louis W. Hill Space Transportation Award to LeRC Director Abe Silverstein for "imagination, technical excellence and leadership which have led to practical application of high energy liquid hydrogen fueled propulsion for space transportation, and for initiating and contributing to the free world's first manned space transportation system"; De Florez Training Award to Edwin A. Link, chairman of General Precision Equipment Corp.'s Link Aviation Div., for "conceiving and implementing the idea of ground-based flight training equipment"; G. Edward Pendray Award to Columbia Univ. professor Robert A. Gross for "original contributions to gas-dynamics and plasma physics especially in the field of shock wave ionization phenomena"; and Lawrence Sperry Award to MSC Flight Director Eugene F. Kranz for "outstanding contributions in directing spacecraft mission planning in developing flight control teams, and as Flight Director of manned Gemini spaceflight missions." (*Av Wk*, 10/30/67, 13; AIAA *Facts*)

October 24: Cosmos *CLXXXIV* was launched by U.S.S.R. into circular orbit of 635-km (394-mi) altitude, 97.14-min period, and 81.2° inclination. Equipment and instruments functioned normally. (GSFC *SSR*, 10/30/67; *SBD*, 10/30/67, 320)

- Two Boosted Arcas I sounding rockets launched from Resolute Bay, Canada, without tracking facilities carried GSFC experiment to obtain D region ionospheric data in vicinity of magnetic north pole. First rocket lost telemetry at seven seconds; recovery search would be attempted if arctic conditions permitted. Second rocket had 255-sec burn (300+

sec planned), so altitude was probably insufficient. Both experiments used ground-based transmission of RF energy to rocketborne receivers (Faraday-rotation experiment). These were first U.S. rocket launchings from this facility. (NASA Rpt SRL)

- NASA's Eighth Annual Honor Awards Ceremony was held in Washington, D.C.

Dr. Roger W. Heyns, Chancellor of Univ. of California, discussed impact of space program on university curricula and research efforts in his keynote address. "Through the stimulation of NASA programs, new fields of study have been created ... and NASA ... [has become] one of the world's most ambitious attempts to pool the talents and learning of thousands. . . .

". . . NASA has been doing pioneering work of immense significance to modern society that has nothing to do with the exploration of space per se. Objectives have been defined, scientific and technical knowledge and talent have been collected, trained and aimed at research objectives and results produced. That is a massive accomplishment of human engineering ... [and] an inspiration to all of us. . . . Just how much . . . is transferable to other situations remains to be seen, but . . . it certainly is not inconceivable that what we have learned, and what we will learn, about public administration from NASA may, in the long pull, mean every bit as much as the scientific and technological developments from the exploration of space itself."

Recipients of Distinguished Service Medal: NASA Associate Administrator Dr. Homer E. Newell; Deputy Associate Administrator for Manned Space Flight Edgar M. Cortright; LaRC Director Dr. Floyd L. Thompson; and former Associate Administrator for Advanced Research and Technology Dr. Raymond L. Bisplinghoff. NASA's second Distinguished Public Service Medal was presented to Dr. Charles S. Draper, director of MIT's Instrumentation Laboratory. Special Group Achievement Award was presented to Canadian government for outstanding success of *Alouette I* satellite (launched Sept. 28, 1962). Exceptional Service Medal: Astronaut Roger B. Chaffee (posthumously); Charles F. Hall, ARC; Donald R. Bellman and Robert D. Reed, FRC; William J. Boyer, Laurence K. Loftin, Jr., James S. Martin, Jr., and Clifford H. Nelson, LaRC; James S. Kramer and H. Warren Plohr, LeRC; Robert H. Gray, KSC; Howard H. Haglund and Robert J. Parks, JPL; Paul G. Marcotte, GSFC; William Cohen, George C. Deutsch, Arthur F. Hood, Joseph B. Mahon, Benjamin Milwitzky, Lee R. Scherer, and William M. Shea, NASA Hq. Recipients of Exceptional Scientific Achievement Award: Michel Bader, Donald E. Gault, and Maurice D. White, ARC; Walter B. Horne, William H. Phillips, and Israel Taback, LaRC; Samuel S. Manson, LeRC; and Eugene M. Shoemaker, USGS. Group Achievement Award: Apollo 204 Review Board; NASA SST Evaluation Team; Lunar Orbiter Spacecraft and Operations Team; and 260-in Solid Motor Project Team. Public Service Award: William Feldman, Eastman Kodak Co.; Robert J. Helberg, Boeing Co.; Mark Sasso, RCA; Robert L. Roderick, Hughes Aircraft Co.; and Richard Cottrell and Paul B. Datner, Aerojet-General Corp. Exceptional Bravery Medal was presented for "courageous and heroic action" in attempting to rescue the Apollo 204 crew to Donald O. Babbitt, Stephen B. Clemmons, James D. Gleaves, Jerry W. Hawkins, and L. D. Reece, NAA; and Henry H. Rogers, KSC. (Text; NASA Release 67–272; Program)

- Commercial service via *Pacific II* comsat, launched by NASA from ETR Sept. 30, would begin Nov. 4, ComSatCorp announced. (ComSatCorp Release 67-52)
- Overall NASA aeronautics program was undergoing "expansion and change to meet the ever growing needs" of air transportation industry and general aviation, NASA Deputy Administrator Dr. Robert C. Seamans, Jr., told the Aero Club in Washington, D.C.: "Less than 1400 direct man-years were applied to our aeronautical effort in 1962 and only $15 million in R&D funding was applied to support their activities. The trend has now been clearly reversed and in [FY 1967] twice the manpower (27.2 directs) and three times the funding ($45.7 million) were applied to NASA's aeronautical program." Although FY 1968 program had not been clarified, NASA planned to increase aeronautical R&D funding to $81.4 million, he said.

    Major aeronautical advances in the past had been exploited by DOD, he noted, which bore development costs and proved out designs before adoption by commercial aviation. Since solutions for many problems could no longer rely heavily on precursor military development, NASA would continue NASA-established role to conduct broadly applicable basic aeronautical research and solve specific problems with reimbursement for special costs. In addition, Dr. Seamans said, NASA would attempt to reduce aircraft noise by modifying engines on existing aircraft, developing quieter engines, and revising operational procedures; cooperate with FAA and industry to solve safety and utility problems of general aviation; and conduct extensive R&D programs for V/STOL, supersonic, and hypersonic aircraft. (Text)
- Fred Friendly, Columbia University professor and consultant to Ford Foundation, delivering Granada Lecture in London, warned that communications satellites' promise of international "common market" of high-quality TV was threatened by commercial and government avarice. In U.S., he said, TV "can make so much money doing its worst, it cannot afford to do its best." Consequence had been that early promise of quality programs had been lost while educational TV, which could fill breach, had been "under-funded, undernourished and under-observed." But even in nations where TV was nationalized, tendency would be for governments to seek profits from it to offset other communications deficits, he said. Thus, he argued, "a promising world village, linked by satellites," could be turned into "an electronic slum," because ambitious corporations and separate governments insisted on going their own ways, concerned more with earnings, national appetites, and images than with what their satellite systems carried. What was required, he suggested, was "an electronic Magna Charta" guaranteeing rights of public to quality TV. (Friendly, *W Post*, 10/25/67, D14)

October 25: Last NASA Arcas sounding rocket in series of seven was launched from Barking Sands, Hawaii, to 36-mi (59-km) altitude, in NOTS experiment to measure incident solar UV irradiance in support of *Ogo IV*'s mission. All experimental objectives were obtained. Rocket and payload performed satisfactorily. (NASA Rpt SRL)
- USAF launched unidentified satellite from WTR using Titan III-B booster; satellite reentered Nov. 5. (*Pres Rep 1967*)
- A 2,000-mph USAF SR-71 reconnaissance aircraft on routine training flight from Beale AFB, Calif., crashed near Lovelock, Nev. Its two crew members ejected safely. Aircraft, built for USAF by Lockheed Aircraft Corp.

as successor to the U-2, was third SR-71 to crash on routine flight; others had crashed Jan. 25, 1966, and April 13, 1967. (AP, *NYT*, 10/27/67, 41)

October 25–27: *Mariner IV*, launched Nov. 28, 1964, responded to a series of commands 56 million mi out in space after three years in orbit. On Oct. 25, the spacecraft was commanded to switch data rate from $8\frac{1}{3}$ to $33\frac{1}{3}$ bps (bits per sec); receive via low-gain antenna; reorient spacecraft so high-gain antenna pointed directly at earth; change to all-engineering data format; and burn midcourse guidance engine for 70 sec. On Oct. 26, commands were sent to actuate planetary scan subsystem and tape recorder. Lines 139 through 199 of picture No. 16 and lines 1 through 59 of picture No. 17 were played back. Initial indications were that photographic data were the same as when first reported July 14, 1965. On Oct. 27, spacecraft responded to commands to test certain redundant equipment—such as transmitter and attitude-control apparatus—and was placed in normal cruise condition by acquiring Canopus. Attitude-control gas in one section of system was thought to be depleted, leaving about 0.2 lb in other section. This would last until early December; when gas was depleted, *Mariner IV* would no longer be able to retain its orientation with respect to the sun and Canopus.

JPL scientists were "delighted" with the firing of midcourse guidance engine, and with transmission of pictures of Mars taken by *Mariner IV* more than two years ago. Scientists reported "a flawless burn . . . [which] proved we can restart engines in space years after we've put them there." Future spacecraft, exploiting this success, could be launched to distant planets where they could take close-up photographs, swing back toward earth, and, on command, transmit pictures.

Mariner IV had taken pictures of Mars from an altitude of 5,500 mi; 21 frames had been exposed and recorded on magnetic tape to be telemetered to earth over a 10-day period. As of 8:00 pm EDT Oct. 27, *Mariner IV* was in its 1,062nd day of flight, 35,604,255 mi (57,299,604 km) from earth, and traveling at 3.91 mps (6.29 kps) relative to earth. (NASA Proj Off; O'Toole, *W Post*, 10/27/67, A3)

October 26: First flight test of Saturn V, designated Apollo 4, was scheduled no earlier than Nov. 7, announced M/G Samuel C. Phillips, Apollo Program Director. "This is a target date," he said. "We are in a very complex learning process and we are going to take all the time we need on this first launch." Launch vehicle and Apollo spacecraft combination was 363 ft tall and weighed 6 million lbs. Thrust of Saturn V's first stage was 7.5 million lbs. Key objective of flight planned for Nov. 7 would be evaluation of Apollo command module heat shield under conditions encountered on return from moon mission. Flight would also test thermal seals for use in new quick-release spacecraft hatch which would be installed on all manned Apollo missions.

Apollo 4 flight plan would call for Saturn V to place spacecraft and launch vehicle 3rd stage (S-IVB) into 117-mi circular orbit. After completing two orbits, 3rd stage would be re-ignited to place spacecraft into orbit with apogee of 10,800 mi. After separation from 3rd stage, service module propulsion system would be fired to raise spacecraft apogee to 11,400 mi. During descent the service module motor would fire again, boosting velocity of spacecraft to about 25,000 mph, the reentry speed of a lunar return flight. (NASA Release 67–274; B *Sun*, 10/26/67, A6)

- GSFC officials completed week of checkout of *Oso IV* and turned on last of nine experiments. All experiments were performing well and spacecraft was transmitting satisfactory data. The 599-lb spacecraft, launched from ETR Oct. 18, had received 1,500 commands. It was 3,000th object in orbit since *Sputnik I* in 1957; 100th satellite orbited in 1967; and 50th satellite orbited by Delta launch vehicle in 53 attempts. (NASA Release 67-273)

- Senate and House adopted conference report on amended NASA FY 1968 appropriations bill (H.R. 12474) totaling $4.6 billion, and legislation was cleared for President's signature. Report approved $3.925 billion for R&D, decrease of $152.1 million from authorization and decrease of $356.5 million from budget; report approved $35.9 million for Construction of Facilities, decrease of $14.6 million from authorization, and decrease of $21.3 million from budget; report approved $628.0 million for Administrative Operations, decrease of $20.2 million from authorization, and decrease of $43.3 million from budget.

  Rep. James G. Fulton (R-Pa.), ranking minority member of House Committee on Science and Astronautics, on floor of House, said that "Voyager and NERVA II programs should be funded this fiscal year. Voyager is the only major planetary exploration program this Nation will have in the 1970's. If this program is abandoned, we will be abdicating our responsibilities. . . . This is particularly significant following the recent Russian Venus IV soft landing on the planet Venus."

  Sen. Gordon L. Allott (R-Col.), ranking minority member of Senate Committee on Appropriations' Space Appropriations Subcommittee, was forced to accept no specific funding for Voyager and Nerva programs (in final version) but said in statement read by Sen. Margaret C. Smith (R-Me.) on floor of Senate that "agreement was reached to add $15.5 million to [R&D] over the House figure." He added that the agreement would provide that NASA's Administrator "may reprogram funds . . . with approval of the appropriate committees of the Congress." He added: ". . . conferees share this belief, that this country must move ahead now in [R&D] leading to interplanetary exploration." (*CR*, 10/26/67, H14036-8, S15405-8)

- Dr. Harvey H. Nininger, leading authority on meteorites and tektites, was honored at Meteoritical Society's 30th annual meeting at ARC by publication of special issue of the *Journal of the Geochemical Society* and by presentation of Society's annual Leonard Medal. First to propose, in 1940, that tektites came from moon, Dr. Nininger later proved that coesite—a quartz-like material created by very high pressures—could be found in meteoroid craters. Presence of coesite was currently being used to determine whether a crater was formed by meteorite impact (ARC Release 67-21)

October 27: NASA Aerobee 150 sounding rocket launched from WSMR reached 99-mi (160-km) altitude in flight to measure, with moderate and high resolution, spectral irradiance of Venus (1,000 to 3,000 Å), and star B Oriones (Rigel). No data were acquired from Venus because tracker was ineffective; Rigel was acquired and five-second exposure sequence was completed. (NASA Rpt SRL)

- U.S.S.R. successfully launched two Cosmos satellites with two boosters: *Cosmos CLXXXV* entered orbit with 888-km (552-mi) apogee, 522-km (324-mi) perigee, 98.7-min period, and 64.1° inclination. *Cosmos CLXXXVI*, launched into orbit with 235-km (146-mi) apogee, 209-km (130-mi) perigee, 88.7-min period, and 51.7° inclination, later docked with *Cosmos CLXXXVIII*. Both spacecraft performed satisfactorily. (*W Post*, 10/29/67, A16; *SBD*, 10/30/67, 320; GSFC *SSR*, 10/31/67)

- Atlas-F carrying unidentified payload exploded shortly after launch by USAF from Vandenberg AFB. (*SBD*, 10/30/67, 316)

October 28: *Cosmos CLXXXVII* was launched by U.S.S.R. into orbit with 210-km (131-mi) apogee, 145-km (90-mi) perigee, 87.8 min period, and 50° inclination. Satellite performed satisfactorily and reentered same day. (GSFC *SSR*, 10/31/67)

October 29: U.S.S.R. would "shortly" launch a manned spacecraft which might land outside Soviet territory—possibly in India—Soviet Ambassador to India Nikolai M. Pagov told reporters in Madras, India, according to AP. Permission to recover spacecraft in India had been requested and granted in early September, he said. Official U.S. sources, noting that several other countries had been approached for such permission, speculated that an unmanned rehearsal for a manned circumlunar flight had been scheduled for late September and postponed because of technical problems. Oct. 27 launches of *Cosmos CLXXXVI* and *Cosmos CLXXXVII*—which were apparently Soyuz-type spacecraft—could be "practice shots" for the planned manned flight. (AP, *W Post*, 10/30/67, A3; Clark, *NYT*, 10/30/67, 54)

- JPL Director Dr. William H. Pickering announced appointment of Fred H. Felberg, Assistant Laboratory Director for Technical Divisions, to the new position of Assistant Laboratory Director for Plans and Programs. He would be succeeded by Jack N. James, Deputy Assistant Director for Flight Projects. In his new position Felberg would be responsible for "planning JPL's total program, including . . . effective distribution of manpower and resources." (JPL Release 460)

October 30: U.S.S.R. launched two Cosmos satellites.

Cosmos CLXXXVIII entered orbit with 276-km (171-mi) apogee, 200-km (124-mi) perigee, 89-min period, and 52° inclination. At 4:20 am, EDT, spacecraft was automatically docked with *Cosmos CLXXXVI* (launched Oct. 27). Tass later announced: ". . . both satellites, equipped with special approach and docking units, carried out a number of complicated maneuvers in space, automatically finding each other, drawing closer, berthing and docking rigidly." It was history's first automatic docking in space and was filmed by TV cameras onboard both satellites. U.S. had successfully accomplished first manned docking March 16, 1966, with *Gemini VII* and Gemini Agena Target Vehicle (GATV); in this docking experiment, astronauts controlled the maneuver. After automatic docking, the two satellites continued in orbit with altitude ranging from 124–171 mi. At 7:50 am, EDT, after completing $2\frac{1}{3}$ orbits, spacecraft were undocked on command from the ground, and placed into different orbits by restarting rocket engines. *Cosmos CLXXXVI*, larger of the two spacecraft, softlanded Oct. 31 after 65 orbits. *Cosmos CLXXXVIII* reentered Nov. 2.

Cosmos CLXXXIX was launched into orbit with 600-km (373-mi) apogee, 535-km (332-mi) perigee, 95.7-min period, and 74° inclination. (Shub, *W Post*, 10/31/67, 1; Kamm, *NYT*, 10/31/67, 1; AP, *W Star*, 10/31/67, A3; Winters, *B Sun*, 11/1/67; GSFC *SSR*, 11/15/67)

- U.S.S.R. announced successful completion of rocket tests in the Pacific northeast of Midway Island to test equipment for spacecraft landings at sea. Series, which began Sept. 2, was first to last as long as scheduled. (UPI, *NYT*, 10/31/67, 44; *M/S Daily*, 10/31/67)

October 30

- Scientific proof that instrumented capsule from U.S.S.R.'s *Venus IV* had impacted on planet Venus Oct. 18 was presented at Moscow news conference by Mstislav Keldysh, President of Soviet Academy of Sciences. Instrumented-capsule-station's transmitters, programmed to broadcast for 100 min, on basis of best previous estimates of pressure in atmosphere of Venus, "stopped transmitting after 94 minutes [exact time for descent]. The station recorded both its own speed and the remaining distance to the surface throughout the descent [and] also recorded the changing atmospheric pressure and recalculations on earth. . . ." Thus, based from this information and other known data, Keldysh confirmed that station stopped transmitting the instant it landed. The question of why radio transmission ended after 94 min—power for 100 min having been provided—was left unanswered.

    Press conference presented additional information. Magnetic field near Venus could not exceed 1/3,000 that of earth's; low concentration of positive ions along *Venus IV*'s route (much less than that at earth atmosphere's top); absence of atomic oxygen in upper Venusian atmosphere; probable dense molecular atmosphere of Venus; large percentage of carbon dioxide (90–95%) in Venusian atmosphere; neutral hydrogen density of 1% of that near earth; nitrogen content of Venusian atmosphere measuring about 7%; and small amounts of argon and other inert gases. Water, at 1/10 to 7/10 of 1%, was reported to exist largely in the planet's cloud layer. Weather was cloudy on Venus, without precipitation. (Shub, *W Post*, 10/31/67, 1; Rausch, *Av Wk*, 11/6/67, 17)

- Cornell Univ. professor Hans A. Bethe, director of Los Alamos Atomic Weapons Laboratory's theoretical physics section during development of atomic bomb and recipient of AEC's 1961 Enrico Fermi Award, was awarded the 1967 Nobel Prize in physics for proving that the sun and other stars produced energy through synthesis of helium from hydrogen. He was cited specifically for "his contributions to the theory of nuclear reaction, especially his discoveries concerning the energy production of stars."

    Nobel Prize in chemistry was presented jointly to George Porter and Ronald G. W. Norrish of U.K. and Manfred Eigen of West Germany for their studies of extremely fast chemical reactions occurring at speeds of one-one-thousandth of a millionth of a second. Nobel Peace Prize was withheld for second consecutive year. (UPI, *W Star*, 10/30/67, 1; Wiskari, *NYT*, 10/31/67, 1)

- AT&T's letter responding to FCC inquiry on need for future communication facilities in Atlantic basin area urged construction of $70.4-million underwater communication cable to southern Europe by 1970's end. It said completion of cable would accommodate rapid growth in transatlantic calls and provide balance between satellite and cable links [see Oct. 4]. ComSatCorp and international telegraph carriers also had received Oct. 4 FCC inquiry. (UPI, *NYT*, 10/31/67, 65)

- Soviet-French space treaty of June 30, 1966, had produced results during September when two Soviet MR–12 sounding rockets carrying French payloads were launched to altitudes of 75 mi and 112 mi over Franz Josef Land in Soviet arctic. Payloads released sodium clouds to measure upper-atmosphere temperatures. (*Av Wk*, 10/30/67, 13)

October 31: Sprint, launched from underground cell as the high-speed, short-range missile for ABM system, completed successful test flight at WSMR when it hit intercept point in sky after several planned sharp maneuvers. Interception simulated actual one for incoming enemy missile and was programmed into computer for 27-ft-long missile. Sprint tests had been conducted since March 29, 1965. (DOD Release 1040-67)

- Georgi Petrov, director of Soviet Academy of Sciences' Space Exploration Institute, claimed that docking of *Cosmos CLXXXVI* and *Cosmos CLXXXVIII*, "for its technical perfection and the possibility it has opened for space exploration, is of greater importance than the American experiment in manual docking . . . [of *Gemini VII* and GATV]. Since the completely automatic linkup, despite the difficulty of its realization, will always be more economical as regards the payload which is so expensive on an orbit, this will make it possible to solve the tasks of helping the crews of space vehicles and open up new prospects for arranging experiments that demand a great weight on an orbit around the earth." (*SBD*, 11/1/67, 3)
- Dr. Ludwig Roth, 58, McDonnell Douglas director of Saturn-Apollo extension program, died in Redondo Beach, Calif. Born in Gross-Gerau, Germany, he had been an associate of Dr. Wernher von Braun in pioneering rocket engineering and space sciences. He had served as Assistant Director of Army Ballistic Missile Agency. (AP, *NYT*, 11/3/67, 43; *World Space Directory*, Fall 1967, 155)

During October: JPL Director Dr. William H. Pickering stressed importance of unmanned space exploration in *Astronautics and Aeronautics:* "During the past decade of space exploration, unmanned spacecraft have demonstrated a remarkable versatility in performing a wide variety of scientific tasks. The success of these robots is due largely to the ingenuity of the designers in building complex but reliable automated devices, capable of withstanding the constraints imposed by the hostile spaceflight environment.

"It is clearly evident that many difficult engineering and scientific problems remain to be solved. The experience of the past decade demonstrates, however, that within the next decade unmanned spacecraft can initiate a detailed surface exploration of our neighboring planets.

"The unmanned exploration of the solar system is technically and scientifically possible. It presents a unique opportunity to step boldly forward into new areas of both technology and science." (Pickering, *A&A*, 10/67, 80-4)

- Congressional role in space program and possibilities for improved international cooperation in space was discussed in *General Electric Forum* interview by Rep. Olin E. Teague (D-Tex.), Chairman of House Committee on Science and Astronautics' Manned Space Flight Subcommittee. Congress' role, Rep. Teague said, was "to examine periodically how well the work is being done and what return we are gaining in new knowledge and practical utility from the space program. . . . Each decision . . . involves the setting of a relative order of priorities for the use of the Nation's resources for science and technology." Expressing conviction that international cooperation in space would increase, he noted that there was "great potential in the use of space for human betterment. . . . More than 60 per cent of the world's people, for instance, are protein-deficient today. If, by earth orbital surveys, in cooperation with the less-developed countries, we can improve that situation, our entire space program will have paid for itself many times over." (*GE Forum*, Autumn 1967, X/3, 11-12; *CR*, 10/19/67, H13719-20)

- Rep. Joseph E. Karth (D-Minn.), Chairman, House Committee on Science and Astronautics' Space Science and Applications Subcommittee, called for "developing a greater public awareness of the benefits of the total space program and . . . driving home the point of how close we are to realizing magnificent achievements in bettering our lives." Writing in *Astronautics and Aeronautics* he said the space program "is no longer insulated from the tough competitive struggle for resources. . . . More bluntly, the NASA part of the national space program no longer has a very high priority." He explained that "scientific and technological programs are the least understood of our national efforts. Tracing national growth and prosperity directly back to research and development is admittedly a tough job, but I am convinced that space technology stimulates social and economic progress." On budget levels through early 1970s, he said, "Compared with earlier forecasts, it seems that a considerably stretched-out program will be characteristic. . . ." In spite of his pessimistic outlook for the present, Rep. Karth said: "This outlook should not cause despair because the long-range prospects for progress are enormous . . . [but] prospects must be sold to the public and top policymakers in terms of what makes sense for the overall benefit of society." (Karth, *A&A*, 10/67, 86–92)
- AFOSR Director of Information Sciences Dr. Harold Wooster, writing on "Basic Research and the Department of Defense," presented his appraisal of what basic research is and why it is a benefit when Government funds support basic research. He stated that the Government had spent $2.1 billion on basic research in FY 1967 while American industry was spending $607 million. Wooster believed that basic research and its results transcend state and national boundaries, and that there were "certain large research tools and facilities" which were too expensive for private industry, or any state or combination of states to build. As an example, he pointed to the National Magnet Laboratory at MIT: "This Laboratory is a national resource already making important direct applied contributions to the defense effort. It would have been impossible to build without Federal financing." (AFOSR *Res Rev*, 10/67, 1–9)
- U.S. scientists, during the next decade, would gain enough experience with meteorological satellites, sensors, and surface station networks, to complete global network for gathering and forecasting weather data, *Space/Aeronautics* reported. This would require a real marriage of "the potential of the meteorological satellite and the computer. . . ." (*S/A*, 10/67, 76)
- U.S.S.R.'s Academician Dr. Anatoli A. Blagonravov, Chairman of Soviet Commission for Exploration and Use of Outer Space, writing in *Astronautics and Aeronautics*, explained that general trend of Soviet space program was similar to that of U.S. program, and refrained from "any forecasts [on] . . . further programs in space research." He said work planning was difficult "because [it] calls for additional checkups or presents prospects of new and unforeseen investigations." He believed "realistically" in prolonged manned space flights, and said there was "no doubt that it will be possible to overcome all the difficulties. . . ." (Blagonravov, *A&A*, 10/67, 70–8)
- AFCRL launched five Nike-Iroquois sounding rockets from Vega Baja, Puerto Rico, so that simultaneous observations of the ionosphere might be made using rocket instruments and the Arecibo radiotelescope. Rock-

ets were designed to reach a peak altitude of 112 mi (180 km). Measurements of temperature, energy distribution, and density of ions and electrons provided basis for calibration and interpretation of ionospheric data acquired by Arecibo radiotelescope. (*OAR Research Review*, 12/67, 6-7)

During October-November: Lunar Orbiter IV, launched May 4, was presumed to have crashed on lunar surface; contact with spacecraft had been lost July 17. (NASA Proj Off)

November 1967

November 1: NASA Aerobee 150 sounding rocket launched from WSMR carried Princeton Univ. payload to 109-mi (176-km) altitude on flight to determine UV spectrum of extremely hot atmosphere and interstellar matter in direction of 05 star Tau Puppis—hottest of bright stars in sky. Excellent film exposure was obtained from the spectrograph; attitude control system stabilized properly. (NASA Rpt SRL)

- NASA announced results from experiment with four "Yolo Wonder" pepper plants subjected to 45 hr of weightless flight aboard *Biosatellite II* (Sept. 7-9) to determine effects of space environment (weightlessness and radiation). Photographs made every 10 min (268 times) in 30 orbits appeared to demonstrate plants depend on gravity for their orientation. Plant leaves had failed to grow in normal, horizontal position, and stems had not held to the vertical position, even with supporting brackets. Disoriented condition of plants was observed when they were delivered to Dr. Samuel Johnson, principal investigator, North American Rockwell Corp. Further evidence of plants' gravity dependence came from wheat seedlings aboard *Biosatellite II;* Dr. Charles Lyon of Dartmouth College experimented with 75 seedlings which, grown from seed in weightlessness, had roots curving upward toward seedling shoots and out to the side. (NASA Release 67-279)

- Continuing Brazil/U.S. space cooperation, experiment to measure airglow would be carried out by National Space Commission (CNAE) of Brazil and NASA, an Aerobee 150 to be launched from Brazil's Barreiro do Inferno range this fall [see Oct. 14-15]. A monochromator would measure day airglow in equatorial regions at altitude of about 130 mi in the range of 2,000 Å. One of two kinds of upper atmosphere's visible radiation (aurora being other), airglow existed when sunlight interacted with upper atmosphere to create visible radiation.

    CNAE would be responsible for preparation and operation of launching range and associated facilities, and for providing personnel for training. NASA would provide Aerobee rocket, payload, and mobile launch facility currently located in Brazil. (NASA Release 67-277)

- ComSatCorp expressed its opposition to AT&T plan to build $70.4-million underwater cable to southern Europe by 1970's end in 92-page response to FCC inquiry on need for future communication facilities in the Atlantic basin area [see Oct. 4 & 30]. James McCormack, Chairman and Chief Executive Officer of ComSatCorp, stated that proposed addition of 720 international communication circuits via new transatlantic cable, designated TAT-5. would be "unnecessary in the public interest and would act as a serious detriment to the development of a really capacious and economical global telecommunications system." He said ComSatCorp planned to orbit Intelsat IV by 1970 to provide at least 5,000 circuits

initially that would accommodate TV, broadband, and other transmissions in serving smaller, developing nations as well as major, technologically advanced countries. (Text; ComSatCorp Release 67-54)
- NASA Administrator James E. Webb appointed L/G August Schomburg (USA, Ret.) as a consultant. Schomburg, whose last duty assignment was Commandant of Industrial College of the Armed Forces, would work in areas of national security, NASA's interfaces with industry and defense, and education in industrial affairs for NASA's field centers. (NASA Release 67-280)

November 2: XB-70 research aircraft, flown at FTC by USAF pilots Col. Joseph Cotton and L/C Emil Strumthal, reached mach 2.52 and 63,500-ft altitude in flight to test stability control and handling qualities. Other test objectives included: handling qualities at mach 1.2 and 40,000-ft altitude; unstarts; radar air speed calibration at mach 1.9 down to 1.2 and 45,000-ft altitude; nose ramp flutter function at mach 1.2 and 32,000-ft altitude, and at mach 2.55 and 64,000-ft altitude. (NASA Proj Off)
- USAF launched two unidentified satellites from WTR on single Thor-Agena D booster; one of two satellites reentered Dec. 2. (*Pres Rep 1967*)
- World's largest optical telescope—Soviet 236.2-in-dia mirror at Leningrad—might be unveiled on Nov. 7, 50th anniversary of Bolshevik Revolution, reported *Space Business Daily*. New telescope was reported to be 82 ft long and to weigh 275 tons; its guidance system would have a precise time-augmented design to guarantee accuracy. U.S. Mount Palomar telescope—world's largest operating mirror—now 20 years old, had diameter of 200 in. (*SBD*, 11/2/67, 11)
- British Univ. of Bristol Prof. H. E. Hinton, speaking to the Royal Society, described a capacity for revival after periods of suspended animation by insects and worm-like parasites with millions of cells. Hinton stated that these living organisms were dried out, kept in suspended animation for years, subjected to extremes of temperature, and then revived no worse for wear. According to Hinton, as long as spatial relations of the molecules were maintained, life activities could be restored to inert organisms whose cells had had no interactions for long periods of time. Injuries inflicted on fly larvae in their inert state, even cutting them into pieces, did not prevent restoration of life—although the larvae, once restored, promptly died of their wounds. Hinton suggested that if complex organisms could be reduced to an inert structure on which passage of time had no effect, infinitely prolonged "calendar" life—important for a space voyage lasting centuries or millennia—was a theoretical, although very distant, possibility. (*W Post*, 11/3/67, A25)

November 3: U.S.S.R. launched *Cosmos CXC* into orbit with 323-km (201-mi) apogee, 195-km (121-mi) perigee, 89.6-min period, and 65.6° inclination. Spacecraft reentered Nov. 11. (GSFC *SSR*, 11/15/67)
- Agreement of Euratom countries with U.S. on a reasonable inspection system had removed "the biggest remaining obstacle" to eventual U.S. and U.S.S.R. agreement on a nuclear nonproliferation treaty [see Aug. 24], according to *New York Times* editorial. A green light had been given U.S. to resume negotiations with U.S.S.R. because principles put forth by five non-nuclear countries "are considered reasonable by the American negotiators and flexible enough to permit common language to be found with the Russians, if Moscow remains interested in completing the treaty." Editorial said Soviet compromise proposal in Sep-

November 3

tember had indicated willingness to let Euratom inspection system continue, subject only to verification by International Atomic Energy Agency in Vienna. Soviet proposal had also implied that controls could center on flow of fissionable materials, to prevent diversion from peaceful to military uses. Soviets had aimed to check new plant designs to make sure the flow of fissionable material could not be concealed. Substantively, no big gap appeared between Soviet interpretation of its September proposal and Euratom countries' desires, although "even after the inspection dispute is resolved, complex issues raised by India and other nuclear-capable countries, including guarantees against nuclear blackmail, will need to be resolved." (*NYT*, 11/3/67, 42)

- NASA announced that research at Ames Research Center indicated a probability of meteoroids' puncturing spacecraft at about six times greater than previous calculations. Tests which simulated meteor entry into atmosphere were conducted in AEC's Hypervelocity Planetary Gas Tunnel. ARC Director H. Julian Allen and assisting scientists Charles Shepard, Howard Stine, and Barrett Baldwin used melting or ablating tests on one stony meteorite and on samples of gabbro—common, meteor-like earth rock—at simulated speeds up to 35,000 mph. Tunnel tests showed that heat of entry into earth's atmosphere melted rock surface, boiled off interior water, and expanded gases occurring in rock. Expansion of exterior rock layer by 10 to 1,000 times converted exterior to foam and acted as a heat shield. Foam was 2 to 20 times less dense than the unheated rock; depending on entry conditions, foam would or would not break off.

  Past experience and theory had suggested that meteors were sandlike and would be remains of broken up comets; in addition, studies of meteor and comet trajectories had supported current theory that comets would be like "dirty snowballs" made mostly of ice, frozen ammonia, and methane with rock and sand additives.

  Opposed to these beliefs, ARC scientists speculated that many more meteors originated in the rocky asteroid belt and that much more solid rock would be contained in luminous clouds of comets. ARC could now prophesy the need for greater protection on long missions where the probability of meteoroid hit would be highest. (NASA Release 67-281)

- GSFC announced award of $3.4-million contract to American Science and Engineering, Inc., for development of x-ray experiment package for first spacecraft in Small Astronomy Satellite (SAS), GSFC-managed program. Spacecraft SAS–A would map sources emitting x-rays outside solar system in collecting information on position, strength, and time variation of emissions. SAS program would also have satellites search for sources radiating in the gamma ray, UV, visible, and IR regions; its data should enable selection of the most interesting stellar sources for detailed study by more sophisticated spacecraft.

  Contract provided for three 140-lb experiment structures: structural and thermal test unit, prototype unit to serve also as flight backup, and flight unit. Johns Hopkins Univ.'s Applied Physics Laboratory would build spacecraft structure and systems, integrate and test total spacecraft before flight. Launch of 330-lb SAS–A was planned for 1970; four-stage Scout would place SAS–A in circular, equatorial orbit at 330-mi altitude. (GSFC Release G-54-67)

- Early findings from *Surveyor V*, launched Sept. 8 and softlanded on moon Sept. 10, were summarized in *Science* by JPL scientists Leonard D. Jaffe and Robert H. Steinbacher. The 18,006 TV pictures taken Sept. 11-24 included views of interior of crater in which *Surveyor V* landed, level surface surrounding crater, star and planet sightings for attitude reference, and solar-corona sequence after local sunset. Alpha-scattering instrument produced data on lunar surface chemical composition—oxygen, silicon, and aluminum identified in preliminary evaluation—and showed general chemical composition to be "similar to that of a silicate of a basaltic type." "Material of basaltic composition at the *Surveyor V* landing site implies that differentiation has occurred in the moon," which summary attributed to "internal sources of heat." Early findings were consistent with hypothesis that "extensive volcanic flows have been responsible for flooding and filling the mare [sea] basins." Erosion experiment, spacecraft landing effects, and other observations indicated soil had significant amounts of fine-grained material and measurable cohesion. *Surveyor V* had provided first direct chemical analysis of lunar surface. Lunar magnet experiment produced results, that "are most similar to a terrestrial plateau basalt with no addition of free iron." (*Science*, 11/3/67, 631-52)

- Secretary of Defense Robert S. McNamara announced U.S.S.R.'s possible development of a Fractional Orbital Bombardment System [see Oct. 16]. While ICBM's followed a ballistic trajectory from launch to impact and had an apogee of 800 mi, FOBS, he said, was fired into a very low orbit, and before first orbit was completed, rocket engine would slow down payload and cause it to drop out of orbit—payload then following a reentry path like reentry of ICBM. He said that "we suspect [they] are pursuing the R&D of a FOBS . . . [and] it is conceivable that they could achieve an initial operational capability during 1968. Because of the low altitude of their orbits, some trajectories of a FOBS would avoid detection by some early warning radars . . . [and] impact point cannot be determined until ignition of the rocket engine [deboosting payload]—roughly three minutes and 500 miles from the target. And the flight time can be as much as 10 minutes shorter than an ICBM."

Accuracy of FOBS would be significantly less than that of ICBM and payload a fraction of that of ICBM, according to McNamara. FOBS' disadvantages had compelled DOD to decide that FOBS "would not improve our strategic offensive posture . . . while development of [FOBS] could be initiated at any time for relatively rapid deployment. . . ."

Anticipating that early warning radars (including BMEWS) could not detect FOBS "because of the low altitude of [FOBS] orbits," McNamara said: ". . . already we are beginning to use operationally over-the-horizon radars which possess a greater capability of detecting FOBS than does BMEWS. These will give us more warning time against a full-scale attack using FOBS missiles than BMEWS gives against a heavy ICBM launch." (Text; DOD Release 1060-67)

Communist-Chinese-oriented ABM system—deployment approved by DOD Sept. 18—was named the Sentinel System, according to Secretary of Defense Robert S. McNamara's announcement. L/G Alfred D. Starbird (USA) was named Army's System Manager for new system; currently Director, DCA, Starbird would assume new position Nov. 15, 1967, and would be responsible for development and timely deployment in a three-pronged organization: Sentinel System Office, Washington, D.C., to be an element of C/Staff's Office, USA; Sentinel System Command, Huntsville, Ala., to develop, procure, and install new system; and Sentinel System Evaluation Agency (SSEA) at WSMR to provide independent evaluation, review, and testing. Existing Nike-X organization, under command of L/G Austin W. Betts (USA), Chief of R&D, would continue (separately from Sentinel System) to perform: R&D, design of equipment to be used for tests of the penetration capabilities of U.S. offensive missiles, and mission of protecting U.S. against large-scale attacks. (DOD Release 1059-67)

- CNO Adm. Thomas H. Moorer officially denied news reports on F-111B program cancellation or drastic change: "The reports may have originated with a manufacturer's brochure received by the Navy." Brochure would have no "official standing," he said. Earlier in the day ABC and Metromedia News reported USN and Congressional sources saying cancellation was official. In Fort Worth, General Dynamics Corp. President Frank W. Davis insisted the F-111 was meeting its objectives and even exceeding them in some cases.

  Navy's highest-ranking admirals had recommended cancellation of F-111B and recasting of major features in a more economic and agile swept-wing fighter, Richard Witkin reported in *New York Times*. Admirals' arguments were listed. Overweight aircraft would perform only one of its two planned missions—defense of fleet against air attack. Air-to-air and air-to-ground missions would be beyond capability of F-111B, and admirals had been convinced that an aircraft capable of all missions could be developed and delivered to the fleet by 1973. To arrive at aircraft they wanted, admirals would scrap wings and fuselage of F-111B; they would keep proven features, such as TF-30 engines, Phoenix air-to-air missiles, advanced radar and electronic gear—new aircraft weighing 52,000 lbs compared with F-111B's 75,000 lbs and costing $4-6 million compared with $8-10 million. Navy's limited budget could not stand the soaring costs (in the light of technical troubles delaying F-111B's delivery on schedule), while potential enemy's development of better aircraft called for competitive U.S. development.

  An aircraft that would keep weight down to 52,000 lbs and exploit the swept-wing and engine features of F-111B had been proposed by McDonnell Douglas Corp. and Grumman Aircraft Engineering Corp.; their promised delivery to the Navy would be less than two years behind that for F-111B, which was three years behind schedule, Witkin reported. (DOD Release 1062-67; Witkin, *NYT*, 11/4/67, 1; AP, *W Post*, 11/4/67, A10)

- At Huntsville, Ala., formal dedication of Toftoy Hall, named for the late M/G Holger N. Toftoy (USA, Ret.), U.S. missile pioneer who died Apr. 19, was made by Mrs. Hazel Toftoy, MSFC Director Wernher von Braun, and Army Missile Command Commander M/G Charles W. Eifler (USA). Toftoy Hall would house Missile Components Dept. of the Army Missile and Munitions Center and School. (*Marshall Star*, 11/1/67, 1)

November 4: NASA Associate Administrator for Manned Space Flight Dr. George Mueller announced revised Apollo mission schedule of six flights in 1968 and five in 1969 using "200-series" (Uprated Saturn) and "500-series" (Saturn V) launch vehicles to test and qualify Command, Service, and Lunar Modules (C&SM and LM). Schedule for 1968 would include: Apollo/Saturn 204 (AS-204), first unmanned test of LM in earth orbit; AS-502, second unmanned flight test of Saturn V and Apollo C&SM; AS-503, third unmanned test of Saturn V and C&SM; AS-206, second unmanned flight test of LM in earth orbit; AS-

205, first manned Apollo flight, a 10-day mission to qualify C&SM for further manned operations; AS-504, first manned Apollo flight on Saturn V, to provide experience with both C&SM and LM, including crew transfer from C&SM to LM and rendezvous and docking. Schedule for 1969 would include five manned flights (AS-505 through AS-509) with the first four programmed as lunar mission development flights or lunar mission simulations—AS-509 being that on which lunar landing would be made. Possibly the lunar landing might be delayed until one of the remaining six Saturn V flights beyond 1969.

All opportunities to accelerate progress in 1969 toward manned flights and a rapid accumulation of manned experience with Apollo/Saturn system would be sought, Mueller said. (NASA Release 67-282)

- NASA appointed Edward M. Shafer, Assistant General Counsel, to become Associate General Counsel. Shafer's principal responsibilities had been in the areas of international programs and communications satellites. He had served as Chairman of NASA Board of Contract Appeals since Jan. 1, 1965. (NASA Ann, 11/8/67)

November 5: Sen. Henry M. Jackson (D-Wash.), Chairman, Joint Committee on Atomic Energy's Military Applications Subcommittee, in informal remarks, characterized U.S.S.R. development of FOBS as "a warning to the world that they are hell-bent on obtaining a superiority over the U.S. in strategic weapons development," and stated: ". . . the significant point is that the Soviet Union has now led the way in the development of three nuclear weapon systems [ICBM, ABM, and FOBS]." Starting Nov. 6, he would preside at Subcommittee hearings on ABM systems, and he said that FOBS would be an important topic for consideration. (UPI, *W Post*, 11/6/67, A9)

- Carl B. Squier, U.S.'s 13th Federally licensed pilot and one of its earliest air daredevils, died at Burbank, Calif. Born in Decatur, Mich., he had worked since 1929 for Lockheed Aircraft Corp., becoming president in charge of sales and working as consultant beyond his retirement. He was a friend of such notables in U.S. aviation as Wiley Post, Charles A. Lindbergh, and Capt. Edward Rickenbacker. (*NYT*, 11/7/67, 41)

- Magnetometer intended as an accurate satellite-borne instrument for measurement of magnetic field in space would be calibrated successfully only through use of discarded 19th-century theodolite, ESSA announced. Discovery of theodolite in storage at ESSA's Fredericksburg Geomagnetic Center would ensure better accuracy for future magnetometers; theodelite would precisely position three coils of wire to improve accuracy. (ESSA Release 67-78)

November 5-6: NASA's 805-lb *Ats III* (ATS-C) (Applications Technology Satellite) was successfully launched from ETR by Atlas-Agena D booster. Primary purpose of launch was to place spin-stabilized spacecraft into geostationary orbit, operate it for at least the first 30 days, and obtain useful data from onboard experiments. Spacecraft entered elliptical transfer orbit with 22,148-mi (35,643-km) apogee, 111-mi (179-km) perigee, 10-hr 28-min period, and 28° inclination. At 16:01:00 GET (Nov. 6) apogee motor was fired to place satellite into near-synchronous orbit with 22,186-mi (35,704-km) apogee, 111-mi (179-km) perigee, 10-hr 28-min period, and 0.536° inclination, where it would be allowed to drift to stationary position over the Atlantic at 47° west longitude.

Ats III was third in series of five satellites designed to improve spacecraft technology, develop long-life control systems, advance spacecraft communications, and improve long-range weather predictions. Five of spacecraft's nine major experiments—Multicolor Spin-Scan Cloud Camera (sscc), Omega Position Location Experiment (ople), Self Contained Navigation System (scns), Image Dissector Camera (idc), and Mechanically Despun Antenna—would be tested in space for first time. sscc would provide first full-disk photos of earth for continuous detailed viewing of short-lived weather phenomena. In addition, spacecraft carried VHF and SHF transponders, resistojet thruster, reflectometer experiment, communications equipment, and engineering experiments to evaluate spacecraft performance. *Ats I* (launched Dec. 6, 1966) had exceeded its test objectives and was still operating flawlessly. *Ats II* (launched April 5, 1967), though adjudged a failure, was transmitting some useful data. *Ats III* was last spacecraft to be launched from ETR's Complex 12 and last ATS to be launched with Atlas-Agena; two remaining spacecraft in series would be launched with Atlas-Centaur. ATS program was managed by GSFC under OSSA direction. (NASA Proj Off; NASA Release 67-276; *W Post*, 11/6/67, A1)

November 6: In its resolution #2260, U.N. General Assembly urged its Committee on Peaceful Uses of Outer Space to elaborate on "agreement on liability for damage caused by the launching of objects into outer space and an agreement on assistance to and return of astronauts and space vehicles." On July 17, Legal Subcommittee of Committee on Peaceful Uses of Outer Space, meeting in Geneva, had reached such provisional agreement. On July 26, 1966, U.S. and U.S.S.R. had agreed in principle that countries were internationally liable for damage caused to other states by objects they launched into outer space.

Other concerns of the resolution were education and training for space exploration, feasibility of communications for direct broadcast from satellites, and applied satellite technology of benefit to man. On Jan. 18, 1966, Committee on Peaceful Uses of Outer Space had proposed a world conference on outer space to be held in New York. This conference should examine the impact of space data on education and communications—issues of major importance even to poor and economically retarded states [see NASA SP-4007, 20]. (Text; NASA SP-4007, 20, 250)

- Joint Committee on Atomic Energy's Military Applications Subcommittee held hearings on ABM systems and U.S. defense against ballistic missile attack; Deputy Secretary of Defense Paul H. Nitze and DOD's Director of Research and Engineering Dr. John S. Foster, Jr., presented DOD position that no system capable of defending against a heavy Soviet attack was now technologically and financially feasible. Foster said: "In 1966, a new threat appeared—the probability that [Communist Chinese] were developing an ICBM. This is obviously vastly different from the Soviet threat . . . [not materializing until early 1970s, and relatively unsophisticated and] . . . against this [Communist Chinese] threat we have high assurance of providing damage denial for the whole country. Furthermore, we think we can maintain this capability against a growing [Communist] Chinese threat at least until the 1980s." (Testimony)
- Rep. Craig Hosmer (R-Calif.), member, Joint Committee on Atomic Energy's Military Applications Subcommittee, during its House hearings on ABM systems, offered reply to Secretary of Defense Robert S. McNamara's announcement on Fractional Orbital Bombardment System

(FOBS): The weapon would "suppress our manned bombers by catching them on the ground with only a 3-minute attack warning . . . destroy the acquisition radar of any [ABM] system we might install . . . [and] destroy our retaliatory ICBM's in their silos." Hosmer noted that Soviets call FOBS by code name Scrag, and said the May 29, 1967, issue of the American Security Council's Washington Report contained an article "written by the exceptionally well-informed Dr. Stefan T. Possony, of the Hoover Institution on War, Revolution, and Peace at Stanford University." Hosmer stated: "A most vital disclosure by Possony is that the Scrag . . . would have a 30-megaton warhead, presumably sufficient to penetrate our hardened ICBM silos, ruin our ICBM missiles and thus destroy our retaliatory capacity. Despite Soviet capabilities with this size warhead and larger . . . McNamara continues to insist that Scrag warhead yields are only in the 1- to 3-megaton range. I believe he does so less with assured knowledge of the yield than he does wishfully, to support his continued claims that we do not need an ICBM system for protection because our missile silos are so strong they cannot be penetrated." (Testimony; *CR*, 11/6/67, H14638–9)

- NASA's Apollo 4 (AS–501), which would be launched by Saturn V in three days, had the power to place astronauts on moon, or telescopes in orbit beyond earth's atmosphere; American capability for major strides toward the unknown had arrived when interest in these endeavors had reached a new and understandable low, Albert Sehlstedt, Jr., reported in Baltimore *Sun*. Along with the new era of less than major interest in space had come the high-priority problems of the Southeast Asia war and the urban riots. *Aviation Week* for Aug. 7 had said "An era ended for the National Aeronautics and Space Administration last week when Congress voted a $234,000,000 cut in the agency's budget authorization for fiscal 1968."

  Many people during recent months had looked toward the achievement of social as well as scientific goals, reappraisement of both being called for by Congress. Because AS–501 would be one of the highest achievements in space, the question was no longer whether man could reach into space, but ". . . . with what urgency it should be done in the light of all the needs on earth." (Sehlstedt, *B Sun*, 11/7/67, 6)

- Detailed information on French space budget for 1968, expected to be approved soon by National Assembly, was reported to total $141 million, increased about $36 million over 1967 budget. It would include $121.5 million for research and operations and $19.3 million for facilities and administration under spending plan being developed by Centre National d'Etudes Spatiales (CNES). Research and operations funds would be divided, $87.8 million for French programs and $33.7 million for ELDO and ESRO projects. About $1 million also had been allocated for bilateral space research projects outside ELDO and ESRO. Major items of budget included: development costs for ELDO Europa 1/PAS launch vehicle and Diamant B launch vehicle; R&D costs for two Roseau radio observatory satellites (scheduled for U.S.S.R. launch in 1971); planning costs for two Symphonie communications satellites (joint French-West German venture with one spacecraft structure group and two avionics groups to be chosen from competing groups in both countries); construction costs for Kourou launch complex in French Guiana; and partial costs of new Toulouse complex for satellite development and balloon sounding projects. Budget also con-

November 6

tained funds for future studies for second phase of France's space program, according to CNES officials. French fifth financial plan committed about $408 million in 1965 for the first phase through 1969. (Fink, *Av Wk*, 11/6/67, 24)

- Some U.S. experts believed Soviets planned further missions: rendezvous and docking of manned Soyuz-class spacecraft in near-earth orbit for long-duration flights; circumlunar flight, fully automated, with animal payload to test reaction to reentering earth's atmosphere at translunar return speeds; manned circumlunar flight; and manned lunar landing. *Aviation Week* stated, "Although there is disagreement among U.S. government and industry on some details, many feel that the Soviets have successfully overcome serious gaps in technology which had slowed their progress. . . ." (*Av Wk*, 11/6/67, 16)

November 6-7: National Academy of Sciences steering committee for Long-Duration Manned Mission Study held first meeting at NAS with UCLA Prof. Donald B. Lindsley as chairman. Committee would study psychological and physiological stresses on astronaut crews traveling in space for up to 700 days. Different parts of overall study would be investigated by small task groups. One would review what is known about the dynamics of small groups. Another would review all available information on psychological, physiological, and performance effects of isolation, confinement, and other stresses. (NAS–NRC–NAE *News Report*, 12/67, 1–2)

November 7: NASA's *Surveyor VI* (Surveyor-F) was successfully launched from ETR by Atlas-Centaur (AC–14) booster on 64-hr lunar intercept trajectory. Primary mission for the 2,223-lb (at launch) spacecraft was to perform a soft-landing on moon and obtain post-landing TV pictures of the lunar surface. As secondary mission spacecraft would determine relative abundance of chemical elements in lunar soil with alpha scattering instrument; obtain touchdown-dynamics, thermal, and radar-reflectivity data; and conduct vernier engine erosion experiment. NASA's *Explorer XXXIV*, in orbit since May 24, would serve as a "solar watchdog," furnishing information on solar radiation for correct interpretation of *Surveyor VI*'s soil analyses.

At 00:04:06 GET *Surveyor VI* with two-burn Centaur separated from Atlas and ignited to reach 100-mi-altitude circular parking orbit where it coasted for 12 min; Centaur then reignited to escape earth orbit and boost spacecraft toward moon. Spacecraft separated from Centaur, deployed its solar panels, and locked on the sun and the star Canopus. Based on a 56-mi target miss estimate, a 1.2-m midcourse maneuver was conducted at 00:18:41 GET to assure precise landing on target in Sinus Medii in center of moon's front face.

Surveyor VI was sixth in series of seven spacecraft designed to prove out design, develop technology of lunar soft-landing, obtain post-landing TV pictures of lunar surface, and provide basic scientific and engineering data in support of Project Apollo. Surveyor program was directed by OSSA Lunar and Planetary Programs Div.; project management was assigned to JPL; Atlas-Centaur launch vehicle was managed to LeRC; and prime contractor for spacecraft development and design was Hughes Aircraft Co. (NASA Proj Off; NASA Releases 67–278, 67–287)

- USAF successfully launched an Advanced Ballistic Re-Entry System (ABRES) module from Vandenberg AFB by Atlas-D booster. (UPI, *W Post*, 11/8/67, A21)

- *Explorer XXVII* and *Vanguard II* had been among the 30–40 artificial satellites to be photographed on a typical night by the three-ton light-gatherer camera at Smithsonian Astrophysical Observatory's Debre Zeit tracking station in Ethiopia. Robert Citron, station manager, described the operation as an "extremely accurate way of measuring distances with the use of geodetic satellites," in a planned program extending into 1972. Debre Zeit, functioning as hub of 12 similar stations in India, Japan, Patagonia, Spain, Brazil, and South Africa, measured earth's size and shape, gravitational pull variations, and atmospheric density changes. Planned progress would include measuring shape of world "down to 33 feet," according to Citron. He explained: "There is a widely accepted theory that the earth's land masses are drifting. In particular, the African continent is said to be drifting away from the Asian land masses." By using planned measuring methods with geodetic satellites, Citron hoped to explore the truth of the theory and measure the drift should the theory be true. (Reuters, *NYT*, 11/7/67, 36)
- President Johnson signed Public Broadcasting Act of 1967 (P.L. 90-129) which would provide new funds for broadcast facilities to give "a wider and . . . stronger voice to educational radio and TV; launch major study of TV's use in Nation's classrooms and potential use throughout the world; and create Corporation of Public Broadcasting to assist stations and producers who aim for the best in broadcasting good music, . . . exciting plays, and . . . reports on the whole fascinating range of human activity." He indicated that he had asked Dr. Milton Eisenhower, Johns Hopkins Univ., and Dr. James Killian, MIT, to serve as members of Corporation's board of directors. He explained that the Corporation would get partial support from Government but "it will be carefully guarded from . . . control."

    The President said: "I believe the time has come to stake another claim in the name of all the people, stake a claim based upon the combined resources of communications. I believe the time has come to enlist the computer and the satellite, as well as television and radio and to enlist them in the cause of education. So I think we must consider new ways to build a great network for knowledge—not just a broadcast system, but one that employs every means of sending and of storing information that the individual can use." (*PD*, 11/13/67, 1530–2)
- Joint Committee on Atomic Energy's Military Applications Subcommittee heard testimony on missile defense from Columbia Univ.'s Professor Philip E. Mosely and RAND Corp.'s senior staff member Thomas W. Wolfe.

    Mosely stated, on subject of nuclear weapons, this philosophy: "In any future period [of U.S.S.R. nuclear equality or superiority over U.S.] . . . we would be prudent to assume that Soviet policy would be tempted to undertake a more extensive, more acute, and more dangerous range of risks [in pursuing] its declared long-range ambition to reshape the world according to its own dogma."

    Wolfe discussed Soviet ABM program, Soviet attitude toward ABM limitation, Soviet ABM activities which relate to overall strategic posture, and political implications of a changing U.S.-Soviet military balance. He saw Soviet leadership seeking "wider options for global intervention," and, for this goal, he looked for a buildup of defensive and offensive nuclear power and extension of the reach of their conventional military power. (Testimony)
- At MSC the TRW Systems Group had received $10.7-million NASA contract

extension to perform spacecraft analysis for Apollo, including studies, technical fact-finding, and investigations for systems in Command, Service, and Lunar Modules. Contract extension would carry through June 1968 and increased total cost of contract to $65.5 million. (MSC Release 67–66)

- The 50th anniversary celebrations of the Bolshevik Revolution had included Moscow display of five new missiles, Bruce Winters reported in Baltimore *Sun*. Two years ago, the Moscow display included three new missiles, one with warhead for "surprise blow on the first or any other orbit around the earth." The 1967 display had included new three-stage ICBM, two-stage IRBM, submarine missile, tactical missile, and antiaircraft missile. (Winters, B *Sun*, 11/8/67, 1)

November 8: NASA's *Ats III*, launched Nov. 5, now in 22,300-mi-altitude synchronous orbit, transmitted to NASA's Rosman, N.C., ground station a color photograph of earth which officials described as excellent in quality. The photograph, processed at GSFC, showed entire earth's disk and details of features as small as two or three miles in size. *Ats III*, third of a series of five satellites managed by OSSA, carried communication, meteorology, navigation, and spacecraft technology experiments.

From its location over the equator at 47° west longitude, the spacecraft had a view of North and South America, part of Africa and Europe, and the southern part of Greenland icecap. Spacecraft's multicolor spin-scan cloud camera obtained color contrast as well as brightness contrast, according to chief experimenter Wisconsin Univ.'s Dr. Verner E. Suomi. He gave this analysis of *Ats III*'s meteorological operations: "From a time series of pictures such as these the United States has, in effect, given South America a means to track their weather far beyond what they are able to do by conventional means. In one single photograph, the equatorial cloud-free band which was seen from *Ats I* [launched Dec. 6, 1966] pictures in the Pacific is also evident in the Atlantic. These pictures are ideal for continuing studies from the heat budget and convective systems. The additional meteorological information in a color photograph is more subtle, but it is there. Low clouds tend to be bluer than high clouds. This has been possible even though adjustments to color balance on the ground are still underway. There are suggestions that the muddy outflow from the Amazon River in Brazil can be seen. We have not yet been able to determine whether or not the Gulf Stream water will be visible. Also, we do not have an example of the colors in the Terminator Zone (sunrise and sunset) to determine if cloud heights can be accurately positioned." (NASA Release 67–286)

- A restructuring of the Nation's space program was outlined by NASA Administrator James E. Webb in testimony on proposed NASA FY 1968 operating plan, before Senate Committee on Aeronautical and Space Sciences. Decisions had reflected fiscal, resources, and time factors. The national fiscal situation required minimum expenditures on space for the next two years. Primary emphasis on completion of Apollo program and other financially important ongoing efforts would be carried out to avoid major losses. Decisions on NASA's space posture for the 1970s would be postponed until "fiscal year 1969 budget or later." According to Webb's planning, "Our allocation of resources in fiscal year 1968 must hold open for later decision the most promising future aeronautical and space options which our past large investments will support."

FY 1968 operating plan figures, presented by Webb, showed NASA's

November 8: The earth from 22,300 miles in space—color photography transmitted by NASA's *Ats III* satellite to Rosman, N.C., ground station beginning Nov. 8. This photo, transmitted Nov. 10, shows portions of North and South America, Africa, and Europe and the Greenland ice cap. The Antarctic continent is under cloud cover. A cold front is moving eastward over the central United States and a tropical storm can be seen at bottom center, with a cold front extending into Argentina.

adjustments of funding under R&D operations for the remainder of FY 1968. Fund reallocation (congressional appropriation figures in parentheses) showed: Apollo Applications, $253.2 (315.5) million; OSSA, $568.1 (528.0) million; OART, $318.7 (301.5) million; OTDA, $275.0 (270.0) million. In more detail, on OSSA, physics and astronomy $141.5 (130.0) million; lunar and planetary, $143.4 (125.0) million; launch vehicle procurement, $141.9 (145.0) million; bioscience, $41.8 (40.0) million; space applications, $99.5 (88.0) million. On OART, electronics systems, $39.2 (35.0) million; space vehicle systems, $35.0 (no change) million; human factors systems, $21.0 (no change) million; basic research, $21.5 (20.0) million; space power and electric propulsion systems, $44.0 (no change) million; nuclear rockets, $54.0 (46.5) million; chemical propulsion, $37.2 (35.0) million; and aeronautics, $66.8 (65.0) million. Sustaining university was $10.0 (no change) million, and technology utilization, $4.0 (no change) million.

Webb presented rather brief outline on present long-range planning for lunar and planetary mission hardware. Limited funds could continue development of flight-qualified engine for NERVA I and would support plans for first Apollo Applications flights in 1970. Two Mars-Mariner flights for 1969 would be the "last in present program." Five Mariner flights in 1970s, two or more Titan III launches in Voyager series in 1973, and Saturn V flight in 1975 could be considered in FY 1969 budget. Webb reasoned that NASA could "use the Atlas-Centaur . . . and the Titan III with transtage, to move out into further planetary operations but with very limited expenditures in 1968 and 1969."

Responding to questions by Senate Committee on Aeronautical and Space Sciences, Webb referred to recent CSC decision on NASA personnel contracts and said he would join with CSC chairman and the Comptroller General in working out pattern for support contracts, the KSC contract providing his guidelines. (Testimony; transcript)

- In a prelaunch press conference at Launch Complex #39, KSC, NASA Deputy Administrator Robert C. Seamans, Jr., referred to high points of Nov. 9 scheduled launch of Apollo 4: ". . . Apollo 4 launch is . . . most difficult and significant milestone to date [and] tomorrow we'll be flight testing the Saturn V for the first time. We'll be flight testing on Apollo spacecraft entering the atmosphere at lunar return speeds for the first time. We'll be live-testing our launch and operational facilities for the first time, and we'll be testing our development and operational team in its first major mission attempt for the first time. I am certain this team . . . will not be found wanting." (Text)

- Deputy Secretary of Defense Paul H. Nitze, in response to statement made Nov. 6 by Rep. Craig Hosmer (R-Calif.), member, Joint Committee on Atomic Energy's Military Applications Subcommittee, during hearings on ABM systems, answered: "First we are taking steps to protect Minuteman against much better Soviet offenses than we actually expect. Secondly, even if any two of the three elements of our second-strike forces—our land-based ICBMs, our submarine-based Polaris and Poseidon, and our strategic bombers—were rendered useless by some unforeseeable disaster, the remaining element could by itself inflict unacceptable damage on the U.S.S.R." He compared effectiveness (number of targets destroyed) of 10 50-kt warheads in a Multiple Independent Re-entry Vehicle (MIRV) configuration to the same effectiveness of one large yield—10 mt—warhead, and stated that "the MIRVs provide much more effective payloads . . . by every relevant criterion of military effectiveness, even though they deliver much less total megatonnage." Response to Hosmer's statement by Nitze had been made at request of Joint Committee on Atomic Energy. (Text)

- FAA, despite qualified objections of air carriers, announced Dec. 15 effective date for imposition of speed limit of 250 kph (288 mph) for all aircraft operating below 10,000 ft mean sea level [see Sept. 1]. FAA believed its action would give pilots more time to "see and avoid" other air traffic in the Nation's most heavily traveled airspace. Larger airlines felt collision avoidance systems and automated terminal approach systems would go further to reduce the growing number of midair collisions; they would have preferred a speed limit from 5,000 ft down to surface. Air Line Pilots Assn. endorsed rule, at the same time making

it plain that they were quite skeptical of its value in collision avoidance. (*W Post*, 11/9/67, M4)
- President Johnson signed H.R. 12474, FY 1968 NASA appropriations bill, designated P.L. 90–131 (NASA LAR VI/119)
- President Johnson signed Executive Order 11381 enlarging Federal Council for Science and Technology to include representatives from DOT, HUD, and Dept. of State for more effective operation. (*PD*, 11/13/67, 1537–8)
- ERC appointed O. Hugo Schuck Director of Office of Control Theory and Application. Schuck had been Visiting Professor of Aeronautics and Astronautics at Stanford Univ. while on leave as Director of Research at Honeywell, Inc. In his new position Schuck would supervise research on application of control theory to complex aeronautics and space vehicle systems. He had been President of the American Automatic Control Council, heading U.S. representation in the International Federation of Automatic Control. (ERC Release 67–37)

November 9: NASA's *Apollo 4* (AS–501) was successfully launched from ETR at 7:00 am EST in first all-up test of three-stage Saturn V launch vehicle, first in-orbit restart of third (S–IVB) stage, and first use of Launch Complex 39 and ground support equipment. Launch vehicle stages performed nominally; orbital insertion, 3rd stage restart, and injection into earth-intersecting coast ellipse were accomplished satisfactorily; Command Module (CM) encountered no functional problems; and splashdown occurred within 10 mi of planned landing point. Third stage ignited to insert Command Service Module (CSM) into orbit with nominal parking apogee, 119 mi; perigee, 114 mi; period, 88.2 min; inclination, 32.6°. After two revolutions in parking orbit, 3rd stage was reignited for simulated translunar injection burn, injected stage and spacecraft into earth intersecting orbit with 10,774-mi apogee. Following stage/CSM separation, Service Propulsion System (SPS) ignited for 16-sec burn, raising apogee to 11,314 mi. Spacecraft was aligned to a specific attitude to achieve thermal gradient across CM heat shield. Attitude, with CM hatch window directly toward sun, was maintained for about 4½ hrs. [CM hatch window had been replaced with instrumented test panel containing simulations of flexible thermal seals and gaps between hatch and surrounding heat shield. Successful performance of thermal seals during reentry heating would qualify seals for use on manned CMs.] SPS was then reignited for 271-sec burn to accelerate spacecraft to the most severe entry conditions that could possibly occur in a lunar-return trajectory. SPS cutoff was followed by CM/SM separation and orientation of CM to entry attitude. Atmosphere reentry at 400,000 ft occurred at flight path angle of $-7.077°$ with inertial velocity of 36,537 fps (24,911 mph). CM landed in mid-Pacific near Hawaii at 3:37 pm, EST, and was recovered by USS *Bennington* 2 hrs 14 min after touchdown.

In significant step toward developing manned lunar landing capability, *Apollo 4*'s flight was the first of two to three missions designed to qualify Saturn V for manned flight, and the first test of the structural integrity and compatibility of launch vehicle and spacecraft. By subjecting CM's heat shield to high heat load and heat rate, NASA was evaluating CM's design adequacy for reentry from lunar missions—primary objective to achieve successful manned lunar mission. Other primary mission objectives were to: confirm launch loads and dynamic characteristics; demonstrate stage separation; verify operation of launch vehicle subsystems and spacecraft subsystems; evaluate performance of space vehicle Emer-

November 9: NASA launches *Apollo 4* into orbit from Complex 39A of Kennedy Space Center in successful first flight test of Apollo/Saturn V space vehicle.

gency Detection System (EDS) in an open-loop configuration; and demonstrate mission support facilities and operations required for launch, mission conduct, and CM recovery.

Saturn V was the most powerful space vehicle developed in U.S. space program; in launch configuration with *Apollo 4*, height was 363 ft, and weight, fully fueled, was 6,220,025 lbs. Saturn V's 1st stage engines produced 7,610,000 lbs of thrust at liftoff, and the launch vehicle placed 278,699 lbs in earth orbit. This launch vehicle could place 96,000 lbs on the moon. Kerosene (RP–1) was the fuel used in the five 1st stage (S–1C) engines, while 2nd stage (S–II) engines used liquid hydrogen. The five S–II stage J–2 engines, developing total thrust of 1,000,000 lbs, were rated for operations to 117-mi altitude and to 935-mi distance downrange. Third stage (S–IVB) J–2 engine, using liquid hydrogen fuel, developed about 200,000 lbs of thrust.

High-quality pictures of the earth were obtained from about one hour before to one hour after apogee on Command Module apogee camera film processed at MSC.

Success for *Apollo 4* (AS-501), according to NASA Associate Administrator for Manned Space Flight Dr. George E. Mueller, had been achieved in terms of "pre-set primary objectives." Heat shield design, S-IVB restart, structural/thermal integrity, compatibility of launch vehicle and spacecraft, and ground support had been proven.

Direction of Apollo Program had been the responsibility of OMSF; development of Apollo spacecraft, flight crew training, and mission control the responsibility of MSC; development of Saturn launch vehicles the responsibility of MSFC; and Apollo/Saturn launch operations the responsibility of KSC. Under overall direction of OTDA, MSFN had been managed by GSFC. (NASA Proj Off; NASA Release 67-275; MSFC Release 67-226)

- Apollo Program Manager M/G Samuel C. Phillips, commented on countdown operations for Saturn/Apollo (AS-501): ". . . I was tremendously impressed with the smooth teamwork that this combined government/multi-industry team put together [and] you could almost feel the will with which it was being carried out. I think that's important to the progress of Apollo . . . perhaps even more so than some of the technical returns we'll get from this mission. . . . Apollo is on the way to the moon." (Text)
- U.S. officials praised successful flight of Apollo 4 (AS-501). President Johnson said that "the whole world could see the awesome sight of the first launch of what is now the largest rocket ever flown. This launching symbolizes the power this Nation is harnessing for the peaceful exploration of space. The successful completion of today's flight has shown that we can launch and bring back safely to earth the space ship that will take men to the moon." NASC's Executive Secretary Dr. Edward C. Welsh said the Nation's people "should be proud of this historic landmark in propulsion and precision, and particularly in the boost given our national space program." Looking toward the future, NASA Administrator James E. Webb praised this "successful demonstration of the devotion and high quality workmanship of over 300,000 men and women in thousands of industrial plants, laboratories, test facilities, universities and government installations [and this] success [which] will permit us to move more rapidly on to a second launch." Dr. Wernher von Braun said, "No single event since the formation of the Marshall Center in 1960 equals today's launch in significance [and] I regard this happy day as one of the three or four highlights of my professional life—to be surpassed only by the manned lunar landing." (*PD*, 11/13/67, 1541; Text; UPI, *NYT*, 11/10/67, 32C; *Marshall Star*, 11/15/67, 1-10)
- Blastoff of Saturn V at ETR produced one of the loudest noises in history, natural or man-made, according to Columbia Univ.'s Lamont Geological Observatory at Palisades, N.Y. Observatory physicist Dr. William Donn labeled U.S. and U.S.S.R. nuclear explosions the only louder manmade sounds and 1883 fall of Great Siberian Meteorite as the only louder natural sounds on record. (AP, *W Post*, 11/10/67, A8)
- President Johnson announced intention to nominate M/G Richard P. Klocko (USAF), Deputy Director, National Military Command System Technical Support, Defense Communications Agency, for promotion to lieutenant general and assignment as Director, Defense Communications Agency. He would replace L/G Alfred D. Starbird (USA) who had been assigned as System Manager, BMDS. (*PD*, 11/13/67, 1539)

November 9-24: NASA's *Surveyor VI*, launched Nov. 7, became fourth U.S. spacecraft to softland on the moon when it touched down at 08:01 EST in Sinus Medii after 64-hr flight and began transmitting first of 30,065 detailed television pictures to JPL Deep Space Facilities, Goldstone, Calif.

Landing sequence began when *Surveyor VI* shifted its normal cruising attitude to position main retrorocket. Triggered by radar, main retromotor slowed spacecraft to 374 mph; retromotor then ejected. Vernier engines cut off at 13 ft above lunar surface and spacecraft landed.

First photos transmitted showed one of spacecraft's landing pads resting on sand-like soil similar to that pictured in previous Surveyor photos. Later pictures confirmed that area was rough and pitted with craters; they revealed a region strewn with rocks up to two feet in diameter and a huge cliff described by USGS scientist Dr. Eugene Shoemaker as "the most rugged feature . . . yet seen on the moon." Efforts to reestablish communications with *Surveyor V*, which had softlanded Sept. 10, for simultaneous communications with *Surveyor VI*, proved futile.

On Nov. 17 *Surveyor VI*'s vernier engines were fired for 2.5 sec, lifting spacecraft to 10-ft altitude before it touched down 8.5 sec later, 8 ft west of original resting point. Purpose of translation—first planned change of location ever accomplished after conclusion of initial flight—was to test lunar surface bearing strength, determine depth of original footprints made by spacecraft, obtain stero-type effect by comparing photos taken of same area from different positions, and study engine erosion effects. Post-translation photos showed cratering and scouring effect on areas under engines and confirmed that alpha-scattering instrument—which had stopped functioning properly after operating successfully for 43 hrs and providing excellent data on chemical composition of lunar soil—had come to rest upside down. Communications with spacecraft were halted Nov. 24 to preserve battery power during cold of lunar night.

Performance of *Surveyor VI* was nearly flawless. *Surveyor I* (launched May 30, 1966), *Surveyor III* (launched April 17, 1967), and *Surveyor V* (launched Sept. 8, 1967) had all softlanded successfully and transmitted photos to earth. *Surveyor II* (launched Sept. 20, 1966) failed to softland because of an ignition failure. Communications with *Surveyor IV* (launched July 14, 1967) were lost seconds before it was scheduled to softland and its condition could not be determined. (NASA Proj Off; NASA Release 67–278; JPL Release; AP, B *Sun*, 11/10/67, 1; Reuters, *W Post*, 11/10/67, 1)

November 10: NASA successfully launched *Essa VI* (TOS–D), sixth meteorological satellite in ESSA's Tiros Operational Satellite (TOS) system, from WTR using three-stage Thrust-Augmented Thor-Delta booster. Primary mission objective was to provide global cloud coverage on a regular, daily basis with six-month-nominal and three-month-minimum spacecraft lifetime. Satellite achieved near-polar, sun-synchronous orbit with 920-mi (1,482-km) apogee, 872-mi (1,402-km) perigee, 114.8-min period, and 102.1° inclination. Wheel orientation maneuver was scheduled for completion during 18th orbit at which time first photos would be programmed and NASA would check out spacecraft before turning its operation over to ESSA.

An advanced version of the cartwheel configuration, 286-lb *Essa VI* carried two Automatic Picture Transmission (APT) camera systems modified with improved magnetic shielding to prevent slight banding visible in previous ESSA APT photos. Cameras would photograph earth's cloud cover and immediately transmit pictures to local APT stations. *Essa VI* would supplement *Essa II* and *Essa IV*, replacing them when they became inoperable.

ESSA financed, managed, and operated TOS system; GSFC was responsible for procurement, launch, and initial checkout of spacecraft in orbit. *Essa I* was launched Feb. 3, 1966; *Essa II*, Feb. 28, 1966; *Essa III*, Oct. 2, 1966; *Essa IV*, Jan. 26, 1967; and *Essa V*, April 20, 1967. All five satellites were still operating. (NASA Proj Off; ESSA Release 67–82)

- Letter to President Johnson from Rep. William F. Ryan (D-N.Y.), member of House Committee on Science and Astronautics, assessed U.S. space program after Apollo/Saturn flight of Nov. 9: U.S. had captured lead in space race with the U.S.S.R. Rep. Ryan urged the President to renounce the space race, and "to make a serious request for the beginning of far-ranging international cooperation to reduce the costs and increase the benefits of space exploration—to remove the wasteful duplication of two great nations." (Text)

November 12: In news release, Sen. Claiborne Pell (D-R.I.), member of Senate Committee on Foreign Relations, called for an international ocean-space treaty to regulate nations' development of deep-ocean riches. In regulating man's behavior in deep-ocean space beyond territorial and continental shelf limits, U.S. was already exploring possible diplomatic initiatives. In U.N. circles, Maltese resolution suggested "marine resources be turned over to an international organization which will supervise their exploitation and distribute the benefits to underdeveloped nations." (Text)

November 13: U.S.S.R. Mikoyan E–266 twin-tailed fighter aircraft—NATO designation Foxbat—averaged 1,804 mph over closed 1,000-km (621-mi) course, carrying two-ton payload. Soviets claimed world speed record. (*Av Wk*, 11/13/67, 35)

- NASA announced establishment of an Astronomy Missions Board, to be chaired by Harvard College Observatory Director Dr. Leo Goldberg. Twelve astronomers and physicists would make up the Board, which would advise NASA through Associate Administrator for Space Science and Applications Dr. John E. Naugle. Executive Director for the Board would be OSSA's Deputy Director of Physics and Astronomy Dr. Henry Smith. The Board would develop and review scientific objectives and general strategy for space astronomy missions of sounding rockets, balloons, and satellites and would recommend designs for missions, telescopes, and other equipment. Excluded from the Board's responsibility would be studies of moon and planets from close range and earth orbital observations of the earth. (NASA Release 67–284)

- DOD would be forced to turn to space technology for early warning, using either manned or unmanned spacecraft to counteract Soviet decision to develop Fractional Orbital Bombardment Systems (FOBS), *Aviation Week* contended. Because they believed that the U.S.S.R. could use FOBS "to foil existing [BMEWS] radars," DOD planners would place initial reliance on HF, over-the-horizon radars. "The Soviet objective appears to be to slash the 15–30 min warning time now available from BMEWS to a mere 3–4 min for a FOBS attack...." Over-the-horizon radars depended on ionospheric reflection which would make them subject to interruption during sunspot activity; further, these radars would be relatively vulnerable to jamming. MOL and, ultimately, continuously manned or unmanned space patrols could receive DOD support if it appeared that the Soviets were succeeding in their plans for penetrating future U.S. missile defenses. (*Av Wk*, 11/13/67, 31)

November 14: Rep. George P. Miller (D-Calif.), Chairman, House Committee on Science and Astronautics, spoke to National Space Club in Washington: "Impressive as our launch vehicles, spacecraft, and scientific instrumentation may be, I have always viewed the space program principally as an investment in people, not hardware." He said principal contribution of the national space effort had been "its stimulation of young people's interest in education [and] much of the credit for the revitalization of American education in recent years must be given to the space program." Scientists and engineers were challenged in keeping up with new knowledge. Universities were made active partners of industry and Government through sustaining university and training grants and laboratory construction, funded by NASA.

Because of current fiscal situation the NASA educational support had been drastically reduced, Rep. Miller noted, and probably would remain at a lower level for some time. Universities had to look to private sources for support previously supplied by Federal Government.

Rep. Miller said he had agreed to serve as chairman of board of trustees for National Space Club's Scientific and Educational Foundation, which would underwrite individual scholarships and fellowships. (Text)

- NASA Aerobee 150 sounding rocket, launched from WSMR, reached 107-mi (172-km) altitude in NRL experiment to obtain stellar spectra in 1,230–1,800 Å far UV wavelength range and photometric data on stellar fluxes in 1,050–1,180 and 1,230–1,350 Å bands. Attitude-control system programming error-pointed instrument 60° from desired targets; only limited data were obtained. (NASA Rpt SRL)

- AFSC's Aeronautical Systems Div., Wright-Patterson AFB, Ohio, had called for industry proposals for concept formulation studies on its first general-purpose, all-weather, air-ground standoff missile, designated X–3. USAF wanted missile with effective operation beyond range of enemy SAMs. Studies would determine specifically what radar or other guidance system would provide desired capability. (*Interavia*, 11/14/67, 1)

- Italy's space effort for 1967–68 was detailed in *Interavia Air Letter*. National research program ($2.72 million) included studies of solar winds, ionosphere's electronic composition, albedo neutrons, x-rays and cosmic rays, solar and galactic gamma radiation, and stellar UV radiation. Other research would cover celestial mechanics, propulsion systems, and telemetry. For international programs, $3.52 million had been approved for Phase B of San Marco project and for launching San Marco C in joint effort with NASA. Supplementary funds ($1.3 million) probably would go for Bologna Univ. and NASA effort using OSO-C to measure x-rays and cosmic rays. Italy's share in 1967 ELDO funding would be $10.2 million, and in ESRO $5.63 million. For ESRO programs, Italian industry would supply antennas and telemetry equipment for tracking network, PCM telemetry equipment, and radars for launching base at Kiruna, Sweden. (*Interavia*, 11/14/67, 3)

- Gen. Curtis E. LeMay (USAF, Ret.) resigned after two years as President of Air Force Historical Foundation. In accordance with Gen. LeMay's recommendation, Gen. Bernard A. Schriever (USAF, Ret.) was unanimously elected to replace him. B/G Monro MacCloskey (USAF, Ret.) was appointed Executive Director. (*AFHF Newsletter*, 12/67, 1)

November 15: NASA X-15 (No. 3) crashed in Mojave Desert when the aircraft "exceeded its structural limitations during the final portions of the reentry maneuver," X-15 Accident Investigating Board suspected. X-15's pilot, Maj. Michael J. Adams, did not use ejection system and was killed, the first fatality in the 191 flights since the three X-15s started their flight operations, June 8, 1959. Usual flight plan called for aircraft to ascend at a sharp angle after being dropped by B-52 aircraft. After engine burnout, plan called for coasting up into high thin air, then dropping back in pancake style into thicker air for reentry. According to a NASA spokesman, the aircraft had risen to about 260,000 ft and encountered trouble on the way down. "There apparently was some sort of control malfunction as the pilot attempted to pull out of his descent," spokesman said. Radar and telemetry data received during flight were excellent, data being received down to an altitude of 60,000 ft.

Speed and altitude records for winged aircraft—4,534 mph, by Maj. William J. Knight (USAF), and 354,200 ft, by NASA's Joseph A. Walker (who had since died in crash of XB-70 No. 2 and F-104 aircraft)—had been set by X-15s. Two crash landings of X-15 aircraft had occurred; pilots had survived and aircraft had been restored to flight status. X-15 No. 3 pilot, Maj. Adams, 37, had finished USAF training in 1952, flown in combat in Korea, and graduated with honors from Aerospace Research Pilots School. Although assigned to MOL program, he had requested transfer to X-15 operations.

Board was continuing attempt to determine the probable cause. (FRC Release 26-67; AP, *NYT*, 11/16/67, 15)

- In message transmitting to Congress annual report on U.S. participation in U.N. for 1966, President Johnson cited successful negotiation of space law treaty as an "outstanding accomplishment . . . which bans weapons of mass destruction from space and calls for peaceful cooperation in its exploration and use." He also cited success of U.N. Development Program in providing economic assistance and approval of charter for U.N. Industrial Development Organization to help new nations create industries best suited to their needs. He commended U.N. representatives' unnoted work to fight ignorance, hunger, and disease, and to promote economic and social development. (*PD*, 11/20/67, 1568-9)

- NASA Administrator James E. Webb, at Chicago meeting of American Petroleum Institute, spoke on space program's technological and social benefits in assessing developing capabilities which would "serve our nation's future needs." He called attention to "special equipment that will automatically monitor ground, underseas cable, or satellite circuits carrying high-speed traffic [and] the entire international communications network structure of the world . . . more valuable today than it was just a few years ago." He also pointed out that "the rocket technology that carried Surveyor to the moon provided astronomers with an improvement in resolution beyond that provided by Ranger by another factor of 1000. That is 1000 times more detail than the Ranger results which themselves were 1000 times better than anything before."

Webb said the U.S.S.R. was "building a rocket bigger than Saturn V, I feel sure." He explained that the Saturn V-Apollo system "permits us to operate out as far as the moon with large payloads of about 100,000 pounds." He compared U.S.S.R. and U.S. exploration of the planet

Venus: "The Russians sent a capsule weighing 845 pounds into the atmosphere of Venus and we made a close fly-by. But in many ways the Mariner [U.S.] data are more complete. It reaches down to about 6 miles from the surface of Venus, while the Venus 4 [U.S.S.R.] data appear to have stopped at about 15 miles up."

Webb explained how NASA's system for making the most of the Nation's technological capabilities uses "a Government-industry-university research base and provides a direct channel for industry to apply new knowledge to industry problems." He referred to work on campuses of 200 universities: ". . . the number of scientists, engineers, researchers, technicians and graduate students participating in NASA activity has grown to about 10,000. Of these, about 1000 are involved as scientific or engineering investigators." Webb concluded that "at any foreseeable budget level NASA will maintain the base of knowledge that will determine the ability of this country to plan, undertake and accomplish successfully new and difficult aeronautical and space objectives." (Text)

- In testimony before House Committee on Interstate Commerce's Transportation and Aeronautics Subcommittee, Rep. John W. Wydler (R-N.Y.), member of House Committee on Science and Astronautics, urged passage of H.R. 3400 to allow Secretary of Transportation to prescribe aircraft noise abatement regulations. Rep. Wydler cited complaints from churches whose religious services had been interrupted; doctors whose patients' mental and physical health had been impaired; schools whose classes had been disrupted; and individuals whose lives had been inhumanly intruded upon by jet aircraft noise. Passage of the bill was imperative, he said, because aircraft noise had become "something that dominates the lives of those who have to live with it to the point where it actually destroys the way in which they are living."

    Secretary of Transportation Alan S. Boyd told Subcommittee that DOT was trying to find a solution to the problem, but he admitted his disbelief that "there will ever be such a thing as a quiet airplane. Despite our far longer experience with the problems of truck noise and railroad noise, we have not been able to produce quiet vehicles in those modes of transportation." However, he believed that "we will be able, by technological and regulatory means, to reduce the impact of aircraft noise exposure for the majority of Americans who are . . . subject to excessive noise exposure." He said bill was "the mechanism by which we can assure future aircraft are substantially quieter than our present generation [of aircraft]." (Transcript)

- Sentinel System—Communist Chinese-oriented ABM system—would be located in accordance with DOD's survey of areas with optimum advantages for the area-defense weapons and their radars; DOD identified the first 10 areas chosen for survey: Albany, Ga.; Chicago, Ill.; Dallas, Tex.; Grand Forks AFB, N. Dak.; New York City; Oahu, Hawaii; Salt Lake, Utah; Seattle, Wash.; Boston, Mass.; and Detroit, Mich. List was not complete, and areas were not final choices.

    Sentinel System would give protection to all U.S. cities and, because of long range of Spartan missile, relatively few batteries could protect entire country against the kind of light and relatively unsophisticated attack that the Communist Chinese might be capable of by the mid-70s, DOD announcement said. (DOD Release 1088-67)

- Fog-dispersal methods would be tested by World Weather, Inc., under Air Transport Assn. contract, with costs defrayed by 33 ATA member airlines.

The $100,000 contract would explore seeding warm fog—above 32° F—with new materials to cause water droplets to combine into larger drops and become precipitation. Materials would be dispensed into fog from aircraft or from ground equipment.

United Air Lines had started successful cold-fog seeding program in 1963, using dry ice pellets, and program now involved several airlines and seeding operations at 17 airports. Warm-fog seeding at Sacramento, Calif., would be technically managed by UAL Meteorology Manager W. Boynton Beckwith; tests would run to Feb. 29, 1968. (ATA Release)

November 16: NASA Nike-Apache sounding rocket, launched from NASA Wallops Station, reached 102-mi (167-km) altitude in Univ. of Maryland experiment to evaluate capabilities and accuracies of pulse and thermal equalization probes as research tools. Experiment also investigated electron energy distribution in normal daytime ionosphere and use of wing-slope techniques with Langmuir probes. Experiment was successful. (NASA Rpt SRL)

- NASA announced Boeing Co. had been awarded incentive fee of $1,811,611 for two missions—*Lunar Orbiter IV*, $1,061,111; *Lunar Orbiter V*, $750,500—completing performance fee portions of Boeing contract and totaling $6,809,053 in such awards for all five Lunar Orbiters. Previous incentive awards—*Lunar Orbiter I*, $1,995,312; *Lunar Orbiter II*, $1,948,725; *Lunar Orbiter III*, $1,053,405—had been earned by Boeing.

  After initial goal to photograph potential moon landing sites had been achieved by first three Lunar Orbiters, NASA's contract modification set a new goal of more advanced scientific missions for *Lunar Orbiter IV* and *V*; Boeing could earn through outstanding performance remaining award fees not earned on previous flights. Modified contract provisions still recognized amounts of original *Lunar Orbiter IV* and *V* award fees.

  NASA's Incentive Award Fee Board for Lunar Orbiter noted that *Lunar Orbiter IV* had photographed moon's "entire front," and *Lunar Orbiter V* had completed photography of "opposite side." (NASA Release 67–285)

- First F–111A USA squadron probably would not be operational in Southeast Asia before November 1968, despite Pentagon "leak" that squadron's scheduled appearance was February 1968, predicted Norman Sklarewitz in *Wall Street Journal*. Of 75 aircraft required for readying of squadron and planning of tactical fighter wing, only two production and six test models had arrived for pilot training at Nellis AFB. At least nine months would be required for pilot training plus several more months for unit training after arrival of sufficient production models at Nellis AFB, he said. (Sklarewitz, *WSJ*, 11/16/67, 14)

- Dr. Donald F. Hornig, Director of OST and science advisor to the President, addressing science writers' seminar in St. Louis, doubted value of scientists' becoming part-time politicians to promote greater public support of scientific activity. He said a scientist who used his skill in explaining complex subjects to laymen was "a much better politician than anyone trying to be an amateur on serious political matters."

  Noting that rate of growth of Federal support of science activities had slowed in recent years, Hornig said $17 billion was still being spent annually on science activities, including $1.8 billion on academic science in universities. (AP, *NYT*, 11/19/67, 2; *Federal Support for Academic Sciences and Other Educational Activities in Univ. and Col.*, July 67)

November 16

- Two scientists, speaking before American Physical Society meeting in New York City, discussed earth's magnetosphere and plasma properties of solar wind. Iowa Univ.'s Director of Physics and Astronomy Depts. Dr. James A. Van Allen believed that particles—protons and electrons—penetrated magnetosphere to populate the outer radiation belt, indicating existence of a peculiar dynamo effect. Los Alamos Scientific Laboratory scientist Dr. A. J. Hundhausen reported Vela bomb-detection satellites observed peculiar solar wind properties—a motion reflecting fluid and electrical characteristics. (Sullivan, *NYT*, 11/17/67, 21)
- West Germany's 1966 budget for space technology and research shown on Finance Minister's summary report revealed that, while $57 million had been available, only $44.3 million had been spent, according to *Interavia Air Letter*. Aerospace industry complained government was so slow to issue contracts that firms themselves had to provide about 50% advance financing. Contracts were not officially approved until the end of each year. (*Interavia*, 11/16/67, 3)
- *Venus IV*'s launch site and launch vehicle had been prepared with assistance from military rocket experts, Deputy Commander Col. Gen. Vladimir Tolubko of Soviet strategic rocket forces reported in U.S.S.R.'s trade-union newspaper *Trud*. Emphasizing role of military specialists in the June 12 launching, Tolubko said they provided guidance during "first 10 minutes of the rocket's flight," and relinquished control to "the command measuring complex." This complex, he wrote, had been controlling, for a decade, "various space apparatuses," presumably both military and scientific vehicles, according to Baltimore *Sun*. "[Complex] receives measuring information and telemetry data from stations in Moscow, Siberia, the Far East as well as from special ships in the Atlantic and Pacific oceans," Tolubko stated. Baltimore *Sun* reported: "Some Western observers believe all of the country's massive lifting rockets fall technically under military control because of their potential as intercontinental missiles. Further command reverts to civilian hands if the rocket is on a scientific mission." (B *Sun*, 11/17/67, A11)
- Soviet journalist Alexander Boiko, in the weekly *Moscow News*, accused Secretary of Defense Robert S. McNamara of "fabrications" about the Soviet orbital bomb system, designated FOBS by DOD. Boiko wrote: "Why is it necessary for McNamara to make up fabrications about the Soviet Union? The U.S.S.R. has always scrupulously adhered to its international obligations [and] we want to feel that others do likewise." Boiko continued: "The U.S.S.R. is most certainly carrying out large-scale space research, but by what right does McNamara connect this with intentions to use outer space for military purposes?" McNamara had referred to U.S.S.R. tests of a system that supposedly could release nuclear bombs from an orbiting vehicle, but had not said experimental vehicles carried even simulated weapons. (Winters, B *Sun*, 11/17/67, 11)

November 16–17: First U.S. high-altitude investigations of aircraft wake turbulence were made by FAA, with assistance from National Aeronautical Establishment (NAE) of Ottawa, Canada, LaRC, and (for radar tracking) NASA Wallops Station. FAA's Convair 880 jet transport cruised at 30,000-ft altitude in restricted airspace southwest of NASA Wallops Station, Va., and trailed colored smoke to outline wake vortices. Two instrumented T-33 jet aircraft, following in wake vortices at distances of 1,000–12,000 ft, recorded wake vortice core diameter, circular velocity, and downward settling during dissipation. Information would be used in evaluating air traffic control standards for separation of enroute aircraft. Although current tests were the first for U.S. in wake turbulence studies at high altitudes, joint LaRC–FAA airport wake turbulence tests had progressed for several years. (FAA Release T 67–24)

November 17: Prototype of West Germany's research satellite Azur had been tested in space simulator at Porz-Wahn, near Cologne. Satellite would explore nature of earth's inner radiation belt, observe polar light zone, and measure spectrum change in solar particles during periods of intense sun-spot activity. Two-phase program would see scientific payloads proved on Nike-Apache and Javelin sounding rockets and satellite launched by Scout booster in NASA-West German cooperative satellite program. (*Interavia*, 11/17/67, 5)

- With 70th edition of *Jane's Fighting Ships*, Sampson Low, Marston & Co., Ltd., had published 1st edition of a new Jane's yearbook, *Jane's Surface Skimmer Systems*, compiled and edited by Roy McLeavy. Aim of new publication was to cover entire field of craft or load carriers that skim across land or sea at interface level. Book was divided into air cushion, hydrofoil, and powerplant sections. (*Interavia*, 11/17/67, 4)

November 18: Soviet references to low-altitude nuclear weapon that would drop on target before completing full orbit of earth appeared to confirm Secretary of Defense Robert S. McNamara's analysis of DOD intelligence reports that U.S.S.R. had a nuclear-warhead missile capable of sudden attack from relatively low altitude. Commemorating U.S.S.R. Rocket and Artillery Forces Day, Strategic Rocket Forces Commander-In-Chief Marshal Nikolai I. Krylov and Col. Gen. Nikolai V. Yegorov described missiles that could deliver "nuclear warheads . . . along ballistic and orbital trajectories." (Anderson, *NYT*, 11/19/67, 15)

November 19: Development of ocean data environmental science services acquisition (ODESSA) system, consisting of unmanned, deep-sea, automated buoys, which would record simultaneously oceanographic and meteorological data, was announced by ESSA. Each buoy would support a group of surface and subsurface electronic sensing packages which would gather and transmit data to central recording station console, either aboard ship or on land. Data could also be recorded at each buoy station on magnetic tape record.

A valuable research tool for oceanographers, the ODESSA system would also have potential use for study of pollution in harbors and estuarine waters and fish habitats. In major role, it would serve ESSA's investigations of the continuous exchange process at interface of air and sea. (ESSA Release 67–80)

- Rear-engine aircraft, built by Sierradyne Corp. and leased by Northrup Corp., was being flight-tested at Dulles International Airport. The 670-lb S–1 experimental aircraft had 87.5-hp engine, 200-mph maximum speed, 20-ft wing span, and 18-ft length. Power plant was placed in fuselage, near aircraft's center of gravity, with nose space left free for equipment. Reduction of cockpit noise, elimination of "propwash," and improved visibility were accomplished by design. (Northrop Corp.; Corrigan, *W Post*, 11/19/67, D8)

- Using $55,000 NASA grant, UCLA had purchased a 24-in telescope with photometer and spectrograph for observatory at Thacher School, Ojai, Calif., *New York Times* reported. Telescope was 15th largest in California and already had produced the most complete spectrographic coverage of stellar explosion of star Nova Delphini. (*NYT*, 11/19/67, 73)

- Development of new 60-mph hydrofoil, designated Burevestnik, had used aircraft turbine engines and water-jet propulsion, Soviet trade-union newspaper *Trud* reported. The Soviet craft had completed its tests to prove that its hardware was much superior to diesel-engine and propeller models previously developed by Soviets. (*NYT*, 11/20/67, 85)

November 20: First Indian-developed rocket, Rohini-75, designed to carry meteorological experiments, was successfully launched from Thumba Equatorial Launching Station (TERLS). (Reuters, *NYT*, 11/21/67, 5)

- NASA Aerobee 150 sounding rocket, launched from WSMR, reached 89-mi (144-km) altitude in American Science & Engineering, Inc., experiment to collect data on celestial x-ray sources for location of sources and flux levels in 1–20 kev range. One strong x-ray source did not appear in data, possibly because of door malfunction. All other data were satisfactory. (NASA Rpt SRL)

- Predicting future trends in aeronautics and space, NASA's Assistant Associate Administrator for Advanced Research and Technology John L. Sloop told American Society for Metals in Rochester, Pa.: "Current events make it clear that many national needs are increasing the pressure on resources and this is being felt over a wide range of activities. . . . It is times like these, however, that give the opportunity for advanced research and technology to open up the range of possible new missions that can be considered in future national decisions [and] we must visualize future mission possibilities and develop the various technologies needed at an optimum pace. Essential to the soundness of this approach is emphasis on technology that has multiple applications."

    Sloop praised "the potential of applications satellites," and said it was "perhaps best gaged by the number of users, domestic and foreign, that are willing to invest their own money. In communications, there are over a dozen proposals for systems that involve an estimated investment of over a billion dollars." (Text)

- NASA named crews for first two manned Apollo/Saturn V flights. Prime crew for AS-504 (first mission), scheduled for 1968: James A. McDivitt, commander; David R. Scott, CM pilot; and Russell L. Schweickart, LM pilot. Backup crew would be Charles Conrad, Jr., commander; Richard F. Gordon, CM pilot; and Alan L. Bean, LM pilot. Prime crew for AS-505 (second mission), scheduled for 1969: Frank Borman, commander; Michael Collins, CM pilot; and William A. Anders, LM pilot. Backup crew would be Neil A. Armstrong, commander; James A. Lovell, CM pilot; and Edwin E. Aldrin, LM pilot. A three-astronaut support team was named for each flight crew: for AS-504—Edgar D. Mitchell, Fred W. Haise, Jr., and Alfred M. Worden; and for AS-505—Thomas F. Mattingly II, Gerald P. Carr, and John S. Bull. (MSC Release; AP, W *Star*, 11/21/67, A2)

- Rep. Joseph E. Karth (D-Minn.), Chairman of House Committee on Science and Astronautics' Manned Space Flight Subcommittee, on floor of House praised NASA's Surveyor program: "It successfully demonstrated the technique of soft-landing [and] . . . with the successfully concluded Lunar Orbiter program, it found and certified manned landing sites. . . ." He said the program had supplied scientists with the first real information on the chemical composition of the lunar surface. In the eastern Sea of Tranquility, for example, *Surveyor V* determined that the surface was made up of the same elements common to the surface of the earth; namely, oxygen, silicon, and aluminum. He concluded:

"Since both Surveyor and its Centaur booster encountered enormous problems in the early stages of development, the eventual success of the program demonstrated the skill, perseverance, and determination of the NASA-industry-university team." (*CR*, 11/20/67, H15619)

- Soviet jet transport, Il-62, carrying delegation of Soviet aviation experts who would negotiate final technical details of proposed Moscow-New York air link with FAA, landed at Dulles International Airport after 11-hr flight from Moscow to Washington.

  As a result of increasing complaints about jet noise and FAA's lack of authority to regulate it, Port Authority of New York had insisted that noise measurement tests be made on Il-62 at Dulles before initiation of direct Moscow-to-New York flights. New York Port Authority had done pioneering work on noise problem.

  Operational safety of Il-62 and alternate landing sites at Washington, D.C., Philadelphia, and Boston had already been approved by FAA for joint use by Pan American Airways and Aeroflot, official Soviet airline. (Hoffman, *W Post*, 11/21/67, A4; *B Sun*, 11/21/67, A8)

- NASA appointed Vincent L. Johnson Deputy Associate Administrator for Space Science and Applications (Engineering). Johnson's former position of Director of Launch Vehicle and Propulsion Programs was filled by Joseph B. Mahon, former Agena Program Manager and Deputy Director of Launch Vehicle and Propulsion Programs. (NASA Release 67-289)

- The U.S. should expect "an early Soviet attempt to send a large payload around the moon and recover it on earth," Rep. George P. Miller (D-Calif.), Chairman, House Committee on Science and Astronautics, said in releasing committee report: "Review of Soviet Space Program," prepared by Dr. Charles S. Sheldon II, Chief of Library of Congress' Science Policy Research Div. This 1957-67 review drew comparisons with the corresponding program in U.S. Rep. Miller commented: "At the same time that space efforts in the United States appear to be slackening, the pace of Soviet space flight has picked up by about 60 percent this year over the corresponding ten months total of last year. The strong technological base the Russians are building through a sustained, long-term commitment to space [assures them] sophisticated and intricate planetary probes . . . unmanned or manned space platforms . . . [application of] aerospace technology systems management to a great variety of industrial and social purposes here on earth."

  Committee report gave "no simple, direct answer" on "whether there is a space race" but suggested "at the present time both the Soviet Union and the United States have put large resources of somewhat similar size into very broadly based programs of great strength."

  Competition in overall space capability, especially on weight lifting and thrust of launch vehicles used by U.S. and U.S.S.R., was illustrated: "In connection with the NASA Administrator's August 1967 pronouncement on the expected large Soviet vehicle [Saturn V class], he made a plea for support of an American nuclear stage to mount on Saturn V in order to leapfrog the Russians. A month later, Leonid Sedov in Moscow stated his belief in the importance of nuclear propulsion for the very large rockets which will soon be needed." (Text; NASA SP-4007, 212; Committee Release, 11/20/67)

- In the last decade, the Nation acquired vast new knowledge, wrote NASA Administrator James E. Webb in *Aerospace Technology*'s Seventh Annual NASA issue, and it entered "a new era, a stirring period of exploration and adventure comparable to the Age of Discovery in the 14th and 15th Centuries." Benefits from our mastery of space, he said, "are opening to us and others new sources of national power and new arenas in which to seek international cooperation as important to our future and the future of the world as sea power and air power have been in the past." He challenged: "Unless we are prepared to move ahead with new programs of [planetary exploration, and work with useful machines and systems in earth orbit] . . . we will be backing out of the Space Age almost before we enter it." (Webb, *Aero Tech*, 11/20/67, 24–5)

- Karl Harr, president of Aerospace Industries Assn., spoke before Metropolitan Baltimore Chamber of Commerce on impact of U.S. space effort and risk that it would be weighed and evaluated in "oversimplified terms." A huge and complex effort should not be judged by superficial aspects alone, forgetting that concentration of "our best industrial, economic, scientific and intellectual resources, organized to an unprecedented degree . . ., has an impact on every aspect of our national life."

Harr noted the strangeness of the unknown medium of space, "for which our ultimate capabilities are yet to be determined." "Today we find ourselves both in midstream in our national space effort and at a crossroads as to future efforts. . . . It is essential that we delay no longer in sorting out that which is valid and enduring, in terms of your total interest and mine, from that which is superficial or transitory." (Text)

November 20–21: Since X–15 No. 3 crash Nov. 15, research missions in X–15 No. 1 and No. 2 had been suspended by FRC until review of flight operating procedures was completed. X–15 No. 1 was awaiting captive flight checkout under wing of B–52 carrier plane. X–15 No. 2 was under repair at North American Rockwell plant following damage to lower stub fin on its tail, but was expected to be flying by February 1968.

Cause of X–15 No. 3 crash was unknown. The pilot, Maj. Michael J. Adams, had difficulty with control booster system; X–15 failed to recover from a dive from extremely high altitude, the pilot losing control above 200,000 ft and regaining it briefly at about 80,000 ft.

Setting unofficial world speed and altitude records, the three X–15s had flown 191 times; No. 3 having had 65 flights. The aircraft were built by North American Rockwell (formerly NAA) under a cooperative program by USAF, USN, and NASA. (NASA Proj Off)

November 21: U.S.S.R. launched *Cosmos CXCI* into orbit with 518-km (322-mi) apogee, 281-km (174-mi) perigee, 92.2-min. period, and 71° inclination. Equipment and instruments functioned satisfactorily. (*Aero Tech*, 12/4/67, 17)

- NASA FY 1968 operating plan was presented by Administrator James E. Webb to House Committee on Science and Astronautics, reflecting accommodation of fund reductions by Congress during recent months— 20% for R&D programs other than Apollo, 50% for construction of facilities, and 7% for administrative operations. He reviewed decisions restructuring NASA effort at least through FY 1969.

The first scheduled Apollo Applications flights of 1970—a "limited number with limited objectives"—would begin with a "precursor training mission for scientists-astronauts with certain orbital experiments, and then proceed with dual launch missions with the Orbital Workshop and the Apollo Telescope Mount." Launch vehicle/spacecraft hardware for follow-on Apollo Applications flights would depend on funding "some time after FY 1969," because of stringent FY 1969 budgetary guidelines.

Future planning for Uprated Saturn—acknowledging necessity for production termination after vehicle No. 216—included FY 1968-funded lead time items "to hold open the option of continued production . . . at the rate of two per year." Webb explained that he held open the option of definite termination of Uprated Saturn production, intending to "preserve competition among the companies involved and between solid, liquid, and, where applicable, nuclear propulsion systems." To preserve competition would cost $3.5 million from FY 1968 funds. He announced NASA–DOD discussions "to consider carefully future national requirements for large launch vehicles, including a possible new booster of the 100,000-pound in earth orbit class."

Webb stressed the need for planetary exploration utilizing both orbiters and landers in a long-term program: "We do not believe that it is the policy of the Nation or the intent of Congress that the United States abandon the field of planetary exploration." Present program still had two Mars-Mariner 1969 flights; continuity depended on NASA's current $143.4-million allocation to lunar and planetary category which would serve as basis for reestablishing future programs in FY 1969 budget.

Development of NERVA I flight-qualified engine could proceed "with the amounts that we and the AEC have available in FY 1968," longer-range planning again dependent on "an affirmative decision" in the Administration's FY 1969 budget for NASA. (Text)

- Operation of *Essa VI* had been turned over to ESSA, NASA announced. (NASA Proj Off)
- Committee on Science, Engineering, and Regional Development had been formed from persons working on problems of science and technology impact for NAS and NAE. Among other projects, Committee would examine "effects on given region of different kinds of institutions—university, not-for-profit laboratories, industrial and government laboratories [and would assess] . . . role of R&D institutions [and] . . . other factors . . . such as risk capital, entrepreneurial skills, and political leadership." The 14-member committee would be chaired by Univ. of Illinois Graduate College Dean Daniel Alpert. (NAS–NRC–NAE *News Report*, 12/67, 8–10)
- DOD announced appointment of Dr. Eberhardt Rechtin, JPL Assistant Director for Tracking and Data Acquisition, to new position of Director, Advanced Research Projects Agency. Rechtin would be responsible for planning, initiating, and directing research and development programs assigned by Director of Defense Research and Engineering Dr. John S. Foster, Jr. (DOD Release 1107–67)
- Compañia Telefónica Nacionale de España (CTNE), Spanish national telephone company, had completed $6.7-million communications satellite ground station at Buitrago (45 mi from Madrid) and would increase its capacity from 36 to 156 channels by 1972. Station had remote-control 82-ft parabolic antenna with klystron amplifier and two 10-kw transmitters to operate in 6,000-megahertz (mhz) band. Station could handle TV relays and could track satellites. (*Interavia*, 11/21/67, 3)

November 22: NASA reported its space engineers at JPL's Space Flight Operations Facility, using Goldstone DSN station, had sent commands to check

Mariner IV's camera and data storage system; spacecraft, launched Nov. 28, 1964, had now traveled about 1.5 billion miles and was circling sun between earth and Mars. *Mariner IV* had been commanded to take TV pictures of black space and record on magnetic tape after long space environment exposure. The spacecraft had responded perfectly. Commands also were sent to switch *Mariner V*'s transmitter from high gain directional to low gain omnidirectional antenna, mission of spacecraft being terminated; *Mariner V*, launched June 14, had completed its Venus mission and was in solar orbit between orbits of Venus and Mercury, about 74 million miles from earth and about 60 million miles from sun. NASA reported that it might be possible to recontact *Mariner V* in Sept. 1968, when spacecraft's antennas would be pointed once again toward earth.

Both Mariners would remain in solar orbit indefinitely; however, because *Mariner IV*'s stabilizing nitrogen gas supply was expected to run out in the next few weeks, spacecraft would drift and solar panels would no longer face the sun. Data reception would end because storage batteries would no longer receive charge. (NASA Release 67-291)

November 23: Cosmos CXCII was launched by U.S.S.R. into circular orbit of 760-km (472-mi) altitude, 99.9-min period, and 74° inclination. Equipment and instruments functioned normally. (*Av Wk*, 12/4/67, 33)

- Detector to measure cosmic dust influx into earth's atmosphere and meteoroid registration would be tested in NASA/German Ministry for Scientific Research (BMWF) cooperative project, part of continuing program first started July 19, 1967. Influx of cosmic dust would be measured by charge emission in detectors (developed by Max Planck Institute of Nuclear Physics, Heidelberg, Germany), which would be launched on NASA-supplied Nike-Apache sounding rockets from ESRO's range in Kiruna, Sweden, May or June 1968. NASA would supply rocket launcher and train Germans in rocket preparation and launching. BMWF would provide two payload packages and ground support equipment. (NASA Release 67-288)

- Program aimed at increasing trained U.S. manpower in oceanography and improving oceanographic applied research and information exchange would be developed by NSF's Office of Sea Grant Programs, according to Director Robert Abel. National Sea Grant College and Program Act, passed Oct. 15, 1966, and funded by FY 1968 $4-million operating budget, would give financial assistance to university development of curricula leading to degrees in ocean engineering and to junior college training programs for technicians in ocean sciences. Program could lead to scientific progress in all fields related to the seas—defense, shipping, agriculture, mining, and weather prediction—and ease "manpower and know-how shortage" slowing U.S. progress in oceanography. (AP, *NYT*, 11/24/67, 30C)

November 24: NASA Nike-Tomahawk sounding rocket was launched from Fort Churchill, Canada, in experiment from Univ. of California at Berkeley to study charged particles in the auroral zone, 1–300 kev electrons, and 1–100 kev protons. Peak altitude was unknown because of loss of telemetry and radar signal. Payload malfunction accounted for failure to obtain complete data. (NASA Rpt SRL)

- Electronic machine was developed by ERC contractor Dr. Huseyin Yilmaz of Arthur D. Little, Inc. Machine visually displayed spoken word profiles on small oscilloscope, with objective of eventually allowing voice commands to maneuver spacecraft. Used to teach deaf and retarded people, machine helped deaf person to correct his speech by "seeing" his voice, and NASA scientists would use this project and similar studies to categorize and encode speech patterns. (ERC Release 67-39)

November 25: Spectrometer aboard NASA's *Oso IV,* launched Oct. 18, had obtained spectral data providing complete picture of corona over whole face of solar disc, Harvard College Observatory Director Professor Leo Goldberg said. Primary purpose of satellite, to obtain high-resolution spectral data, had been achieved. Goldberg said that 4,000 UV photos of the sun, made above earth's atmosphere, would provide three-dimensional information on sun's structure, temperature, and density, and on how they vary with height. Previously, scientists could study solar flares only by observing them at the sun's rim during eclipse. Goldberg said interpretation of UV photos, still being transmited to earth, should also be related to unsolved problems about sun's chemical composition and in turn to theories of sun's origin and evolution. (Weil, *W Post,* 11/26/67, A6; AP, W *Star,* 11/26/67, A17)

- U.S.S.R. successfully launched *Cosmos CXCIII.* Orbital parameters: apogee, 354 km (220 mi); perigee, 203 km (127 mi); period, 89.9 min; inclination, 65.7°. Satellite reentered Dec. 3. (AP, *NYT,* 11/26/67, 46; GSFC *SSR,* 12/15/67)

- National Policy Panel of U.N. Assn. of the U.S.A. published report "Stopping the Spread of Atomic Weapons." Report favored proposed treaty ending spread of nuclear weapons, urged U.S. to open its peaceful nuclear reactors to inspection by IAEA, recommended specific international facilities and arrangements for handling explosions for peaceful purposes, asked for a fairer balance between nuclear and nonnuclear nations, and called for inclusion of U.N. guarantee (based on U.S. and U.S.S.R. power) against "nuclear blackmail." Panel report's "underlying theme" was that some nations might refuse to sign if the two major powers avoided treaty terms designed to meet concern of nonnuclear states that "special units be created within International Atomic Energy Agency to assure that technical, industrial innovations in civilian nuclear energy be made available on a pooled, non-discriminatory basis." (Finney, *NYT,* 11/26/67, 12; Marder, *W Post,* 11/26/67, 1)

- Dr. Jack E. Froehlich, 46-year-old scientist and rocket pioneer, was drowned with his nine-year-old son, Mark, in a boating accident. He had been president of National Engineering Science Co. and, during 1959, had been at JPL as a project manager for *Explorer I. (Pasadena Star-News,* 11/27/67)

November 26: U.S.S.R.'s long-held superiority in large boosters had given it an advantage over U.S. in nuclear warhead protection, according to Pentagon sources. The advantage would become significant both offensively and defensively because of present developments in ABM defenses—developments designed to destroy incoming warheads in outer space with intensive x-ray emissions. Soviet scientists could easily add more shielding to their warheads to halt x-ray bombardments from U.S. antimissile missiles, article noted, but U.S. was handicapped by weight limitations of its lower-thrust boosters. "Some officials," article said, "are concerned that the hardness of Russian warheads will lessen seriously the protective capability of the Pentagon's planned [ABM systems] . . . [and] that U.S. offensive missiles now need more protection . . . to survive Russian anti-missile explosions and reach their targets."

U.S. answer might lie in new lightweight composite materials under development; made by winding and inorganically bonding filaments of such materials as carbon, silica, and graphite, they would strengthen heat shield at little penalty in weight. (B *Sun*, 11/27/67, A1)

November 27: In apparent sharp increase in testing, U.S.S.R. would launch fourth 1967 test series of "carrier rockets" into the Pacific between Nov. 28 and Dec. 30, according to Tass announcement. Aircraft and ships were warned not to enter either of two 129-km (80-mi) -wide target areas northwest and south of Midway Island, a U.S. possession, between noon and midnight local time each day, Tass said. UPI quoted observers on Soviet aims: to test new and bigger Soviet rockets needed to send a man to the moon and bring him back, and to expand recent Soviet experiments on the landing of space capsules at sea. UPI said series would include launchings of 3,000-4,000 mi on down-range course from the Soviet Union, change from maximum of 6,000 mi for past launchings. Past announcements by U.S.S.R. on three 1967 "carrier-rocket" series had termed all three "successful." (UPI, W *Star*, 11/27/67, A5)

- Former Vice President and current leader for Republican Presidential nomination, Richard M. Nixon, on TV said it was "very likely . . . some of those appropriations [for the space program] are excessive." Referring to the space program as "a sacred cow," he called for the President to be "a balance wheel" in evening off the pressure from "vested interests." He defended a congressman's right when representing an area with important space business to fight for more funds. (*SBD*, 11/29/67, 140)

November 27–29: American Astronautical Society's Astronautics International Conference was held in New York City.

Coupling a plea for constructive realism with a denunciation of U.S.S.R.'s mere lip service toward international space cooperation, NASA Assistant Administrator for International Affairs Arnold W. Frutkin discussed ways for understanding "international space cooperation." No Russian attended the conference, two speakers and one cosmonaut declined invitations. Frutkin explained that cooperation meant "joint projects, jointly undertaken, jointly carried out, and jointly profitable."

Realistic projects practicing space cooperation on an international scale were cited by Frutkin to be: NASA's program; ESRO/ELDO complex for combining "ten European nations in spacecraft development and seven nations in large vehicle development;" and INTELSAT satellite telecommunications consortium, including "some sixty nations in day-to-day operational space communications traffic."

He included in U.S. benefits from the NASA cooperative program "gold flow [for] foreign purchase of space hardware and services," Canadian ionospheric investigations, West German barium cloud technique, Italian atmospheric density measurements, and French global wind circulation studies.

Frutkin referred to ESRO's program as "a true pooling of resources [which] goes well beyond the NASA cooperative program. In ESRO, "common facilities have been planned, financed and operated on an international basis by an international governmental agency."

MSC Director Dr. Robert R. Gilruth discussed manned space flight "from Mercury to Apollo." He said Mercury defined man's survivability in space, Gemini defined man's operational capability, and Apollo would be first in utilizing space for manned exploration. "To progress in ten short years from exploring survivability to utilization on the most difficult exploration mission ever undertaken implies a rate of change in technical capability which has no real parallel in history," he said. Deep space and lunar operations would be faced as "two new major dimensions to our space capabilities." Space explorers would be coping with the need for huge amounts of energy, extremely precise navigation to hit target and reentry corridor, and radiation protection that could include the intensities generated by solar flares. Lunar explorers would require a vehicle designed to operate entirely outside earth's atmosphere, descend to and ascend from lunar surface, and propel reliably while exhibiting "relative sophistication of a throttleable engine."

Man returning from first lunar exploration trip would have accomplished "an age-old dream [of visiting] his nearest neighbor in space [and found] a key to the riddle of Earth's relation to its nearest neighbor [and] a chance of understanding the creative forces that formed his own home—Earth." "For the first time he will have soared beyond the confines of Earth's gravity and felt the pull of a foreign planet [and] personally viewed our whole planet as a small bluish ball only four times the span of the lunar disk. These experiences may well produce profound changes in man's attitude towards himself and his world comparable to those wrought by Galileo's dramatic demonstration of the Copernican theory, changes which should tend to inspire man to place the affairs of his very small house in order." (Text)

NASA's Apollo Applications Program (AAP) Director Charles W. Mathews listed basic objectives for AAP: long-duration space flights of men and systems based on unique capabilities of man, habitability, biomedical and behavioral considerations, and systems development; scientific investigations in earth orbit based on solar astronomy, earth observations, and stellar astronomy; applications in earth orbit based on meteorology, earth resources, and communications; and extended lunar exploration. "The activities involved in [AAP] represent major steps in the utilization of our space capability. The results . . . can serve to establish the direction of future space exploration and applications. In particular, increased knowledge on the effective integration of men into the total system should accomplish much in determining the character, systems configurations and operational approach in future programs. The ability to capitalize on the large investments already made in the Apollo Program affords the opportunity to carry on this work in [AAP] in an efficient and economical manner." (Text)

November 28: SST design refinements reflecting 1966 recommendations made by 240-man Government and airline evaluation team were announced by Boeing Co. Vice President H. W. Withington. Prototype refinements were still to be reviewed by FAA and described in detail to the 26 airlines which had reserved SST deliveries. Changes included addition of canard surfaces, incorporation of direct-lift control, lengthening fuselage by 12 ft, and provision of more convenience in passenger boarding and deplaning. Withington cited progress in component and subsystem testing and in wind-tunnel investigations. The first metal (titanium alloy billets) for airliner prototypes was under production. Major contractors were preparing for work on assemblies and Boeing was making ready facilities for construction of the two prototypes. A titanium fabrication facility would be completed as part of the Boeing Development Center and would include a furnace large enough for heat-treating sheets of titanium 70 ft long and 10 ft wide. The full-scale demonstration mockup, constructed in 1966, had recently been dismantled to make room for a dimensionally accurate engineering mockup to be constructed at Boeing Co. (Boeing Release)

- Dr. Jerome B. Wiesner—MIT Provost, longtime member of President's Science Advisory Committee, and in 1961–64 Special Assistant to the President—expressed "The Case Against an Anti-Ballistic-Missile System" in *Look*. He believed arguments were "overwhelmingly against building" an anti-Chinese ABM system. "I do not believe that a really effective antimissile system is remotely possible for either the U.S. or the Russians. And even if the Russians could develop one, and a truly effective defense against our SAC bombers as well, our installing an ABM system would not restore our powers of deterrence. Only improvements in our own offensive-missile force . . . could achieve this." Nor did he believe ABM system would give "either complete or lasting protection against Chinese missiles." He was "convinced . . . we must rely on our known ability to retaliate devastatingly in case of nuclear attack" and must "accept and live with a 'deterrent balance.' " (Wiesner, *Look*, 11/28/67, 25–7)
- NASA announced closing of its Western Support Office, Santa Monica, Calif., reduction in size of its Pasadena Office, and plans to determine manpower requirements for continuing functions in Los Angeles area. NASA centers or NASA Hq. would perform functions eliminated from the two west coast offices. Work at the offices had included contract administration, financial management and disbursement, public affairs, personnel services, procurement, and administration of NASA contract with CIT for JPL's operation. (NASA Release 67–292)
- JPL Director Dr. William H. Pickering appointed William H. Bayley, General Manager of Deep Space Network (DSN), as Assistant Laboratory Director for Tracking and Data Acquisition. In new post, Bayley would direct worldwide DSN which JPL had operated for NASA. Bayley succeeded Dr. Eberhardt Rechtin, who had taken leave of absence from JPL to accept post of Director of DOD's Advanced Research Projects Agency. (JPL Release 461/11–67)
- Gen. James Ferguson, AFSC Commander, presented Distinguished Service Medal for outstanding service to M/G John L. Zoeckler, his Deputy C/S Systems. Ferguson cited Zoeckler's "strategic use of management techniques to eliminate serious scheduling problems inherent in the F–111 program." Zoeckler had managed System Program for F–111 aircraft system at ASD, Wright-Patterson AFB, from Oct. 14, 1963, until Sept. 1, 1967. (AFSC Release 209.67)

November 29: Australia launched her first satellite, *Wresat I* (Weapons Research Establishment Satellite), from Woomera Rocket Range, becoming seventh nation to place a satellite in space. Satellite—107-lb, 5-ft-long cone $2\frac{1}{2}$ ft in dia at its base—went into elliptical orbit with 1,249-km (776-mi) apogee, 170-km (106-mi) perigee, 98.9-min period, and 83.3° inclination. U.S. had provided three-stage Redstone launch vehicle. First attempt, on previous day, had aborted because failure of air conditioner to eject had prevented ignition.

Wresat I, joint project of Australia's Adelaide Univ. Physics Dept. and the Dept. of Weapons Research at Salisbury, was sending back data on interaction of solar radiation, particularly x-rays and ultraviolet, with earth's outer atmosphere. NASA provided satellite acquisition and data reception. (*SBD*, 11/30/67, 149; GSFC *SSR*, 11/30/67; GSFC Historian; *Aero Tech*, 12/4/67, 17; *Interavia*, 12/67, 1842)

- Secretary of Defense Robert S. McNamara would take new post as President of World Bank with approval of President Johnson, McNamara said. He explained that President George Woods of World Bank had offered him position Apr. 18 but, because of deep obligation to serve U.S. President, his decision had been deferred. President Johnson had informed him in October that nomination to succeed Woods would soon have to be made. Agreement that post would go to McNamara had been reached, but he would stay on job "at least long enough into next year to complete the work on the military program and financial budget for fiscal year 1969." (DOD Release 1130-67; *NYT*, 11/30/67, 16)

- Assistant Secretary of Defense for International Security Affairs Paul C. Warnke, in hearings before Senate Committee on Foreign Relations, presented DOD views on ocean space activities by nations. He considered S. Joint Res. 111, introduced by Sen. Norris Cotton (R-N.H.), ranking minority member of Senate Committee on Commerce, and S. Res. 172 and 186, introduced by Sen. Claiborne Pell (D-R.I.), member of Senate Committee on Foreign Relations. He cited primary concern and interest of DOD to be U.S. use of the "ocean environment for purposes of maintaining or enhancing our national security."

    Pointing out deep ocean exploitation involved "unsettled areas" of international law—one area being "the appropriate breadth of the continental shelf"—Warnke cited the 1958 Geneva Convention on the Continental Shelf and noted that congressional policy "in this field is manifested in the Marine Resources and Engineering Development Act of 1966."

    He stated that DOD supported the need for comprehensive studies on ocean exploration and exploitation reflected in Senate Joint Resolution 111. On security grounds, he specifically objected to Section IV of S. Res. 186 on "use of seabed and subsoil of ocean space for peaceful purposes only" until thorough studies could be completed leading to "a sound judgment consistent with security" of U.S. Further objection was voiced by Warnke on a "sea guard" which S. Res. 186 called for; his argument cited "the difficult political and military problems inherent in the organizing of international peacekeeping or enforcement agencies or forces." (Testimony)

- NASA Nike-Apache sounding rocket, launched from Sonmiani, Pakistan, carried payload to peak altitude of 175 km (109 mi) to obtain atmospheric wind and temperature data up to 135 km (84 mi). Grenade and chemical cloud techniques were used by experimenters, Dr. D. Rees, London, England, and Mr. M. Rahmatullah, Karachi, Pakistan. Rocket performed satisfactorily; grenade-explosion timing sequence was late. (NASA Rpt SRL)

- Reduction in total number of positions throughout NASA would approximate 5%, because of budget cuts, NASA announced. NASA would eliminate 1,700 jobs; it now employed about 32,000 persons. MSFC would lose 700 employees and JPL 550. Other reductions included: MSC, 189; GSFC, 96; LaRC, 246; LeRC, 191; ERC, 225; ARC, 79; Wallops Station, 21; Space Nuclear Propulsion Office, 2; and NASA Hq, 142. KSC would be the only NASA field center to add personnel in 1968, a total of 211. (*Aero Tech*, 11/20/67, 3; von Braun, *Marshall Star*, 11/29/67, 1; *W Post*, 11/30/67, A9; UPI, *NYT*, 12/1/67, 5; *Av Wk*, 12/4/67, 31)

November 29

- NASA awarded General Dynamics Corp. a one-year, $21.1-million, cost-plus-award-fee contract to provide management and engineering services for Centaur launch vehicle, ending Sept. 30, 1968. This vehicle had been used successfully to launch six Surveyor spacecraft and would continue in use for lunar and planetary as well as earth orbital missions. LeRC was Centaur project manager. (LeRC Release 67-72)

November 30: Sen. Henry M. Jackson (D-Wash.), Chairman, Senate Committee on Armed Services' Nuclear Safeguards Subcommittee, reported from Senate floor that "Nuclear Test-Ban Treaty safeguards [were] being supported and implemented in a satisfactory manner." He said that recent Soviet and Chinese missile and nuclear developments had formed "serious challenge to the strategic superiority of U.S. power on which our defense planners have counted to maintain political stability and to keep the peace." Sen. Jackson listed two prime requirements to maintain U.S. strategic superiority. First would be strategic offensive capability to penetrate Soviet ABM defenses using new generations of land-based ICBMs and nuclear submarines with better missiles and using MIRVs. Second requirement would be the best ABM defense to protect U.S. retaliatory second-strike force, to safeguard American people, and to account for allies' needs. He cited seriousness of Soviet's doubling ICBMs, developing orbital nuclear bombs, and deploying ABM system around Moscow, especially when combined with growing Communist Chinese nuclear threat. (Text)

- DOD was studying contract performance under its letter contract with Pratt & Whitney for F-111 engines. Orr Kelly, reporting in Washington *Star*, stated, "The Defense Dept. has ordered a high-level investigation of [Pratt & Whitney operations] in an effort to cut [F-111 engine costs], Pentagon sources said today." A firm price had not been set—final contract negotiations would begin in early 1968—but negotiation and fact-finding teams, headed by Asst. Secretary of Navy for Installations and Logistics Graeme C. Bannerman, Chief of Naval Material Adm. I. J. Galantin, and Procurement Control and Clearance Div. Chief Gordon W. Rule would continue to work for final contract containing "target costs, profits and prices to the government." Rising costs were reflected in history of F-111 engine procurement: 1961 cost estimate, $270,000; 1965 Pratt & Whitney ceiling price for British government, $400,000; 1966 Pratt & Whitney letter contract estimates to U.S. Government, $700,000 to $750,000. Article stated that negotiations on price of engine were "not related to the quality;" further, negotiations would not delay aircraft's development, according to a Navy official. DOD sources emphasized price differences could not be taken "as a full measure of the rise in price because there have been significant improvements in the engine, giving it greater thrust, since 1961." After letter contract was signed, Naval Air Systems Command had asked the Performance Technology Corp. to make an independent investigation of Pratt & Whitney operations. At issue were 2,053 TF-30 engines manufactured in four-year period 1967–1970. (Kelly, W *Star*, 11/30/67, 1)

- AFSC announced that inertial navigation systems of several contractors were receiving first-phase ground testing before second-phase flight testing in jet cargo test-bed aircraft, in AMSA program for advanced development of components for next generation of manned bombers. Central Inertial Guidance Test Facility (CIGTF) of AFMDC was testing systems in specially built van that made daily trips on "uneven" public highways.

After ground and flight tests, hardware would be returned to AFMDC for evaluation of "new navigation techniques in an actual flight environment." (AFSC Release 188.67)

- New U.S.S.R. commission would scientifically study reported UFOs, reported *Newport News Press*. DOD's contractor to investigate UFOs, Dr. Edward U. Condon of Univ. of Colorado, would invite head of new Soviet commission, retired Air Force Gen. Porfiri A. Stolyarov, to be his guest. Condon would offer Stolyarov access to all U.S. information, hoping for full exchange of data and close working relationship. U.S.S.R. had in past dismissed UFOs as "American hoax," now did not reject possibility that all UFOs could still be explained away, but warned against outright dismissal. (*Newport News Press*, 11/30/67, 4)

- Dr. Alan T. Waterman, 75, physicist and for 12 years Director of National Science Foundation, died at the National Institutes of Health. During WW II, he had served on the National Defense Research Committee and in the Office of Scientific Research and Development. More recently, he had been a member of NASA's Historical Advisory Committee. President Johnson said: "The American people mourn the passing of a foremost man of science and of human purpose, Dr. Alan T. Waterman. Our Government has lost a trusted counselor. As Chief Scientist of the Office of Naval Research and as first Director of the National Science Foundation, he left an indelible stamp of achievement on one of the most vital areas in American life. He will be missed. But succeeding generations will be wiser for his skill and richer for the foresight that marked his long career." (*PD*, 12/8/67, 1648; *W Post*, 12/2/67, B6)

- Two MSFC scientists had received awards from Alabama Section of American Institute of Aeronautics and Astronautics. Physicist William C. Snoddy was presented Hermann Oberth award for contributions to Explorer, Pegasus, and Apollo/Saturn programs. Deputy Director, Technical, Eberhard F. M. Rees was given General Holger N. Toftoy award for outstanding technical management during 1947–67. (*Space Propulsion*, 11/30/67, 147)

- HYCAT, an organo-iron compound synthesized by United Technology Center as a new solid propellant catalyst, was described as nonexplosive and completely safe compound having low volatility. Dr. David Altman, UTC President, stated that "HYCAT is actually safer than some of the less effective agents now used in operational missiles." New catalyst test firings were reported and research indicated that its performance was completely predictable, storage being possible for years with no impairment. (*Space Propulsion*, 11/30/67, 141)

During November: While cost-cutting had taken place for NASA program for FY 1968, financial pressures on Congress had not prevented $0.4- billion increase for DOD space program—total FY 1968 DOD funding reaching record-breaking $2 billion, according to *News Front*. Giving support to "the new momentum in military space," it said were "major gains in space technology [that] have put a score of defense projects on a much firmer footing." These projects included reliable launch vehicles, communications satellites, refined reconnaissance and early warning sensors, and Secretary-of-Defense-endorsed MOL. "Space has become a demonstrably important place from which to perform militarily useful functions," *News Front* reported, and space law treaty had made no mention of "outlawing the so-called 'spy-in-the-sky' satellites." (*News Front*, 11/67)

- "The Impact of the Space Program," in November *Liturgical Arts*, presented views of Dr. T. W. Adams, Socio-political Specialist with NASA's Office of Policy: "Possibly the most psychologically important fact that could come out of space exploration is the confirmation that there are other planets with intelligent life [and] . . . it is entirely conceivable that within perhaps a century, earth-men will have communicated with intelligent beings in other star systems. For this possibility . . . it is surely not overstating to say the Space Age is bound to have an increasing impact on all elements of today's mobile society." (*CR*, 12/15/67, H17285)
- Gen. James Ferguson, Commander, AFSC, writing in November *Air Force and Space Digest*, characterized AFSC responsibilities: "Our research and development charter and our assigned mission literally cast the Systems Command in the role of architect for the future Air Force [and] the operational commands in the Air Force share substantially in the Systems Command mission of responsiveness. . . . While the current focus of attention is on Southeast Asia we are being careful to assure the continued viability of our strategic missile forces, which has served—and must continue to serve—as a restraint, an ever-present warning, and a prime deterrent." Gen. Ferguson also noted establishment of new Director of Laboratories. (*AF/SD*, 11/67, 90–3)
- Method for obtaining a three-dimensional view of clouds photographed two-dimensionally by weather satellites was reported by AFCRL's John H. Conover and Itek Corp.'s Ronald J. Ondrejka. Photos used were from *Nimbus II* mission.

    The two photos used in this stereo-pair arrangement had to be taken from different points, in this method long used in photo reconnaissance studies. With three-dimensional satellite-photograph presentation, absolute cloud heights could be measured to within a theoretical accuracy of about one km. (*OAR Res Rev*, 11/67, 11–2)

December 1967

December 1: General aviation (nonairline) pilots in 1966 flew 104,706 aircraft a record 3.3 billion mi in 21 million hrs, consuming 512 million gal of gasoline and jet fuel, FAA reported. The 10% increase in number of active general aviation aircraft over the 95,442 in 1965 was largest yearly increase since 1946. Miles flown increased by 30% and hours by 26%. Most active segment of general aviation operations was business flying, which accounted for 33% of total hours and 46% of total miles. (FAA Release 67–84)

- At dedication of Georgia Institute of Technology's Space Science and Technology Center, NASA Administrator James E. Webb reviewed NASA-Georgia Tech relationships: "In the first year of the NASA Traineeship program, Georgia Tech was one of the ten original schools. It is one of five schools . . . that started NASA's new program of Systems Design Traineeships in September. . . . Thirty-eight NASA regular trainees and five design trainees are enrolled at Georgia Tech this fall.

    "The . . . Center we are dedicating is one of 36 such facilities that NASA has helped to build and support on university campuses . . . in a very real sense a new and valuable kind of 'capital asset' that will pay dividends to our nation—and to the states in which they are located—for many years to come. Here . . . for example, in the past few years NASA has invested $6.5 million for research, training, and the facility dedicated today. . . .

    "Even under our reduced budget NASA is sponsoring sustaining research here at a level of some $300,000 annually. . . . In project research, NASA has funded a cumulative total of some $2.1 million at this university. . . . The total picture of cooperation . . . would also include a cooperative student program . . . started with . . . NACA, in 1952. There are currently 31 Georgia Tech cooperative students at [MSC] and six at [LaRC]. More than 125 Georgia Tech alumni are employees of [MSC], and 60 alumni are at [LaRC]. . . ." (Text)
- Frank J. Magliato, Special Assistant to the NASA Administrator, was appointed NASA Executive Secretary, replacing John R. Biggs, who became Special Assistant to the Associate Administrator for Advanced Research and Technology. (NASA Ann, 12/1/67)
- ERC Director James C. Elms announced the appointment of Dr. Richard J. Hayes, Chief of ERC's Space Guidance Laboratory, as ERC Assistant Director. In his new position Dr. Hayes would become one of five Assistant Directors and would manage activities of two of the center's nine laboratories. He would be responsible for research in guidance and control of future aeronautical and space vehicles. (ERC Release 67–40)

December 2: 25th anniversary of first controlled, self-sustaining nuclear reaction, achieved by the late Enrico Fermi at Univ. of Chicago in 1942.

December 2

At two-day meeting at Univ. of Chicago sponsored by AEC and the International Atomic Energy Agency, AEC Chairman Dr. Glenn T. Seaborg said U.S. nuclear plants currently generated 2.8 million kw of electrical power. By 1980, he said, figure would reach 150 million kw and by 2000, 700 million kw.

President Johnson, stressing significance of the anniversary, said: "Throughout history, man has struggled to find enough power—enough energy—to do his work in the world. He domesticated animals, he sold his brother into slavery, and enslaved himself to the machine—all in the desperate search for energy. . . . By learning the secret of the atom, we have given mankind, for the first time in history, all the energy he can possibly use." Urging acceptance of a nonproliferation treaty, he noted that countries with a single reactor, while generating electricity, could produce enough plutonium to make dozens of bombs every year. Even if their purpose were peaceful, he said, "the fact remains that the secret diversion of even a small part of the plutonium they create could soon give every nation power to destroy civilization. . . ." To prove U.S. desire to halt spread of nuclear weapons he announced that "when such safeguards are applied under the treaty, the [U.S.] will permit the International Atomic Energy Agency to apply its safeguards to all nuclear activities in the [U.S.]—excluding only those with direct national significance." (Sullivan, *NYT*, 12/3/67, 1; *PD*, 12/11/67, 1650–1)

- Trendex poll conducted for Thiokol Chemical Corp. in six American cities indicated that interest in the national space program had dropped to lowest in more than four years. Some 41% of persons questioned thought the Government was spending too much on the space program; 83% thought more attention should be given to water pollution, 82% to air pollution, 68% to training unskilled workers, 51% to beautification program, 45% to antipoverty drive, and 26% to space exploration. Competition to achieve first manned lunar landing was favored by 51%, dropping from 77% in September 1965, and opposed by 35%. Most significant change of attitude, Trendex said, was increase in number of respondents who did not know whether they favored the competition—14%, compared to 1% in 1963. (*NYT*, 12/3/67, 28; Thiokol PIO)

- FAA announced establishment of a one-year study (Jan. 1–Dec. 31, 1968) of causes of near midair collisions. To encourage full reports vital to study's success, FAA said that no action would be taken against any person involved in a reported near collision during the period of study. In addition, FAA would withhold the report and the identity of persons from public disclosure at the reportee's request. (FAA Release 67–85)

- Moscow Aviation Institute professor Feliks Zigel had called for a "joint effort of all the scientists of the world," to determine the nature of unidentified flying objects (UFOs), according to Soviet press agency Novosti. Group of which he was a member had had 200 reports of sightings, Zigel said. Typical of these was "a luminous orange-colored crescent flying with its outward bend forward. Its surface is only a little duller than that of the moon. The horns of the crescent throw out jets, sometimes with sparks. The outer contour of the crescent is sharp and the inner contour blurred and wavy." The main task, Zigel asserted, was to organize a systematic study of the phenomena from astronomical and meteorological observatories to determine whether the objects were of protoplasmic or extraterrestrial origin. "Unfortunately, certain scientists both in . . . [U.S.S.R. and U.S.] deny the very existence of the problem instead of helping to solve it. . . . The u.f.o. phenomenon is a challenge to mankind. It is the duty of scientists to take up this challenge, to disclose the nature of the u.f.o. and to establish the scientific truth." (Kamm, *NYT*, 12/10/67, 70)

December 3: U.S.S.R. launched *Cosmos CXCIV* into orbit with 333-km (207-mi) apogee, 205-km (127-mi) perigee, 89.7-min period, and 65.7° inclination. Equipment and instruments functioned satisfactorily, and spacecraft was successfully recovered Dec. 11. (*W Post*, 12/6/67, B2; *Interavia Air Letter*, 12/14/67, 6; GSFC *SSR*, 12/15/67)

- "Evaluation of Apollo 4 mission data . . . continues to confirm initial reports that Spacecraft 017 met all flight objectives without problems," NASA announced. Detailed systems analyses were still in process, but evidence to date indicated spacecraft systems operated properly during Nov. 9 mission and met all specifications.

  First Service Propulsion System (SPS) burn had occurred as scheduled; second SPS burn, however, was 13 sec longer than planned, because of a switchover to ground control after burn was started by onboard guidance and navigation system. Although review of the burn was still being conducted, NASA had already determined that there had been no failure in onboard systems involved. Because of the longer-than-planned SPS burn, spacecraft's reentry velocity was 0.0058% higher than expected; actual velocity was 24,913 mph, compared with nominal 24,772 mph. Increased velocity caused higher Command Module (CM) maximum heat rate—620 BTUs per sq ft per sec instead of planned 586 BTUs. Maximum rate expected on lunar return was 480 BTUs. Maximum g on reentry was 7.3, compared with expected 8.33 g, because of a shallower reentry flight-path angle.

  Cabin pressure remained between 5.6 and 5.8 psi during entire mission, indicating negligible leakage rate. Cabin air temperature remained stable at 60° F during orbit and increased to 70° F during reentry. Structural performance of the spacecraft and Lunar Test Article 10-R (a simulated lunar module) during launch and boost phase was satisfactory. Earth landing system functioned as planned; all parachutes inflated properly, and recovery aids deployed and operated normally. Heat shield performance was good; maximum char thickness was 3/4 in, and charring of crew compartment heat shield was less than expected. Fuel cell and cryogenic subsystems functioned normally, with fuel cells producing potable sterile water and demonstrating excellent load-sharing and thermal-control capability. CM and SM reaction-control systems, electrical power subsystem, and spacecraft sequential devices operated normally, with all functions occurring at scheduled times.

  Guidance and control system and the mission control programmer performed properly. Range-to-go at drogue parachute deployment calculated was 2.2 nm. Comparisons with measured landing point indicated better than predicted performance. All communications objectives were accomplished. Each MSFN station, Apollo tracking ship *Vanguard*, and at least two of the Apollo/Range Instrumentation Aircraft established two-way communications with the spacecraft as planned. (NASA Release 67-294; Hines, *W Star*, 12/3/67, A20; *SBD*, 12/5/67, 172-3)

- As of Nov. 15, U.S. had dropped 1,630,500 bombs on North and South Vietnam since July 1965—twice the tonnage dropped during the Korean War, and three times that dropped in the Pacific Theater during World War II. (Wilson, *W Post*, 1/5/68, 4)

- William Littlewood, a former vice president of American Airlines and developer of the DC-3 aircraft, died at age 69. Recipient of numerous awards—including Wright Brothers Medal, Flight Safety Foundation Medal, and Guggenheim Medal—Littlewood had been a member of NACA, Flight Safety Foundation, Cornell Aeronautical Laboratory board of directors, U.K. Royal Aeronautical Society, and Canadian Aeronautics and Space Institute. (*NYT*, 12/5/67, 45; *W Star*, 12/5/67, B4)

December 4: NASA's total procurement in FY 1967 fell to $4.651 billion—8% below FY 1966—and procurement actions dropped to 283,000—11% less than FY 1966—because of budget reductions, *Aerospace Technology* reported. NASA's "Annual Procurement Report, Fiscal Year 1967" disclosed that 83% of net dollar value went directly to business firms; 4% to education and other nonprofit institutions; 5% to Cal Tech for operating JPL; and 8% to other Government agencies. NASA continued its policy of making greater use of incentive contracts, awarding 81 new incentive contracts and converting 11 cost-plus-fixed-fee contracts to incentive awards during FY 1967. North American Rockwell Corp. led in dollar amount of NASA awards, receiving 25.46% of total. (*Aero Tech*, 12/4/67, 46)

- Thomas O'Toole, writing in the *Washington Post*, reviewed NASA's "growing role in the Vietnam war." Both NASA and DOD were reluctant to disclose the size of NASA's defense effort, he charged, "but it's known that [OART] is spending between $4 million and $5 million a year directing the efforts of 100 scientists and engineers to tasks vital to the Vietnam war." Results of NASA research, he said, included a new acoustic detector able to locate mortars by measuring ground vibrations, a more-steerable parachute, aircraft engines too quiet to be heard during approach, a scissors-like sling for helicopter rescues, and a helicopter which did not make a chopper noise. Although NASA willingly participated in defense research, its authorization to do so was "fuzzy," he said, based only on a section of the 1958 National Aeronautics and Space Act which directed NASA to make available to defense organizations "discoveries that have military value or significance." NASA was "nervous" about its military role, he asserted, because of its unclear authorization, fear of congressional inquiry, and fear of international reaction.

Sen. Strom Thurmond (R-S.C.) later attacked O'Toole's article on the Senate floor: "One might expect that if . . . a story of this sort were to be dug up, the article would express gratification that the taxpayer's money was being used on behalf of the American people . . . [and not purport to be] shocked that NASA projects help our national objective. . . . I am firmly of the conviction that anything that can be developed technically to save a single American life in Vietnam, especially by any individual or organization that is being paid by American taxpayers, belongs freely to our soldiers. . . .

"The article is obviously a patently calculated leak. The informed details it provides on NASA's research and technical work proves this. . . . This is a shocking thing, that is not a matter of press freedom, or even of press license, but plain aid and comfort to the enemy. . . . Publishers and editors should accept their responsibility to prevent such . . . news coverage and to prevent, too, the violation of security classifications on subjects that involve adversely the preservation of the lives of American fighting men." (O'Toole, *W Post*, 12/4/67, A1; NASA LAR VI/128)

December 4

- French experimental aerotrain Le Zinc reached 215 mph in test on 4.2-mi track at Gometz-la-Ville, France, claiming new world record for track vehicles. Previous 195 mph record had been set by a French Aerotrai Co. locomotive. Le Zinc, a half-size experimental model of an aerotrain designed by Jean Berlain, rode on a 1/10-in.-thick air cushion over a concrete track. Full-scale model, driven by turbo-propeller, would carry 80 passengers up to 250 mph. (Hess, *NYT*, 12/5/67, 1)
- *Izvestia* announced completion of one of the world's largest atomic research reactors, designed to produce and test materials that could provide heat and power for long-duration space flights. Material Investigations Reactor (MIR), located in Melekess reactor development center in Volga River Valley, reportedly could generate 100,000 kw of heat at full power from fuel elements containing 90% uranium 235. Designed to produce an intense flux of neutrons to irradiate test materials rapidly, MIR could attain maximum neutron flow of 500,000 billion neutrons per sq cm per sec. Highest flux attained to date was 600,000 billion, achieved by AEC's Savannah River reactor in 1965–1966 demonstration. (Shabad, *NYT*, 12/10/67, 72)
- Michael Getler, in *Aerospace Technology* editorial, cited U.K.'s lack of "willingness to invest at home in the pursuit of new technology needed to compete and to grow," as a major reason for devaluation of the British pound [see Nov. 18]. Comparing French and U.K. economies Getler noted that France had invested heavily in her own industry to support a large-scale military buildup and an expensive national space program. "In contrast to French willingness to pay for the cost of [R&D], the British have tended to treat their big companies solely as manufacturers. . . .

    "While the British have purchased [U.S.] Polaris missiles . . . the French have built their own. The British . . . now are buying U.S.-built F–4s and F–111s, while the French continue to design and build newer versions of their own Mirage family, and are now exporting them. . . .

    "The British national space program is virtually non-existent despite a growing realization . . . [by] Parliament that investment in a properly scaled effort can be of great importance in developing [British] technology. . . . There is no suggestion that the British take on a multibillion dollar effort, but there certainly are grounds to support investment in a $100 million program . . . that would be directed at developing satellites with some obvious commercial applications."

    Although U.K.'s purchases of U.S. goods were "good business for American industry," Getler noted, "the long range risk to this country is that an old and comparably faithful ally will be weakened in the process. Eventually, the British will tire of buying U.S. weapons . . . [but will have] no R&D base . . . to fall back. . . ." (Getler, *Aero Tech*, 12/4/67, 58)

December 5: USAF launched *OV III–6* research satellite from Vandenberg AFB by Scout booster into orbit with 275-mi (439-km) apogee, 254-mi (408-km) perigee, 93-min period, and 90.6° inclination. The 219-lb satellite, designed for 256-day lifetime, carried two main experiments: mass spectrometers and ion density gauges to measure density, temperature, and composition of the upper atmosphere and its variation with latitude. (*Pres Rep 1967*; GSFC *SSR*, 12/15/67; *SBD*, 12/5/67, 176)

- USAF launched an unidentified satellite from Vandenberg AFB toward polar orbit using Titan III-Agena D booster. (UPI, *W Post*, 12/6/67, A24)

- NASA Aerobee 150A sounding rocket launched from NASA Wallops Station carried 300-lb payload containing two white rats to 85-mi (137-km) altitude in first of four experiments to study behavior in an artificial gravity field and determine minimum level of gravity needed by biological organisms during space flight. Two arms of payload were extended after rocket burnout, producing a centrifuge with artificial gravity levels between 0.35 and 1.65 g. During five minutes of free fall, rats selected their own gravity levels by walking along tunnel runway. Data on their movement and position were telemetered to ground stations. Payload impacted 70 mi downrange in the Atlantic; no recovery was attempted. (WS Release 67–36; NASA Release 67–290; NASA Rpt SRL)
- Aerobee 150 sounding rocket launched by NASA from WSMR carried 14-in telescope with new STRAP III stellar pointing control to 96.8-mi (155-km) altitude in Johns Hopkins Univ. experiment to measure vacuum spectral emission lines from atmospheres of Venus, Jupiter, and Regulus. STRAP III (used for first time) rocket and other instruments performed satisfactorily. (NASA Rpt SRL)
- NASA Nike-Tomahawk sounding rocket was launched from Churchill Research Range, Canada, in experiment on vertical and horizontal variation of auroral light emissions. Peak altitude was not obtained because of heavy fog effect on radar tracking. Rocket and experiment performance was satisfactory. (NASA Rpt SRL)
- ELDO's Europa I rocket was launched from Woomera Rocket Range but was automatically destroyed in mid-air after French Coralie 2nd stage failed to ignite. U.K.'s Blue Streak 1st stage performed satisfactorily. Purpose of suborbital flight test was to evaluate performance of the Coralie and separation of the three stages. (AP, *NYT*, 12/6/67, 16; *SBD*, 12/7/67, 189)
- MSFC awarded IBM Corp. a $1,292,218 supplemental contract for configuration management of 27 instrument units for Saturn V and Uprated Saturn I. Agreement, effective through June 1970, brought total value of contract to $200,644,441. (MSFC Release 67–236)
- Gerald J. Mossinghoff, Director of the Office of Legislative Planning at U.S. Patent Office, was named director of Congressional Liaison Div. in NASA's Office of Legislative Affairs. (NASA Release 67–297)

December 5–6: New York Academy of Sciences held ceremonies marking its 150th anniversary.

Dr. Hilliard V. Paige, General Electric Co. Vice President and general manager of Missile and Space Div., compared role of the space program in stimulating technological development to that of World Wars I and II. World War I, he said, marked real beginning of American aviation activities. World War II led to current U.S. leadership in production of jet aircraft. Space program was current counterpart of this stimulus, he noted, but program was declining because of crippling budget and program cuts.

Former Astronaut M. Scott Carpenter (Cdr., USN) described his experiences living in Sealab II [see Aug. 28, 1965] and plans for a larger Sealab III in 1968. In addition to Sealab experiment, series of special vehicles would be tested, he said: Deep Submergence Rescue Vehicle (DSRV-1) to rescue men from disabled submarines resting on ocean floor; Deep Submergence Search Vehicles (DSSV) to conduct research on ocean floor, collecting and surfacing small objects; Nuclear Powered Deep Submergence Research and Ocean Engineering Vehicle (NR-1) which would run along ocean floor on wheels; and Large Object Salvage System (LOSS) to develop a variety of tools, from surface craft with strong lifting ability to equipment enabling swimmers to operate effectively in deep water. (Sullivan, *NYT*, 12/6/67, 35; 12/24/67, E3; NY Academy of Sciences PIO)

December 6: NASA Aerobee 150 sounding rocket launched by NASA from WSMR carried GSFC payload to 131-mi (210-km) altitude in experiment to measure spectral irradiance of Gamma Velora and Zeta Puppis with a UV stellar spectograph, star tracker, and modified attitude-control system (STRAP). Rocket and instruments performed satisfactorily. (NASA Rpt SRL)

- News briefing on results of Lunar Module (LM) flammability tests, conducted as a result of Jan. 27 Apollo fire, was held at MSC. Using a full-scale boilerplate mockup of LM cabin interior fabricated specifically for the test program by Grumman Aircraft Engineering Corp., fires were deliberately set in 41 tests to investigate the flammability and propagation properties of all interior LM materials, contractor fire-hazard fixes, crew equipment and stowage areas, and apparent propagation paths. Although results indicated that most LM areas were adequately protected against propagation of an accidental fire, four materials which under specific conditions might act as flame propagation media were ordered replaced. In addition, beta cloth containers for food and flight data would be modified to enclose the stored items completely; flight data file and crew procedure notebooks would be redesigned to minimize their flammability; and procedures would be developed to control location and quantity of loose items and to combat onboard fire. Similar tests would be conducted in the Command Module (CM). (Transcript; AP, B *Sun*, 12/7/67, A3)

- British Defence Ministry denied report by London *Evening Standard* that U.K. would cancel its agreement with U.S. to purchase 50 F-111 variable sweep-wing aircraft because devaluation of the pound would increase cost of the aircraft. Devaluation of the pound from $2.80 to $2.40 Nov. 18 added one sixth of original cost to total, but arrangements were reportedly being made to adjust the contract, possibly by increasing U.S. purchases of U.K. military equipment to help offset aircraft cost. (Flowers, B *Sun*, 12/7/67, A11)

- U.K. was considering halting its activities at Woomera Rocket Range because of large expense due to distance between England and the Australian range and because only a few of its missile projects required such an extensive facility, *Space Business Daily* reported. (*SBD*, 12/6/67, 181)

- Dr. Herbert Friedman, Naval Research Laboratory Superintendent of Atmosphere and Astrophysics Div., received 1967 Rockefeller Public Service Award ($10,000) from Princeton Univ. President Robert F. Goheen. Award was given in the field of science, technology, and engineering. Scientist had specialized in astrophysics from first experiments in rocket astronomy (revealing strength and pattern of UV and x-rays 50 mi above atmosphere) down to present service, which also included post of Chief Scientist of Hulburt Center for Space Research. (Lobsenz Public Relations Co. Release, 11/13/67; *Science*, 12/29/67, 1655)

- Jerome F. Lederer, NASA Manned Space Flight Safety Director, received

Flight Safety Foundation's Distinguished Service Award for distinguished service in achieving aircraft uses under safer conditions. (NASA Hq WB, 50)

December 7: Javelin sounding rocket launched by NASA from Churchill Research Range reached 500-mi (805-km) altitude in GSFC-Southwest Center for Advanced Studies experiment to study "ionosphere-protonosphere transition region at L values beyond the plasma pause." Rocket and instrumentation, including a magnetic mass spectrometer, two Langmuir probes, a magnetometer, and a Lunar Aspect Sensor, performed satisfactorily. (NASA Rpt SRL)

- NASA reported that RAM C-1 spacecraft had met all mission objectives and officially classified the mission a success. Spacecraft was launched Oct. 19 to assess the effectiveness of water addition technique in alleviating spacecraft reentry blackout. Effects of water injection in reducing RF attenuation and blackout were observed during 50% of total VHF blackout period (approximately two thirds of total water-addition time). During last one-third of water injection period, no signal recovery was observed. NASA concluded that water addition was effective in reducing attenuation of the X-band signal down to 130,000-ft altitude and that reentry communications problem could be significantly reduced by utilization of higher transmission frequencies. (NASA Proj Off)

- NASA established Aerospace Safety Advisory Panel to review safety studies and operations plans, report on them to the NASA Administrator and advise him of any hazard in facilities and proposed operations, and perform other duties. Panel of chairman, vice chairman, and seven other members would review, evaluate, and advise on all elements of NASA's safety program. (NMI-1156.14)

- GSFC scientist Dr. Norman F. Ness reported new findings on moon's nature and environment. Data obtained from *Explorer XXXV* in lunar orbit indicated that: (1) no supersonic shock front preceded moon to impede flow of solar wind toward the lunar surface; (2) a sizable empty cavity or solar wind void existed behind the moon away from the sun; (3) moon had practically no magnetic field, and therefore no complex magnetosphere, surrounding radiation belts, or lunar ionosphere; and (4) moon's average electrical conductivity was low (under 1,800° F). Findings did not appear to create any additional hazards to Apollo manned lunar landing program. *Explorer XXXV*, launched into lunar orbit from ETR July 22, to study interaction of solar wind with the moon and investigate the lunar environment, was first satellite to orbit the moon without having a midcourse maneuver capability and the only operating satellite currently in lunar orbit. (NASA Release 67-295)

- Four students from UCLA would spend 60 days in simulated space cabin at McDonnell Douglas Corp.'s Santa Monica, Calif., plant in early 1968 under $200,000 NASA contract for research in biotechnology and human factors. Test, which would take place in a closed system using recirculated water and oxygen, would be longest ever attempted with crew subsisting in long-duration flight conditions. Crew members, currently undergoing training for test, would have only radio contact with outside but would be under close observation by medical personnel at all times. (NASA Release 67-296)

- NASA contract activity:

    Negotiations were initiated with McDonnell Douglas Corp. on a $21-million contract for Improved Delta launch support services. Contract would cover 20 launchings from WTR and ETR during 21-mo period beginning Jan. 1, 1968, and would provide for inspection, checkout, and actual launch.

Negotiations were concluded with North American Rockwell Corp. on a $812-million, cost-plus-fixed-fee contract for continuation of Apollo Command and Service Module Program from Dec. 4, 1966, through program completion. Contract covered engineering, design, manufacture, analysis, and testing for Apollo spacecraft and related equipment, including production of four additional spacecraft through S/C 115A; contained improved plans for quality, reliability, assurance, and safety; and provided for an award fee based on achievement of specified management objectives. (NASA Releases 67-298, 67-299)

- R/A Rawson Bennett II (USN, Ret.), former chief of naval research (1955-1961) and head of Project Vanguard, died at his home in Arlington, Va. (W Star, 12/9/67, A23)

December 8: Maj. Robert H. Lawrence, Jr. (USAF), first Negro selected for a mission in the Nation's space program, was killed during routine training flight when his F-104 aircraft crashed on landing at Edwards AFB. Copilot Maj. Harvey J. Royer (USAF), chief of operations at USAF's Aerospace Research Pilot School, was injured. USAF appointed a board of officers to investigate the crash. Selected June 30, to train for DOD's Manned Orbiting Laboratory (MOL) program, Major Lawrence was ninth astronaut—first in DOD astronaut program—to be killed in an accident and fifth to die in an air crash; three others had died in Jan. 27 Apollo fire and one in an automobile accident. Memorial services were conducted at Edwards AFB and in Chicago. (*W Post*, 12/9/67, A1; AP, *NYT*, 12/10/67, 44; B *Sun*, 12/11/67, A15)

- Apollo Command Module (CM) No. 009, launched on a ballistic mission Feb. 25, 1966, and successfully recovered, would be modified by North American Rockwell Corp. and reused in series of dry-land impact tests at MSC. It was first of four previously used CMs to undergo the tests. (NAR Release NL-15)

- Robert F. Thompson, Assistant Manager of MSC's Apollo Applications Program Office since its establishment in July 1966, was appointed Manager of that program. Post had been vacant since April 1967 when MSC Deputy Director George M. Low, who had been Acting Manager, became Manager of the Apollo Spacecraft Program Office. (MSC Release 67-73; MSC *Roundup*, 12/22/67, 1)

Kurt Jung, member of West Germany's Bundestag Defense Committee, said West Germany had allotted $25 million to purchase 220 F-4 Phantom jet aircraft from McDonnell Douglas Co., to replace ill-fated Lockheed F-104G Starfighter aircraft that had been worn out in use or lost in air crashes. (AP, B *Sun*, 12/9/67, 1)

December 9: USAF launched an unidentified satellite from Vandenberg AFB using a Thor-Agena D booster. (*Aero Tech*, 12/18/67, 10)

- International Atomic Energy Agency (IAEA) would conduct first worldwide computerized exchange of nuclear data under $100,000 pilot program, Kathleen Teltsch reported in the *New York Times*. U.S., U.S.S.R., and EURATOM would supply indexes of new atomic literature, patents, university theses, and conference papers which IAEA would program for a computer. System was expected to store more than 100,000 entries, each including title, author's name, key words in English, and an abstract. (Teltsch, *NYT*, 12/11/67, 27)

December 10: Aerobee 150 sounding rocket was launched by NASA from WSMR, carrying ARC payload to 99.2-mi (159.5-km) altitude to verify performance of SPARCS (Solar Pointing Aerobee Rocket Control System) and map the flight path magnetic field. Rocket and instrumentation performed satisfactorily. (NASA Rpt SRL)

- AEC successfully detonated a nuclear explosion 4,240 ft underground in Farmington, N. Mex., creating a 160-ft-dia cavity. No radioactivity was released. The $4.7-million, 26-kiloton Project Gasbuggy explosion, first commercial nuclear explosion in Plowshare program for peaceful uses of atomic energy, was an attempt to unlock natural gas deposits trapped in rock formation too dense to be tapped by normal drilling methods. Analyses would be conducted for several months to analyze gas flow, determine reasons for any increases in potential production, and check for presence of radioactivity in gas. Test was sponsored jointly by AEC, El Paso Natural Gas Co., and Bureau of Mines. (AEC Release K-281; *WSJ*, 12/11/67, A1, A2; Smith, *NYT*, 12/11/67, 1)

- JPL scientists Dr. Dimiter I. Tchernev and Dr. George W. Lewicki had developed recording method which used pulsed laser beam to record magnetic spots (bits) on thin magnetic film, retaining several hundred times more data per square inch of film than was generally stored on present magnetic film or tape. Success of new process might permit storage of data for 500 photos on interplanetary spacecraft on a one-inch-square strip of film; retention of 10 million bits of scientific data per square inch of film by onboard computers; and substantial reduction in computer size. Project was being funded by NASA at rate of $175,000 a year. (JPL Release 463)

- Boeing Co. President William M. Allen received Los Angeles Chamber of Commerce's 1967 Kitty Hawk Memorial Award for distinguished civilian achievement in aviation. Award for distinguished achievement in military aviation was presented to Air Force Academy Commandant Col. Robin Olds. (*Av Wk*, 12/25/67, 68; Boeing PIO)

December 11: NASA Deputy Administrator Dr. Robert C. Seamans, Jr., testifying before the House Committee on Merchant Marine and Fisheries' Subcommittee on Oceanography, summarized NASA's activities in national marine science program. NASA's activities, primarily exploratory, were aimed at ascertaining what potential benefits could be gained in oceanography through the application of space techniques and equipment, he said. Major areas included: (1) investigation of the applicability of existing space technology to problems of oceanography; (2) studies of remote sensing of oceanographic phenomena, including analysis of data obtainable from space missions not originally conceived for oceanographic purposes such as Gemini, Tiros, and Nimbus photos; and (3) development and test of remote sensors for viewing and discrimination of oceanographic phenomena.

Spacecraft, Dr. Seamans said, "will not supplant surface, subsurface, and airborne research and monitoring systems; rather they will supplement them. The absolute accuracy, three-dimensional capability, and selective resolution of these conventional systems will continue to be necessary and must be utilized as a part of the overall system to their best advantage. However, Earth-orbiting spacecraft have unique advantages for obtaining certain types of marine data as follows: repetitive worldwide coverage, greater frequency of observation (once or twice daily for polar orbiting satellites and continuous for geostationary satellites),

ability to observe remote areas at will, and an 'all-weather' capability for certain parameters." (Testimony; *SBD,* 12/20/67, 257)
- Model No. 001 of delta-wing Anglo-French Concorde supersonic aircraft was publicly displayed in Toulouse, France. The 132-passenger, 1,450-mph, aluminum aircraft was scheduled to make its first flight Feb. 28 and enter commercial service in 1971—three years earlier than 277-passenger, 1,800-mph U.S. SST. U.S. airlines accounted for more than half of the 74 orders already placed. (Garrison, *NYT,* 12/12/67, 1; B *Sun,* 12/12/67, 1; *NYT,* 12/24/67, D33)
- Boeing Co. revealed details of design proposals for Airborne Warning and Control System (AWACS) being evaluated by USAF. Proposed designs, result of a one-year study conducted by Boeing, used a Boeing 707 jet aircraft to house extensive radar and communications equipment for dual-purpose capability in air defense and as a tactical flying command post. Main radar antenna was housed in a 30-ft-dia rotodome on top of aircraft's vertical stabilizer to assure excellent radar performance and minimize reduction of aircraft performance. (Boeing Release S-9051)
- "Our national security depends upon progress in science and technology as much as or more than any other aspect of our national strength," Dr. Alexander H. Flax, Assistant Secretary of the Air Force for Research and Development, said at Univ. of Tennessee Space Institute in Tullahoma. "The Federal budget for research and development test and evaluation [RT&DE] has experienced a steady and phenomenal growth over the past twenty-five years . . . from 7/10s percent in 1941 to almost 3% in 1965. Most recently, we have been through a period of highly accelerated growth in which the initiation of a massive space program added to a rising curve of cost for development, and acquisition of ballistic missiles resulted in RDT&E growth rates . . . of 20% per year. . . . This trend, if it were to continue from the level of '66 for ten years, would result in a Federal RDT&E budget almost as large as the total budget of today by that time." (NASA LAR VI/129)
- *Philadelphia Inquirer* editorial on Dec. 8 death of Maj. Robert H. Lawrence, Jr., (USAF): "[His death] deprives the Nation of another highly trained scientist who had been selected for future service in space. . . .

 "With the toll of astronauts now at nine victims, it is significant that not one has been lost in space itself. . . . Our astronauts are not only men who have extensive educational qualifications but who have equally extensive experience as test pilots and similar extremely hazardous activities. . . .

 "They may put their lives on the line willingly, but the Nation should strive to safeguard them as much in their non-space activities as . . . in space." (*P Inq,* 12/11/67, 10)

December 11–14: Some 100 representatives of Government and industry attended Orbital Workshop design meeting at MSFC to discuss structures, mechanical systems, propulsion, instrumentation, communications, crew station, and electrical systems. A five-day Orbital Workshop mockup review would be held in late January 1968 with a McDonnell Douglas Corp. mockup containing recent design concepts. (NASA Release 67–239)

December 12: President Johnson, on tour of Michoud Assembly Facility, reaffirmed support for U.S. space leadership. "Ten years ago, we could put scarcely 100 pounds into orbit about the earth. Today we can orbit 285,000 pounds. That is progress. That is something we are proud of . . . something we are doing together.

December 12: "We will never evacuate the frontiers of space to any other nation," President Johnson pledges during tour of Michoud Assembly Facility. On the platform in front row from left are Astronaut Walter Cunningham, Astronaut Walter M. Schirra, Jr., the President, NASA Administrator James E. Webb, Mrs. Webb, Gov. John J. McKeithen, and NASA Associate Administrator for Administration Harold B. Finger. In the back row, behind and to the right of Webb, are Gen. Edmund F. O'Connor, Director of Industrial Operations, MSFC; Dr. George N. Constan, manager of Michoud; H. D. Lowrey, president of Chrysler Corp. Space Division; and J. K. Swearingen, project manager for Ling-Temco-Vought Range Systems Division.

"In the 9 years since I first introduced in the . . . Senate the Aeronautics and Space Act of 1958, we have seen the power of our rocket engines increase 50 times—from 150,000 to 7,500,000 pounds of thrust in the engines that you build here.

"Not long ago we had to stand by and watch other countries accomplish what we could not accomplish. I will never forget the days of Sputnik 1 and Sputnik 2, and the real concern. . . .

"We were the most scientifically advanced nation on the face of the earth but we did not launch man's first earth satellite.

"We were backward because we did not choose to adventure. We did not choose to have vision. We did not choose to look forward. Now let us remember that our future achievements—or our future failures—will depend on how far ahead we choose to look and how far ahead we choose to think.

"If we think second, and if we look third, then we are going to wind up not being first.

"... we will advance in space to the extent that our people and their representatives are prepared for us to advance and are prepared to pay the cost of that advance. We may not always proceed at the pace we desire. I regret . . . that there have been reductions and there will be more. There have been interruptions, and I hope that we have had all we can take. But I do have faith and confidence in the American people.

"We will not surrender our station. We will not abandon our dream. We will never evacuate the frontiers of space to any other nation." (*SBD*, 12/13/67, 218; *PD*, 12/18/67, 1696–7)

- Two sounding rockets were launched from NASA Wallops Station. Nike-Apache carried Univ. of Colorado experiment to 60-mi (96.4-km) altitude to obtain a vertical profile of the density of nitric oxide in region at 47–75-mi (75–120-km) altitude with a scanning UV monochromator. Payload was first to utilize Tomahawk clamshell nose cone. Instruments performed satisfactorily, but late ignition of rocket's 2nd stage caused 12-mi (20-km) loss in apogee. Nike-Cajun carrying GSFC-instrumented payload used grenade-explosion technique to measure temperature, pressure, density, and wind in region 22–59-mi (35–95-km) altitude in coordination with nitric oxide experiment. Rocket performed satisfactorily, reaching 68-mi (109.6-km) altitude. The near-simultaneous launches were intended to permit analysis of theory relating nitric oxide concentration to temperature and absorption of radio energy. (NASA Rpt SRL)
- USAF awarded Martin Marietta Corp. a $4,392,994 increment to a previously awarded contract for design, development, fabrication, and delivery of Titan III boosters. (DOD Release 1169–67)

December 13: NASA's *Pioneer VIII* (Pioneer C), third in series of five spacecraft designed to provide continuing measurements over the solar cycle at widely separated points in interplanetary space, was successfully launched from ETR by Thrust-Augmented Thor-Delta booster into orbit around the sun. Orbital parameters: aphelion, 1.0080 astronomical units (au) or 101.2 million mi (162.8 million km); perihelion, 0.9892 au or 91.5 million mi (147.2 million km); period, 386.6 days; and inclination, 0.057°. Primary mission objective of the 145-lb, drum-shaped satellite was to collect scientific data on interplanetary phenomena, including magnetic field, plasma, and cosmic ray measurements for two or more passages of solar activity. As secondary mission *Pioneer VIII* would: (1) investigate continuously characteristics of the magnetosheath, investigate the geomagnetospheric tail, and acquire data when a highly significant solar event occurred; (2) refine primary determinations of earth and moon masses, the astronomical unit, and osculating elements of earth's orbit; (3) provide synoptic study of solar-interplanetary relations; and (4) provide target for checkout and training of MSFN equipment and operations personnel by launching a Test and Training Satellite (TTS) as secondary payload. The 40-lb *Tts I*, carried pickaback on 2nd stage, was successfully ejected after 3rd-stage burnout, and entered orbit with 303-mi (488-km) apogee, 187-mi (301-km) perigee, 92-min period, and 33° inclination. *Pioneer VIII* separation, boom deployment, and initial solar orientation occurred as planned; all seven experiments—including an interplanetary dust detector and a radio propagation experiment—functioned satisfactorily; and on Dec. 15 spacecraft successfully completed maneuver which orientated the spin axis perpendicular to ecliptic plane.

Pioneer VI (launched Dec. 16, 1965) and *Pioneer VII* (launched Aug. 17, 1966) were currently 144 million mi and 68 million mi, respectively, from earth and were still returning excellent data. Pioneer program was managed by ARC under OSSA direction. (NASA Releases 67-293, 67-310; NASA Proj Off; *W Post*, 12/14/67, A17; AP, B *Sun*, 12/14/67, 1)

- Dr. John S. Foster, Jr., Director of Defense Research and Engineering, discussed importance of defense R&D and problem of choosing among new technological developments available. Speaking before the Dallas Chapter of the Assn. of the U.S. Army and the Dallas Council on World Affairs, he noted that over 80% of all the scientists and engineers who had ever lived were alive today; scientific and technical manpower had increased 400%, compared to 50% for labor force; and $\frac{1}{3}$ of all U.S. R&D supported national defense. Current defense R&D funding was $8 billion.

  Discussing selection of proposed projects, he described a new tool and procedure called the Development Concept Paper, "a short, single document . . . [which] requires that . . . parties agree to a common document that states the issues, the assumptions, the alternative courses of action available, the pros and cons for each." With this type of presentation, he said, the Secretary of Defense could more easily and more objectively decide whether to approve or reject a project, and if approved, the paper could be transmitted to project's implementers as a guide. This new tool, he indicated, would be used in future to avoid repetition of past inappropriate decisions in which precious resources were wasted, such as air-launched Skybolt missile and B-70 aircraft which proved too vulnerable after $2 billion had been spent on their development; and C5-A subsonic transport aircraft, currently being produced for operation in 1969, whose production had been delayed since late 1950's pending development of a system to utilize it. From such cases, Dr. Foster noted, "we have repeatedly learned that money is not enough. That enthusiasm is not enough. That technical innovation is not enough. We need critical analysis of purposes, missions, threats, and requirements. We also need realistic tested advanced alternatives produced by our investment in research. And to gain wise management decisions, we must join these together in constructive confrontation." (Text)

- U.S.S.R. and Communist-bloc countries—including Bulgaria, Hungary, East Germany, Mongolia, Poland, Romania, and Czechoslovakia—had agreed to conduct "a program of joint launchings of sputniks and rockets," Evert Clark reported in the *New York Times*. Yugoslavia had declined admission to the group, choosing instead to join INTELSAT. U.S.S.R. reportedly had constructed or would construct tracking stations in Cuba, U.A.R., Mali, and "some other countries of Asia and Africa." Since U.S.S.R.'s land mass stretched over $\frac{2}{3}$ of earth's circumference, U.S.S.R. had less need for tracking stations in other countries than U.S., and assigned little or no responsibility to nation where station was located. NASA, however, had bilateral agreements with more than 70 nations covering joint activities from student exchanges to satellite launches. (Clark, *NYT*, 12/14/67, 26C)

December 14: NASA *Surveyor V* spacecraft on the moon responded to turn-on commands from Goldstone, Calif., station at beginning of its fourth lunar day. Spacecraft had operated during its second lunar day but had failed to respond during third. Battery was still functioning, and camera and alpha-scattering instrument were responding to commands. (NASA Proj Off; *W News,* 12/15/67, 5)

- NASA's *Oso IV,* launched from ETR Oct. 18, had achieved its primary and secondary objectives and was continuing to operate satisfactorily. Although anomalies occurred in some of the nine experiments, and three were turned off, all of the pointed experiments had operated as planned for more than one solar rotation, and the wheel experiments had obtained useful data. (NASA Proj Off)

- *Surveyor VI* briefing was held at NASA Hq.

    Dr. Leonard D. Jaffe, JPL Surveyor Project Scientist, said *Surveyor VI* had discovered or confirmed that: (1) lunar maria were covered primarily with very fine, 0.001-in-dia particles but also with "some coarse particles and rocks up to a yard across"; (2) particles stuck together weakly and were not suspended above lunar surface; (3) radar return was from surface and not from subsurface layer; (4) strength of lunar maria was three to eight pounds psi at one- to two-inch depth; (5) composition of rock was basic and basaltic with very little meteoritic iron on the surface; (6) surface was extremely uniform; and (7) particles were cohesive and adhesive, but there was "no need for concern" about manned spacecraft's sinking or being covered by sticky soil in maria sites investigated.

    Simultaneous operation of *Surveyors V* and *VI* would permit measurement of moon's "wobble" or physical libration by obtaining spacecraft's range and range rate, Dr. John A. O'Keefe, Assistant Chief of GSFC's Laboratory for Theoretical Studies, explained. "The moon has a sort of a nose [Sinus Medii, landing site of *Surveyor VI*] that sticks out in front. The earth's gravity takes hold of that nose and wiggles it . . . back and forth as the earth slightly moves either way." Although scientists had a fairly good idea of the size of the "nose" and how firmly earth could grab it, they did not know how the moon responded to the pull, he said. Response depended on whether lunar mass was concentrated toward the center, in which case it would yield easily, or toward the outside, in which case moon's inertia would be much stronger. Dr. O'Keefe also showed *Surveyor VI* photos of lunar surface lit by sunshine. In one, corona was traceable out to a distance of 30 solar radii from the sun—furthest ever traced. Another revealed persistence of a line of light after the instant of true sunset. Line, which had appeared less clearly in previous Surveyor photos, he said, could be explained either as a layer of 0.001-in-thick particles or as lunar material floating a short distance from surface because of electrostatic charges.

    Surveyor VI mission represented "the completion of two major milestones of lunar surface exploration—the investigation and certification of four prime candidates for future manned landings and the scientific evaluation of four . . . widely separated regions in the moon's equatorial belts," Surveyor Program Manager Benjamin Milwitzky asserted. To maximize the scientific value of the last Surveyor mission in the series, NASA had established a group of scientists to study and recommend possible landing sites for Surveyor G. As a result of group's recommendations, NASA had selected Tycho crater, south of the equator, as primary target, and Fra Mauro, north of the equator, as an alternate. Highland Tycho area was "risky as a landing site," he said, because it was covered by ridges, grooves, and rocks, but it would provide scientific data for comparison with data from previous lowland maria landing sites. Sur-

December 14

veyor G, scheduled for launch Jan. 7, 1968, would be equipped with surface sampler, TV cameras, alpha-scattering instrument, and minor payload items such as magnets on footpads and surface sampler scoop and special mirrors. (Transcript; Reistrup, *W Post,* 12/15/67, A20; *W Star,* 12/15/67, A7)

- Stanford Univ. School of Medicine biochemists Drs. Arthur Kornberg and Merhven Goulian successfully synthesized a simple, biologically-active form of DNA (deoxyribonucleic acid) from inert materials. The manmade DNA virus—a basic molecule of life—was capable of reproducing itself and generating new viruses. Discovery could eventually make it possible to create artificial viruses that could attack diseases such as cancer and leukemia; modify genes and produce specific biological change; and cure hereditary defects such as diabetes, hemophilia, and mental retardation.

  President Johnson, speaking at Smithsonian Institution ceremony marking 200th anniversary of *Encyclopædia Britannica,* cited creation of active DNA as "a spectacular breakthrough in human knowledge. . . . These men have unlocked a fundamental secret of life. It opens a wide door to new discoveries in fighting disease and building much healthier human beings." (*W Post,* 12/15/67, A10; Randal, *W Star,* 12/15/67, 1; O'Toole, *W Post,* 12/15/67, A1; *PD,* 12/18/67, 1712-5)

- Three Soviet scientists confined in a small chamber for 70 days suffered hallucinations and "psychological changes," Tass announced. Scientists, who subsisted on "dehydrated meat, cottage cheese, various tinned food and concentrates," lost weight, and their muscle tone deteriorated despite daily exercise on a bicycle stand. All three recovered. (*W Post,* 12/15/67, A20; *SBD,* 12/15/67, 236)

December 15: NERVA (Nuclear Engine for Rocket Vehicle Application) reactor (NRX–A6) was successfully ground-tested by NASA and AEC at Jackass Flats, Nev., achieving major goal of one-hour duration for nuclear rocket technology program and demonstrating that rocket reactors could be operated for periods longer than those required for most space missions. Test, in which reactor operated for 60 min at 1,100-mw design power, was sixth in series to obtain additional data on reactor characteristics under extended operation duration in joint NASA–AEC Rover program. (AEC Release K–285; *Aero Tech,* 1/1/68, 13)

- U.S.S.R. announced successful completion of rocket tests in second Pacific area near Midway Island to test equipment for spacecraft landings at sea. Tests in other areas had been completed Dec. 13. Series, which began Nov. 28, had been scheduled to last until Dec. 30. (UPI, *NYT,* 12/14/67, 16; *SBD,* 12/18/67, 246)

- Univ. of Iowa professor Dr. James A. Van Allen, writing in *Science,* warned against "abandoning *in situ* study of the planets to the Soviet Union." Citing important contributions made by unmanned planetary probes, he noted that intelligible telemetry signals had been received from *Mariner IV* from ranges as great as 320 million km and that the spacecraft was still operating properly 36 mo after launch. Entire *Mariner V* mission, he said, "was conducted with such precision and competence as to draw cheers from even the most hardened professionals." Although successful Mariner missions had been accomplished at a cost of less than 2% of NASA budget, Congress had failed to provide for preparatory work for any specific planetary mission beyond Mariner Mars '69, he said. "Even more devastating is the reluctance of NASA to forcefully request

December 15: The Nuclear Rocket Development Station at Jackass Flats, Nevada, as the NERVA nuclear rocket reactor, NRX–A6, is operated at or above design power of 1,100 mw energy (55,000-lb thrust) for 60 min.

adequate funding for such work . . . despite the existence of well-conceived programs of great scientific potential. . . . The basic technology is available, and a rich diversity of feasible experiments has been proposed. . . ." Despite evidence that U.S.S.R.'s planetary exploration plans were both ambitious and increasingly competent, he said, U.S. "is now allowing its own high competency in planetary exploration to decay. . . ." (*Science*, 12/15/67)

The *New York Times* later commented: "Professor Van Allen is entirely right in his complaint . . . that [U.S.] is spending a ridiculously small sum . . . on planetary exploration by unmanned rockets.

". . . the cost-effectiveness in scientific terms of the nation's planetary program is far greater than that of the Apollo program for a manned flight to the moon, the nearly bottomless well into which the vast majority of NASA's funds are now going. The existing allocation of funds is scientifically irrational, explicable principally in terms of the propaganda and public relations returns envisaged from 'winning' the moon race." (*NYT*, 12/19/67, 46)

December 16: Cosmos CXCV was successfully launched by U.S.S.R. Orbital parameters: apogee, 375 km (233 mi); perigee, 211 km (131 mi); period, 90.1 min; and inclination, 65.7°. Spacecraft reentered Dec. 23. (GSFC *SSR*, 12/30/68; *SBD*, 12/19/67, 251)

- U.N. Committee on the Peaceful Uses of Outer Space approved space rescue treaty proposed by U.S. and U.S.S.R. Draft was submitted to General Assembly for final approval. (*NYT*, 12/17/67, 1)

December 17: First flight of USAF RF-111A reconnaissance aircraft was successfully conducted at Fort Worth, Tex. Reconnaissance equipment and flight-test instruments were checked out during three-hour flight. RF-111A was fourth F-111 model to be flight tested; other versions—F-111A tactical fighter, FB-111A strategic bomber, and USN's F-111B—had logged more than 6,500 flying hrs. Of 1,016 supersonic flights made by F-111 models, 213 had exceeded mach 2. (AFSC Release 210.67; DOD Release 1182-67)

- The 64th anniversary of first powered flights by Orville and Wilbur Wright from Kitty Hawk, N.C. Proclamation issued by President Johnson to commemorate their achievements cited "their inventive genius [which] revolutionized transportation, and gave rise to great new industries that have strengthened America's defense and economy."

 Igor I. Sikorsky, consulting engineer at United Aircraft Corp.'s Sikorsky Aircraft Div., received the Wright Brothers Memorial Trophy at Aero Club's annual Wright Brothers Memorial Dinner in Washington, D.C. (NAA Release)

- Walter Sullivan, writing in the *New York Times*, listed proposed projects to use atomic explosions for commercial applications being considered by AEC. Project Sloop, a 20-kiloton, $13,175,000 explosion at 1,200-ft depth, would shatter a large body of copper ore in southeastern Arizona. Highly diluted acid would then be pumped into the broken rock to convert the copper into a soluble compound which could be pumped out. Project Bronco would shatter a northeastern Colorado shale oil deposit beyond reach of conventional extraction methods. Heat from the blast would liquefy the oil so that it would flow into the central cavity for extraction. Projects Dragon Trail and Ruleson would use nuclear explosion to carve a huge gas storage area beneath a Renovo, Pa., site. Other proposals included use of explosions to break through impermeable rock formations in Arizona so that rain water could sink in and be stored underground and to dig canals, harbors, and mountain passes. According to AEC price list, charge for 10-kiloton explosion would be $350,000 and for a two-megaton explosion, $600,000, excluding charges for safety studies, site preparation, transportation and emplacement of the devices, and support services. (Sullivan, *NYT*, 12/12/67, E7; *Plowshare*)

December 18: USNS *Mercury*, third Apollo insertion-injection tracking ship, left General Dynamics Corp.'s Quincy, Mass., shipyard after successfully completing final contractor instrumentation systems tests. Ship docked in Hoboken, N.J., where it would undergo inspection and instrumentation adjustments before range testing and final preparations in Florida for Apollo 503 mission. *Mercury*, her two sister ships *Vanguard* and *Redstone*, and reentry ships *Huntsville* and *Watertown* were part of National Range Instrumentation Ship pool. They would be equipped with special antennas for communications between Apollo astronauts and mission control via Atlantic and Pacific satellites. For Apollo missions they would be operated as part of GSFC-operated MSFN, which supported Apollo with tracking, command, and voice and telemetry communications via land lines, microwave, submarine cables, and comsats. Operation of Apollo ships was assigned to WTR; crew was provided by Military Sea Transport Service, and technical instrumentation crew by USAF. (NASA Release 67–305)

- Rep. William F. Ryan (D-N.Y.) charged that the House Committee on Science and Astronautics and the Senate Committee on Aeronautical and Space Sciences had defaulted on their responsibility to report to the American public about the Congressional hearings which had been held on the Jan. 27 Apollo accident. In a statement inserted in the *Congressional Record*, Rep. Ryan said that since Congress seemed reluctant "to face some unpleasant truths and to ask some hard questions that need to be asked," he had prepared his own report based on the hearings and on the hundreds of letters, documents, and personal commentaries he had received. Much of this correspondence, he said, which cited deficiencies and negligence which were both intentionally ignored and unintentionally overlooked, was made in confidence for fear of reprisals by NASA or NASA-contractor employers.

  Rep. Ryan concluded that Congress overfunded and underexamined the validity of major technological programs and the success and integrity of their administration. Having an excellent space program, he said, "is essential both to our pride and to our progress . . . [but] it is essential as well that we throw off the shackles of the space race and declare our intention to make rational plans designed to meet the needs of America not distorted by the conceptions and value judgments of another nation. It is essential that we examine our national priorities with the utmost care and make certain that the space program is neither made a handy villain by its critics nor a handy saviour by its spokesmen. We must establish reliability and make a sober and objective appraisal of the space program's role in our society." Congress, he asserted, ought to: (1) demand that NASA adopt new standards of candor, present regular status reports, and clarify its budget presentations, and that a panel of independent experts be created to provide objective technical evaluation of NASA management and planning procedures; (2) relinquish romanticism and evaluate each program in terms of specific objectives and social value; (3) revise its attitudes about hearings on the space program, being sure to call both friendly and hostile witnesses; and (4) establish a realistic view so that the space program could be assigned its proper place in national priorities and receive whatever funds necessary. (*CR*, 12/18/67, A6320–9; Ryan's Off)

- ELDO Technical Director Dr. W. H. Stephens announced that ELDO would develop Europa III following development of Europa II. Europa III booster would have Europa I 1st stage (Blue Streak) and liquid hydrogen-oxygen 2nd stage with two 14,000-lb-thrust rocket engines. Configuration would be able to boost 6,000-lb payload into low earth orbit. Using high-energy 3rd stage, new launch vehicle would be able to place 1,200 lbs into synchronous orbit; adding two strap-on solid rocket motors rated at 420,000 lbs thrust each, launch vehicle could place 4,000 lbs into synchronous orbit.

Low-thrust, very-high-specific-impulse systems were being considered, Stephens said, for use in transferring payloads from low to high orbits, and for accelerating to escape speeds. Using Dragon reactor technology, a small reactor could also be employed in such a system. (S/P, 12/18/67, 150)

- Several Latin American nations, including Argentina, Brazil, and Mexico, were interested in making extensive use of Kourou space research base, French Guiana, according to Centre National d'Etudes Spatiales (CNES) sources. New equatorial range would be fully operational in 1969, with first segment handling sounding rocket firings early in 1968. Initial plans for Kourou facility had included three launch pads, but CNES was already making plans to expand the facility. One new pad would handle Scout and Thor-Able vehicles, subject to agreement with U.S. (*Space Pro,* 12/18/67, 150)
- Paul E. Cotton, Director of Management Operations, in NASA's OMSF, was appointed director of OART's Programs and Resources Div. Cotton would replace William E. Hanna, Jr., who had been named Director of the Bureau of Data Processing and Accounts in the Social Security Administration. (NASA Release 67–304)
- *New York Times* editorial on space rescue treaty approved by U.N. Committee on the Peaceful Uses of Outer Space Dec. 16: "[Treaty] . . . is a substantial advance, but what must remain worrisome is the attitude toward such mishaps that will be shown [by] countries refusing to approve the pact. The largest of the abstainers, of course, is likely to be China.

"Even without this gap, the treaty has limitations. The suspicious will note that . . . the launching nation has no right to join in the search until and unless requested to do so by the authorities governing the area where the ship has landed. Nor does the treaty define with precision how long the latter authorities may hold a space vehicle and its crew before returning them . . . [or specify] payment for damages caused by errant spacecraft.

"The important fact, however, is not that the rescue treaty has deficiencies. Most treaties do. Far more significant is the demonstration it has provided of intimate and fruitful Soviet-American cooperation for achievement of a common end. . . ." (*NYT,* 12/18/67)

December 19: U.S.S.R. launched *Cosmos CXCVI* into orbit with 887-km (551-mi) apogee, 225-km (140-mi) perigee, 95.5-min period, and 45° inclination. Equipment and instruments performed satisfactorily. (AP, *NYT,* 12/20/67, 35; *SBD,* 12/20/67, 258; GSFC *SSR,* 12/30/67)

- Uprated Saturn I (SA-211) 1st stage was successfully static-fired for 35 sec at MSFC by Chrysler Corp. personnel. Stage would be shipped to Michoud Assembly Facility for post-firing checkout after completion of a 145-sec, full-duration captive firing. (MSFC Release 67–243)
- JPL scientists continued efforts to reestablish contact with *Surveyors V* and *VI* resting on the moon. *Surveyor V* (launched Sept. 8) had responded

immediately to turn-on commands during its fourth lunar day Dec. 14 and later transmitted 67 recognizable 200-line pictures. Following unsuccessful attempts to reposition the Antenna/Solar Panel Positioner on Dec. 16, transponder signals were lost, and complete loss of signal occurred. *Surveyor VI* (launched Nov. 7) had also responded to turn-on commands Dec. 14, but telemetry lock-up was poor, and erratic signals were lost completely after 2½ hrs. Attempts to reactivate both spacecraft for simultaneous communications and additional data would continue until end of lunar day Dec. 22. (NASA Proj Off; AP, *NYT*, 12/20/67, 29)

- USA–USAF–USN X–22A V/STOL research aircraft designed and built by Bell Aerosystems Co. under USN contract completed its 100th successful test flight at Niagara Falls (N.Y.) International Airport. The four-engine, dual-tandem, ducted-propeller aircraft had made 239 vertical and 100 short takeoffs and 245 vertical and 94 short landings. During 40.9 hrs of actual flying time 54 vertical/horizontal transitions had been conducted. Present aircraft was second of two models in X–22A program; first model crashed Aug. 8, 1966, after completing 15 flights. Second model would be tested through 1968 and would then be delivered to a tri-service group at Patuxent (Md.) Naval Air Test Center for further evaluation. (Bell Release 92; *NYT*, 12/25/67, 42)

- In its final 1967 session U.N. General Assembly unanimously endorsed new space rescue treaty and urged signature by all countries. Primary provisions: (1) immediate notification when astronauts landed in other countries; (2) assistance to these astronauts, including rescue efforts on the high seas; (3) safe and prompt return of the astronauts; and (4) notification and return of objects launched into outer space. Treaty would be open to all nations and would become effective when ratified by U.S., U.S.S.R., U.K., and two other countries.

  President Johnson, praising the agreement, said: "I hope that this agreement will help to ensure that nations will assist astronauts in the event of accidents or emergency. The agreement would carry forward the purpose of this administration to promote international cooperation in the peaceful uses of outer space." (Teltsch, *NYT*, 12/17/67, 1; Estabrook, *W Post*, 12/20/67, A20; UPI, *NYT*, 12/21/67, 29; *SBD*, 12/21/67, 263; *PD*, 1/1/68, 1763)

- NASA announced establishment of Apollo Lunar Exploration Office within OMSF to increase effectiveness of directing Apollo lunar exploration and planning for post-Apollo lunar exploration. New office, headed by former Lunar Orbiter Manager Lee R. Scherer under general direction of Apollo Program Director M/G Samuel C. Phillips (USAF), would consist of Flight Systems Development and Lunar Science Divisions. It would be responsible to OSSA for scientific aspects of Apollo lunar exploration; OSSA would review operating plans, science SR&T tasks, mission objectives and plans, and scientific payloads and principal investigators for specific missions. (NASA Ann, 12/19/67)

- NASA was negotiating with McDonnell Douglas Corp. on a $10-million, fixed-price-incentive contract for 10 improved 2nd stages for the Delta launch vehicle. Procurement for new stages would include modified versions, larger in diameter than original Delta, to be mated to new elongated Thor 1st stages which would be incorporated in the Delta configuration beginning in 1968. (NASA Release 67–306)

December 20: France's first variable-sweep-wing jet aircraft, the Mirage G, flew 1,500 mph in public debut at Istres, France, military airport. Dassault-built aircraft was world's fourth variable-sweep-wing model and first built in Western Europe; others were American F-111, and Soviet Sukhoi and Mikoyan. (*W Post*, 12/21/67, A31)

- United Aircraft Corp. turbotrain, powered by gas turbine engine similar to those used in jet aircraft, reached 170 mph during test run in New Jersey, setting new speed record for passenger train in U.S. Train, designed for future passenger service between Boston and New York City, would be operated by New Haven Railroad under DOT contract. (*W Post*, 12/21/67, A7)

- NASA and Italian Space Commission (ISC) agreed to launch jointly Italy's third San Marco satellite (San Marco C) from mobile launcher in Indian Ocean off the coast of Kenya, Africa. Spacecraft would measure air density by continuously monitoring spacecraft's drag forces, investigate ionospheric characteristics which interfered with long-range radio transmission, and conduct other upper-atmosphere experiments in 1969 and 1970. San Marco project, provided for under May 31, 1962, cooperative agreement, was a mutual program of NASA and ISC with no exchange of funds. NASA supplied launch vehicle and provided personnel training and tracking and data acquisition services. ISC was responsible for design, fabrication, and testing of payload and for launching of satellite built by Centro Ricerche Aerospaziali (CRA) of the Univ. of Rome. *San Marco I* was launched December 14, 1964, and *San Marco II*, April 26, 1967. (NASA Release 67-303; *A&A*, 67)

- *NASA Thesaurus* (SP-7030) was published by NASA to provide a standardized list of terms for indexing and retrieving documents in the NASA scientific and technical information system. The three-volume set, containing approximately 15,000 indexing terms and scope notes, subject categories, and a cross-reference for each term, would be available through Superintendent of Documents, GPO. (NASA Release 67-308)

- New reconnaissance camera system for use in Vietnam on RF-4C aircraft was undergoing tests at Wright-Patterson AFB, Ohio, to determine camera's operating capability and devise techniques to enhance low-altitude aerial photography capability. Developed by AFSC's Aeronautical Div., fully automated camera could take six black and white or color pictures per second during daytime or nighttime reconnaissance and could provide stereo coverage at extremely low altitudes and supersonic speeds. (AFSC Release 192.67)

- NASA announced appointment of Bob P. Helgeson, Deputy Manager of AEC's Hanford Project in Richland, Wash., as NASA Director of Safety. He would report to Associate Administrator for Organization Harold B. Finger and would be responsible for "developing and implementing safety programs throughout NASA . . . [with] broad authority over all safety activities, including those involving manned and unmanned flights in aeronautics and space, and ground and test research operations." Helgeson would replace George D. McCauley, who retired. (NASA Release 67-307; UPI, *W Post*, 12/21/67, A2)

- Ben W. Hersey, Chief of KSC's Personnel Office, announced that more than 200 additional permanent employees would be hired for KSC's Civil Service staff by July 1968. First consideration would be given to personnel at MSFC and other NASA centers where personnel reductions had occurred. (KSC Release 443-67)

- Secretary of Interior Stewart L. Udall had invited a number of distinguished U.S. scientists to serve on an advisory committee to evaluate how widespread sonic booms expected from SST might destroy the national resource of tranquility, David Hoffman reported in the *Washington Post*. Discussions would include effect of sonic boom on wildlife, national parks, and Indian adobe structures in the Southwest. (Hoffman, *W Post*, 12/20/67, A1)

December 21: USAF launched an experimental reentry vehicle by Atlas-F booster from Vandenberg AFB as part of Advanced Ballistic Reentry Systems (ABRES) program. (*Aero Tech*, 1/1/68, 13)

- Communications with NASA's *Mariner IV* Mars probe, launched Nov. 28, 1964, were terminated because of greatly reduced telemetry availability. Spacecraft had responded to series of commands Oct. 25–27 and had continued operating satisfactorily until it exhausted its gas supply in the attitude control system Dec. 7 and began to pitch, roll, and yaw. On Dec. 10 and 11 *Mariner IV* recorded 83 micrometeoroid hits which apparently induced severe perturbations in pitch and yaw causing degraded signal strength. During its 3.06-yr lifetime, *Mariner IV* had traveled more than 1.5 billion mi in heliocentric orbit, taken world's first closeup photos of Mars from 5,500-mi altitude, and returned excellent data. (NASA Proj Off; AP, B *Sun*, 12/23/67; *W Post*, 12/23/67, A5)

- FAA proposed that all civil turbojet aircraft be required to carry altitude warning devices to signal pilots visually and aurally when they were approaching or deviating from a preselected altitude. Signals were preset when device was installed and activated at a point within a prescribed band as aircraft approached a preset altitude. Device was aimed primarily at alerting pilots' sense of altitude awareness, particularly during high-speed climb and descent maneuvers. (FAA Release 67-87)

- Hughes Aircraft Co. had completed negotiations with the government of Brazil on a $3.8-million contract for construction of a comsat earth station at Itaborai, near Rio de Janeiro. Brazilian Telecommunications Co. (EMBRATEL) would spend an additional $1 million or more for site preparation and buildings. Station, which would initially handle TV and 120 two-way telephone conversations via INTELSAT satellites over the Atlantic, would be completed within one year. (Hughes Release; Montgomery, *NYT*, 12/22/67, 63)

- Aerospace industry sales in 1967 would total $27.3 billion—a 13% increase over 1966 sales—and in 1968 would reach $29.2 billion, Aerospace Industries Assn. (AIA) predicted. Commercial transport aircraft production and DOD procurement had increased significantly in 1967, while NASA sales had declined from $4.9 billion to $4.1 billion. (AIA Release 67-55)

- Existence of a secret three-station system which would have enabled SAC to broadcast final messages to its surviving bomber and missile forces in event of a general nuclear attack on U.S. was revealed by Howard Silber in the *Omaha World-Herald*. System, operational since July 1963, had completed its mission and was being superseded by newer devices. Located near West Point, Wisner, and Tekamah, Neb., each constantly manned station had consisted of a guard house and three trailer-vehicles housing launch control equipment, electronics equipment, and a Blue Scout Junior rocket. The three-stage rockets were to have been launched in sequence so their messages could be sounded at specific prearranged intervals at altitudes of more than 100 mi. (*Omaha W–H*, 12/21/67, 1; AP, *NYT*, 12/22/67, 31)

December 22: ERC awarded $99,151 to TRW Systems, Inc., and $99,370 to RCA for research on a satellite system to provide more efficient navigation and air traffic control for aircraft and ships in North Atlantic area in 1975. Both companies would conduct eight-month studies to determine best technical approach, identifying related problems and requirements. (FRC Release 67–47)

- MSFC awarded Mason-Rust Co. a one-year, $8,990,826 contract extension for continued provision of services—including transportation, safety and security, supply, communications, and custodial services—at Michoud Assembly Facility. Extension brought total cost-plus-award-fee contract to $39,073,652. (MSFC Release 67–246)
- U.S. R&D funding in 1968 would total about $26.5 billion, $700 million (3.5%) more than 1967 estimate, Battelle Memorial Institute (BMI) said in its annual forecast. BMI noted that "significantly, for the first time since reliable figures for total research and development expenditures became available, it is estimated that for [CY] 1968 the increase in Federal spending on research in the social sciences will be greater than the increase in the physical sciences." This trend, forecast said, was partly because of reductions in growth of military, space, and atomic energy programs and because of increasing national concern with education, health, urban, employment, and welfare problems. (AP, *NYT*, 12/23/67, 7; BMI PIO)
- ERC Director James C. Elms announced appointment of Dr. D. M. Warschauer as Chief of ERC's Component Technology Laboratory. Dr. Warschauer had been Manager of Itek Corp.'s Physics Laboratory since 1965. (ERC Release 67–45)
- J. V. Reistrup, in the *Washington Post*, reported that the Senate Committee on Aeronautical and Space Sciences had prepared a draft report on Apollo accident hearings which would be published in early 1968. The report, revealed as a result of Rep. William Ryan's (D-N.Y.) Dec. 18 criticisms on Congress' failure to produce such a report, was said to be critical of NASA but to have reaffirmed Apollo lunar-landing goal. (Reistrup, *W Post*, 12/22/67, A1)
- U.S.S.R. team to study UFO (unidentified flying object) sightings would consist of 18 scientists and Air Force officers supported by 1,000 field observers, *Time* reported [see Nov. 30]. Flurry of UFO sightings had been reported in recent weeks by presumably reliable Aeroflot and military pilots, who usually described them as sickle-shaped. (*Time*, 12/22/67, 21)

December 24: Communist China exploded her seventh nuclear device at Lop Nor test site in Sinkiang Province. Low yield of test marked reversal of trend since first test Oct. 16, 1964. Yield size had increased progressively, culminating with June 17, 1967, explosion of hydrogen bomb. Test was not announced by Communist Chinese officials, but was detected by AEC. (AEC Release K–289; Finney, *NYT*, 12/25/67, 1)

- Robert Walker, writing in the *New York Times*, reviewed new applications for industrial use of gold. In telephone industry, gold was used for coating electromechanical switchgear and for transmitter domes; in computers, for coating metal tapes and electroplating printed circuits; and in automotive industry, for connecting alternator-regulator systems. To aerospace industry, one of gold's most important characteristics was that it reflected heat and light. Gold was used to coat the "umbilical cord" which connected Astronaut Edward H. White II to *Gemini IV* spacecraft (June 3, 1965); to cover shrouds that contracted very hot jet aircraft engine parts; and to coat satellites so that they reflected more than 90% of sun's heat. By adding oxides to gold coatings space scientists used the gold to absorb infrared energy which could be converted to electrical energy. (Walker, *NYT*, 12/24/67, 1F)

December 25: Washington *Evening Star* editorial on space rescue treaty approved by U.N. Dec. 19: "International cooperation in space has made another advance . . . 'prompted by sentiments of humanity' in an era when manned flights to the moon and beyond will require all countries to be on the alert for possible tragic misadventures.

"There has been some grumbling in the U.N. over the fact that there is as yet no provision in international law for the payment of damages a country might suffer as a result of spatial mishaps—for example, the crashing of a spaceship on a populated area or on some vital economic or military installation. This matter, however, can and will be taken care of in a special treaty that almost certainly will be worked out in the coming year." (*W Star*, 12/25/67)

December 26: U.S.S.R. successfully launched *Cosmos CXCVII*. Orbital parameters: apogee, 505 km (314 mi); perigee, 220 km (137 mi); period, 91.5 min; and inclination, 48.5°. (GSFC *SSR*, 12/30/67); *Aero Tech*, 1/1/68, 13)

- First flight test of Apollo Lunar Module (LM), designated Apollo 5, was scheduled no earlier than Jan. 17, 1968, NASA announced. The 31,700-lb LM, one of three modules comprising the Apollo spacecraft, was designed to carry two astronauts from lunar orbital flight to a landing on the moon, then back to the Apollo spacecraft in lunar orbit. It would be launched on the 6-hr 30-min test flight by 1.6-million-lb-thrust Uprated Saturn I (AS-204) originally scheduled for first manned Apollo mission in February 1967, and would not be recovered. Primary objective of the unmanned Apollo 5 flight would be verification of LM's flight readiness for manned operations in space. Test would flight qualify LM descent stage propulsion engine, including restart; ascent stage propulsion engine; systems; structures; and staging. (NASA Release 67-313; UPI, *NYT*, 12/27/67, 15; AP, *W Post*, 12/27/67, A4)

- In response to FCC inquiry on future communications in Atlantic area—currently served by two comsats and four cables—ComSatCorp President Joseph V. Charyk and AT&T Vice President Richard R. Hough presented their proposals in separate letters to FCC Chairman Rosel H. Hyde. Hough, advocating a fifth underwater cable, said that although AT&T supported the idea of a global comsat system, all of the international carriers believed the cable project was "the right way to add to our capabilities at this time." The best service, he said, could be provided by "a balanced, integrated network of both cables and satellites."

Charyk charged that the cable would not be economically feasible: ". . . when a new cable is laid to carry traffic that could otherwise be carried at no extra cost by satellite facilities which must be established in any case, the ultimate customer must bear the cable cost as an addition to the already committed costs of service." As traffic was diverted from comsats to a cable, a greater share of satellite revenue requirement would have to be carried by routes not served by the cable, primarily less developed countries with smaller traffic streams. U.S., he said, was committed to develop the INTELSAT system. To ensure that U.S. met this commitment, "the advanced [comsat] system must be in operation in the time frame we have proposed for INTELSAT IV. It is essential that a vigorous advanced satellite development progam be under way during 1968." (Texts; Smith, *NYT*, 12/29/67, 43)

December 27: U.S.S.R. launched Cosmos CXCVIII into orbit with 281-km (175-mi) apogee, 265-km (165-mi) perigee, 89.8-min period, and 65.1° inclination. Equipment and instruments performed satisfactorily. (GSFC *SSR*, 12/30/67)

- NASA's *Ats III*, launched from ETR Nov. 5, had successfully met its mission objective of operating for 30 days and obtaining useful data from onboard experiments. Resistojet thruster had experienced valve malfunction, but thrust data were still being obtained. Multicolor Spin-Scan Cloud Camera successfully transmitted 123 photos before it was turned off because of an arcing problem. Omega Position Location Experiment was successfully operated in communications tests with aircraft in flight over the Atlantic Ocean, and all other experiments except Self-Contained Navigation System, which was scheduled to be operated in spring 1968, had been successfully operated. (NASA Proj Off)

- Apollo/Saturn V 3rd stage was flown from McDonnell Douglas' Sacramento, Calif., plant to KSC via Super Guppy aircraft. Other stages for third Apollo/Saturn V mission had been transported to KSC by barge. Boeing-built 1st stage, aboard NASA barge *Poseidon*, and NAR-built 2nd stage, aboard *Point Barrow*, had reached KSC Dec. 26. IBM instrument unit would be flown to KSC from IBM's Huntsville, Ala., plant via Super Guppy Dec. 29. Stages would be assembled in KSC's Vehicle Assembly Building (VAB). (KSC Release 445-67; MSFC Release 67-245)

- ComSatCorp awarded J. W. Bateson Co., Dallas, Tex., a $7.9-million contract for construction of a new research laboratory near route 70-S in Montgomery County, Md., 30 mi from ComSatCorp's Washington, D.C., headquarters. Construction would begin immediately. (ComSatCorp Release 67-56; *W Post*, 12/28/67, F9)

- President Johnson was expected to ask Congress for the smallest civilian space budget in six years for FY 1969, Evert Clark reported in the *New York Times*. According to his sources, the FY 1969 request would probably be close to $4 billion. DOD's space budget request, however, was expected to be higher than record $2 billion requested for FY 1968, chiefly because USAF's MOL would require more development funds. (Clark, *NYT*, 12/28/67, 6)

December 28: NASA Wallops Station awarded Doyle and Russell, Inc., a $1,233,644, fixed-price contract for construction of a new launch complex. Work, to be completed within one year, would include construction of an assembly shop, launch pad and mobile shelter, launch terminal building, liquid fuel storage area, launch control building, and substation enclosure. (WS Release 67-38)

- Soviet scientists had compiled a structural and geological map of the lighted side of the moon to aid in selection of lunar landing sites, Tass announced. (Reuters, *W Post*, 12/29/67, D6)

December 29: NASA announced personnel changes, effective Jan. 1, 1968. Adm. W. F. Boone (USN, Ret.), Assistant Administrator for Defense Affairs, would retire after five years of service with NASA. He would continue to serve as a part-time consultant. Functions of Admiral Boone's office would be included in new Office of Department of Defense and Interagency Affairs, headed by Gen. Jacob E. Smart (USAF, Ret.). Gen. Smart, Assistant Administrator for Policy, would be replaced by Deputy Associate Administrator for Advanced Research and Technology Dr. Alfred J. Eggers, who would also continue as a Special Assistant to the NASA Administrator on aerospace safety.

NASA Administrator James E. Webb, expressing his appreciation for Admiral Boone's service, praised the "effective cooperation developed between [DOD] and NASA in the . . . buildup . . . of large launch vehicles and spacecraft required for a lunar landing and exploration . . . in large measure due to [his] leadership. . . ." (NASA Release 67-314)

- ERC Director James C. Elms announced appointment of Dr. Richard M. Head, Manager of ERC's Aeronautics Programs Office, as ERC Chief Scientist. In his new position Dr. Head's responsibilities would include research in ERC aeronautics and space programs and on solar flares. (ERC Release 67-44)
- Univ. of Brussels glaciologist Dr. Edgard E. Picciotto was shipping four tons of snow from Plateau Station in Antarctica to Brussels for analyses to determine chemistry and rate of fall of microscopic dust particles from space. Study, supported by NSF, would be conducted in cooperation with Ohio State Univ. and Belgian Atomic Center. Since the Antarctic snow was not contaminated by urban air pollution and never melted, extraterrestrial dust remained lodged as it fell in layers of snow. Plateau Station area, Dr. Picciotto said, was "the only place on earth you can retrieve atmospheric precipitation from before the industrial revolution and the atomic bombs." Ideally, the dust should have been collected in space, but space collection was too expensive and would be contaminated by spacecraft. Dr. Picciotto would extrapolate annual worldwide total from fallout rate at Plateau Station. Preliminary results suggested a 100,000-ton annual worldwide rate. (Reinhold, *NYT*, 1/9/68, 4)

December 30: Among 12 scientists named by President Johnson to receive the 1967 National Medal of Science were Edwin H. Land, President of the Polaroid Corp.; Igor I. Sikorsky, former engineering manager of United Aircraft Corp.'s Sikorsky Aircraft Div.; and George B. Kistiakowsky, Harvard Univ. chemistry professor. Presentations would be made in early 1968. (*PD*, 1/6/68, 11)

December 31: AEC Chairman Dr. Glenn T. Seaborg reviewed highlights of AEC's 1967 activity. "Outstanding . . . was the trend by electric utilities to go nuclear. During the year, utilities announced plans for at least 32 nuclear power plants and ordered 30 reactors . . . [representing] a capacity of more than 24 million kilowatts. . . . The nation enters the New Year with 15 operable nuclear power plants and 21 under construction. The trend points to a generating capacity by nuclear plants of between 120,000 and 170,000 net electrical megawatts by the end of 1980, or 25 percent of the total national capacity." Other activities included: continued R&D with DOD for strong nuclear defense capability; testing of first model of a cardiac pacemaker; limited distribution of a purified multiple vaccine which provided immunity to several different flu viruses with one inoculation; continued development of food irradiation program; sixth successful test of NERVA; and successful Project Gasbuggy nuclear explosion. In addition, AEC conducted research which provided numerous spin-off benefits to industry and sponsored lectures, demonstrations, and exhibits. (AEC Release K-286; AP, *W Post,* 12/31/67, D14)

During December: A preliminary report on *Surveyor V*, compiled by OSSA, recounted spacecraft's performance. "*Surveyor V* will be remembered as having first performed one of man's most extraordinary technical feats, a remotely conducted analysis of the chemical constituents on an extraterrestrial body in our solar system." The compilation included information on principal scientific results, TV observations, astronomy, lunar surface parameters, and lunar theory and processes. (Text, NASA SP-163)

- JPL scientist Sidney J. Slomich, writing in *Astronautics and Aeronautics*, emphasized need for national purpose to apply the highly advanced technology of space. "New impetus would be gained through the establishment of a coordinating national organization, governmental or private, to take advantage of the potential of advanced technology through modern communication techniques. The situation calls for a comprehensive national effort, with leadership coming from the highest levels of government to break down further interdisciplinary, interagency, organizational, and psychological and occupational barriers and to check out in detail a wide variety of potential applications like those cited above, relating them to the country's and the world's most pressing problems. . . . Emphasis should be on the application of the vast array of existing techniques, not on the development of new ones, and on near-term objectives—no longer, say, than two or three years. Government agencies, the universities, publicly and privately supported research centers . . . must be banded together into a cooperative effort devoted to applying to the problems of our age the many solutions the new high technology makes possible. There is needed . . . a public or private national organization, a research center, and, above all, a national purpose to bring this about." (Slomich, *A&A,* 12/67, 54–8)

- Claude Witze, in *Air Force and Space Digest*, claimed that USAF pilots testing F-111A were enthusiastic about the aircraft's potential. "These . . . pilots consider the F-111A weapon system the greatest single technological jump designed for their mission since the wedding of the jet engine and modern avionics. The F-111A, they predict, will let them hit tactical targets harder, with greater accuracy, and at longer ranges than any other airplane. . . ." Accelerated F-111A testing and training program begun at Nellis AFB, Nev., in July had accumulated encouraging statistics, he said: "By September, the new wing had set an unprecedented record. During that month, the five planes flew a total of 304.1 hours, an average utilization rate of 60.8 hours per aircraft. In October, the month in which the first production model was delivered . . . it was 59.7 hours per aircraft. The stated requirement for the F-111A is thirty hours per aircraft. The best previous records set at Nellis on other aircraft have been in the area of thirty-eight hours a month per aircraft . . . with systems far less complex than those of the F-111A." In addition, he noted, F-111A had set "an extraordinary record for safety. Far fewer aircraft have been lost than USAF experienced in previous similar programs." (*AF/SD,* 12/67, 45–55; *CR,* 12/5/67, S1788–90)

- Writing in *Astronautics and Aeronautics*, Hudson Institute scientists Herman Kahn and Anthony J. Wiener listed possible events and discoveries that could occur by the year 2000. List, according to the authors, showed "that as a result of the long-term trends toward accumulation of scientific and technological knowledge and the institutionalization of change through research, development, innovation, and diffusion, many important new things are likely to happen in the next few decades." List included: extension of life expectancy to more than 150 yrs; antigravity, practical use of gravity waves; interstellar travel; substantial lunar or planetary bases or colonies; and discovery of extraterrestrial life and, possibly, communication with extraterrestrial intelligence. (Kahn, Wiener, *A&A*, 12/67, 28–48)

- Calculation of effective radiated power by measurement of signal strength from eleven orbiting military communications satellites was accomplished by AFCRL; two satellites exhibited below standard response. Measurements, made at 401 megacycles, were carried out at the 84-ft parabolic antenna facility, Sagamore Hill Radio Observatory, Mass. (*OAR Res Rev*, 12/67, 4)

- Development of a laser system for detecting small variations in moon's motion, and for measuring earth-moon distances to within two meters was reported by AFCRL. A laser beam would be directed at a reflector placed on lunar surface; reflected light would be detected. (*OAR Res Rev*, 12/67, 5)

During 1967: In 1967, U.S. orbited 87 spacecraft; U.S.S.R., 67; France, 2; Italy, 1; Australia, 1; and U.S.-U.K., 1. U.S. total included 30 launches with 61 payloads for DOD and 20 successful NASA missions in 22 attempts—highest percentage of successes in one year to date. NASA had one satellite as a secondary payload.

Highlighting NASA space achievements were closeup photographs of the moon provided by Surveyor and Lunar Orbiter spacecraft; soft-landing on the moon by *Surveyor III, V,* and *VI*; successful Venus flyby by *Mariner V*; and preparation of a UV chart of sun's surface temperature from *Oso IV* photos of the sun. *Surveyor III, V,* and *VI* radioed more than 50,000 closeup photos of the moon and conducted on-site chemical analyses of the lunar soil, confirming its being similar to basaltic rock on earth. First launch from lunar surface was accomplished when *Surveyor VI*'s retrorockets were fired, lifting spacecraft and translating it to another site eight feet away. NASA launched *Lunar Orbiter III, IV,* and *V,* successfully completing the five-mission program and obtaining detailed photos of entire lunar surface—with 100 times the detail possible with earth-based telescopes—including eight prospective Apollo manned landing sites. *Mariner V* passed within 2,600 mi of Venus and radioed data on Venusian atmosphere and effects of solar wind.

Ground tests in Project Apollo were continued in preparation for three-man Apollo/Saturn space flight planned for 1969. Following flash fire in which three astronauts died Jan. 27 during ground test at KSC, intensive studies were conducted to devise new safety measures. Nearly all combustible materials were removed from the spacecraft, and the hatch was redesigned for fast escape. Apollo program climaxed Nov. 9 with successful maiden flight of Saturn V which orbited unmanned *Apollo 4* spacecraft.

Applications satellites launched included three INTELSAT comsats for ComSatCorp and three ESSA meteorological satellites for ESSA. *Ats II* provided first continuous high-quality photos of earth from geostationary orbit, and *Biosatellite II* provided new knowledge on effects of radiation and weightlessness on life forms. Scientific achievements included orbiting of *Oso III* and *IV*, *Ogo IV*, *Explorer XXXIV* and *XXXV* satellites, and *Pioneer VIII* interplanetary spacecraft.

Some 180 meteorological sounding rockets and 105 scientific sounding rockets were launched, and RAM C-1 spacecraft was launched on suborbital mission to measure effectiveness of water-injection concept in maintaining communications during reentry.

Some 12 flights of X-15 rocket research aircraft were conducted, including record-setting 4,534 mph (mach 6.72) flight by Maj. William J. Knight (USAF). Following crash Nov. 15 in which first fatality in 191 X-15 tests occurred, flights were suspended pending a review of the accident. NASA-USAF flight research continued with 14 flights of XB-70 supersonic aircraft. President Johnson authorized FAA to proceed with SST prototype construction, contracts were signed with Boeing Co. and General Electric Co., and construction was begun. Research programs were initiated for wing-structure designs for hypersonic aircraft which could make sustained flights at 5,300 mph (mach 8), and studies were undertaken to reduce aircraft noise. M2-F2 lifting-body vehicle crashed May 10, during its second flight of the year.

DOD program included orbiting of two Vela nuclear-detection satellites and 11 IDCSP comsats; static test-firing of Titan IIIM 1st stage; and beginning of construction of MOL launch complex at Vandenberg AFB.

Among the highlights in propulsion technology was the third and final test firing of the 260-in-dia solid-propellant rocket motor, producing 5.8 million lbs of thrust. One test of NERVA nuclear reactor was conducted in which NRX-A6 was operated at full design power for 60 min, achieving major goal of one-hour-duration operation. Phoebus 1B was operated at 1,500 mw for 30 min, and Phoebus 2C-F underwent series of systems tests with conventional fuel.

Tracking, communications, and data acquisition reached greatest volume in 1967 with a large number of unmanned satellites and space probes. MSFN stations conducted tracking operations with Lunar Orbiter spacecraft as targets after completion of missions' photo-acquisition phases, and new Apollo Manned Space Flight Network was completely checked out using *Tts I*, launched as secondary payload with *Pioneer VIII*. At end of 1967, NASA's Office of Tracking and Data Acquisition was simultaneously supporting 50 flight programs for NASA; six in cooperative international programs; and 12 for DOD.

U.S.S.R. launched 67 payloads, including 61 Cosmos satellites, one Venus probe, one Soyuz spacecraft, and three Molniya I comsats. *Venus IV* landed on the Venusian surface and transmitted atmospheric data; *Soyuz I*, after 25-hr manned flight, crashed on landing, killing the pilot; and *Cosmos CLXXXVI* and *Cosmos CLXXXVIII* unmanned spacecraft achieved world's first automatic docking in space. (NASA Release 67-301; *Space Flight Record 1958-1967*; NASA Proj Off)

- International space events were highlighted by ratification on Oct. 10, 1967, of space law treaty, approved by U.N. on Dec. 19, 1966, and endorsement on Dec. 19, 1967, of a new treaty providing for rescue of astronauts. Draft of space rescue treaty, negotiated by the Legal Subcommittee of the U.N. Committee on the Peaceful Uses of Outer Space, would become effective following ratification by five nations, including U.S., U.S.S.R., and U.K.

NASA's international program included orbiting of U.K.'s *Ariel III* and

ESRO's *Esro II* and assistance with orbiting of Italy's *San Marco II*—first satellite to be launched from a platform at sea. Some 124 scientists from 26 countries participated in research at NASA centers; scientific sounding rockets were launched from six sites in cooperation with Argentina, Brazil, West Germany, India, Japan, and Norway; and by the end of 1967, 84 countries had cooperated with U.S. in space research activities.

Kenya, Korea, Peru, and Tanzania joined the International Telecommunications Satellite Consortium (INTELSAT), bringing total membership to 60.

Australia successfully launched her first satellite, *Wresat I*. (NASA Release 67–301; EH; ComSatCorp PIO; NASA *International Programs*)

- Rep. Joseph E. Karth (D-Minn.), Chairman, House Committee on Science and Astronautics' Space Science and Applications Subcommittee, in interview with *Aerospace Management,* said: "The Space Program got off the ground ten years ago on a surge of apprehension and sensationalism. This year it became very clear that public acceptance or rejection will be a key factor in shaping the U.S. objectives in space for the years ahead. Having ushered in the second decade of space exploration, both NASA and the aerospace industry should now review their 'ledgers' to make the best use of their achievements to date, as well as learn from the mistakes and false-starts of the past." He predicted: "I have no doubts that before this century is over, man will set foot on the near planets [although] it seems more reasonable to me to first find out more about the other planets by means of automated robots [and] I will be disappointed if . . . we don't couple the great capabilities of the Saturn V vehicle with those of a Voyager spacecraft." He added: "However, be they the merits of the Voyager concept or the potential benefits of Earth Resources Satellites, the public and Congress must be honestly convinced of their values if space efforts are to be supported." (Karth, *Aerospace Management*, Fall/Winter 67, 5–9)

- High probability of mission success for Saturn project was born of sound engineering judgment built on extensive experience, and supported by modern analytical techniques, Special Assistant in Manufacturing Problems to MSC Apollo Spacecraft Program Mgr. (former MSFC Deputy Director) Dr. Eberhard F. M. Rees said in interview with *Aerospace Management*. He summarized the factors which had enhanced the probability: "Considering that one Saturn V, including the launching costs, amounts to over 150 million, it is not hard to see why the test launchings preceding manned flights must be limited to only a few. What this limitation has done then is to stress more than ever before the critical need for intensive engineering, almost a compulsive meticulousness in manufacturing assembly, quality control, and most extensive testing on the ground, such as component qualification testing, and subsystem and system testing." (Rees, *Aerospace Management*, Fall/Winter 67, 23–27)

- Managerial and technical challenges of the Mariner project were highlighted by JPL scientist Dan Schneiderman in interview with *Aerospace Management:* "We have tried to recognize and correct our deficiencies, to retain the proven practices in the screening and use of reliable hardware, also to retain the integrity of our experienced manpower resources." He added: "The basic challenge in planetary exploration is that of designing and building autonomous systems that can operate unattended for long periods. This is a quality and reliability challenge, augmented by cost and weight limitations, plus a launch-window constraint. In the face of these constraints, we can't buy in large quantities, nor be extensively experimental. We have instead concentrated on implementing a tight quality assurance." (Schneiderman, *Aerospace Management,* Fall/Winter 67, 17–21)

- Highlights of AEC's annual report to Congress for 1967: major system testing for NERVA I and its reactor was planned; post-operation examination revealed no life-limiting problems in ground test unit of SNAP–10A after operation in simulated space environment for 417 days; Project Gasbuggy, first joint industry-government Plowshare experiment, investigated use of underground nuclear explosion to increase productivity in New Mexico natural gas field; two AEC-instrumented Vela satellites for nuclear test detection were launched, raising total now in orbit to eight; 25 underground nuclear tests were conducted. (*Major Activities in the Atomic Energy Programs,* January–December 1967)

Appendix A

SATELLITES, SPACE PROBES, AND MANNED SPACE FLIGHTS

A CHRONICLE FOR 1967

The following tabulation was compiled from open sources by Dr. Frank W. Anderson, Jr., Deputy NASA Historian. Sources included the United Nations Public Registry; the *Satellite Situation Report* issued by the Operations Control Center at Goddard Space Flight Center; public information releases of the Department of Defense, NASA, ESSA, and other agencies, as well as those of the Communications Satellite Corporation. Russian data are from the U.N. Public Registry, the *Satellite Situation Report*, translations from Tass News Agency, statements in the Soviet press, and international news services' reports. Data on satellites of other foreign nations are from the U.N. Public Registry, the *Satellite Situation Report*, government announcements, and international news services' reports.

It might be well to call attention to the terms of reference stated or implied in the title of this tabulation. This is a listing of payloads that have (a) orbited; (b) as probes, ascended to at least the 4,000-mile altitude that traditionally has distinguished probes from sounding rockets, etc.; or (c) conveyed one or more humans into space, whether orbit was attained or not. Furthermore, only flights that succeeded—or at least can be shown by tracking data to have fulfilled our definition of satellite or probe or manned flight—are listed. Date of launch is referenced to local time at the launch site. An asterisk by the date marks those dates that are one day earlier in this tabulation than in listings which reference to Greenwich Mean Time. A double asterisk by the date marks those dates of Soviet launches which are a day later in this compilation that in listings which reference to Greenwich time.

World space activity in 1967 continued at about the same pace as 1966, confirming the leveling-off process which has now seen world space activity at roughly the same level for three years following several years in which activity approximately doubled each year. There was a slight rise in total successful launches—127 against 116 in 1966—and a slight rise in total payloads orbited—159 against 143 in 1966, still under the record 160 of 1965. The difference between launches and payloads is of course accounted for by the multiple-payload launches (DOD is the principal user of this system, with 9 multiple launches orbiting a total of 31 payloads and with as many as 8 payloads on one Titan IIIC vehicle; NASA had one multiple launch of two payloads against none in 1966; the U.S.S.R. again had none).

Of the 1967 world total, the U.S. launched 55 boosters carrying 87 payloads (compared with 92/100 in 1966), the U.S.S.R. launched 67 (compared with 43 in 1966), followed by France with 2, and Australia, Italy, and U.S.-U.K. with 1 each. Of the U.S. total, DOD accounted for 30 launches and 61 payloads (compared with 42/70 in 1966); 4 of NASA's total were non-NASA missions—*Intelsat II–B, Intelsat II–C, Intelsat II–D,* and *Ariel III.* The U.S. and U.S.S.R. each had one orbital test flight in its manned spaceflight program (as opposed to the U.S. 5 Gemini flights in 1966), the Soviet flight—*Soyuz I*—being manned and the pilot being killed in landing. Each of these nations had one planetary flight. The big difference was in lunar flights: the U.S. unmanned lunar exploration program climaxed with 7 flights in 1967 while the U.S.S.R. had none (compared with the U.S.S.R.'s 5/4 edge in 1966 and 5/2 bulge in 1965).

As we have cautioned in previous years, the "Remarks" column of these appendixes is never complete because of the inescapable lag behind each flight of the analysis and interpretation of results.

Launch date	Name	International designation	Vehicle	Payload data	Apogee (st. mi.)	Perigee (st. mi.)	Period (minutes)	Inclination (degrees)	Remarks
Jan. 11	*Intelsat II-B* (United States)	1967-1A	Thor-Delta	Total weight: 192 lbs. (in synchronous orbit; 357.5 lbs. at liftoff, including 165 lbs. of solid-rocket fuel). Objective: Place satellite and apogee motor into proper transfer orbit, provide tracking and telemetry and backup calculations through the transfer orbit so the satellite can be injected into synchronous orbit. Payload: 56″ (dia.) x 26½″ cylindrical satellite, with apogee motor and whip antennas beneath and sleeve rising above and covering antennas; 2 frequency translation mode repeaters; 4 traveling wave tube amplifiers; control system; 2 batteries; 12,756 n-on-p solar cells.	22,904 After apogee motor firing 22,442	185 22,235	654 1,436	26 1.42	Thrust-Augmented Thor-Delta vehicle put *Intelsat II-B* into good transfer orbit; on 1/14 ComSatCorp took over control of the satellite and fired apogee motor, putting satellite into synchronous orbit over the Pacific. Commercial operations began 1/26. Still in orbit, still transmitting.
Jan. 14	DOD Spacecraft (United States)	1967-2A	Thor-Agena D	Total weight: Not available. Objective: Develop spaceflight techniques and technology. Payload: Not available.	227	123	90	80.08	Reentered 2/2/67.
Jan. 18	*IDCSP VIII, IX, X, XI, XII, XIII, XIV, XV* (United States)	1967-3A, B, C, D, E, F, G, H	Titan IIIC	Total weight: 800 lbs. (100 lbs. for each satellite). Objective: Establish an interim defense communication satellite system. Payload: 36″ x 32″ symmetrical 24-face polyhedron satellite, spin stabilized; 40-watt solar cell array; x-band transmitter.	20,963–21,261	20,836–20,927	1,330–1,343	.08–.18	The 8 spacecraft went into near synchronous orbit, joining 7 other IDCSP satellites launched 6/16/66. Still transmitting.
Jan. 19	Cosmos *CXXXVIII* (U.S.S.R.)	1967-4A	Not available	Total weight: Not available. Objective: Continuation of Cosmos scientific satellite series. Payload: Not available.	182	119	89.2	65	Reentered 1/27/67.

Launch date	Name	International designation	Vehicle	Payload data	Apogee (st. mi.)	Perigee (st. mi.)	Period (minutes)	Inclination (degrees)	Remarks
Jan. 26[1]	Cosmos CXXXIX (U.S.S.R.)	1967-5A	Not available	Total weight: Not available. Objective: Continuation of Cosmos scientific satellite series. Payload: Not available.	130	89		50	Reentered 1/26/67.[1]
Jan. 26	Essa IV (United States)	1967-6A	Thor-Delta	Total weight: 290 lbs. Objective: Place an Automatic Picture Transmission satellite of the TOS series into sun-synchronous orbit with a local equator crossing time between 9:00 a.m. and 9:20 a.m., so that APT pictures can be furnished to APT ground stations anywhere in the world regularly. Payload: 22″ x 48″ 18-sided polygon, with 18″ receiving antenna and 4 22″ transmitting whip antennas; containing 2 APT cameras, FM transmitters, 2 spin-control systems (magnetic coil; small solid-propellant rockets), 2 infrared horizon sensors; 8 solar and terrestrial radiation sensors; 63 nickel-cadmium batteries; 9,100 n-on-p solar cells.	888	822	113	102	Essa IV was put into near-polar orbit by Thrust-Augmented Thor-Delta from WTR. All systems functioned normally and ESSA reported photos were of excellent quality. Still in orbit; still transmitting.
Feb. 2	DOD Spacecraft (United States)	1967-7A	Atlas-Agena D	Total weight: Not avaliable. Objective: Develop spaceflight techniques and technology. Payload: Not available.	543	503	101.4	98.8	Reentered 2/12/67.
Feb. 4[1]	Lunar Orbiter III (United States)	1967-8A	Atlas-Agena D	Total weight: 850 lbs. Objective: Acquire from lunar orbit detailed photographic information on various lunar surface areas of interest to Apollo and Surveyor as landing sites, and improve knowledge of the moon.	[Lunar orbit:] 1,117 [Final lunar orbit:] 1,149	130 33	215 209	20.93 20.85	Lunar Orbiter III orbited the moon, photographed all 12 Apollo sites, but a failure in a photo subsystem prevented readout of low resolution photos of the 6 easterly sites; there were high-resolution photos of

ASTRONAUTICS AND AERONAUTICS, 1967

Date	Name	Designation	Launch Vehicle	Payload/Objective	Weight (lbs)	Perigee	Period	Inclination	Remarks
				these. Photographed soft-landed *Surveyor I*. With *Lunar Orbiter II*, acted as tracking target to exercise Apollo network on lunar-distance tracking. Was intentionally crashed onto the moon 10/9/67.					
Feb. 7	Cosmos CXL (U.S.S.R.)	1967-9A	Not available	Payload: 5'0" x 5' (dia.) (when deployed, 18'6" along the antenna booms and 12' across solar panels) conical spacecraft, with body containing attitude control system, retromotor, S-band transmitter, dual-lens (24" and 3" focal length) camera system, 2 radiation dosimeters; high and low gain transmitters; 20 pressurized cells for detecting micrometeoroid flux; 4 solar paddles with a total of 10,856 n-on-p solar cells. Total weight: Not available. Objective: Continuation of Cosmos scientific satellite series. Payload: Not available.	150	106	88.48	51.7	Reentered 2/9/67.
Feb. 8	DOD Spacecraft	1967-10A	Thor-Burner II	Total weight: Not available. Objective: Develop space-flight techniques and technology. Payload: Not available.	543	503	101.4	98.8	Still in orbit.
Feb. 8	*Diademe I* (France)	1967-11A	Diamant	Total weight: 50 lbs. Objective: Orbit geodetic satellite which uses three different systems of geodetic measurement as checks against each other. Payload: 19.7'' (dia.) x 7.8'' circular body, with a single antenna rising from the center of the top side, flanked by four antennas canted outward; outer surface mostly covered with silver-coated quartz laser reflectors; 4 stubby solar-array paddles extend outward from bottom; transmitters controlled in frequency by ultrastable oscillator.	833	363	104	40	Launched from Hammaguir Range, *Diademe I* had unexplained low apogee (1,118 mi apogee was intended), but could still perform 60% of intended mission. Third satellite to be launched by France, *Diademe I* would check geodetic measurements from Doppler shift against ones from laser beams against photographs of the satellite against a star background. Still in orbit.
Feb. 8	Cosmos CXLI (U.S.S.R.)	1967-12A	Not available	Total weight: Not available. Objective: Continuation of Cosmos scientific satellite series. Payload: Not available.	214	180	89.8	72.9	Reentered 2/16/67.

399

ASTRONAUTICS AND AERONAUTICS, 1967

Launch date	Name	International designation	Vehicle	Payload data	Apogee (st. mi.)	Perigee (st. mi.)	Period (minutes)	Inclination (degrees)	Remarks
Feb. 14	Cosmos CXLII (U.S.S.R.)	1967–13A	Not available	Total weight: Not available. Objective: Continuation of Cosmos scientific satellite series. Payload: Not available.	846	113	100.3	48.4	Reentered 7/5/67.
Feb. 15	Diadème II (France)	1967–14A	Diamant	Total weight: 50 lbs. Objective: Orbit geodetic satellite which uses 3 different systems of geodetic measurement as checks against each other. Payload: 19.7″ (dia.) x 7.8″ circular body, with a single antenna rising from the center of the top side, flanked by 4 antennas canted outward; outer surface mostly covered with silver-coated quartz laser reflectors; 4 stubby solar-array paddles extend outward from bottom; transmitters controlled in frequency by ultrastable oscillator.	1,170	368	110.4	39.96	Launched from Hammaguir Range, Diadème II had lower apogee than the 1,300-mi one planned. Fourth satellite launched by France and the last from the Hammaguir Range (which would revert to Algeria on 7/1/67). Diadème II would check geodetic measurements from Doppler shift against ones from laser beams against photographs of the satellite against a star background. Still in orbit.
Feb. 22	DOD Spacecraft (United States)	1967–15A	Thor-Agena D	Total weight: Not available. Objective: Develop spaceflight techniques and technology. Payload: Not available.	228	107	90.1	48.4	Reentered 3/11/67.
Feb. 24	DOD Spacecraft (United States)	1967–16A	Titan IIIB	Total weight: Not available. Objective: Develop spaceflight techniques and technology. Payload: Not available.	242	96	89.8	106.9	Reentered 3/6/67.
Feb. 27	Cosmos CXLIII (U.S.S.R.)	1967–17A	Not available	Total weight: Not available. Objective: Continuation of Cosmos scientific satellite series. Payload: Not available.	188	127	89.5	65	Reentered 3/7/67.

ASTRONAUTICS AND AERONAUTICS, 1967

Date	Name	Designation	Launch Vehicle	Payload/Objective	Weight (lbs)				Remarks
Feb. 28	Cosmos CXLIV (U.S.S.R.)	1967-18A	Not available	Total weight: Not available. Objective: Continuation of Cosmos scientific satellite series. Payload: Not available.	388	388	96.92	81.2	Still in orbit. After Cosmos CLVI was orbited, it and Cosmos CXLIV were revealed to be experimental weather satellites.
Mar. 3	Cosmos CXLV (U.S.S.R.)	1967-19A	Not available	Total weight: Not available. Objective: Continuation of Cosmos scientific satellite series. Payload: Not available.	1,327	137	108.6	48.4	Reentered 3/8/68.
Mar. 8	Oso III (United States)	1967-20A	Thor-Delta	Total weight: 627 lbs. Objective: Continue Oso I and II studies of the sun and its effects on earth's upper atmosphere by obtaining high-resolution spectral data from pointed sun experiments during major portion of 1 solar rotation. Payload: Top part of spacecraft a 22"-radius semi-circular sail continuously pointed at the sun with 3 experiments, 1,860 n-on-p solar cells; lower part a 44" x 9" 9-sided revolving wheel containing 5 non-pointed experiments, controls, telemetry, recorder, and batteries.	354	336	96	33	Oso III was launched into good orbit and experiments all functioned; by 5/15/67 it had exceeded its primary objective by factor of 2. Still in orbit, still transmitting.
Mar. 10	Cosmos CXLVI (U.S.S.R.)	1967-21A	Not available	Total weight: Not available. Objective: Continuation of Cosmos scientific satellite series. Payload: Not available.	193	118	89.2	51.5	Reentered 3/18/67.
Mar. 13	Cosmos CXLVII (U.S.S.R.)	1967-22A	Not available	Total weight: Not available. Objective: Continuation of Cosmos scientific satellite series. Payload: Not available.	196	123	89.5	65	Reentered 3/21/67.
Mar. 16	Cosmos CXLVIII (U.S.S.R.)	1967-23A	Not available	Total weight: Not available. Objective: Continuation of Cosmos scientific satellite series. Payload: Not available.	271	171	91.3	71	Reentered 5/7/67.

Launch date	Name	International designation	Vehicle	Payload data	Apogee (st. mi.)	Perigee (st. mi.)	Period (minutes)	Inclination (degrees)	Remarks
Mar. 21	Cosmos CXLIX (U.S.S.R.)	1967-24A	Not available	Total weight: Not available. Objective: Continuation of Cosmos scientific satellite series. Payload: Not available.	185	154	89.8	48.4	Reentered 4/7/67.
Mar. 22	Cosmos CL (U.S.S.R.)	1967-25A	Not available	Total weight: Not available. Objective: Continuation of Cosmos scientific satellite series. Payload: Not available.	232	128	90.1	65.7	Reentered 3/30/67.
Mar. 22	Intelsat II-C (United States)	1967-26A	Thor-Delta	Total weight: 192 lbs. (in synchronous orbit; 357.5 lbs. at liftoff, including 165 lbs. of solid rocket fuel). Objective: Place satellite and apogee motor into proper transfer orbit, provide tracking and telemetry and backup calculations through the transfer orbit so the satellite can be injected into synchronous orbit. Payload: 56" (dia.) x 26½" cylindrical satellite, with apogee motor and whip antennas below and sleeve rising above and covering antennas; 2 frequency translation mode repeaters; 4 traveling wave tube amplifiers; control system; 2 batteries; 12,756 n-on-p solar cells.	23,114 After apogee motor firing: 22,250.	183 22,221	660.56 1,436	26.56 .804	Intelsat II-C was put into good transfer orbit by Thrust-Augmented Delta vehicle; on 3/25 ComSatCorp took over control and fired apogee motor, putting satellite in synchronous orbit over the Atlantic. Commercial operations began 4/6/67. With Intelsat II-B in synchronous orbit over the Pacific, ComSatCorp now had improved model (240 channel) comsats over both oceans. In addition to regular commercial traffic, they would relay Apollo traffic during flights. Still in orbit, still transmitting.
Mar. 24	Cosmos CLI (U.S.S.R.)	1967-27A	Not available	Total weight: Not available. Objective: Continuation of Cosmos scientific satellite series. Payload: Not available.	391	391	97.1	56	Still in orbit.
Mar. 25	Cosmos CLII (U.S.S.R.)	1967-28A	Not available	Total weight: Not available. Objective: Continuation of Cosmos scientific satellite series. Payload: Not available.	318	176	92.2	71	Reentered 8/5/67.

Date	Name	Designation	Launch Vehicle	Payload / Objective	Weight (lbs)	Perigee/Apogee (mi)	Period (min)	Inclination (°)	Remarks
Mar. 30	DOD Spacecraft (United States)	1967-29A	Thor-Agena D	Total weight: Not available. Objective: Develop spaceflight techniques and technology. Payload: Not available.	246	114	90.2	85	Agena 2nd-stage engine failed on restart; payload did not achieve intended 6,900-mi orbit. Reentered 4/17/67.
Apr. 4	Cosmos CLIII (U.S.S.R.)	1967-30A	Not available	Total weight: Not available. Objective: Continuation of Cosmos scientific satellite series. Payload: Not available.	181	126	89.3	64.6	Reentered 4/12/67.
Apr. 5	Ats II (United States)	1967-31A	Atlas-Agena D	Total weight: 715 lbs. Objective: Place spacecraft in medium-altitude orbit of about 6,900 mi and obtain design and operational data on gravity-gradient stabilization; also obtain data from its experiments for at least 30 days. Payload: 56" (dia.) x 72" cylindrical satellite; cylinder is in 3 segments, the 2 end ones faced with total of 22,344 solar cells, the middle one (24" tall) thermally insulated and containing experiments; gravity-gradient stabilization system has 4 main booms, each 127' long, forming an X, and damper boom of 24 5'-long sections orthogonal to main booms; 8 telemetry antennas are mounted on top rim; 7 applications and technology experiments—AVCS cameras, omnidirectional detector, VLF detector, cosmic radio noise measurement, solar-cell damage measurement, thermal coatings, albedo experiments—and 6 scientific experiments—electron spectrometer, particle telescope, 4 electric-field measurement devices; 4 telemetry transmitters; 2 command receivers; 2 batteries.	6,033	101	219.7	28.3	Agena engine failed to ignite for 2nd burn, so Ats II did not achieve intended 6,900-mi circular orbit. This limited usefulness of gravity-gradient experiment and of most other experiments; although all experiments returned data, mission was adjudged a failure. Still in orbit, still transmitting.
Apr. 8	Cosmos CLIV (U.S.S.R.)	1967-32A	Not available	Total weight: Not available. Objective: Continuation of Cosmos scientific satellite series. Payload: Not available.	144	116	88.5	51.6	Reentered 4/10/67.

ASTRONAUTICS AND AERONAUTICS, 1967

Launch date	Name	International designation	Vehicle	Payload data	Apogee (st. mi.)	Perigee (st. mi.)	Period (minutes)	Inclination (degrees)	Remarks
Apr. 12	Cosmos CLV (U.S.S.R.)	1967-33A	Not available	Total weight: Not available. Objective: Continuation of Cosmos scientific satellite series. Payload: Not available.	178	126	89.2	51.8	Reentered 4/20/67.
Apr. 13	DOD Spacecraft (United States)	1967-34A	Scout	Total weight: Not available. Objective: Provide a navigation satellite capability. Payload: Not available.	673	665	106.3	90.2	Still in orbit.
Apr. 17	Surveyor III (United States)	1967-35B	Atlas-Centaur	Total weight: 2,200 lbs. (weight at launch, including 1,444-lb. retromotor, propellants, etc.; weight of Surveyor landed on the moon, 620 lbs.). Objective: Softland on moon within the Apollo zone and east of Surveyor I landing site; obtain post-landing TV photos of lunar surface; manipulate lunar surface with surface sampler and observe effects with TV camera. Payload: 10'-high x 14' (around 3 extended landing gear) spacecraft, consisting of triangular aluminum frame to which are attached: a mast supporting rotatable planar array antenna and solar panel (with 3,960 solar cells); 2 folding booms deploying conical omnidirectional antennas; thermal compartment housing 2 transmitters, 2 receivers, silver-zinc battery, thermal compartment housing decoder and signal processing equipment; 2 altitude-radar antennas; survey TV camera; extendable surface sampler; retromotor; 3 vernier motors.	Softlanded on moon				Surveyor III was launched into earth orbit, then put into lunar-transfer trajectory by Centaur stage. Softlanded 4/19 in eastern part of Oceanus Procellarum, though bounding 3 times in landing; during 1st lunar day Surveyor III took 6,315 TV pictures, operated surface sampler to confirm lunar bearing strength of 3 to 8 psi, firm enough to support Apollo lander or astronaut. Since spacecraft was at 14° angle on inside slope of a crater, it was able to photograph the earth in color and a solar eclipse. Survived 1 lunar night but could not be raised for communication after 5/23/67.

404

ASTRONAUTICS AND AERONAUTICS, 1967

Date	Name (Country)			Objective/Payload				Remarks
	(United States)			Objective: Place and operate Advanced Vidicon Camera System in sun-synchronous orbit with a local equator crossing time between 3:00 p.m. and 3:20 p.m., so that daily AVCS pictures of the entire earth can be obtained regularly. Payload: 22″ x 42″ 18-sided hatbox-shaped polygon, with 18″ receiving antenna and 4 22″ transmitting whip antennas; containing 2 AVCS cameras, FM transmitters, 2 spin-control systems (magnetic coil; small solid-propellant rockets), 2 infrared horizon sensors; 8 solar and terrestrial radiation sensors; 63 nickel-cadmium batteries; 9,100 n-on-p solar cells.		110.0	101.5	ESSA V was put into near-polar orbit by Thrust-Augmented Thor-Delta from WTR. All systems functioned normally; after extensive postlaunch evaluation, the satellite was turned over to ESSA to operate on 5/8/67. Still in orbit, still transmitting.
Apr. 23	Soyuz I (U.S.S.R.)	1967-37A	Not available	Total weight: Not available. Objective: Test manned spacecraft, continue biological experiments. Payload: Not available.	139	88.1	51.4	Cosmonaut Vladimir A. Komarov was killed 4/24 when his spacecraft reentered prematurely—on 18th orbit after 26 hrs 45 min of flight time—when tumbling of spacecraft apparently caused parachute to foul and fail to slow descent sufficiently. Cosmonaut Komarov was 1st man to die while in spaceflight.
Apr. 26	San Marco II (Italy)	1967-38A	Scout	Total weight: 285 lbs. Objective: Orbit a satellite to measure air density in altitude range between about 120 mi and 200 mi and electron content between spacecraft and ground and in locality of spacecraft. Payload: 20″ spherical satellite consisting of a heavy sphere contained in a much lighter one as a device for measuring air density; 4 battery packages; telemetry and command radio; 4 19″ antennas for telemetry, 2 retractable 100′ antennas for the electron experiment.	465	136	94	2.89	San Marco II was launched into good equatorial orbit by U.S.-furnished Scout vehicle handled by Italian crew. Launching took place from sea platforms in Indian Ocean off Kenya, Africa, the 1st satellite to be launched from a sea platform. Experiments performed well. Reentered 10/14/67.

Launch date	Name	International designation	Vehicle	Payload data	Apogee (st. mi.)	Perigee (st. mi.)	Period (minutes)	Inclination (degrees)	Remarks
Apr. 27	*Cosmos CLVI* (U.S.S.R.)	1967-39A	Not available	Total weight: Not available. Objective: Continuation of Cosmos scientific satellite series. Payload: Not available.	398	364	96.9	81.19	*Cosmos CLVI* was later revealed to be a meteorological satellite; together with *Cosmos CXLIV* it would form experimental system known as "Meteor," would obtain data on ½ of earth's surface each 24 hours.
Apr. 28	*Vela Hotel VII* (United States) and *Vela Hotel VIII*	1967-40A and 40B	Titan IIIC	Total weight: 508 lbs. (for both Vela satellites). Objective: Conduct research on techniques of nuclear-test detection. Payload: 56" x 46" cylindrical 26-sided polyhedron, consisting of reaction wheel plus gas jets for stabilization, 120-watt solar cell array.	A: 69,972 B: 71,216	67,799 66,805	6,655.1 6,671.8	32.1 33.1	Five spacecraft launched with single booster. Both Velas still in orbit.
	and *Ers-XVIII*	1967-40C		Total weight: 20 lbs. Objective: Measure natural radiation background in space. Payload: 11" octahedron; spin stabilized, random orientation; 5.5-watt solar cell array.	69,123	5,338	2,831.3	32.8	Still in orbit.
	and *Ers-XX*	1967-40D		Total weight: 14 lbs. Objective: Correlate differences in friction reaction of materials in space and in vacuum chambers. Payload: 11" octahedron; spin stabilized, random orientation; 5.5-watt solar cell array.	69,122	5,338	2,831.2	32.8	Still in orbit.

	Ers-XXVII	1967-40E		Total weight: 20 lbs. Objective: Provide single-station readout of x-ray (0.5-14Å), electron (.04-33 Mev), and proton (1-80 Mev) detectors. Payload: 11" octahedron; spin stabilized, random orientation; 5.5-watt solar cell array.	69,122	5,388	2,831	32.8	Still in orbit.
May 4	Lunar Orbiter IV	1967-41A	Atlas-Agena D	Total weight: 850 lbs. Objective: Obtain from high-inclination lunar orbit a broad, systematic photo survey of major portion of lunar surface to increase scientific knowledge and to provide basis for detailed study by subsequent orbital and landing missions. Payload: 5'6" x 5' (dia.) (when deployed, 18'6" along the antenna booms and 12' across solar panels) conical spacecraft, with body containing attitude control system, retromotor, S-band transmitter, dual-lens (24" and 3" focal length) camera system, 2 radiation dosimeters; high and low gain transmitters; 20 pressurized cells for detecting micrometeoroid flux; 4 solar paddles with total of 10,856 n-on-p solar cells.	[Lunar orbital data:] 3,797	1,681	721	85.5	Lunar Orbiter IV went into elliptical lunar orbit, photographed 99% of earth side of moon, raised to 80% the total coverage of hidden side of the moon by the 4 Lunar Orbiters. Photo readout was completed 6/1; on 7/17 contact with spacecraft was lost; from orbital tracking data Lunar Orbiter IV is presumed to have crashed onto the moon in Oct.-Nov. 1967.

Launch date	Name	International designation	Vehicle	Payload data	Apogee (st. mi.)	Perigee (st. mi.)	Period (minutes)	Inclination (degrees)	Remarks
May 5	Ariel III (United States-U.K.)	1967-42A	Scout	Total weight: 198 lbs. (in orbit; 226 lbs. at launch, including separation system). Objective: Place U.K. satellite in desired orbit, provide tracking and telemetry support; U.K. to conduct radiation and noise measurements in earth's atmosphere. Payload: 35" x 23" (dia.) 12-sided cylinder with cone on top surmounted by turnstile antenna; 4 booms, when deployed, point outward and downward, support solar-cell trays (these cells plus the ones mounted on the satellite total 8,000), and 2 experiments; telemetry system; 1 nickel-cadmium battery; transmitter-receiver; tape recorder; 5 experiments measuring molecular oxygen, galactic noise, vlf radiation, ionisation and temperature, and terrestrial radio noise.	373	306	95.6	80	Third U.K.-U.S. satellite launching was successful; orbit was good; all experiments operated and returned good data. Still in orbit, still transmitting.
May 9	DOD Spacecraft (United States)	1967-43A	Thor-Agena D	Total weight: Not available. Objective: Develop spaceflight techniques and technology. Payload: Not available.	492	114	94.3	85	Two payloads launched with single booster. 43A reentered 7/13/67.
	and DOD Spacecraft	1967-43B		Total weight: Not available. Objective: Develop spaceflight techniques and technology. Payload: Not available.	501	346	98.4	84.9	43B still in orbit.

Date	Name	Designation	Launch Vehicle	Description				Remarks
May 12	Cosmos CLVII (U.S.S.R.)	1967-44A	Not available	Total weight: Not available. Objective: Continuation of Cosmos scientific satellite series. Payload: Not available.	184	126	89.4	51.3	Reentered 5/20/67.
May 15	Cosmos CLVIII (U.S.S.R.)	1967-45A	Not available	Total weight: Not available. Objective: Continuation of Cosmos scientific satellite series. Payload: Not available.	528	528	100.7	74	Still in orbit.
May 17	Cosmos CLVIX (U.S.S.R.)	1967-46A	Not available	Total weight: Not available. Objective: Continuation of Cosmos scientific satellite series. Payload: Not available.	37,655	236	1,173	51	Still in orbit.
May 17	Cosmos CLX (U.S.S.R.)	1967-47A	Not available	Total weight: Not available. Objective: Continuation of Cosmos scientific satellite series. Payload: Not available.	127	88		49.6	Reentered 5/18/67.
May 18	DOD Spacecraft (United States)	1967-48A	Scout	Total weight: Not available. Objective: Develop spaceflight techniques and technology. Payload: Not available.	678	667	106.9	89.5	Still in orbit.
May 22	Cosmos CLXI (U.S.S.R.)	1967-49A	Not available	Total weight: Not available. Objective: Continuation of Cosmos scientific satellite series. Payload: Not available.	213	127	89.8	65.7	Reentered 5/30/67.
May 22	DOD Spacecraft (United States) and	1967-50A	Atlas-Agena D	Total weight: Not available. Objective: Develop spaceflight techniques and technology. Payload: Not available.	151	102	89.3	91.5	Two payloads launched with single booster. 50A reentered 5/30/67.
	DOD Spacecraft	1967-50B		Total weight: Not available. Objective: Develop spaceflight techniques and technology. Payload: Not available.	121	99	88.3	91.4	50B reentered 5/27/67.

Launch date	Name	International designation	Vehicle	Payload data	Apogee (st. mi.)	Perigee (st. mi.)	Period (minutes)	Inclination (degrees)	Remarks
May 24	*Explorer XXXIV* (United States)	1967-51A	Thor-Delta	Total weight: 163 lbs. Objective: Place spacecraft into orbit with 90,000 mi or more apogee, investigate the region between the shock front and the magnetosheath and obtain data on the cislunar environment. Payload: 28" (dia.) x 8" octagonal spacecraft, with 4 telemetry antennas extending from the top and 2 7' booms deploying flux-gate magnetometers and 4 solar paddles extending from the sides; telemetry system and transmitter; 6,144 n-on-p solar cells; silver-cadmium battery pack; 11 scientific experiments, including 7 devices for measuring energetic particles, 2 for measuring energetic particles and plasma, 1 for measuring plasma, and 1 for measuring magnetic fields.	133,131	150	6,360	67.2	*Explorer XXXIV* was launched into good highly elliptical orbit by Thrust-Augmented Thor-Delta vehicle from WTR; 10 of 11 experiments returned data; mission was adjudged a success when spacecraft returned data in excess of 3 months. Still in orbit, still transmitting.
May 24 [1]	*Molniya I-5* (U.S.S.R.)	1967-52A	Not available	Total weight: Not available. Objective: Develop and further improve a satellite radio and TV communication system. Payload: Satellite with transmitter, command system, orientation system, orbit correction device, power supply.	19,672	285	715	64.8	*Molniya I-5* was thought to be a replacement for *Molniya I-8*, which prematurely reentered 3/17/67. Still in orbit.
May 31	DOD Spacecraft (United States)	1967-53A, B, C, D, E, F, G, H	Thor-Agena D	Total weight: Not available. Objective: Develop spaceflight techniques and technology. Payload: Not available.	572-578	568-576	103.5	69.9	Eight payloads launched with single booster. All still in orbit. 53O and D were later identified as *Gravity Gradient 4* and *6*.

ASTRONAUTICS AND AERONAUTICS, 1967

Date	Name	Designation	Launch Vehicle	Objective/Payload	(col)	(col)	(col)	(col)	Remarks
June 1	Cosmos CLXII (U.S.S.R.)	1967-54A	Not available	Total weight: Not available. Objective: Continuation of Cosmos scientific satellite series. Payload: Not available.	174	125	89.2	51.8	Reentered 6/9/67.
June 4	DOD Spacecraft (United States)	1967-55A	Atlas-Agena D	Total weight: Not available. Objective: Develop spaceflight techniques and technology. Payload: Not available.	157	148	89.3	104.8	Reentered 6/12/67.
June 5	Cosmos CLXIII (U.S.S.R.)	1967-56A	Not available	Total weight: Not available. Objective: Continuation of Cosmos scientific satellite series. Payload: Not available.	329	162	93.1	48.4	Reentered 10/11/67.
June 8	Cosmos CLXIV (U.S.S.R.)	1967-57A	Not available	Total weight: Not available. Objective: Continuation of Cosmos scientific satellite series. Payload: Not available.	199	126	89.5	69.4	Reentered 6/14/67.
June 12	Venera IV (U.S.S.R.)	1967-58A	Not available	Total weight: 2,438 lbs. Objective: Place spacecraft in orbit to planet Venus. Payload: Not available.	Burned up in Venusian atmosphere.				Venera IV (Venus IV) burned up in thick Venusian atmosphere after having ejected an instrumented capsule that parachuted to a landing on Venus. Data from capsule held the atmosphere to be almost wholly carbon dioxide (1.5% oxygen and water vapor), temperature range from 104–536°F, no noticeable magnetic field, no radiation belt, but with faint hydrogen corona. Signals ceased at impact, 10/18/67.
June 12	Cosmos CLXV (U.S.S.R.)	1967-59A	Not available	Total weight: Not available. Objective: Continuation of Cosmos scientific satellite series. Payload: Not available.	958	131	102.1	81.9	Reentered 1/15/68.

ASTRONAUTICS AND AERONAUTICS, 1967

Launch date	Name	International designation	Vehicle	Payload data	Apogee (st. mi.)	Perigee (st. mi.)	Period (minutes)	Inclination (degrees)	Remarks
June 14	*Mariner V* (United States)	1967-60A	Atlas-Agena D	Total weight: 542 lbs. Objective: Conduct a flyby of Venus and obtain scientific information that complements and extends information obtained by *Mariner II* relating to the origin and nature of Venus and of its environment. Payload: 9'6''-high x 18' (cruise position, with solar panels and antennas deployed)-structure; basically a 50'' octagon with 4 solar panels extended horizontally and a pole low-gain antenna and a dish high-gain antenna extending from the top; transmitters; control system; 5 experiments to measure interplanetary and Venusian phenomena; 16,700 solar cells; 33-lb. silver-zinc battery.	0.74 au	0.58 au	194.6 (days)	1.39	*Mariner V* flew past Venus at 2,544-mi altitude on 10/19/67 and continued traveling in heliocentric orbit. All experiments returned good data, showed evidence of interaction of solar wind with Venus, very weak interplanetary field if any, no radiation belts, a temperature at extreme outer edge of Venusian atmosphere of 700°F (compared with 1,300°F for comparable edge of earth's atmosphere), detected hydrogen corona as strong as earth's, and provided refined data on relative masses of earth/moon and Venus. Still in heliocentric orbit, still operating.
June 16	*Cosmos CLXVI* (U.S.S.R.)	1967-61A	Not available	Total weight: Not available. Objective: Continuation of Cosmos scientific satellite series. Payload: Not available.	359	175	92.9	48.4	Reentered 10/25/67; provided solar-flare data for continuous 3-mo period.
June 16	DOD Spacecraft (United States)	1967-62A	Thor-Agena D	Total weight: Not available. Objective: Develop spaceflight techniques and technology. Payload: Not available.	175	105	90	80	Two payloads launched with single booster. 62A reentered 7/20/67.
and	DOD Spacecraft	1967-62B		Total weight: Not available. Objective: Develop spaceflight techniques and technology. Payload: Not available.	329	326	94.9	80.2	62B reentered 10/22/68.

ASTRONAUTICS AND AERONAUTICS, 1967

Date	Name	Designation	Launch Vehicle	Objective/Payload					Remarks
June 17	Cosmos CLXVII (U.S.S.R.)	1967-63A	Not available	Total weight: Not available. Objective: Continuation of Cosmos scientific satellite series. Payload: Not available.	369		175	92.9	48.4	Reentered 6/25/67.
June 20	DOD Spacecraft (United States)	1967-64A	Titan IIIB	Total weight: Not available. Objective: Develop spaceflight techniques and technology. Payload: Not available.	175		88	89.9	111.5	Reentered 6/30/67.
June 29	Egrs IX (United States) and	1967-65A	Thor-Burner II	Total weight: Not available. Objective: Provide an aid to the geodetic survey of the earth sphere. Payload: Not available.	2,457		2,378	172.1	89.8	Two payloads launched with single booster. 65A still in orbit.
	Aurora I	1967-65B		Total weight: 40 lbs. Objective: Study the charged particles that precipitate in the upper atmosphere and cause auroras. Payload: Rectangular box structure containing scientific instrumentation; solar-cell array.	2,451		2,366	172.1	89.8	65B still in orbit.
July 1	IDCSP XVI, XVII, XVIII (United States) and	1967-66A, B, C	Titan IIIC	Total weight: 300 lbs. (100 lbs. for each satellite). Objective: Establish an interim defense communication system. Payload: 36" x 32" symmetrical 24-face polyhedron satellite, spin stabilized; 40-watt solar-cell array; x-band transmitter.	20,833–20,848		20,497–20,555	1,308.8–1,311.7	7.1–7.2	Six spacecraft launched with single booster. 66A, B, and C were placed in near-synchronous orbit constellation along with 14 of 15 previously launched IDCSP satellites. Still in orbit.
	Dats and	1967-66D		Total weight: Not available. Objective: Test and evaluate a mechanically despun antenna system for a spin-stabilized spacecraft. Payload: Not available.	20,857		20,594	1,313.6	7	66D still in orbit.

413

Launch date	Name	International designation	Vehicle	Payload data	Apogee (st. mi.)	Perigee (st. mi.)	Period (minutes)	Inclination (degrees)	Remarks
July 1 —Con.	*Lee V*	1967-66E		Total weight: Not available. Objective: Provide test data on feasibility of a tactical communication satellite operating with various types of ground antennas. Payload: Not available.	20,853	20,609	1,315.4	7	66E still in orbit.
	and								
	Dodge I	1967-66F		Total weight: 430 lbs. Objective: Operate and evaluate a passive 3-axis satellite with gravity-gradient attitude control at near-synchronous altitude. Payload: Truncated octahedron; gravity-gradient and flywheel stabilization; solar-cell array.	20,926	20,685	1,319	7	66F still in orbit.
July 4	*Cosmos CLXVIII* (U.S.S.R.)	1967-67A	Not available	Total weight: Not available. Objective: Continuation of Cosmos scientific satellite series. Payload: Not available.	167	124	89	52	Reentered 7/12/67.
July 14	*Surveyor IV* (United States)	1967-68A	Atlas-Centaur	Total weight: 2,287 lbs. (weight at launch, including 1,444-lb. retromotor, propellants, etc.; weight of Surveyor lander at moon, 620 lbs.). Objective: Softland on moon in Sinus Medii, obtain post-landing TV photos of lunar surface. Payload: 10'-high x 14' (around 3 extended landing gear) spacecraft, consisting of triangular frame to which are attached: a mast supporting	Landed on moon.				*Surveyor IV* had excellent launch and flight to moon. Initial trajectory was so accurate—spacecraft would have landed within 90 mi of aiming point—that midpoint correction was postponed 24 hrs and achieved accuracy within 5 mi. Lunar descent was begun 36 min prior to landing; at less than 3 min prior to landing all communication with the spacecraft was suddenly lost.

Date	Name	Cospar Designation	Launch Vehicle	Payload/Objective	Weight (lbs)			Remarks
July 17	Cosmos CLXIX (U.S.S.R.)	1967-69A	Not available	rotatable planar array antennas and solar panel (with 3,960 solar cells); 3 folding booms deploying conical omnidirectional antennas; extendable surface sampler; 2 mirrors; thermal compartment housing decoder and signal processing equipment; 2 altitude radar antennas; survey TV camera; retromotor; small bar magnet attached to footpad. Total weight: Not available. Objective: Continuation of Cosmos scientific satellite series. Payload: Not available.	129			Spacecraft could have completed soft landing but probably crashed. Cause for communications loss could not be determined.
July 19	Explorer XXXV (United States)	1967-70A	Thor-Delta	Total weight: 230 lbs. Objective: Place spacecraft in either a captured lunar orbit or a geocentric orbit with apogee near or beyond lunar distance, obtain data on the interplanetary plasma and the interplanetary magnetic field. Payload: 28″ (dia.) x 8″ octagonal spacecraft, with 4 telemetry antennas and a 21″-long retromotor extending from the top and 2 7′ booms deploying fluxgate magnetometers and 4 solar paddles extending from the sides; telemetry system and transmitter; 7,660 solar cells; silver-cadmium battery pack; 7 scientific experiments, including 3 on energetic particles, 1 on plasma, 2 on magnetic fields, and 1 on micrometeoroid flux.	[Lunar orbital data:] 4,760.	497	690	88	Reentered 7/17/67.
								Explorer XXXV's retromotor was fired on 7/22, slowing the spacecraft enough to allow it to be captured into lunar orbit. By 8/9/67 it had orbited the moon 33 times, had been adjudged a success, and its data had disproved theory that the moon had captured interplanetary magnetic field lines and formed a lunar magnetosphere. Still in orbit, still transmitting.
July 25	DOD Spacecraft (United States)	1967-71A	Thor-Agena D	Total weight: Not available. Objective: Develop spaceflight techniques and technology. Payload: Not available.	378	243	94.5	75.1	Still in orbit.

Launch date	Name	International designation	Vehicle	Payload data	Apogee (st. ml.)	Perigee (st. ml.)	Period (minutes)	Inclination (degrees)	Remarks
July 27	*OV I-86* (United States) and	1967-72A	Atlas	Total weight: Not available. Objective: Develop space-flight techniques and technology. Payload: Not available.	431	346	95.5	101.6	Two payloads launched with single booster. 72A still in orbit.
	OV I-12	1967-72D		Total weight: Not available. Objective: Develop space-flight techniques and technology. Payload: Not available.	431	346	95.5	101.6	72D still in orbit.
July 28	*Ogo IV* (United States)	1967-73A	Thor-Agena D	Total weight: 1,240 lbs. Objective: Obtain data on latitude-dependent phenomena from an attitude-stabilized platform and for a period of perigee rotation from the Northern Hemisphere across the Arctic pole into the Southern Hemisphere (about 50 days). Payload: 67" x 32" x 31" rectangular parallelepiped spacecraft, containing many of the experiments and the subsystems such as telemetry, attitude control, temperature control, nickel-cadmium batteries; deployed from the ends of the spacecraft are 2 22' booms and 4 4' booms, all of which mount experiments; deployed from the sides of the spacecraft are 2 large solar paddles, from the end of one of which extends a 60' experimental antenna; total of 20 experiments in and deployed from the spacecraft, including ones for atmospheric and ionospheric measurements, solar radiation measurements,	564	256	97.9	86.01	*Ogo IV* achieved good orbit, received data from all 20 experiments; by 10/6/67 it had passed its 50-day operating requirement with 18 experiments still functioning. Still in orbit, still transmitting.

Date	Name	Cospar designation	Launch vehicle	Payload/Objective				Remarks
July 31	Cosmos CLXX (U.S.S.R.)	1967-74A	Not available	radio measurements, magnetic field measurements, and cosmic ray experiments. Total weight: Not available. Objective: Continuation of Cosmos scientific satellite series. Payload: Not available.			50	Reentered 7/31/67.
Aug. 1	Lunar Orbiter V (United States)	1967-75A	Atlas-Agena D	[Lunar orbital data:] 3,734. Total weight: 850 lbs. Objective: Place 3-axis-stabilized spacecraft in high-inclination lunar orbit and obtain photos of selected scientifically interesting areas on lunar surface and supplemental photos of candidate Apollo landing sites. Payload: 5'6" x 5' (dia.) (when deployed, 18'6" along the antenna booms and 12' across solar panels) conical spacecraft, with body containing attitude-control system, retromotor, S-band transmitter, dual-lens (24" and 3" focal length) camera system, 2 radiation dosimeters; high and low gain transmitters; 20 pressurized cells for detecting micrometeoroid flux; 4 solar paddles with a total of 10,856 n-on-p solar cells.	121	510	85	Lunar Orbiter V achieved elliptical orbit around the moon, photographed 23 previously unphotographed areas on the moon's far side, the 1st photo of the "full earth," 36 sites of scientific interest, and 5 Apollo sites. With this mission, Lunar Orbiters had photographed substantially the entire lunar surface at about 10 times the resolution possible from earth and had provided large and small scale coverage of all Apollo landing sites. Also each spacecraft had remained in lunar orbit for a time to provide meteoroid-impact data, information on lunar gravity, and a valuable means of exercising the Apollo tracking net with actual problems of lunar orbit. Lunar Orbiter V was crashed onto the face of the moon 1/31/68.
Aug. 7	DOD Spacecraft (United States)	1967-76A	Thor-Agena D	Total weight: Not available. Objective: Develop spaceflight techniques and technology. Payload: Not available.	117	89.8	79.9	Reentered 9/1/67.
Aug. 8	Cosmos CLXXI (U.S.S.R.)	1967-77A	Not available	Total weight: Not available. Objective: Continuation of Cosmos scientific satellite series. Payload: Not available.	90		50	Reentered 8/8/67.

Launch date	Name	International designation	Vehicle	Payload data	Apogee (st. mi.)	Perigee (st. mi.)	Period (minutes)	Inclination (degrees)	Remarks
Aug. 9	Cosmos CLXXII (U.S.S.R.)	1967-78A	Not available	Total weight: Not available. Objective: Continuation of Cosmos scientific satellite series. Payload: Not available.	187	125	89	52	Reentered 8/17/67.
Aug. 16	DOD Spacecraft (United States)	1967-79A	Titan IIIB	Total weight: Not available. Objective: Develop spaceflight techniques and technology. Payload: Not available.	277	77	90.1	111.5	Reentered 8/23/67.
Aug. 23	DOD Spacecraft (United States)	1967-80A	Thor-Burner II	Total weight: Not available. Objective: Develop spaceflight techniques and technology. Payload: Not available.	556	517	102.1	98.9	Still in orbit.
Aug. 24	Cosmos CLXXIII (U.S.S.R.)	1967-81A	Not available	Total weight: Not available. Objective: Continuation of Cosmos scientific satellite series. Payload: Not available.	328	174	92	71	Reentered 12/17/67.
Aug. 31	Cosmos CLXXIV (U.S.S.R.)	1967-82A	Not available	Total weight: Not available. Objective: Continuation of Cosmos scientific satellite series.	24,699	311	715	64.5	Reentered 12/30/68.
Sept. 7	Biosatellite II (United States)	1967-83A	Thor-Delta	Payload: Not available. Total weight: 955 lbs. (including 275-lb. experiment capsule). Objective: Conduct biological experiments on effects of weightlessness and controlled gamma radiation aboard an attitude-controlled spacecraft with gravity effect below 10⁻⁵ for 3 days and recover experiment capsule. Payload: 72″ × 56″ (dia. at base) cylinder-cone adapter section, covered with thermal insulating material, and con-	202	187	90.8	33.5	Biosatellite II was placed in good orbit by 2-stage Thrust-Augmented Delta; all experiments and the life-support system operated. Communications difficulties and prospects of bad weather in the recovery area led to decision to cause spacecraft reentry on orbit 30 (after 45 hrs. of weightlessness) rather than orbit 46 (after 70 hrs. of weightlessness) as planned. Spacecraft re-

Date					
			taining attitude-control system (nitrogen jets, IR sensors, magnetometer), telemetry (transmitter, 2 receivers, decoders, silver-zinc batteries, and power controller); antennas, 40" (dia. at base) reentry vehicle, containing experiment capsule with 13 biological experiments; strontium-85 radiation source and dosimeter; life-support system; separation and entry systems; de-orbit telemetry transmitter, programmer, tracking beacon, tape recorder; recovery system.		entered, was caught in midair by a recovery aircraft over the Pacific. Most of the experiments suffered only minor loss of data because the flight period was shortened. Quick-look results indicated that gravity does control direction of plant growth; bacteria had increased 20 to 30% faster in weightless state, suggesting that astronauts on future flights of years' duration might need artificial gravity or drugs as protection against too rapid growth of human cells.
Sept. 8	*Surveyor V* (United States)	1967-84B	Atlas-Centaur	Total weight: 2,218 lbs. (weight at launch, including 1,444-lb. retromotor, propellants, etc.; weight of Surveyor lander at moon, 619 lbs.) Objective: Softland on moon in Mare Tranquillitatis; obtain post-landing TV pictures of lunar surface; conduct vernier-engine erosion experiment. Payload: 10'-high x 14' (around 3 extended landing gear) spacecraft, consisting of triangular frame to which are attached: a mast supporting rotatable planar array antenna and solar panel; 2 folding booms deploying conical omnidirectional antennas; alpha scattering experiment; thermal compartment housing decoder and signal processing equipment; 2 altitude radar antennas; survey TV camera; retromotor; small bar magnet attached to footpad.	Softlanded on moon.	*Surveyor V* was launched into good parking orbit and then into good lunar transfer trajectory; leaking valve in helium tank lost some helium pressure, forced change in lunar landing procedure, retromotor having to fire longer and vernier engines being turned on only in the last few thousand feet. Spacecraft landed safely in Mare Tranquillitatis on 9/10; during its 1st lunar day it took 18,006 pictures, more than *Surveyor I* and *Surveyor III* combined. Survived lunar night, took another 1,048 photos of poorer quality. Alpha-scattering experiment showed chemistry of lunar soil to be that of basaltic rock; vernier engine firing produced observable erosion. Spacecraft was shut down 11/1/67.

ASTRONAUTICS AND AERONAUTICS, 1967

Launch date	Name	International designation	Vehicle	Payload data	Apogee (st. mi.)	Perigee (st. mi.)	Period (minutes)	Inclination (degrees)	Remarks
Sept. 11	Cosmos CLXXV (U.S.S.R.)	1967-85A	Not available	Total weight: Not available. Objective: Continuation of Cosmos scientific satellite series. Payload: Not available.	240	130	92	72.9	Reentered 9/19/67.
Sept. 12	Cosmos CLXXVI	1967-86A	Not available	Total weight: Not available. Objective: Continuation of Cosmos scientific satellite series. Payload: Not available.	982	126	102.5	81.9	Reentered 3/3/68.
Sept. 15	DOD Spacecraft (United States)	1967-87A	Thor-Agena D	Total weight: Not available. Objective: Develop space-flight techniques and technology. Payload: Not available.	245	94	89.9	80.1	Reentered 10/4/67.
Sept. 16	Cosmos CLXXVII (U.S.S.R.)	1967-88A	Not available	Total weight: Not available. Objective: Continuation of Cosmos scientific satellite series. Payload: Not available.	166	125	89.1	51.7	Reentered 9/24/67.
Sept. 19	Cosmos CLXXVIII (U.S.S.R.)	1967-89A	Not available	Total weight: Not available. Objective: Continuation of Cosmos scientific satellite series. Payload: Not available.	193	85	88.6	50	Reentered 9/19/67.
Sept. 19	DOD Spacecraft (United States)	1967-90A	Titan IIIB	Total weight: Not available. Objective: Develop space-flight techniques and technology. Payload: Not available.	214	76	89.1	106	Reentered 9/30/67.
Sept. 22	Cosmos CLXXIX (U.S.S.R.)	1967-91A	Not available	Total weight: Not available. Objective: Continuation of Cosmos scientific satellite series. Payload: Not available.	98	88	87.3	49	Reentered 9/22/67.

ASTRONAUTICS AND AERONAUTICS, 1967

Date	Spacecraft	Designation	Launch vehicle	Objective/Payload	Weight (lbs)	Perigee/Apogee	Period (min)	Inclination	Remarks
Sept. 25	DOD Spacecraft (United States)	1967-92A	Not available	Total weight: Not available. Objective: Provide navigation satellite capability. Payload: Not available.	693	648	106.7	89.2	Still in orbit.
Sept. 26	Cosmos CLXXX (U.S.S.R.)	1967-93A	Not available	Total weight: Not available. Objective: Continuation of Cosmos scientific satellite series. Payload: Not available.	218	128	90.1	72.9	Reentered 10/4/67.
Sept. 27	Intelsat II-D (United States)	1967-94A	Thor-Delta	Total weight: 192 lbs. (in synchronous orbit; 357.5 lbs. at liftoff, including 165 lbs. of solid-rocket fuel). Objective: Place satellite and apogee motor into proper transfer orbit, provide tracking and telemetry and backup calculations through the transfer orbit so the satellite can be injected into synchronous orbit. Payload: 56″ (dia.) x 26½″ cylindrical satellite, with apogee motor and whip antennas below, and sleeve rising above and covering antennas; 2 frequency translation mode repeaters; 4 traveling wave tube amplifiers; control system; 2 batteries; 12,756 n-on-p solar cells.	23,056 After apogee motor firing: 22,240.	186 22,236	656.8 1,436	26.4 .73	Thrust-Augmented Thor-Delta booster put Intelsat II-D into good transfer orbit; on 9/30 ComSatCorp took over control and fired apogee motor, putting satellite into synchronous orbit over the Pacific. Commercial communications operations began on 11/4/67; Intelsat II-D would also carry Apollo traffic during flights. Still in orbit, still transmitting.
Oct. 3	Molniya I-8 (U.S.S.R.)	1967-95A	Not available	Total weight: Not available. Objective: Develop and further improve a satellite radio and TV communication system. Payload: Satellite with transmitter, command system, orientation system, orbit correction device, power supply.	24,774	308	715	64.96	Reentered 3/4/69.
Oct. 11	DOD Spacecraft (United States)	1967-96A	Thor-Burner II	Total weight: Not available. Objective: Develop spaceflight techniques and technology. Payload: Not available.	534	418	100.1	99.1	Still in orbit.

ASTRONAUTICS AND AERONAUTICS, 1967

Launch date	Name	International designation	Vehicle	Payload data	Apogee (st. ml.)	Perigee (st. ml.)	Period (minutes)	Inclination (degrees)	Remarks
Oct. 11	Cosmos CLXXXI (U.S.S.R.)	1967-97A	Not available	Total weight: Not available. Objective: Continuation of Cosmos scientific satellite series. Payload: Not available.	536	415	101.1	98.16	Reentered 10/18/67.
Oct. 16	Cosmos CLXXXII (U.S.S.R.)	1967-98A	Not available	Total weight: Not available. Objective: Continuation of Cosmos scientific satellite series. Payload: Not available.	221	131	89.9	65	Reentered 10/24/67.
Oct. 18	Cosmos CLXXXIII (U.S.S.R.)	1967-99A	Not available	Total weight: Not available. Objective: Continuation of Cosmos scientific satellite series. Payload: Not available.	132	90	89.9	49	Reentered 10/18/67.
Oct. 18	Oso IV (United States)	1967-100A	Thor-Delta	Total weight: 605 lbs. Objective: Obtain high-resolution spectral data (within the range 1 Å–1,350 Å) from the pointed solar experiments during a major portion of a solar rotation. Payload: Top part of spacecraft a 22″-radius semicircular sail continuously pointed at the sun with 8 experiments, 1,860 n-on-p solar cells; lower part a 44″ x 9″ 9-sided revolving wheel containing 6 nonpointed experiments that scan the sun every 2 sec., controls, telemetry, recorder, batteries.	369	337	95.8	32.99	Oso V was launched into extremely good circular orbit; experiments all functioned and returned good data. By 1/3/68 it had achieved its primary objective. Still in orbit, still transmitting.
Oct. 22	Molniya I-7 (U.S.S.R.)	1967-101A	Not available	Total weight: Not available. Objective: Develop and further improve a satellite radio and TV communication system. Payload: Satellite with transmitter, command system, orientation system, orbit correction device, power supply.	24,693	283	714	64.7	Still in orbit.

422

ASTRONAUTICS AND AERONAUTICS, 1967

Date	Name	Designation	Launch vehicle	Objective/Payload	Weight (lbs)	Period (min)	Inclination	Remarks
Oct. 24	Cosmos CLXXXIV (U.S.S.R.)	1967-102A	Not available	Total weight: Not available. Objective: Continuation of Cosmos scientific satellite series. Payload: Not available.	394	394	97.1	81.2	Still in orbit.
Oct. 25	DOD Spacecraft (United States)	1967-103A	Titan IIIB	Total weight: Not available. Objective: Develop spaceflight techniques and technology. Payload: Not available.	247	78	89.5	111.4	Reentered 11/5/67.
Oct. 27	Cosmos CLXXXV (U.S.S.R.)	1967-104A	Not available	Total weight: Not available. Objective: Continuation of Cosmos scientific satellite series. Payload: Not available.	552	324	98.7	64.1	Reentered 1/14/69.
Oct. 27	Cosmos CLXXXVI (U.S.S.R.)	1967-105A	Not available	Total weight: Not available. Objective: Achieve automatic docking in space. Payload: Not available.	146	130	88.7	51.7	On 10/30/67, Cosmos CLXXXVI achieved automatic docking with Cosmos CLXXXVIII. World's 1st automatic docking in space was photographed by cameras in both spacecraft. Cosmos CLXXXVI reentered 10/31/67.
Oct. 28	Cosmos CLXXXVII (U.S.S.R.)	1967-106A	Not available	Total weight: Not available. Objective: Continuation of Cosmos scientific satellite series. Payload: Not available.	131	90	87.8	50	Reentered 10/28/67.
Oct. 30	Cosmos CLXXXVIII (U.S.S.R.)	1967-107A	Not available	Total weight: Not available. Objective: Achieve automatic docking in space. Payload: Not available.	171	124	89	52	On 10/30/67, Cosmos CLXXXVIII achieved automatic docking with Cosmos CLXXXVI. World's 1st automatic docking in space was photographed by cameras in both spacecraft. Cosmos CLXXXVIII reentered 11/2/67.
Oct. 30	Cosmos CLXXXIX (U.S.S.R.)	1967-108A	Not available	Total weight: Not available. Objective: Continuation of Cosmos scientific satellite series. Payload: Not available.	373	332	96.7	74	Still in orbit.

ASTRONAUTICS AND AERONAUTICS, 1967

Launch date	Name	International designation	Vehicle	Payload data	Apogee (st. mi.)	Perigee (st. mi.)	Period (minutes)	Inclination (degrees)	Remarks
Nov. 2	DOD Spacecraft (United States) and	1967-109A	Thor-Agena D	Total weight: Not available. Objective: Develop spaceflight techniques and technology. Payload: Not available.	255	114	90.4	81.5	Two payloads launched with single booster. 109A reentered 12/2/67.
	DOD Spacecraft	1967-109B		Total weight: Not available. Objective: Develop spaceflight techniques and technology. Payload: Not available.	326	282	94.3	81.6	109B reentered 3/28/69.
Nov. 3	Cosmos CXC (U.S.S.R.)	1967-110A	Not available	Total weight: Not available. Objective: Continuation of Cosmos scientific satellite series. Payload: Not available.	201	121	89.6	65.6	Reentered 11/11/67.
Nov. 5	ATS III (United States)	1967-111A	Atlas-Agena D	Total weight: 798 lbs. (in synchronous orbit, after kick-motor firing; 1,674 lbs. before motor firing). Objective: From synchronous orbit, flight-test and obtain data from the application and technology experiments; operate spacecraft for at least the 1st 30 days. Payload: 55" (dia.) x 54" cylindrical satellite, the sides of which are covered with 24,320 n-on-p solar cells; from the top center extends a mechanically despun microwave antenna and from the top rim extend 8 whip antennas for the telemetry and command system; from the	22,148 After apogee motor firing: 22,186.	21,953	628 1,422	28 .536	ATS III was put into synchronous orbit and all experiments returned data except for the navigation experiment, which would not be exercised until spring 1968. On 1/2/68 ATS III was adjudged successful. Still in orbit, still transmitting.

424

Date	Name	Designation	Launch Vehicle	Payload/Findings	Remarks
Nov. 7	Surveyor VI (United States)	1967-112A	Atlas-Agena D	Total weight: 2,223 lbs. (weight at launch, including 1,444-lb. retromotor, propellants, etc.; weight of Surveyor lander at moon, 621 lbs.). Objective: Softland on moon; obtain post-landing pictures of lunar surface; conduct vernier-engine erosion experiment. Payload: 10' high x 14' (around 3 extended landing gear) spacecraft, consisting of triangular frame to which are attached: a mast supporting rotatable planar array antenna and solar panel; 2 folding booms deploying conical omnidirectional antennas; thermal compartment housing decoder and signal processing equipment; 2 altitude radar antennas; survey TV camera; retromotor; small bar magnet attached to footpad. bottom center extends the nozzle of the apogee motor and from the bottom rim extend 8 dog-leg antennas for the VHF experiment; 9 major experiments—mechanically despun antenna, linear VHF communication transponder, multicolor spin scan cloud camera, image dissector cameras, self-contained navigation experiment, resistojet thruster, reflectometer experiment; 4 telemetry transmitters; 2 command receivers; 2 batteries.	Softlanded on moon. Surveyor VI made a flawless flight to the moon, softlanded in Sinus Medii on 11/9; during its 1st lunar day, transmitted to earth 30,065 TV pictures of lunar surface and star surveys. Alpha scattering equipment operated 43 hrs. On 11/17 the 3 vernier engines were restarted and for the 1st time a spacecraft was lifted off the moon; spacecraft landed some 8 ft from its starting point. Attempts to revive spacecraft during 2nd lunar day were only partially successful, ending on 12/14/67.

Launch date	Name	International designation	Vehicle	Payload data	Apogee (st. mi.)	Perigee (st. mi.)	Period (minutes)	Inclination (degrees)	Remarks
Nov. 9	Apollo 4 (United States)	1967-113A	Saturn V	Total weight: 278,699 lbs. (weight in earth orbit, including S-IVB stage and command and service modules). Objective: Demonstrate integrity and compatibility of launch vehicle and spacecraft; demonstrate stage separation; demonstrate spacecraft heat-shield performance at lunar-return speeds; demonstrate support facilities and operations needed for launch, conduct of mission, and recovery of spacecraft. Payload: 120′-long S-IVB/IU/Apollo adapter/command module/service module/boilerplate lunar module combination; cameras; telemetry.	119	114	88.2	32.6	Apollo 4 was launched from Launch Complex 39 at ETR, went into earth orbit; after 2 revolutions the S-IVB was fired again, lifting the command and service modules to peak altitude of 11,240 mi; the service module propulsion system accelerated the command module to lunar-return speed of 36,537 fps. Apollo 4 command module reentered, landed 9 mi from aiming point in Pacific, was picked up by U.S.S. Bennington. This 1st flight of the Saturn V launch vehicle was a complete success; all stages performed well, the command module and its heat shield withstood the temperatures of lunar-return reentry. At least one more flight was planned to man-rate the Saturn V.
Nov. 10	Essa VI (United States)	1967-114A	Thor-Delta	Total weight: 286 lbs. Objective: Provide global cloud coverage daily for a nominal spacecraft life of 6 mos, minimum life of 3 mos. Payload: 22″ x 42″ 18-sided hatbox-shaped polygon, with 18″ receiving antenna and 4.22″ transmitting whip antennas; containing 2 APT camera systems, FM transmitters, 2 spin-control systems (magnetic coil; small solid-propellant rockets), 2	921	872	114.8	102.139	Essa VI was put into near-polar orbit by Thrust-Augmented Thor-Delta from WTR. All systems functioned normally; after extensive postlaunch evaluation, the satellite was turned over to ESSA to operate on 11/21. Still in orbit, still transmitting.

Date	Name		Designation	Launch vehicle	Remarks	Weight (lbs)	Inclination	Period (min)	Notes
Nov. 21	Cosmos CXCI (U.S.S.R.)		1967-115A	Not available	Infrared horizon sensors; 8 solar and terrestrial radiation sensors; 63 nickel-cadmium batteries; 9,100 n-on-p solar cells. Total weight: Not available. Objective: Continuation of Cosmos scientific satellite series. Payload: Not available.	322	174	92.2	71	Reentered 3/2/68.
Nov. 23	Cosmos CXCII (U.S.S.R.)		1967-116A	Not available	Total weight: Not available. Objective: Continuation of Cosmos scientific satellite series. Payload: Not available.	472	472	99.9	74	Still in orbit.
Nov. 25	Cosmos CXCIII (U.S.S.R.)		1967-117A	Not available	Total weight: Not available. Objective: Continuation of Cosmos scientific satellite series. Payload: Not available.	220	127	89.9	65.7	Reentered 12/8/67.
Nov. 29	Wresat I (Australia)		1967-118A	Redstone	Total weight: 107 lbs. Objective: Measure interaction of solar radiation, particularly x-rays and ultraviolet, with the earth's atmosphere. Payload: 60" x 30" (dia. at base) cone-shaped satellite, with radiation measuring instruments, telemetry, battery.	776	106	98.9	88.3	Wresat I (Weapons Research Establishment Satellite), 1st satellite launched by Australia, went into elliptical polar orbit from Australia's Woomera Rocket Range, returned scientific data, U.S. provided satellite acquisition and data reception. Reentered 1/10/68.
Dec. 3	Cosmos CXCIV (U.S.S.R.)		1967-119A	Not available	Total weight: Not available. Objective: Continuation of Cosmos scientific satellite series. Payload: Not available.	207	127	89.7	65.7	Reentered and spacecraft recovered 12/11/67.
Dec. 5	OV III-6 (United States)		1967-120A	Scout	Total weight: 219 lbs. Objective: Obtain data on the density, temperature, and composition of the upper atmosphere and on the atmosphere's variation with latitude. Payload: Not available.	275	254	93.0	90.6	Reentered 3/9/69.

Launch date	Name	International designation	Vehicle	Payload data	Apogee (st. mi.)	Perigee (st. mi.)	Period (minutes)	Inclination (degrees)	Remarks
Dec. 5	DOD Spacecraft (United States)	1967-121A	Titan IIIB	Total weight: Not available. Objective: Develop spaceflight techniques and technology. Payload: Not available.	226	87	90.1	109.5	Reentered 12/16/67.
Dec. 9	DOD Spacecraft (United States)	1967-122A	Thor-Agena D	Total weight: Not available. Objective: Develop spaceflight techniques and technology. Payload: Not available.	142	104	88.4	81.6	Reentered 12/25/67.
Dec. 13	Pioneer VIII (United States) and	1967-123A	Thor-Delta	Total weight: 145 lbs. Objective: Obtain scientific data on interplanetary phenomena, including magnetic field, plasma, and cosmic rays, in heliocentric orbit for a period covering 2 or more passages of solar activity centers; investigate continuously the magnetosheath at large distances from the bow shock wave; provide a target for MSFN checkout of equipment and training of personnel by launching a Test and Training Satellite (TTS) as secondary payload. Payload: 37" (dia.) x 35" cylindrical spacecraft, with a boom protruding from the top containing communications antennas; the sides covered with solar cells except for a narrow band in which are located the experiments and 3 booms, 2 for orientation jets and 1 for the magnetometer; data storage; transmitter; batteries; total of 7 experiments.	1.0880 au	0.9892 au	386.6 (days)	.057	Pioneer VIII was launched by Thrust-Augmented Thor-Delta from ETR. As Delta 2nd stage burned out, TTS I pickaback satellite was ejected by a spring, went into earth orbit; Pioneer VIII continued on into heliocentric orbit slightly farther out from the sun than the orbit of earth. All experiments functioned and returned data. Still in orbit, still transmitting.

ASTRONAUTICS AND AERONAUTICS, 1967

	Tts I	1967-123B		Total weight: 40 lbs. Objective: See *Pioneer VIII*. Payload: 11″ on-a-side octahedron, with the top apex supporting an S-band antenna and the sides covered with solar cells; S-band transmitter, receiver; batteries.	301	182	92.3	32.9	See *Pioneer VIII*. Reentered 4/28/68.
Dec. 18	Cosmos *CXCV* (U.S.S.R.)	1967-124A	Not available	Total weight: Not available. Objective: Continuation of Cosmos scientific satellite series. Payload: Not available.	233	131	90.1	65.7	Reentered 12/23/67.
Dec. 19	Cosmos *CXCVI* (U.S.S.R.)	1967-125A	Not available	Total weight: Not available. Objective: Continuation of Cosmos scientific satellite series. Payload: Not available.	351	140	95.5	45	Reentered 7/7/68.
Dec. 26	Cosmos *CXCVII* (U.S.S.R.)	1967-126A	Not available	Total weight: Not available. Objective: Continuation of Cosmos scientific satellite series. Payload: Not available.	314	137	91.5	48.5	Reentered 1/30/68.
Dec. 27	Cosmos *CXCVIII* (U.S.S.R.)	1967-127A	Not available	Total weight: Not available. Objective: Continuation of Cosmos scientific satellite series. Payload: Not available.	175	165	89.8	65.1	Still in orbit.

[1] Local time; one day later by Greenwich time.

Appendix B

CHRONOLOGY OF MAJOR NASA LAUNCHINGS

JANUARY 1, 1967, THROUGH DECEMBER 31, 1967

This chronology of major NASA launchings in 1967 is intended to provide an accurate and ready historical reference, one compiling and verifying information previously scattered over several sources. It includes launchings of all rocket vehicles larger than sounding rockets launched either by NASA or under "NASA direction" (e.g., NASA provided vehicles, launch facilities, and performed the launches for ComSatCorp's three INTELSAT II launches in 1967). NASA sounding rocket launches are published annually by the Goddard Space Flight Center Historian in *Goddard Projects Summary: Satellites and Sounding Rockets*.

An attempt has been made to classify the performance of both the launch vehicle and the payload and to summarize total results in terms of primary mission. Three categories have been used for evaluating vehicle performance and mission results—successful (S), partially successful (P), and unsuccessful (U). A fourth category, unknown (Unk), has been provided for payloads where vehicle malfunctions did not give the payload a chance to exercise its main experiments. These divisions are necessarily arbitrary, since many of the results cannot be neatly categorized. Also they ignore the fact that a great deal is learned from missions that may have been classified as unsuccessful.

Date of launch is referenced to local time at the launch site. Sources used were all open ones, verified where in doubt from the project offices in NASA Headquarters and from the NASA Centers. For further information on each item, see Appendix A of this volume and the entries in the main chronology as referenced in the index. Prepared January 1968 by Dr. Frank W. Anderson, Jr., Deputy NASA Historian (EH).

ASTRONAUTICS AND AERONAUTICS, 1967

Date	Name (NASA Code)	General mission	Launch vehicle (site)	Performance Vehicle	Performance Payload	Performance Mission	Remarks
Jan. 11	*Intelsat II-B*	Communications satellite, commercial.	Thrust-Augmented Thor-Delta (DSV-3E) (ETR)	S	S	S	Placed in synchronous orbit over Marshall Is. on 1/14/67. Launched by NASA for ComSatCorp; NASA would lease portion of communications capacity for Apollo support. Was nicknamed *Lani Bird II*, because available for commercial service 1/27/67.
Jan. 26	*Essa IV* (TOS-B/APT)	Operational weather satellite.	Thrust-Augmented Thor-Delta (DSV-3E) (WTR)	S	S	S	Replaced *Essa II* in TOS system of weather satellites; carried 2 Automatic Picture Transmission (APT) systems.
Feb. 4	*Lunar Orbiter III* (Lunar Orbiter C)	Lunar probe, orbital.	Atlas-Agena B (ETR)	S	S	S	Placed in lunar orbit 2/8, then into close-in lunar orbit; took 211 medium- and high-resolution photos of Apollo landing sites and other lunar features. 30 photos were lost when readout ended on 3/2/67 because of spacecraft equipment failure.
Mar. 8	*Oso III* (OSO-E)	Scientific satellite, solar.	Thor-Delta (DSV-3C) (ETR)	S	S	S	Placed in excellent circular orbit. All 9 experiments operated and returned data.
Mar. 22	*Intelsat II-C*	Communications satellite, commercial.	Thrust-Augmented Thor-Delta (DSV-3E) (ETR)	S	S	S	Placed into synchronous orbit over the west coast of Africa 3/25/67. Launched by NASA for ComSatCorp; NASA would lease portion of communications capacity for Apollo support. Was nicknamed *Canary Bird*.
Apr. 5	*Ats II* (ATS-A)	Applications technology satellite.	Atlas-Agena D (ETR)	U	U	U	*Ats II* was to have entered 6,900 mi. circular orbit, but Agena engines failed to reignite and the tumbling spacecraft remained in elliptical transfer orbit.
Apr. 17	*Surveyor III* (Surveyor C).	Scientific lunar landing probe.	Atlas-Centaur (ETR)	S	S	S	Soft-landed on moon 4/19/67 after 3 bounces. Transmitted 6,315 detailed photos of lunar surface. Surface sampler confirmed bearing strength of 10 psi for the lurain.
Apr. 20	*Essa V* (TOS-C)	Operational meteorological satellite.	Thrust-Augmented Thor-Delta (WTR)	S	S	S	Launched by NASA for ESSA, *Essa V* was 5th in ESSA's Tiros Operational Satellite (TOS) system. The cartwheel configuration satellite carried 2 AVCS cameras to provide 24-hr. global weather coverage. 1st cloud cover photos transmitted on 4/24/67. NASA turned over the satellite to ESSA for operation on 5/8/67.
May 4	*Lunar Orbiter IV* (Lunar Orbiter D)	Lunar probe, photographic.	Atlas-Agena D (ETR)	S	S	S	Transmitted 163 high- and medium-resolution photos of lunar surface, including coverage of 99% of moon's front face and much of the back face; high-inclination orbit permitted broad systematic photography of lunar surface. Readout of photos completed 6/1.

432

Date	Name	Purpose	Launch vehicle			Results
May 5	Ariel III (UK-E)	Scientific earth satellite, atmospheric.	Scout (WTR)	S	S	S	Third U.S.-U.K. cooperative space project; NASA launched satellite, provided tracking and telemetry support.
May 24	Explorer XXXIV (IMP-F)	Scientific earth satellite.	Thrust-Augmented Delta (WTR)	S	S	S	In highly elliptical polar orbit (113,131-mi. apogee/150-mi. perigee); fifth IMP in series. All 11 onboard experiments were operating normally.
May 29	Esro II-A	Scientific earth satellite.	Scout (WTR)	U	Unk	U	Did not achieve orbit because of failure of NASA Scout launch vehicle; cause of failure still being investigated. First of 2 planned ESRO satellites in cooperative NASA-ESRO project, it contained 7 experiments for solar astronomy and cosmic ray studies.
June 14	Mariner V (Mariner E/Venus 67)	Scientific Venus probe.	Atlas-Agena D (ETR)	S	S	S	Excellent launch, would have brought Mariner V within 42,000 mi. of Venus on 10/18/67; successful midcourse maneuver on 6/19/67 altered flight path to pass within 2,480 mi. of Venus on 10/19/67, returned good data.
July 14	Surveyor IV (Surveyor D)	Scientific lunar landing probe.	Atlas-Centaur (ETR)	S	U	U	Performance throughout 66-hr. lunar-intercept trajectory was excellent, but on 7/16 all communications were lost with Surveyor IV seconds before attempt at softlanding on the moon. Softlanding was unlikely. Last of Centaur direct-ascent missions in Surveyor project.
July 19	Explorer XXXV (IMP-E)	Scientific interplanetary probe.	Thrust-Augmented Thor-Delta (ETR)	S	S	S	Traveled to moon on direct-ascent trajectory; on 7/21 retromotor slowed spacecraft enough to permit lunar capture; went into elliptical (4,760-497-mi.) lunar orbit, returned data on radiation at lunar distance.
July 28	Ogo IV (OGO D)	Scientific geophysical satellite.	Thrust-Augmented Thor-Agena D (WTR)	S	S	S	By 9/17 Ogo IV had transmitted 19,000 hrs. of data and 18 of 20 experiments still operated; completed requirements of primary mission on 9/25.
Aug. 1	Lunar Orbiter V (Lunar Orbiter E)	Lunar probe, orbital.	Atlas-Agena D (ETR)	S	S	S	Took 424 photos of lunar surface, completed readout 8/27; substantially filled gaps in lunar coverage, provided detailed coverage of 36 scientific-interest sites and 5 Apollo sites.
Sept. 7	Biosatellite II (Biosatellite B)	Life science satellite.	Thrust-Augmented Thor-Delta (ETR)	S	S	S	Reentered in 30th orbit on 9/9, one day early because of faulty communications; capsule was recovered in midair by USAF aircraft in Pacific only 15 mi. from aiming point. Specimens were intact and analysis indicated definite effects from weightless state.
Sept. 8	Surveyor V (Surveyor E)	Scientific lunar landing probe.	Atlas-Centaur (ETR)	S	S	S	Softlanded in lunar Sea of Tranquility on 9/10 in spite of leak in vernier-engine gas supply. Transmitted 18,006 photos during 1st lunar day; soil test confirmed basaltic character of lunar soil, similar to earth's. Spacecraft survived 1st lunar night.

See footnote at end of table.

Date	Name (NASA Code)	General mission	Launch vehicle (site)	Performance Vehicle	Performance Payload	Performance Mission	Remarks
Sept. 27	Intelsat II-D	Communications satellite, commercial.	Thrust-Augmented Thor-Delta (ETR)	S	S	S	Launched by NASA for ComSatCorp. Apogee motor was fired 9/30, achieved synchronous orbit for positioning over Western Pacific to supplement and back up Intelsat II-B.
Oct. 18	Oso IV (OSO D)	Scientific solar satellite.	Thrust-Augmented Thor-Delta (ETR)	S	S	S	Oso IV achieved extremely good circular orbit; all experiments returned good data, achieved its primary objective.
Oct. 19	RAM C-1 (RAM C-A)	Reentry-physics probe.	Scout (WI)	S	S	S	RAM C-1 attained maximum reentry speed of 25,168 fps at 245,672 ft., landed 10 mi. from aiming point NE of Bermuda. Water-injection experiment did reduce length of blackout.
Nov. 5	Ats III (ATS-C)	Applications technology satellite.	Atlas-Agena D (ETR)	S	S	S	Ats III achieved synchronous orbit; all experiments returned good data except the navigation experiment, which would not be turned on until Spring 1968.
Nov. 7	Surveyor VI (Surveyor F)	Scientific lunar landing probe.	Atlas-Agena D (ETR)	S	S	S	Surveyor VI softlanded in Sinus Medii; during 1st lunar day transmitted 30,065 TV photos of lunar surface and star surveys. On 11/17 vernier engines were restarted and for 1st time a spacecraft was lifted off the surface of the moon, landing 8 ft. away.
Nov. 9	Apollo 4 (AS-501)	Launch vehicle and spacecraft development.	Saturn V (ETR)	S	S	S	Apollo 4 was launched into earth orbit; S-IVB fired again and lifted payload to peak apogee of 11,240 mi; then service module propulsion system powered command module to lunar-reentry velocity of 36,537 fps. Command module landed 9 mi. from aiming point in Pacific, was picked up by U.S.S. Bennington. First flight of Saturn V was a major success.
Nov. 10	Essa VI (TOS-D)	Operational weather satellite.	Thrust-Augmented Thor-Delta (WTR)	S	S	S	Essa VI was launched into near-polar orbit, subjected to postlaunch evaluation, and turned over to ESSA to operate on 11/21.
Dec. 13	Pioneer VIII (Pioneer C) and Tts I	Scientific probe, sun-orbiting; and piggyback test and training satellite.	Thrust-Augmented Thor-Delta (ETR)	S	S	S[2]	As Delta 2nd stage burned out, Tts-1 pickaback satellite was spring-ejected into earth orbit to exercise Manned Space Flight Network; Pioneer VIII continued on into heliocentric orbit slightly farther from sun than earth's orbit. All experiments returned good data.

[1] Local time; one day later by Greenwich time.
[2] Provided the pointing experiments continue to return data for most of one rotation of the sun.

Appendix C

ABBREVIATIONS OF REFERENCES

Listed here are abbreviations for sources cited in the text. This list does not include all sources provided in the chronology, for some of the references cited are not abbreviated. Only those references which appear in abbreviated form are listed below. Abbreviations used in the chronology entries themselves are cross-referenced in the Index.

A&A	AIAA's magazine, *Astronautics & Aeronautics*
A&A 67	NASA's *Astronautics and Aeronautics 1967* [this publication]
ABC	American Broadcasting Company
AEC Release	Atomic Energy Commission News Release
Aero Tech	*Aerospace Technology* magazine (formerly *Technology Week*)
AFFTC Release	Air Force Flight Test Center News Release
AFHF Newsletter	*Air Force Historical Foundation Newsletter*
AFOSR Release	Air Force Office of Scientific Research News Release
AFRPL Release	Air Force Rocket Propulsion Laboratory News Release
AFSC *Newsreview*	Air Force Systems Command's *Newsreview*
AFSC Release	Air Force Systems Command News Release
AF/SD	*Air Force and Space Digest* magazine
AFSSD Release	Air Force Space Systems Division News Release
AIA Release	Aerospace Industries Association News Release
AIAA *Facts*	American Institute of Aeronautics and Astronautics' *Facts*
AIAA *News*	American Institute of Aeronautics and Astronautics' *News*
AIAA Release	American Institute of Aeronautics and Astronautics News Release
AIP *News*	*American Institute of Physics News*
AP	Associated Press
ARC *Astrogram*	NASA Ames Research Center's *Astrogram*
ARC Release	NASA Ames Research Center News Release
Atlanta J/C	*Atlanta Journal and Constitution* newspaper
Av Daily	*Aviation Daily* newsletter
Av Wk	*Aviation Week and Space Technology* magazine
B News	*Birmingham News* newspaper
B *Sun*	Baltimore *Sun* newspaper
Can Press	Canadian Press
CBS	Columbia Broadcasting System
C Daily News	*Chicago Daily News* newspaper
C&E News	*Chemical & Engineering News* magazine
ComSatCorp Release	Communications Satellite Corporation News Release
CR	*Congressional Record*
CSM	*Christian Science Monitor* newspaper
CTNS	Chicago Tribune News Service
C Trib	*Chicago Tribune* newspaper
DOD Release	Department of Defense News Release
DOT Release	Department of Transportation News Release
EH	NASA Historical Staff (Code EH)
ERC Release	NASA Electronics Research Center News Release

ESSA Release	Environmental Science Services Administration News Release
FAA Release	Federal Aviation Administration News Release
FonF	*Facts on File*
FRC Release	NASA Flight Research Center News Release
FRC X-Press	NASA Flight Research Center's *FRC X-Press*
GE Forum	*General Electric Forum* magazine
Goddard News	NASA Goddard Space Flight Center's *Goddard News*
GSFC Release	NASA Goddard Space Flight Center News Release
GSFC *SSR*	NASA Goddard Space Flight Center's *Satellite Situation Report*
H Chron	*Houston Chronicle* newspaper
J/Armed Forces	*Journal of the Armed Forces* magazine
JPL *Lab-Oratory*	Jet Propulsion Laboratory's *Lab-Oratory*
JPL Release	Jet Propulsion Laboratory News Release
KC Star	*Kansas City Star*
KC Times	*Kansas City Times* newspaper
KSC Release	John F. Kennedy Space Center, NASA, News Release
Langley Researcher	NASA Langley Research Center's *Langley Researcher*
LaRC Release	NASA Langley Research Center News Release
LA Times	*Los Angeles Times* newspaper
LeRC Release	NASA Lewis Research Center News Release
Lewis News	NASA Lewis Research Center's *Lewis News*
Marshall Star	NASA George C. Marshall Space Flight Center's *Marshall Star*
M Her	*Miami Herald* newspaper
MJ	*Milwaukee Journal* newspaper
M News	*Miami News* newspaper
MSC Release	NASA Manned Spacecraft Center News Release
MSC *Roundup*	NASA Manned Spacecraft Center's *Space News Roundup*
M/S Daily	*Missile Space Daily* newsletter
MSFC Release	NASA George C. Marshall Space Flight Center News Release
M Trib	*Minneapolis Tribune* newspaper
NAA *News*	National Aeronautic Association *News*
NAC Release	National Aviation Club News Release
NANA	North American Newspaper Alliance
NAR Release	North American Rockwell Corp. News Release
NAR *Skywriter*	North American Rockwell Corp. *Skywriter*
NAS Release	National Academy of Sciences News Release
NASA Ann	NASA Announcement
NASA Hq PB	NASA Headquarters Personnel Bulletin
NASA Hq *WB*	NASA Headquarters' *Weekly Bulletin*
NASA Int Aff	NASA Office of International Affairs
NASA LAR VI/8	NASA Legislative Activities Report, Vol. VI, No. 8
NASA Proj Off	NASA Project Office
NASA Release	NASA (Headquarters) News Release
NASA Rept SRL	NASA Report of Sounding Rocket Launching
NASA SP-4006	NASA Special Publication #4006
NASC Release	National Aeronautics and Space Council News Release
NAS–NRC Release	National Academy of Sciences-National Research Council News Release
NAS–NRC–NAE *News Report*	National Academy of Sciences-National Research Council-National Academy of Engineering
Natl Obs	*National Observer* magazine
NBC	National Broadcasting Company
NGS Release	National Geographic Society News Release
NMI–	NASA Management Instruction–
NN	NASA Notice
N News	*Newark News* newspaper
NSC Release	National Space Club News Release
NSF Release	National Science Foundation News Release
NYT	*New York Times* newspaper
NYTNS	New York Times News Service
Omaha W–H	*Omaha World-Herald* newspaper

O Sen	*Orlando Sentinel* newspaper
PD	National Archives and Records Service's *Weekly Compilation of Presidential Documents*
P EB	Philadelphia *Evening Bulletin* newspaper
PMR Release	USN Pacific Missile Range News Release
Pres Rep 1967	*Report to the Congress from the President of the United States*, U.S. Aeronautics and Space Activities, 1967
P SB	Philadelphia *Sunday Bulletin* newspaper
SA	*Scientific American* magazine
S/A	*Space Aeronautics* magazine
SBD	*Space Business Daily* newsletter
SciServ	Science Service
SD Union	*San Diego Union* newspaper
SE Post	*Saturday Evening Post* magazine
S/F	*Space Flight* magazine
SF Chron	*San Francisco Chronicle* newspaper
S/P	*Space Propulsion* newsletter
SR	*Saturday Review* magazine
St. Louis P-D	*St. Louis Post-Dispatch* newspaper
Testimony	Congressional testimony, prepared statement
Text	Prepared report or speech text
Transcript	Official transcript of news conference or congressional hearing
UPI	United Press International
US News	*U.S. News and World Report* magazine
USGS Release	U.S. Geological Survey News Release
WH Release	White House News Release
WJT	*World Journal Tribune* newspaper
W News	*Washington Daily News* newspaper
W Post	*Washington Post* newspaper
WSJ	*Wall Street Journal* newspaper
WS Release	NASA Wallops Station News Release
W Star	Washington *Evening Star/Sunday Star* newspaper

INDEX AND LIST OF ABBREVIATIONS AND ACRONYMS

A

A-1 (French satellite), 41
AA. See Apollo Applications.
AAS. See American Astronautical Society.
ABC. See American Broadcasting Company.
Abel, Robert, 354
Abelson, Dr. Philip H., 80
Aberdeen (Md.) Proving Ground, 113
ABM. See Antiballistic missile system.
Abraham, Karl, 42
Abrahamson, Maj. James A. (USAF), 196
ABRES. See Advanced Ballistic Reentry System.
Accelerator, 11, 49, 188
Acelerometer, 183
AC Electronics Div. (General Motors Corp.), 43
Accident, 278
  aircraft, 177, 188, 207, 234, 286, 293, 309, 371
    F-104, 180, 371, 373
    F-111A, 122, 149, 311
    F-111B, 122, 149, 161, 312
    P-1127, 277
    SR-71, 5, 111, 318
    X-15, 345, 352, 392
    XB-70, 180
  Apollo AS-204, 152, 167, 171
    cause, 36, 40, 42, 51, 52, 71, 91, 101, 103, 107, 110, 244
    damage, 27
    fire, 21–23, 26, 36
    investigation, 25, 33, 34, 36, 41, 49, 53, 86, 121, 122, 131–132, 158, 161–162, 171, 204, 211
    report, 47, 51–52, 90, 101–102, 105–106, 110, 114, 118, 128, 139, 145, 148, 381, 386
    schedule, 65, 70, 145
    standards, 139, 187, 391
    U.S.S.R. statement, 26, 34
  astronaut, 177, 293, 371
  automobile, 177, 293
  Brooks AFB, 29, 49, 83–84
  cosmonaut (see also *Soyuz I*), 41, 130
  M2-F2 lifting-body vehicle, 147, 225–226, 392
  missile, 3
  safety practices, 82, 99, 112–113, 145, 294
  Saturn V, 13

Accident—Continued
  U.S.S.R. spacecraft, 82, 293
    Soyuz I, 124, 130–131, 134, 139–140, 143, 149, 158, 175, 213, 239, 291, 392
Adams, Dr. Mac C., 69, 73, 137, 153
Adams, Maj. Michael J. (USAF), 83, 129, 188, 254, 345, 352
Adams, Dr. T. W., 362
Adelaide Univ., 358
ADSCP. See Advanced Defense Satellite Communications Project.
Advanced Ballistic Reentry System (ABRES), 181, 202, 302, 334, 385
Advanced Defense Satellite Communications Project (ADSCP), 32, 71
Advanced Research Projects Agency (ARPA), 2–3, 168, 353, 358
Advanced Vidicon Camera System (AVCS), 120
Advisory Committee on Supersonic Transport (President's), 84
Aebersold, Paul C., 168–169
AEC. See Atomic Energy Commission.
AEDC. See Arnold Engineering Development Center.
Aero Club of Washington, 89, 318, 380
Aerobee (sounding rocket)
  150
    failure, 268
    infrared astronomy, 60
    magnetic field studies, 372
    meteorite debris, 240
    micrometeoroid sampling, 177, 293
    solar astronomy, 125, 146, 285, 289, 293, 311
    stellar data, 95–96, 140, 344
    ultraviolet astronomy, 60, 78, 173, 326, 369
    upper atmosphere data, 48, 146, 191, 214–215, 232, 236–237, 285, 320, 368
    x-ray astronomy, 201–202, 253, 289, 350
  150 (Mod I)
    x-ray astronomy, 98
  150-A
    biological studies, 368
    ultraviolet astronomy, 72, 83
    upper atmosphere data, 72, 83
Aeroflot, 170, 185, 351, 386
Aerojet-General Corp., 14, 61, 130, 184, 189–190, 201, 281, 317

Aeronautics, 16, 54, 56, 86, 126, 143, 164, 183, 209, 237, 298
  anniversary, 183, 290–291, 380
  award, 66, 96, 153, 154, 164, 172, 179, 181, 212, 232, 285, 369–370, 372
  employment, 112
  funds for, 17, 54, 73, 318, 336–337
  general aviation, 16, 184, 255, 363
  laboratory, 183, 188, 290
  noise abatement. See Noise, aircraft.
  research (see also X-15, X-22A, and XB-70), 12, 13–14, 19, 54, 67, 73, 94, 137, 142–143, 184, 185, 188, 218, 261, 291, 318, 348, 350, 364
Aeronautics and Space Engineering Board, 195
Aeronomy satellite, 11, 75, 92–93
Aerospace Art, Hall of, 109
Aerospace Corp., 18, 142
Aerospace Industries Assn. (AIA), 19, 30, 109, 112, 184, 352, 385
Aerospace industry, 1, 36, 109, 112, 215, 225, 245, 385
Aerospace Research Pilots School, 345, 371
Aerospace Safety Advisory Panel (proposed), 99, 281
Aerospace Security Analysts, 109
Aerospike rocket engine, 246–247
Aerotrain, 367
AFA. See Air Force Assn.
AFCRL. See Air Force Cambridge Research Laboratories.
AFETR. See Air Force Eastern Test Range.
AFOSR. See Air Force Office of Scientific Research.
Africa, 41, 335, 376
AFRPL. See Air Force Rocket Propulsion Laboratory.
AFSC. See Air Force Systems Command.
Agena (booster), 279
Agency for International Development (AID), 173
Agnew, Dr. Harold, 77
Agreement
  astronaut, 133
  France-Quebec, 161
  international, 48, 88–89, 101, 211, 216, 218, 270, 327–328, 376
  NASA-Dept of Commerce, 42
  -Dept. of Interior, 193
  -ESRO, 160
  -Italy, 384
  -Smithsonian Institution, 72
  -USAF, 74–75, 92, 112, 190
  U.K.-France, 16
  -U.S.S.R., 286
  U.S.-U.K., 93, 369
  -U.S.S.R., 9, 41, 85, 299, 327
  -U.S.S.R.-Australia, 64
  -North Vietnam, 309–310
Agriculture, Dept. of., 167
AIA. See Aerospace Industries Assn.
AIAA. See American Institute of Aeronautics and Astronautics.

AIAA/AAS Space Forum, 4
AIAA Aerospace Sciences Meeting, 19
AID. See Agency for International Development.
Aiken, Sen. George D., 32
Aiken, William S., Jr., 242
Air cushion vehicle, 152, 349
Air Force Academy, 153, 372
Air Force Assn. (AFA), 77, 273
Air Force Cambridge Research Laboratories (AFCRL), 83, 195, 225, 285, 289, 324–325, 362, 391
Air Force Eastern Test Range (AFETR), 196
Air Force Historical Foundation, 344
Air Force Logistics Command, 59
Air Force Medical Corps, 164
Air Force Office of Scientific Research (AFOSR), 149
Air Force Rocket Propulsion Laboratory, (AFRPL), 63–64
Air Force Systems Command (AFSC), 15, 163, 178, 362
  Aeronautical Systems Div., 344, 384
  aircraft, 268, 269
  all-weather landing system, 300
  Avionics Lab., 18–19
  award, 136, 213, 358
  Ballistic Systems Div. (BSD), 199
  booster, 62
  camera, aircraft, 384
  contract, 8, 168
  experiment, 18–19, 182
  missile, 344
  National Range Div., 200
  organization, 35
  personnel, 87, 230
  Space and Missile Systems Organization (SAMSO), 199
  Space Physics Laboratory, 227
  Space Systems Div. (SSD), 199
Air Line Pilots Assn., 338–339
Air Line Traffic Assn. "Man of the Year" award, 181
Air pollution, 52, 182, 189
Air traffic control, 154, 161, 260, 284, 300, 338, 364, 385
Air Transport Assn. of America (ATA), 77, 89, 161, 224, 312, 346–347
Airborne Warning and Control System (AWACS), 373
Airbus (European commercial aircraft), 219–220, 270
Aircraft (see also individual aircraft, such as B-52, X-15, etc.), 1, 19, 70, 93, 106, 143, 166, 260, 269, 347, 369, 380
  accident, 5, 111, 122, 149, 161, 177, 180, 188, 207–208, 234, 277, 286, 293, 309, 311–312, 318–319, 345, 352, 371, 373, 392
  air show, 116, 156, 162, 166, 203, 211, 218
  bomber, 61, 111, 149, 269, 365, 380
  cargo, 268, 304
  carrier, 111, 313
  collision study, 110, 364

Aircraft—Continued
  communications tests, 177, 250
  delta-wing, 60–61, 373
  fighter, 16, 111, 149, 161, 164, 180, 237, 239–240, 269, 289, 295, 307, 330, 343, 347, 371, 373, 380
  foreign, 156, 177, 184, 343, 351, 369, 371, 384
  general-aviation, 188, 304, 309, 349, 363
  helicopter, 16, 111, 164, 174, 262, 268, 285, 366
  hypersonic, 8, 17, 61, 182, 392
  interceptor, 307
  navigation, 271–272
  noise. See Noise, aircraft.
  personal, 13
  rear-engine, 349
  reconnaissance, 5, 111–112, 303, 380, 384
  record, 267, 295, 304, 343, 345
  regulations, 260, 276–277, 307, 338, 385
  research (see also X-15, X-22A, and XB-70), 109, 188, 349, 364, 383, 392
  safety, 276, 292, 338, 346–347, 348, 384
  tracking, 195, 207–208, 300
  training, 347, 390
  transport (see also Supersonic transport), 11, 12, 17–18, 33–34, 60–61, 130, 338, 366
    air-bus, 219–220, 270
    jet, 1, 4, 12, 64, 156, 184, 237, 244, 260, 261, 268, 271–272, 348, 351, 373
    military, 268
    STOL, 244
  variable-sweep-wing, 13, 16, 239–240, 383–384
  V/STOL, 4, 146, 244, 318, 330, 383
  VTOL, 203
Aircraft carrier, 111, 313
Airlines, 170–171, 177, 185, 256, 285, 312, 366
  air traffic control, 161, 260, 300, 338–339
  fares, 224, 312
  jet aircraft, 100
  safety, 154, 276–277, 284
  statistics, 304
  supersonic aircraft, 33–34, 39, 176, 211, 237, 261–262, 357–358
Airports, 137, 199, 254, 256, 279, 314
  Beautification award, 179
  charges, 312
  noise, 34, 250, 256
  site selection, 247
  statistics, 84, 230, 304
Alabama, Univ. of, 180
Albania, 23
Albany, Ga., 346
Albert, Rep. Carl, 132
Albuquerque, N. Mex., 180
Alcock, G. E. D., 216
Aldrin, L/C Edwin E., Jr., (USAF), 12, 16, 175, 267, 300–301, 350
Alexander, Charles C., 10
Alexander, George, 27, 191
Alexander, Holmes, 147

Algeria, 197
All-weather landing system (AWLS), 300
Allen, H. Julian, 328
Allen, Dr. Joseph P., 233
Alliance for Progress, 111
Alliluyeva, Mrs. Svetlana, 143
Allis-Chalmers Manufacturing Co., 49
Allnutt, Robert F., 174, 250
Allott, Sen. Gordon L., 320
Alouette I (Canadian satellite), 197, 317
Alouette II, 191, 197
Alpert, Dean Daniel, 353
ALSEP. See Apollo Lunar Surface Experiments Package.
Althouse, Edwin, 191
Altman, Dr. David, 361
Amazon River, 336
AMC. See U.S. Army Missile Command.
American Airlines, 366
American Astronautical Society (AAS), 4, 134, 184, 209, 356
American Automatic Control Council, 339
American Broadcasting Company (ABC), 112–113, 114, 200, 330
American Chemical Society, 109
American Geophysical Union, 93, 115
American Helicopter Society, 172
American Institute of Aeronautics and Astronautics (AIAA), 79, 227, 241, 289
  Alabama Section, 243, 361
  award, 316
  meeting, 4, 19, 36, 55–56, 61, 139, 178, 215, 315–316
American Institute of Physics, 26
American Marketing Assn., 48
American Personnel and Guidance Assn., 86
American Petroleum Institute, 345–346
American Physical Society, 27, 35, 348
American Rocket Society (ARS), 79
American Science and Engineering, Inc., 16–17, 184, 328, 350
American Security Council (ASC), 206, 333
American Society for Engineering Education, 180
American Society for Metals, 350
American Society of Magazine Editors, 97
American Society of Mechanical Engineers (ASME), 251
American Telephone and Telegraph Co. (AT&T), 88, 97, 118, 293, 322, 326, 387
Ames, Milton B., Jr., 76
Ames Research Center (ARC), 143, 180, 240, 359
  award, 232, 317
  probe, 265, 375
  research, 50, 109, 177, 183, 240, 328, 372
Ammonia, 189
AMR. See Atlantic Missile Range.
AMSE. See American Society of Mechanical Engineers.
Amundsen, Roald, 112
Anaheim, Calif., 315
Anders, William A., 85, 350

Anderson, Sen. Clinton P., 13, 25, 33, 63, 115, 152, 195, 216–217, 274
Anderson, Rep. William R., 275
Anderton, Capt. H. L. (USN, Ret.), 204
Andoeya, Norway, 258, 260, 268
Andover, Me., 89, 134, 232–233, 282
Andrews AFB, Md., 126
Andrews, Rep. George W., 174
Ann Arbor, Mich., 87–88, 313
Anniversary, 378
 aeronautics, 380
 Government, 1, 80, 220–221, 284
 nuclear reaction, 363–364
 satellite, 27
 USAF, 220
 U.S.S.R., 9, 107, 269, 273, 291, 309, 327, 336, 349
Antarctic Treaty, 303
Antarctica, 2, 6, 26
Antenna, 56, 70, 78, 89–90, 95, 121, 189, 240–241, 248–249, 332, 333
Antigua Island, 226
Antiballistic missile (ABM) system, 9, 206, 221, 267, 332, 338, 346, 355–356
 Nike X, 7, 201, 274, 292, 296, 297–298, 322–323, 329–330
 research program (Project Sparta), 121
 U.S., 21, 41, 69, 85, 129, 206, 274, 275, 284–285, 297, 322–323, 329–330, 332, 333, 346, 355, 358, 360
 U.S.S.R., 21, 27, 41, 85, 104, 267–268, 331, 355–356, 358, 360
Antimissile missile system. See Antiballistic missile (ABM) system.
APL. See Applied Physics Laboratory.
Apollo (program), 48–49, 114–115, 125, 136, 143, 161–162, 229, 330–331, 333, 356–357, 371, 380, 381, 383, 386, 391
 astronaut. See Astronaut.
 communications, 282, 365
 criticism, 57, 63, 72–73, 87, 89, 94, 101–102, 102–103, 128, 132, 147–148, 171, 178, 225
 facilities, 36, 70, 192–193, 244–245, 339–340
 fire. See Accident, Apollo AS–204.
 funds for, 17, 54–55, 59–60, 81–82, 87, 144–145, 150, 216–217, 227, 246, 248, 380
 landing site, 8–9, 99, 305, 334
 landing technique, 149, 179, 305
 launch (see also Apollo spacecraft), 111, 125, 179, 189, 296–297, 299, 319
  AS–206, 36
  AS–501 (Apollo 4), 36, 191, 254, 298, 333, 339, 340, 365, 391
  AS–502 (Apollo 5), 36, 205
  AS–504, 350
  AS–505, 350
 management, 36, 53–54, 72–73, 87, 90, 94, 101–102, 103, 105–106, 114, 116, 128, 139, 144–145, 161–162, 167, 168, 175, 211, 219, 248, 300, 341, 383
 plans for, 13, 17, 36, 38, 53–54, 81–82, 85, 89, 144–145, 147–148, 149, 242, 298, 300, 319, 330–331, 383, 386

Apollo—Continued
 progress, 7, 40, 65, 68, 70, 72–73, 97–98, 114–115, 130, 139, 179, 221, 226–227, 254, 279–280, 339–340
 test, 21–23, 40, 45–46, 53–54, 111, 112–113, 179, 205, 232, 246, 272, 274, 279–280, 319, 330–331, 339–340, 371, 387, 391
 tracking, 70, 79, 242, 278, 365, 381, 392
 training, 76–77, 122, 212, 330–331
Apollo (spacecraft), 4, 48–49, 63, 72–73, 105, 110, 114, 138, 148–149, 150, 221, 275–276, 279, 302–303
 Command Module (CM), 23, 53–54, 57, 105, 144–145, 175, 189, 216–217, 335–336, 356, 371
 control, 216, 365
 escape device, 52, 145, 179, 205
 fire prevention, 49, 52, 102–103, 144, 187, 247, 302–303, 369
 heat shield, 205, 272, 341, 365
 launch, 36
  Apollo 4 (AS–501), 191, 254, 298, 333, 339, 340, 365, 391
  Apollo 5 (AS–502), 205, 387
 launch vehicle. See Saturn.
 Lunar Module (LM), 145, 187, 189, 200, 231, 293, 298, 312, 335–336, 369, 387
 materials, 51, 105–106, 111, 112, 145, 148–149, 187, 246, 369
 quality control, 138, 294
 recovery, 111
 Service Module (SM), 24–25, 90, 144–145, 175, 189, 246, 319, 335–336, 371
 test, 23, 57, 86, 106, 144, 179, 200, 205, 232, 279–280, 312, 330–331, 365, 371, 387, 391
Apollo 4 (AS–501) (spacecraft), 36, 191, 254, 298, 333, 339–340, 365, 391
Apollo 5 (AS–502) (spacecraft), 36, 205, 387
Apollo Applications (AA) program, 10, 92, 194, 242, 275–276
 criticism, 192
 funds for, 17, 49, 54–55, 129, 192–193, 199, 220, 225, 235, 246, 283, 338
 management, 126, 178, 283–284, 371
 plans for, 10, 21, 54–55, 71, 178, 199, 260, 338, 357
 recovery system, 168
Apollo AS–204 Review Board
 appointment, 25
 investigation of accident, 33, 34, 102–103
  criticism, 62
  report, 101–102, 105–106, 110, 114, 139
   interim, 36, 46–47, 51–52
   recommendations, 145, 187, 246
Apollo Lunar Surface Experiments Package (ALSEP), 169–170
Apollo Telescope Mount (ATM), 21, 30, 55, 81, 220
Applications Technology Satellite (ATS), 17, 30, 55, 189, 242, 331–332, 336

Applied Physics Laboratory (APL) (Johns Hopkins Univ.), 83, 219, 328
APT. See Automatic Picture Transmission.
Aquanaut, 37, 122, 232
ARA. See Associates for Radio Astronomy.
Arbeitsgemeinschaft Airbus, 219–220
ARC. See Ames Research Center.
Arcas (sounding rocket), 93, 231, 237, 248, 270, 272, 276, 301, 311, 313, 316, 318
Arcasonde (meteorological rocket), 311
Archangel, U.S.S.R., 259
Arecibo Ionospheric Observatory, Puerto Rico, 249, 263, 324–325
Argo D–4 (sounding rocket). See Javelin.
Argentina, 307, 382, 393
Ariel III (UK–E) (scientific satellite), 140, 392–393
Aristarchus (moon crater), 1
Arizona, 8, 380
Arizona, Univ. of, 132, 166
Arlington National Cemetery, 29, 176
Arlington, Va., 371
Armed Forces Communications and Electronics Assn., 178, 181
Armed Forces Reserves Joint Assembly, 3
Armed Forces Unification Act, 220
Arms Control and Disarmament Agency, 48
Armstrong, Neil A., 350
Armstrong, Warren E., 217
Army Air Corps, 126
Army Corps of Engineers, 278
Arnold Engineering Development Center (AEDC), 7, 79, 216, 246, 292
Arnoldi, Dr. Louis B., 59
ARPA. See Advanced Research Projects Agency.
ARS. See American Rocket Society.
AS–204 (booster), 81, 387
AS–206 (booster), 81
ASC. See American Security Council.
Ascension Island, 89
Asia, 41, 376
Asimov, Dr. Isaac, 15
ASME. See American Society for Mechanical Engineers.
Associated Universities, Inc., 316
Associates for Radio Astronomy (ARA), 188, 249
Assn. of the U.S. Army, 376
Astor, Vincent, Foundation, 64
Astro Space Labs, Inc., 218–219
Astrobee 1500 (sounding rocket), 50
"Astromouse," 4
Astronaut (see also Cosmonaut; Extravehicular activity), 4, 26, 31, 49, 79, 141, 166, 205, 232, 284, 288, 368–369
  Apollo mission, 17, 22–23, 36, 45–46, 84–85, 97–98, 105, 111, 112, 115, 148–149, 161–162, 196, 230, 232, 238–239, 244–245, 350

Astronaut—Continued
  contract, life story, 133
  crew assignment, 84–85, 97–98, 144–145, 148–149, 350
  death, 22–23, 25, 26–27, 29, 33, 36, 42–43, 46–47, 49, 51–52, 57, 58, 68, 102–103, 122, 143, 161–162, 177, 293, 371, 373
  memorial services, 27, 29, 180
  hazards, 46, 53–54, 102–103, 105–106, 107–108, 143, 144, 145, 213, 373
  honors, 41, 45, 58, 69, 86, 175, 178, 180, 207, 267, 294, 300–301, 317
  memorial, 27, 58, 60, 69, 85, 142, 284
  performance, 17, 56–57, 213, 334
  physiology (see also Space biology), 12, 36, 39, 47, 56, 64, 101–102, 113–114, 213, 284, 288, 334
  press conference, 148–149, 244–245
  rescue, 22–23, 111, 125–126, 205, 392
  scientist-astronaut, 1, 51–52, 94, 230, 233, 234–235, 244
  selection, 1, 52, 94, 230, 233, 234, 244
  speech, 16, 49, 97
  training, 1, 12, 46, 52, 94, 196, 288, 294
  TV appearance, 112, 161
Astronauts Memorial Commission, 58
Astronomy (see also individual planets; Radioastronomy; Star), 16, 17, 21, 31, 45, 166, 181, 188, 236, 240, 250–251, 253, 343, 344, 378–379
  optical, 5, 17, 21, 35, 46, 78, 82, 181, 182–183, 216, 237, 250–251, 252
  solar, 66, 83, 112, 115, 154–155, 191–192, 206, 227, 236, 287, 288, 303–304, 307, 308, 311, 357
  stellar, 89, 192, 216, 344, 356–357
  ultraviolet, 60, 72, 200, 270, 272, 276, 301, 311, 313, 318, 326, 358–359, 369
  x-ray, 17, 66, 142, 201–202, 288, 289, 328
Astrophysics, 16, 17, 27
ATA. See Air Transport Assn. of America.
AT&T. See American Telephone and Telegraph Co.
Athena (missile), 80
Atlanta, Ga., 139–140
Atlantic II (communications satellite). See *Intelsat II–C*.
Atlantic Ocean, 7, 28, 37, 78, 174, 250, 293, 336, 368, 385, 387
Atlas (booster), 62, 163, 221
  D, 202, 334
  F, 302, 321, 385
Atlas (missile), 3, 46
  F, 235
Atlas-Agena (booster), 17–18
  launch
    Agena-D
      Ats II, III, 98, 331–332
      Lunar Orbiter III, V, 38, 138, 229
      Mariner V, 186–187
      unidentified, 36, 161, 175

Atlas-Centaur (booster), 68, 207, 209–210, 265, 332
  AC–12, 113
  AC–14, 334
ATM. See Apollo Telescope Mount.
Atmosphere (see also Ionosphere), 18–19
  artificial, 29, 38, 39, 51, 52, 53, 64, 105, 107–108, 118, 149, 259
  meteorological experiments grenade, 359
  upper, study of, 17, 55–56, 74, 75, 91, 109, 157, 212, 214–215, 272, 285, 311, 314–315, 316–317, 324–325, 359, 367, 375
Atomic bomb, 48
Atomic Energy Commission (AEC) (see also NASA–AEC Space Nuclear Propulsion Office, NERVA, Rover, SNAP, and Vela programs), 15, 168–169, 170, 228, 263, 285, 384, 386, 394
  accelerator, 188
  accomplishments, 363–364, 389–390
  anniversary, 1, 363–364
  award, 48, 332
  budget, 17, 207
  contract, 61, 281
  cooperation, 267, 378
  Hypervelocity Planetary Gas Tunnel, 328
  Los Alamos Laboratory, 77
  nuclear explosion, 372, 380
  nuclear reactor, 206–207, 367
  nuclear test, 2–3, 372
  space nuclear power, 50–51, 378
ATS. See Applications Technology Satellite.
Ats I (Applications Technology Satellite), 36–37, 89–90, 91, 177, 270, 332, 336
Ats II (ATS–A), 98, 173, 332
Ats III (ATS–C), 331–332
Atwood, Harry N., 209
Atwood, J. Leland, 105, 138–139
Atwood, Lee, 23
Auburn Univ., 180
Aurora borealis, 38
Australia, 7, 41, 47, 68, 102, 153, 157, 176, 233, 263
  Dept. of Weapons Research, 358
    international cooperation, 64, 79, 182, 241
  launch, 3, 368, 369, 391, 393
  tracking station, 64, 79, 265
Automatic Picture Transmission (APT), 20, 29, 270, 342–343
Avco Corp., 172
AVCS. See Advanced Vidicon Camera System.
Aviation Hall of Fame, 298
Aviation-Space Writers' Assn., 153
Avions Marcel Dassault, 383–384
AWACS. See Airborne Warning and Control System.
Awards, 23–24, 85, 149, 190, 217–218
  civic, 70, 72, 85, 105, 136, 154, 164, 237, 267, 275, 372, 380

Awards—Continued
  Government, 39, 45, 48, 54, 129, 153, 164, 179, 195–196, 207, 213, 358, 369, 389
  institutions, 6, 10, 72, 128–129, 133, 136, 253–254, 309, 369
  society
    aeronautics, 96, 172, 181, 232, 285
    astronautics, 41, 136, 215, 361
    engineering, 66, 84, 212
    labor, 175
    management, 86, 361
    science, 19, 174
AWLS. See All-weather landing system.
Azur (West German satellite), 349

B

Babbitt, Donald O., 317
Babcock, Horace W., 181
Bader, Michel, 317
Bahamas, 209
Baikonur, U.S.S.R., 124
Baker, Erwin, 217
Baker, Robert, 171
Baldwin, Barrett, 328
Baldwin, Hanson, 203, 239–240
Ball, Robert S., Memorial Award, 154
Ballantine, Stuart, Medal, 309
Balloon, 96, 227
  Echo I, 241
  France, 8, 177, 333–334
  NASA, 18, 238, 241, 281
  tracking of, 89
  Voyager parachute tests, 179, 222
Baltimore, Md., 63, 206
"Bambi" (antimissile system), 169
Bannerman, Graeme C., 360
Barbados, West Indies, 247
Barking Sands, Hawaii, 270, 272, 276, 301, 311, 313, 318
Baron, Thomas R., 94, 122
Barreira do Inferno Range, Brazil, 89, 326
Barstow, Calif., 180
Bartley, Airman 2/C William F., Jr., 29
Bates, Dr. F. C., 245
Bateson, J. W., Co., 388
Battelle Memorial Institute (BMI), 386
Baxter, L/C William D. (USAF), 25
Bay of Lunar Landing (moon), 56
Bay of Lunik (moon), 56
Bayley, William H., 358
Baylor College of Medicine, 233
Bayne, James M., 54
Beale AFB, Calif., 111–112, 318–319
Bean, Alan L., 350
Bechtel Corp., 170–171
Beckwith, W. Boynton, 347
Beecher, William, 121–122, 201, 221, 267
Belew, Leland F., 92
Belgian Atomic Center, 389
Belgium, 251–252, 389
Belgrade, Yugoslavia, 253–254, 279
Bell Aerospace Corp., 231
Bell Aerosystems Co., 13, 76–77, 96, 146, 383
Bell Telephone Laboratories, 142

Bellman, Donald R., 317
Belyayev, Col. Pavel (U.S.S.R.), 166, 175
Bendix Corp., 81–82, 169–170, 193, 226, 312
Bendix Field Engineering Co., 179
Benedict, Howard, 154, 204–205
Bennett, Ralph K., 289
Bennett, R/A Rawson, II (USN, Ret.), 371
Benton, Ark., 262
Bergen, William B., 136
Berkeley, Calif., 217
Berkner, Dr. Lloyd V., 175–176
Berlain, Jean, 367
Berlin, Germany, 29
Berlin, Technical Univ. of, 280
Bermuda, 311
Berry, Dr. Charles A., 40, 108, 136
Bethe, Prof. Hans A., 322
Bethpage, N.Y., 187, 279
Betts, L/C Austin W. (USA), 329–330
Biggs, John R., 38, 75, 299, 363
Bigler, Stewart, 137
Bingham, Charles F., 176
Bioastronautics Conference, 243
Biosatellite (program), 47
Biosatellite I, 47, 68, 263
Biosatellite II, 263–265, 313, 326, 391–392
Bioscience. See Space biology.
Bishop, Jerry, 31
Bisplinghoff, Dr. Raymond L., 129, 135, 317
Black Arrow (rocket), 3
Blacksburg, Va., 243
Blagonravov, Dr. Anatoly A., 101, 292, 324
Blair, G. Richard, 56
Blamont, Prof. Jacques-Emile, 253–254
Bleymaier, B/G Joseph S. (USAF), 87
Blue Scout-Junior (rocket), 385–386
Blue Streak (U.K. rocket), 233, 368
BMWF. See Federal German Ministry for Scientific Research.
BMI. See Battelle Memorial Institute.
BOB. See Budget, Bureau of.
Boeing Co., 158, 372, 373
  award, 294, 317
  booster, 61, 144–145, 220, 232
  contract, 23–24, 36, 47, 88, 133, 137, 144–145, 192, 193, 201, 220, 347, 392
  Lunar Orbiter III, IV, 202, 347, 392
  supersonic transport, 5, 31–32, 130, 176, 357–358
Boeing 707 (jet aircraft), 2, 143
Boeing 2707 (supersonic aircraft), 237
Boelsche medal, 70
Boiko, Alexander, 348
Bollerud, B/G Jack (USAF), 173
Bologna Univ., 344
Bond, Dr. Richard, 117
Bonestell, Chesley, 109
Bonn, Univ. of, 312
Boodley, Lewis E., 56
Boone, Adm. William F. (USN, Ret.), 388–389
Booth, Eugene T., 109

Bordeaux, France, 47
Borman, L/C Frank (USAF), 25, 85, 128, 350
  congressional testimony, 115
  interview, 112, 161
Boston College, 86, 174, 256, 351
Boston, Mass., 209, 346, 384
Boston Univ. School of Medicine, 15
Boulder, Colo., 206
Bow, Rep. Frank T., 8
Bowdoin College, 224
Bowhill, S. A., 212
Boyer, William J., 317
Boyd, Secretary of Transportation Alan S., 8, 35, 90–91, 95, 97, 188, 200, 203, 346
Boynton, Melbourne, Award, 136
Bracewell, R. N., 37
Bradford, Dr. Smith A., 252
Bradley, Gen. Mark (USAF, Ret.), 281
Brauer, Dr. R. W., 122–123
Brazil, 89, 248, 303, 307, 326, 335, 336, 382, 393–394
Brazilian Space Commission (Comissão Nacional de Atividades Espaciais (CNAE), 189, 326
Brazilian Telecommunications Co. (EMBRATEL), 385
Brewer, Frank G., Trophy, 96
Brewster Flat, Wash., 134
Bridgeport, Conn., 262
Brinkley, Rep. Jack, 18
Bristol, Univ. of, 327
British Astronomical Society, 216
British Interplanetary Society, 73, 259
Bronco, Project, 380
Bronstein, Vitali, 308
Brooke, Sen. Edward W., 32
Brookings Institution, 69
Brooks AFB, Tex., 36, 39, 49, 83, 233
Brown and Root-Northrop Corp., 205
Brown Engineering Co., 112
Brown, Dr. Harold, 111, 223, 286, 294
Brunenkant, Edward J., 170
Brunswick, Me., 224
Brussells, Univ. of, 389
Bryce Canyon National Park, Utah, 8
Bryson, Reid A., 143
Bubble chamber, 216
Buckley, Edmond C., 79
Budget, Bureau of (BOB), 254, 263
Buitrago, Spain, 353
Bulgaria, 376
Bull, Lt. John S. (USN), 350
Bullock, Robert O., 19
Bureau of Commercial Fisheries, 287
Bureau of Sport Fisheries and Wildlife, 193
Burevestnik (U.S.S.R. hydrofoil), 350
Burger, Col. Robert J. (USAF), 195
Burgess, E. H., 5
Burroughs, Richard Hansford, Test Pilot Award, 6
Byrd, Sen Harry F., Jr., 32

C

C-5A Galaxy (cargo transport), 11, 268.
CAB. See Civil Aeronautics Board.
Cable, underwater, 293, 322, 326, 387–88
Cabot, Godfrey L., Aviation Award, 288
Caditi (Communist Chinese sounding rocket), 314
Caldara, M/G Joseph D. (USAF, Ret.), 309
Calder, Nigel, 212, 315
California, 126, 202–203, 251, 269, 279, 309
California Institute of Technology (Cal Tech), 16, 35–36, 82, 87, 118, 162, 188, 261, 366
Cal Tech. See California Institute of Technology.
California, Univ. of, 13–14, 72, 126, 217, 233, 317, 354
   Los Angeles (UCLA), 209, 316, 334, 349, 370
Calverton, N.Y., 122
Calvo Rodes, Gen. Rafael (Spain), 1
Cambridge, Mass., 29, 54, 190, 278, 312
Camera, 260–261, 331–332, 384
Cameron, Air Commodore Neil (U.K.), 267–268
Canada (see also Churchill Research Range), 66, 78, 166, 170, 191, 245–246, 247, 317, 348
   Defence Research Board, 197
   Defence Research Telecommunications Establishment, 85, 110
Canadian Aeronautics and Space Institute, 366
Canadian National Telecommunications, 170
Canadian Pacific, 170
Canary Islands, 89
Canberra, Australia, 79, 121, 212, 242
Candy, M. P., 216
Canham, Edwin D., 234
Cannon, Sen. Howard W., 185
Canopus (star), 38, 113, 138, 295, 334
Canyon de Chelly National Monument, Ariz., 8
Cape Kennedy, Fla., 26, 33, 49, 57, 64, 226, 279
Carbon dioxide, 37–38, 113–114
Carbon monoxide, 53–54
Carco Electronics Co., 55
Carnegie Commission on Educational Television, 107
Carpenter, Cdr. M. Scott (USN), 12, 113, 232, 368–369
Carr, Maj. Gerald P. (USMC), 350
Carter, M/G Wendell E. (USAF), 230
Case Institute of Technology, 150
Casteau, Belguin, 204
CAT. See Clear air turbulence.
Catalytic Construction Co., 170
Cayey, Puerto Rico, 232–233, 269
Celebration of the Prelude to Independence (Williamsburg, Va.), 166–167
Centaur (booster), 59, 68, 350–351, 360

Central Intelligence Agency (CIA), 8, 41, 204, 228, 293
Centre National d'Études Spatiales (CNES), 221–222, 333–334, 382
Centro Ricerche Aerospaziali (CRA), Rome, 228, 384
CERN. See European Center for Nuclear Research.
Cernan, LCdr. Eugene M. (USN), 144
CETS. See European Conference on Communications Satellites.
China, Communist
   hydrogen bomb, 190
   missile, 8, 201, 228, 296, 332, 346, 358, 360
   nuclear test, 190, 205, 386
   satellite, 14, 314
   sounding rocket, 314
   space law treaty, 23, 26
   space program, 314
Chaffee, LCdr. Roger B. (USN)
   Apollo (AS-204) accident, 22–23, 25, 26, 27, 36, 46–47, 101
   burial, 29
   honors, 317
   memorial, 69, 85
Chanute, Octave, Award, 184
Chapman, Dr. John H., 85
Chapman, Dr. Philip K., 233, 234
Charyk, Dr. Joseph V., 114, 146, 387–388
Cheyenne, Wyo., 90
Chicago, Ill., 35, 98, 122–123, 345, 346, 371
Chicago, Univ. of, 363–364
Chico, Calif., 238
Childs, Marquis, 33
Chile, 307
Chlorophyll, 13–14
Christensen, Everett E., 39
Christmas Islands, 270
Chrysler Corp., 196, 239, 382
Churchill Research Range, Canada (see also Ft. Churchill, Canada), 25, 28, 67, 80, 110, 142, 177, 368, 370
CIA. See Central Intelligence Agency.
Citizens League Against the Sonic Boom, 190
Citron, Robert, 335
City University, London, 109
Civil Aeronautics Board (CAB), 256, 286
Civil Service Commission (CSC), 338
Clark, Evert, 31, 80, 215, 247, 304, 376
Clark, Dr. John F., 74
Clark, Sen. Joseph S., 220, 297
Clarke, Arthur C., 73
Clarksburg, Md., 255
Clear air turbulence (CAT), 90
Clemmons, Stephen B., 317
Cleveland Engineering Society, 97–98
Cleveland, Ohio, 38
Cloud, 114, 362
CM. See Command Module.
CNAE (Comissão Nacional de Atividades Espaciais). See Brazilian Space Commission.

CNES. See Centre National d'Études Spatiales.
Cockburn, Sir Robert, 291
Cocoa Beach, Fla., 212
Coesite, 320
Cohen, William, 317
Coleman, Herbert J., 236
College Park, Md., 284
College Station, Tex., 272
Collier, Robert J., Trophy, 164, 237
Collins, Maj. Michael (USAF), 85, 350
Collision, aircraft, 110, 271–271, 338–339, 364
Cologne, W. Germany, 349
Colorado, 8, 12, 380
Colorado, Univ. of, 282, 361
Columbia Univ., 233, 335
Columbus, Ohio, 45
Comet, 251–252, 328
Command Module (CM), 90, 168, 178, 189, 216–217, 335–336, 339, 340, 371
Command Service Module (CSM), 175, 339
Commerce, Dept. of, 34, 42, 73, 84, 132, 143, 162, 263
Comisión Nacional de Investigación es del Espacio (CONIE), 1
Committee on Institutional Cooperation, 249
Committee on Space Research (COSPAR), 87, 93, 212, 236
Communications, 63, 96, 100, 101, 107, 117, 170–171, 236, 241, 242, 251, 262, 276, 335
  deep space, 38, 80, 138, 229, 238, 266, 267, 341–342, 353–354, 358
  global, 41, 80, 180–181, 193–194, 326–327, 387–388
  international, 21, 66, 71, 85, 88–89, 90, 121–122, 133–134, 170, 204, 218, 318
  laser use in, 10, 18–19, 221–222
  military use, 71, 182, 199, 204, 224, 236
  underwater, 326–327, 387–388
Communications satellite (see also individual satellites: *Intelsat I* (*Early Bird I*), *Molniya I–6*, etc.), 215, 249, 282, 290, 313, 318, 387–388, 392–393
  agreement, 101, 218, 276
  charges, 30, 100, 222–223
  cooperation, 193–194, 255, 262
    international, 98–99, 133–134, 326–327, 333–334
  ground station, 101, 143, 146, 184, 202–203, 204, 353, 385
  launch
    Initial Defense (IDCSP), 199
    Intelsat II–B (*Pacific I*), 7
    Intelsat II–C (*Atlantic II*), 83
    Intelsat II–D (*Pacific II*), 215, 282
    Molniya I–5, 165
    Molniya I–6, 290
    Molniya I–7, 313
  military, 32, 178, 199, 204, 213, 250, 361, 391–392

Communications satellite—Continued
  system
    domestic, special purpose, 95, 96, 117, 222–223, 276
    foreign, 170, 204, 218, 255, 277, 290, 313, 314, 393
    global, 133–134, 146, 193–194, 241, 250, 277, 282, 318, 326, 387–388, 392
    transmission, 133–134, 182, 193–194, 204, 262
  use of, 32, 95, 182, 213, 241, 249–250, 262, 276, 318, 326, 385–386
Communications Satellite Corp. (ComSatCorp), 80, 293, 322, 391–392,
  Atlantic II. See *Intelsat II–C*.
  contract, 13, 86, 88–89, 202–203
  Early Bird I. See *Intelsat I*.
  FAA, service to, 236
  FCC regulation, 6, 114, 118, 276, 326–327, 387
  ground station, 118, 126, 130, 202–203, 222–223, 232–233, 269
  Intelsat I (*Early Bird I*), 7–8, 83, 100, 157, 262, 282
  Intelsat II–A, 7, 83
  Intelsat II–B (*Pacific I*), 7, 21, 41, 82, 133–134, 157, 193–194, 282
    transmission, 7, 21, 41, 133–134, 193–194
  Intelsat II–C (*Atlantic II*), 42, 76, 83, 85, 89, 282
    transmission, 83
  Intelsat II–D (*Pacific II*), 146, 215, 282
  INTELSAT III, 157
  INTELSAT IV, 249–250, 293, 326–327
  laboratory, 255
  rates, 93, 222–223, 236, 318
  revenues, 285
  satellite program, 6, 7, 13, 21, 42, 80, 83, 88–89, 95, 225, 232–233, 276, 318, 326–327, 387, 391–392
  services, 7–8, 95, 114, 134, 222–223, 236, 269, 276, 282, 318, 326, 387–388
  test, 21
Compañía Telefónica Nacionale de España (CTNE), 353
Computer, 31, 70, 104, 153, 197, 292, 324, 371
ComSatCorp. See Communications Satellite Corp.
Concord, Mass., 227
Concorde (U.K.-France) supersonic transport, 2, 116, 215
  cost, 230
  production schedule, 5, 66, 154, 185, 269, 373
  sales price, 220, 224
Condon, Dr. Edward U., 282, 361
Conference on Exploration and Peaceful Uses of Outer Space, 38–39, 46
Congress
  Apollo AS–204 accident, 131, 132, 386
  communications satellite, 6, 241
  F–111, 239–240, 330

Congress—Continued
  Conference Committee on NASA FY 1968 Authorization, 204, 213, 225, 226, 227, 230
  Joint Committee on Atomic Energy, 8, 188, 228, 267
    Subcommittee on Military Applications, 331, 332, 338
  NASA Semiannual Report, 150
  NASA budget, 13, 17–18, 18, 46, 211, 215, 226, 255–256, 257, 296, 297, 299, 333, 356, 388
  research and development, 310
  science, 313
  SST program, 31–32, 35, 171–172, 302
Congress, House of Representatives, 69, 128, 139, 183, 194, 208, 211, 213, 215, 249, 275, 281, 283, 291, 292, 295, 313, 314, 323
  bills introduced, 33, 69, 85, 99, 181
  bills passed, 67, 194, 207, 212, 237, 250, 320
  Committee on Appropriations, 207, 246, 248
    Subcommittee on Dept. of Agriculture, 167
    Subcommittee on Dept. of Defense, 169, 174
    Subcommittee on Dept. of Transportation, 203
    Subcommittee on Independent Offices, 243
  Committee on Armed Services, 49, 206
    Subcommittee on Appropriations, 140
  Committee on Government Operations, 254, 261
    Subcommittee on Military Operations, 88
    Subcommittee on Research and Technical Programs, 231
  Committee on Interstate Commerce, Subcommittee on Transportation and Aeronautics, 346
  Committee on Merchant Marine and Fisheries, Subcommittee on Oceanography, 372
  Committee on Science and Astronautics, 165, 178, 185, 238, 278, 310, 313, 346, 351
    Apollo AS–204 accident, 33, 65, 103, 381
    NASA budget, 54, 59, 68, 156, 177, 192–193, 320
    Panel on Science and Technology, 19
    Subcommittee on Advanced Research and Technology, 76, 78, 84, 120
    Subcommittee on Manned Space Flight, 72, 150, 323, 350
    Subcommittee on NASA Oversight, 51, 90, 95, 102, 105, 107, 112, 115, 122, 147, 278
    Subcommittee on Space Science and Applications, 82, 84, 117–118, 158, 255–256, 393

Congress, House of Representatives—Continued
  Committee on Science and Astronautics—Continued
    Subcommittee on Science, Research, and Development, 181, 200–201, 230, 277–278, 314
  Committee on Ways and Means, 297
  Seminar on Technology Assessment, 277–278
Congress, Senate, 40, 48, 60, 126, 128, 156, 179, 185, 188, 194, 195, 221, 229, 289, 314, 323, 366
  bills introduced, 75–76
  bills passed, 194–195, 207, 225, 230, 284, 293–294, 295, 297, 320
  Committee on Aeronautical and Space Sciences, 25, 32, 64, 274
    Apollo AS–204 accident, 25, 33, 40, 51–52, 105, 110, 114–115, 130, 138–139, 145, 147–148, 152, 216–217, 381, 386
    NASA budget, 13, 20, 121, 192, 336–337
    Subcommittee on Space Science and Applications, 324
  Committee on Appropriations, 284, 291, 293–294, 295, 324
    Subcommittee on Dept. of Defense, 21
    Subcommittee on Independent Offices, 220, 235, 282–283
    Subcommittee on Space Appropriations, 320
    Subcommittee on Transportation, 284
  Committee on Armed Services, 21
    Preparedness Investigating Subcommittee, 294–295, 296
    Subcommittee on Nuclear Safeguards, 360
  Committee on Commerce, 8, 21, 33–34, 73, 107, 359
    Subcommittee on Aviation, 256
  Committee on Foreign Relations, 65, 71, 108, 297–298, 323, 343, 359
    Subcommittee on Disarmament, 41, 146–147
  Committee on Labor and Public Welfare, Special Subcommittee on Scientific Manpower Utilization, 19, 72–73
  Committee on Public Works, Subcommittee on Air and Water Pollution, 73
  resolutions introduced, 58
Congressional Record, 33–34, 103, 185, 290–291, 294–295, 314, 350–351, 381
CONIE. See Comisión Nacional de Investigación es del Espacio.
Connes, Dr. Pierre, 251
Connes, Mrs. Pierre, 251
Connolly, V/A T. F. (USN), 174
Conover, John H., 362
Conrad, LCdr. Charles E., Jr. (USN), 136, 350
Contract (see also under agencies, such as NASA, USAF, etc.)

Contract—Continued
  cost-plus-award-fee, 81–82, 170–171, 194, 205, 312, 360
  cost-plus-fixed-fee, 47, 49, 82, 188, 220, 292
  cost-plus-incentive-fee, 112, 168, 174–175, 312
  fixed price, 129, 159–160, 195, 220, 221
  fixed-price-incentive-fee, 126, 149, 161, 383
  incentive-fee, 23–24, 347
  multiple-incentive-fee, 196
  study, 13, 86, 88, 179–180, 201, 255, 386
Convair 880 (jet transport), 348–349
Convair 990 (jet aircraft), 166
Cooper, Maj. Gordon L. (USAF), 128
Cooper, John Cobb, 217
Cooper, M/G Paul T. (USAF), 199
Copeland, R., 251–252
Coralie (rocket engine), 224, 233, 368
Cornell Aeronautical Laboratory, 366
Cornell Univ., 25, 151, 188, 248–249, 322
Coronagraph, 146, 293
Corporation of Public Broadcasting (proposed), 252, 335
Corporation for Public Television (proposed), 107
Corpus Christi, Tex., 70
Cortright, Edgar M., 3, 30, 67–68, 74, 117, 300, 317
Cosmic dust, 354
Cosmonaut, 1–2, 14, 62, 82, 101, 194, 260, 277, 288, 291
  accident, 41, 124, 125, 126, 128, 130, 131, 134, 139–140, 143, 149, 158, 239, 293
  anniversary, 80, 107
  memorial, 161, 213
  Paris International Air and Space Show, 166, 175, 181–182
  training, 253
Cosmonautics Day, 107
Cosmos (U.S.S.R. satellite program), 218
Cosmos CXXXVIII (U.S.S.R. satellite), 12, 392
Cosmos CXXXIX, 19, 304
Cosmos CXL, 40
Cosmos CXLI, 41
Cosmos CXLII, 46
Cosmos CXLIII, 53
Cosmos CXLIV, 54, 60, 77–78
Cosmos CXLV, 61
Cosmos CXLVI, 70, 80
Cosmos CXLVII, 70
Cosmos CXLVIII, 77
Cosmos CXLIX, 82
Cosmos CL, 83
Cosmos CLI, 86
Cosmos CLII, 87
Cosmos CLIII, 97
Cosmos CLIV, 101, 129
Cosmos CLV, 107
Cosmos CLVI, 129
Cosmos CLVII, 152
Cosmos CLVIII, 154
Cosmos CLIX, 157–158
Cosmos CLX, 157–158, 304
Cosmos CLXI, 161
Cosmos CLXII, 173
Cosmos CLXIII, 176
Cosmos CLXIV, 180
Cosmos CLXV, 183
Cosmos CLXVI, 188, 288, 314–315
Cosmos CLXVII, 189
Cosmos CLXVIII, 201
Cosmos CLXIX, 210–211, 304
Cosmos CLXX, 225, 239, 304
Cosmos CLXXI, 237, 239, 304
Cosmos CLXXII, 238
Cosmos CLXXIII, 252
Cosmos CLXXIV, 258
Cosmos CLXXV, 268
Cosmos CLXXVI, 268
Cosmos CLXXVII, 272
Cosmos CLXXVIII, 275
Cosmos CLXXIX, 278
Cosmos CLXXX, 281
Cosmos CLXXXI, 299
Cosmos CLXXXII, 304
Cosmos CLXXXIII, 309
Cosmos CLXXXIV, 316
Cosmos CLXXXV, 320–321
Cosmos CLXXXVI, 320–321, 323
Cosmos CLXXXVII, 321
Cosmos CLXXXVIII, 320–321, 323
Cosmos CLXXXIX, 321
Cosmos CXC, 327
Cosmos CXCI, 352
Cosmos CXCII, 354
Cosmos CXCIII, 355
Cosmos CXCIV, 365
Cosmos CXCV, 380
Cosmos CXCVI, 382
Cosmos CXCVII, 387
Cosmos CXCVIII, 388
Cosmos, Minn., 162
COSPAR. See Committee on Space Research.
Cotton, Col. Joseph (USAF), 9, 11, 26, 28, 174, 265, 327
Cotton, Sen. Norris, 359
Cotton, Paul E., 382
Cottrell, Richard, 317
Coughlin, William, 161–162
CRA. See Centro Ricerche Aerospaziali.
Cray, Douglas W., 250
Crossfield, A. Scott, 237
Cryogenics, 136, 142–143
CSC. See Civil Service Commission.
CSM. See Command Service Module.
CTNE. See Compañia Telefónica Nacionale de España.
Cuba, 3, 23, 376
Cummings, Robert L., 244
Cunningham, R. Walter, 23, 84–85, 144, 244–245
Cutler-Hammer, Inc., 202–203, 244
Cygnus (constellation), 16–17
Czechoslovakia, 192, 376

D

Daddario, Rep. Emilio Q., 200–201, 230, 277–278, 313
Dallas, Tex., 80, 134, 284, 346, 388
Dallas Council on World Affairs, 376
Dana, William H., 128, 157, 193, 214, 292
Dartmouth College, 326
Datner, Paul B., 317
DATS. See Despun Antenna Test Satellite.
Daunt, Dr. John G., 142–143
David, Heather M., 1–2
Davies, Merton E., 87
Davis, Frank W., 330
Davis, Dr. Jesse H., 142–143
Davis, L/C Leighton I. (USAF), 195–196, 200
Dawson, James P., 37
Dayton, Ohio, 269, 298
DC–3 (airliner), 366
DC–8 Super 63 (jet airliner), 64, 166
Dean, Ambassador Sir Patrick, 23, 298
Deauville, France, 164–165
Debre Zeit, Ethiopia, 335
Debus, Dr. Kurt H., 40–41, 135, 158–159
Deep Space Network (DSN) (NASA), 38, 229, 267, 341–342, 353–354, 358
Deep Submergence Rescue Vehicle (DSRV–1), 368–369
Deep Submergence Search Vehicle (DSSV), 368–369
Deep Submergence Systems Project (DSSP), 232
Defense Communications Agency, 32
Defense, Dept. of (DOD), 27, 58, 228, 348, 359, 366
  Advanced Research Projects Agency (ARPA), 2–3, 168, 353, 358
  aircraft, 111, 149, 318–319, 385
  budget, 17, 18, 140, 361, 388
  communications satellite system, 11–12, 32, 71, 204, 213, 236, 261, 392
  cooperation, 111, 152, 213, 279, 293, 307, 389–390
  Manned Orbiting Laboratory. See Manned Orbiting Laboratory.
  missile program, 21, 273, 292, 296, 297–298, 329, 332–333, 343–344
  nuclear detection satellite, 129, 133, 392
  personnel, 182, 234, 353, 359
  R&D, 123, 152, 318, 376, 388–389
  space program, 140, 205, 286, 296, 343–344, 391, 392
  university program, 13
Defense Documentation Center, 197–198
De Florez Training Award, 316
De France, Dr. Smith J., 232
De Gaulle, President Charles (France), 90, 166, 257–258
Delphinius (constellation), 216
Dembling, Paul G., 299
Delta (booster) (see also Thor-Delta), 66–67, 70, 319–320, 383
  Thrust-Augmented, 214, 282
  Thrust-Augmented Improved, 7, 42, 76–83, 163, 263

Dementyev, Peter V., 46
Denisov, Dr. V., 296–297
Denver, Colo., 153, 209
Denver Research Institute (DRI), 271
Despun Antenna Test Satellite (DATS), 199
Detroit, Mich., 11, 81, 346
Deutsch, George C., 317
Development Concept Paper, 376
Diademe I (D–1C) (French geodetic satellite), 41, 47
Diademe II (D–1D), 47
Diamant (French booster), 41, 47, 333
Diamonds (meteoritic), 92
Diapason I (D–1A) (French satellite), 41
Dicke, L. M., 280
Dicke, Dr. Robert H., 17, 249
Direct Lift Control (DLC), 143
Dirksen, Sen. Everett M., 9–10
Disarmament, 42, 48, 86, 252–253, 303
Distinguished Flying Cross, 207
Distinguished Service Medal, 164, 195–196, 358
Distinguished Service Medal (NASA), 63
Djakarta, Indonesia, 184
DLC. See Direct Lift Control.
DNA (deoxyribonucleic acid), 378
Dobrov, Dr. G. M., 153
Dobrynin, Ambassador Anatoli F., 23, 298
Docking, 21, 60, 124, 321, 334, 392
Documentation, Inc., 284
DOD. See Defense, Dept. of.
Dodge (DOD gravity experiment satellite), 199, 219
Dollfus, Dr. Audouin, 5
Doolittle, Gen. James H. (USAF, Ret.), 153
Doppler effect, 41, 47
Dorman, Dr. Berhardt L., 75
Dorn, Rep. William Jennings Bryan, 69, 292
DOT. See Transportation, Dept. of.
Douglas Aircraft Co., Inc. See McDonnell Douglas Corp.
Douglas, Col. William K. (USAF), 136
Douglas DC–8 (jetliner), 2
Dow Chemical Co., 170–171
Downey, Calif., 36, 38, 63, 102, 108, 116, 148–149
Downing, Rep. Thomas N., 291
Doyle and Russell, Inc., 388
Dragon Tail, Project, 380
Draper, Dr. Charles S., 15, 253–254, 317
Drexel Institute of Technology, 190
DRI. See Denver Research Institute.
Dryden, Dr. Hugh L., 150, 177
Dryden, Hugh L., Memorial Fund, 94
DSN. See Deep Space Network.
DSRV. See Deep Submergence Rescue Vehicle.
DSSP. See Deep Submergence Systems Project.
DSSV. See Deep Submergence Search Vehicle.
Dublin, Ireland, 312

DuBridge, Lee A., 240
Dudley Observatory, 234
Duke, Charles, 86
Dulles International Airport, 262, 349
Durban, South Africa, 46

E

E-266 (U.S.S.R. aircraft), 295, 343
Eaker, L/G Ira C. (USAF, Ret.), 113, 262
Early Bird I (communications satellite). See *Intelsat I*.
Earth
  continental drift, 182
  gravity, 85–86, 357
  photographs of, 190, 219, 245, 258–259, 336, 342, 391
  shape, 245
  temperature, 143, 182
Earth Landing System (ELS), 205
Earth Photographs from Gemini III, IV, and V (atlas), 190
Earth Reentry Module (ERM), 125
Earth Resources Observation Satellite (EROS), 15, 30, 393
Earthquakes, 85–86
East Coast Laboratory, Fla., 132
Eastern Airlines, 237
Eastern Test Range (ETR) (see also Cape Kennedy and Kennedy Space Center).
  launch, 21, 47, 208, 214, 226, 370, 388
    contract, 126, 370–371
  vehicle
    Delta, 370
    Minuteman, 62
    satellite
      Atlas-Agena D, 98, 331–332
      Thor-Delta, 66, 307, 319–320
      Thrust-Augmented Delta, 282
        Thrust-Augmented Improved, 7, 83, 264
      Titan III-C, 11, 129, 199
    space probe
      Atlas-Agena D, 38, 138, 185, 229
      Atlas-Centaur, 113, 209, 265
      Saturn, 298, 339–340
      Thrust-Augmented Delta, 214
      Thrust-Augented Thor-Delta, 375
Eastman Kodak Co., 317
Echo I (communications satellite), 241
Eckert, Dr. W. J., 63
Eckhardt, Rep. Bob, 18
Eckman, Dr. Philip K., 87
Eclipse, solar, 112, 119
Economic Development Administration, 250
Edison Electric Institute, 117
EDS. See Emergency Detection System.
Education, 11, 16, 26, 51, 52, 80, 227, 285, 297, 309, 317, 332, 344
Edwards AFB, Calif., 197
  flight
    F-104, 371
    F-111A, 239–240
    M2-F2 (lifting-body vehicle), 147, 225–226

Edwards AFB—Continued
  flight—continued
    SR-71, 5
    X-15, 83, 128, 143, 157, 193, 196, 214, 254, 289–290, 292
    XB-70, 3, 9, 11, 174, 193, 253
  sonic boom tests, 231
Eggers, Dr. Alfred J., 281, 389
Eglin AFB, Fla., 3, 10, 111, 240
Ehrlich, Eugene, 134, 207, 271
Eifler, M/G Charles W. (USA), 330
Eigen, Manfred, 322
Einstein, Dr. Albert, 17
Einstein, Albert, Award, 72
Eisele, Maj. Donn F. (USAF), 23, 45, 85, 144, 244
Eisenhower, President Dwight D., 25
Eisenhower, Dr. Milton, 335
El Centro, Calif., 205
El Paso Natural Gas Co., 372
ELDO. See European Launcher Development Organization.
Electric propulsion, 73, 77
Electro Mechanical Research Corp., 218
Electro-Optical Systems, Inc., 13, 27
Electron, 237, 272, 348
Electron II (U.S.S.R. satellite), 288
Electronics, 10, 48, 55, 73, 137–138, 163, 238, 271–272, 278
Electronics and Aerospace Systems Technical Convention, 307
Electronics Industries Assn., 137–138
Electronics Research Center (ERC) (NASA), 69–70, 191, 359
  contract, 184, 242, 278, 386
  facilities, 98, 278
  navigation satellite system, 242, 386
  personnel, 54, 86, 174, 201, 363
Ellis V/A William E. (USN), 313
Elms, James C., 99, 201, 363, 386, 389
ELS. See Earth Landing System.
Ely, Nev., 195
EMBRATEL. See Brazilian Telecommunications Co.
Emergency Detection System (EDS), 339–340
Emme, Dr. Eugene M., 315
Employment, 68, 69, 259, 285
EMSC. See European Ministerial Space Conference.
Encyclopaedia Britannica, 378
Ends, Dr. Earl J., 110
Engine (see also individual engines, such as F-1, M-1, etc.), 315
  aircraft, 255, 286, 349, 360
    gas turbine, 66, 172, 384
    jet, 296, 261
    supersonic transport, 130, 261
    turbofan, 207, 250
  chemical, 49, 102
  electric, 52, 73, 102
    solar-electric, 77, 102
  nuclear (see also NERVA), 11, 54, 64–65, 102, 281, 351, 353

Engine—Continued
  rocket, 32, 63–64, 73, 77, 189–190, 217–218, 231, 246
  vernier, 266–267, 342
Engineers, 39, 231, 259, 278, 285, 313
Engins Matra, 197
England, Dr. Anthony W., 233
Engle, Capt. Joseph H. (USAF), 212
Environmental Science Services Administration (ESSA)
  budget, 17–18
  cooperation, 269, 287
  Geomagnetic Center, 331
  Institutes for Environmental Research, 15
  satellite, 20, 42, 120, 125, 143, 189, 270, 342, 353, 391–392
  Solar Disturbance Forecast Center, 206
Epstein, Julius, 293
Erb, R. Bryan, 133
ERC. See Electronics Research Center.
Erikson, Leif, Day, 281
ERM. See Earth Reentry Module.
E.R.N.O., 197
Ers-XVIII (research satellite), 129
Ers-XX, 129
Ers-XXVII, 129
Esro I (ESRO satellite), 160
Esro II, 112, 160, 168, 392–393
ESSA. See Environmental Science Services Administration.
Essa I (meteorological satellite), 20, 120, 343
Essa II, 20, 120, 343
Essa III, 20, 120, 343
Essa IV, 20, 42, 120, 343
Essa V (TOS-C), 120, 125, 143, 189, 270, 343
Essa VI (TOS-D), 342–343
Esso Production Research, 244
ESTEC. See European Space Technology Center.
Ethiopia, 335
ETR. See Eastern Test Range.
EURATOM. See European Atomic Energy Commission.
Europa I (ELDO booster), 224, 233, 368
Europa I/PAS, 333–334
Europa II, 255, 381–382
Europa III, 381–382
Europe, 41, 52, 59, 70, 73, 194
European Atomic Energy Community (EURATOM), 327–328, 371
European Center for Nuclear Research (CERN), 216
Eshleman, V. R., 208
ESRANGE (ESRO rocket range), 94
ESRO. See European Space Research Organization.
European Conference on Communications Satellites (CETS), 302
European Institute of Science and Technology (proposed), 164–165
European Launcher Development Organization (ELDO), 224, 233, 255, 302, 333–334, 344, 368
European Ministerial Space Conference (EMSC), 218
European Space Conference, 302
European Space Research Organization (ESRO), 38, 66–67, 94, 112, 168, 276, 302, 333–334, 344, 356, 392–393
  TD–1, 197
  TD–2, 197
European Space Technology Center (ESTEC), 302
European Travel Commission, 312
Eurosat (communications satellite system), 255
Eurospace, 255
Eutechnics, 15
EUV. See Extreme ultraviolet radiation.
Evans, A. J., 138
Evans, M/G Harry L. (USAF), 87
Ewing, Cortez A. M., Foundation, 141
Ewing, Cortez A. M., Lecture, 140–141
Exceptional Service Medal (NASA), 196
Exhibit, 181–182, 203, 211, 218
Explorer (program), 361
Explorer I (U.S. satellite), 28, 355
Explorer XXVII, 335
Explorer XXVIII, 163–164
Explorer XXIX (GEOS-A), 41–42
Explorer XXXI, 191
Explorer XXXII (AE-B), 11, 92–93
Explorer XXXIII (IMP-D), 10, 163–164, 214
Explorer XXXIV (IMP-F), 163–164, 174, 214, 265, 334, 391–392
Explorer XXXV (IMP-E), 214, 238, 370, 391–392
Explosion
  nuclear, 2–3, 137
  quasar, 16
Expo '67, 158, 170
Extraterrestrial life, 5, 6, 9, 13–14, 37, 38, 53, 61, 109, 166, 271, 362
Extravehicular activity, 41, 142
Extreme ultraviolet radiation (EUV), 72
Ezra, Dr. Arthur A., 197

F

F–1 (rocket engine), 61, 194
F–4 (Phantom II) (fighter aircraft), 164, 367, 371
F–5 (supersonic fighter aircraft), 307
F–100 (Super Sabre), 12
F–104 (Starfighter) (aircraft), 112, 345, 371
F–104G, 371
F–111 (supersonic fighter), 213, 292, 358
  contract, 149
  criticism, 289
  U.K., 93, 367, 369
  USAF, USN support, 111, 174, 294–295
F–111A, 162, 166, 239–240, 347, 380
  contract, 149
  crash, 122, 311–312
  NASA, 13
  USAF support, 111, 390

F-111B, 330, 380
  contract, 149
  crash, 122, 161, 174, 311-312
  USN support, 111, 174, 307
F-111C, 149
F-111K, 149
FAA. See Federal Aviation Administration.
Faget, Maxime, 25, 136
Fairbanks, Alaska, 120, 174-175
Fairmont, Minn., 279
Fanfani, Amintore, 59
Fargo, N.D., 153
Farmington, N. Mex., 372
Farnborough, U.K., 222, 230
Farnsworth, Clyde H., 204
Fascell, Rep. Dante B., 85
FB-111A (supersonic bomber), 380
FCC. See Federal Communications Commission.
Federal-Aid Airport Program, 279
Federal Aviation Administration (FAA), 33-34, 236, 285-286, 351
  air traffic control, 84, 260, 307, 338, 348-349, 364
  airports, 230, 256
  appropriations, 284
  award, 96, 179, 181
  contract, 133, 392
  cooperation, 20, 34, 69, 348-349
  landing system, 300
  noise, aircraft, 8, 34, 84, 106-107
  regulations, 89, 260, 276-277, 307, 338-339
  "Report to Industry" meeting, 185
  statistics, 64, 84, 230, 363
  transport, supersonic (see also Supersonic transport), 2, 39, 50, 69, 104, 157, 245, 357-358
    design and development, 5, 48, 176, 392
Federal Communications Commission (FCC), 114, 293, 322, 387
  approvals, 146
  briefs filed with, 96-97, 276, 326-327
  ComSat Corp., 13, 86, 88-89, 93, 95, 202-203, 222, 232-233, 276, 326-327, 388
  criticism of, 6
  ground stations, 130, 146, 232-233
  requests to, 93, 95, 130
Federal Council for Science and Technology, 339
Federal Electric Co., 170-171, 194
Federal Mediation and Conciliation Service, 299
Federal Power Commission (FPC), 52
Fédération Aéronautique Internationale, 174, 295
Federation of American Societies for Experimental Biology, 122-123
Fedorov, Dr. Yevgeny K., 113-114
Fedotov, Alexandr, 295
Felberg, Fred H., 321
Feldman, William, 317
Fels Institute of Local and State Government, 163

Feoktisov, Konstantin Petrovich, 175, 291
Ferguson, Gen. James (USAF), 358, 362
Fermi, Enrico, Award, 48
Ferri, Dr. Antonio, 61, 64
Field Enterprises Educational Corp., 133
Finger, Harold B., 5, 75, 102, 384
Fink, Donald E., 218, 255
Finland, 78
Fire
  Apollo AS-204. See Accident, Apollo AS-204.
  escape system, 40
  Lunar Module, 187, 200
  pressure chamber (Brooks AFB), 29, 49, 83-84
Fire, Project, 290-291
Fjeldbo, G., 208-209
Flammability test, 369
Flax, Dr. Alexander H., 174, 373
Flight Research Center (FRC) (NASA), 13, 74-75, 92, 99, 112, 195, 317, 352
Flight Safety Foundation, Inc., 6, 173, 309, 315, 366, 369-370
Flight Test Center (FTC), Edwards AFB, Calif., 254
Florida, 3, 10
Florida State Univ., 233
Flory, Donald A., 37
FOBS. See Fractional orbital bombardment system.
Fog dispersal, 346-347
Ford Foundation, 96, 107, 276, 318
Ford, Rep. Gerald R., 132
Ft. Belvoir, Va., 229
Ft. Churchill, Canada (see also Churchill Research Range), 28, 33, 42, 354
Ft. Monmouth, N.J., 197-198
Fort Worth, Tex., 330
Foster, Dr. John S., Jr., 146-147, 154, 169, 332, 353, 376
Foulois, M/G Benjamin D. (USA, Ret.), 126
Fowler, Dr. William A., 35-36
Foxbat (U.S.S.R. fighter), 343
FPC. See Federal Power Commission.
Fractional Orbital Bombardment System (FOBS), 329, 331, 332-333, 343, 348
France, 166, 181-182, 204, 239, 251, 299, 367
  aircraft, 12, 16, 219-220, 270
    Concorde (France-U.K. supersonic transport), 2, 154, 215, 220, 230, 269, 373
  balloon, 8, 177, 333
  cooperation, 16, 23, 161, 168, 216, 218, 219-220, 257-258, 270, 333, 356
    U.S.S.R., 257-258, 261, 322, 333-334
  hydrogen bomb, 152, 177, 194
  launch
    satellite, 16, 41, 47
    sound rocket, 65
  missile program, 41, 93, 197
  satellite, 16, 41, 47, 50, 161, 218, 333-334

France—Continued
  space program, 5, 218, 257-258, 291, 333-334, 391
  submarine, nuclear-powered, 90
Franken, Dr. Peter, 168
Frankfurt, Germany, 262
Franklin Award, 10-11
Franklin Institute, 210, 309
Fra Mauro (lunar landing site), 377
FRC. See Flight Research Center.
Frederick, Md., 279
Freeman, Secretary of Agriculture Orville L., 81
Freitag, Capt. Robert F. (USN, Ret.), 87
French Guiana, 47, 334, 382
Friedman, Dr. Herbert, 39, 236, 369
Friedman, Capt. Robert (USN), 102
Friendly, Prof. Fred W., 107, 318
Friendship 7, 49
Fritz, John M., 180
Froelich, Dr. Jack E., 355
Frosch, Robert A., 307
Frunze Military Academy, 50
Frutkin, Arnold W., 135, 160, 356
FTC. See Flight Test Center.
Fubini, Dr. Eugene G., 84
Fuel, 363
  ammonia, 189
  hydrogen, 60
  kerosene, 340
  liquid, 215, 217
  liquid hydrogen, 316, 340
  liuid hydrogen-oxygen, 382
  liquid oxygen-kerosene, 61
  slush hydrogen, 61
  solid, 50, 63, 189-190, 382
Fulton, Fitzhugh, 3, 11, 19, 28, 126, 193, 253, 298
Fulton, Rep. James G., 55, 67, 195, 314, 320
FW-4A (rocket engine), 76

G

Gabriel, David S., 155
Gagarin, Col. Yuri A. (U.S.S.R.), 80, 101, 107, 128, 158, 166, 182
Galantin, Adm. I. J. (USN), 360
Gamma ray, 61, 262, 308, 313
Gamma Velora (star), 369
GAO. See Government Accounting Office.
Garbarini, Robert F., 158
Garber, Ruvin, 88
Garrett Corp., 281
Gas turbine, 66, 172, 384
Gasbuggy, Project, 394
Gault, Donald E., 317
GCA Corp. (Geophysics Corp. of America), 28, 91, 216, 237, 238
Geer, E. Barton, 25
Gemini (program), 7, 28, 58, 196, 356
Gemini (spacecraft), 164, 166
Gemini IV (flight), 387
Gemini XII (flight), 16, 175
Gemini Summary Conference, 34

General aviation, 184, 188, 309, 363
General Dynamics Corp., 59, 88, 105, 149, 311, 330, 360, 381
General Electric Co.
  Apollo spacecraft report, 138
  contract, 99, 129, 133, 168, 188, 192
  meteorites, 92
  Missile and Space Div., 60, 368
  Re-Entry Systems Div., 264
  supersonic transport engine, 5, 32, 48, 130, 133, 392
  XB-70 accident, 180
General Electric Forum, 132, 286, 323
General Motors Corp., AC Electronics Div., 43
General Precision Equipment Corp., 316
General Services Administration (GSA), 235
Geneva Convention on the Continental Shelf, 359
Geneva, Switzerland, 92, 114, 216, 252
Geodetic satellite, 41, 47, 221, 335
Georgia Institute of Technology, 151, 363
Georgia, Univ. of, 309
Geotechnology, 16
Germantown, Md., 281
Germany, East, 192
Germany, West, 70, 268
  aircraft, 270, 371
  cooperation, 189, 216, 218, 219, 270, 276, 302, 333, 349, 354, 356, 393
  Ministry of Scientific Research (BMWF), 189, 219, 354
  radiotelescope, 312
  satellite, 218, 276, 333, 349
  sounding rocket, 189, 349, 354, 356, 393
  space program, 219, 348
  tracking station, 29
Gernsback, Hugo, 247-248
GET: ground elapsed time
Getler, Michael, 211, 242-243, 367
Giacconi, Dr. Riccardo, 16-17
Gibson, Col. M. B. (USAF), 178
Gilruth, Dr. Robert R., 23, 27, 34, 136, 150, 356-357
Givens, Maj. Edward G. (USAF), 177, 180
Gleaves, James D., 317
Glenn, Col. John H., Jr. (USMC, Ret.), 27, 49, 112
Glennan, Dr. T. Keith, 150, 316
Gluhareff, Michael E., 262
Goddard Institute for Space Studies, 16
Goddard Memorial Dinner, 76
Goddard, Robert H., 177
Goddard, Robert H., Award, 19
Goddard, Robert H., Memorial Trophy, 72
Goddard Space Flight Center (GSFC), 180, 197, 359
  aeronomy satellite, 142
  award, 218, 317
  contract, 60, 176, 328
  facilities, 18, 179
  management, 278, 281, 332, 341
  Memorial Symposium, 73-74
  personnel, 38, 218, 250, 254, 270
  Real Time Computer Complex, 312

Goddard Space Flight Center—Con.
 satellite monitoring, 242, 274
  Ats II, 331–332
  Ats III, 336
  Biosatellite II, 265
  Echo I, 241
  Essa IV, 20, 120, 343
  Explorer XXXIII, 10
  Observatory class, 66, 307–308, 319–320
 sounding rocket experiments, 142, 173
  astronomical, 125, 140, 303, 369, 389
  atmospheric data, 25, 28, 67, 91, 97, 103, 106, 116, 117, 140, 151, 155, 191, 237–238, 258, 268, 272, 289, 316, 370, 375
  magnetic field, 238
Goheen, Robert F., 369
Gold, 386–387
Gold, T., 238
Goldberg, Ambassador Arthur J., 23, 65
Goldberg, Dr. Leo, 343, 355
Goldstone Tracking Station, 118, 194, 212, 229, 242, 342, 353, 376
Goldwater, Barry, 216
Gometz-la-Ville, France, 367
Goodell, Rep. Charles E., 69
Goodrich, Joseph A., 217
Goodyear Aerospace Corp., 189
Gordon, Max, 311
Gordon, Mitchell, 2
Gordon, LCdr. Richard F., Jr. (USN), 136, 350
Gordon, T. J., 280
Gore, Sen. Albert, 65, 71
Gould, Jack, 240
Goulian, Dr. Merhven, 378
Government Accounting Office (GAO), 13, 47–48, 168, 175
"Governor's Award" (Ohio), 45
Graduate Research Center of the Southwest, 176
Graham, Mayor Milton, 179
Graham Engineering Corp., 205
Granada Lecture, 318
Grand Forks AFB, N. Dak., 346
Grants, 51, 64, 132, 151, 288, 306–307, 349
Gray, Robert H., 317
Green Bank, W. Va., 155, 251
Green, Leon, Jr., 189
Green River, Utah, 80
Green Valley, W. Va., 232
Greenland, 336
Grimwood, James M., 10
Grissom, L/C Virgil I. (USAF)
 award, 207
 death, 22, 25, 26, 27, 29
  cause, 36, 47, 101, 107
 memorial to, 58, 60, 69, 85, 181
Gross, Prof. Robert A., 316
Groton, Conn., 267
Ground elapsed time: GET
Grumman Aircraft Engineering Corp., 122, 168, 187, 279, 301, 330, 369
GSA. See General Services Administration.

GSFC. See Goddard Space Flight Center.
Guerin, Pierre, 252
Guggenheim Aerospace Laboratories, 61
Guggenheim, Daniel and Florence, Aviation Safety Center (Cornell Univ.), 173, 188
Guggenheim, Daniel and Florence, Award, 253–254
Gulf of Mexico, 244
Gurney, Rep. Edward J., 58
Guss, Dr. Donald E., 38
Gyroscope, 183

H

H–1 (rocket engine), 196
Hachenberg, Prof. O., 312
Haddon, M. Carl, 36
Hage, George H., 158, 201, 302
Haglund, Howard, 317
Haise, Fred W., Jr., 350
Halaby, Najeeb E., 12, 34
Haley Astronautics Award, 41
Hall, Charles F., 317
Hall, Rep. Durward G., 49
Hall of Fame for Great Americans (New York Univ.), 142
Hall of Science (proposed), 170
Hamilton AFB, Calif., 280
Hamilton Standard Div., United Aircraft Corp., 284
Hammaguir Range, Algeria, 5, 16, 41, 47, 65, 93, 197
Hammarskjold, Knut, 6
Hanford Project (AEC), 284
Hann, H. Frank, 203
Hanna, William E., Jr., 382
Harbour Island, 209
Harford, James J., 289
Harkins, R. Roger, 282
Harmon, Clifford B., Trust, 267
Harmon International Aviation Trophy, 267, 300
Harmon, Airman 3/C Richard G., 29
HARP. See High Altitude Research Project.
Harper, Charles, 11, 138
Harr, Dr. Karl G., Jr., 19, 30, 109, 352
Harris poll, 225
Hartke, Sen. Vance, 58
Harton, Erskine E., Jr., 212
Harvard College Observatory, 284, 355
Harvard Univ., 389
Haute Provence Observatory, 251
Hawaii, 7, 203
Hawaiian Telephone Co., 88, 101
Hawker Siddeley Aviation Co., 220, 277
Hawker Siddeley Dynamics, Ltd., 197
Hawkins, Jerry W., 317
Haworth, Leland J., 230, 248
Hayes International Corp., 112
Hayes, Dr. Richard M., 363
Hayward, Calif., 279
Head, Dr. Richard M., 155, 191–192, 389
Hearth, Donald P., 55, 300
Hedrick, B/G Walter R., Jr., (USAF), 87

Heeschen, Dr. David, 156
Helberg, Robert J., 317
Helgeson, Bob P., 384
Helicopter, 262
   Atlantic flight, 166, 174
   military, 16
   record, 174
   rescue use, 166
   spacecraft recovery use, 111
Helium, 35–36, 39, 201, 259
Heller, Gerhard B., 309
Hello, Bastian, 136
Helms, J. Lynn, 269
Helms, Richard, 8
Henize, Dr. Karl G., 233
Heos-A (interplanetary physics satellite), 66–67
Hereford, Tex., 279
Herres, L/C Robert T. (USAF), 196
Hersey, Ben W., 384
Herzfeld, Dr. Charles M., 168
Hess, Dr. Wilmot N., 46
Hewes, Donald, 288
Heyns, Dr. Roger W., 217, 317
HH–3E (helicopter), 166, 174
High Altitude Research Project (HARP), 247
Highwater, Quebec, 247
Hill, Louis W., Space Transportation Award, 316
Hines, William, 62, 64–65, 95, 121, 178, 221, 247
Hinton, Prof. H. E., 327
HL–10 (lifting-body vehicle), 190
Hoboken, N.J., 142–143, 381
Hoffman, David, 385
Hoffman, S. K., 294
Holifield, Rep. Chet, 62, 71
Holland, Sen. Spessard L., 58, 220
Holloman AFB, N. Mex., 12, 39
Hollomon, Dr. J. Herbert, 73, 162
Holmberg, Mervin, 122
Holmes, Adm. Ephraim P. (USN), 313
Holmes & Narver, Inc., 126
Holmquest, Dr. Donald L., 233
Holt, Prime Minister Henry E. (Australia), 79
Homestead, Fla., 189–190
Honest John-Nike (rocket), 146, 191
Honeysuckle Creek Tracking Station, Australia, 79
Honeywell, Inc., 59
Hood, Arthur F., 317
Hootman, Dr. James A., 12
Hoover Institution on War, Revolution, and Peace, 333
Horne, Walter B., 317
Hornig, Dr. Donald F., 84, 165, 215, 347
Horowitz, Dr. N. H., 117
Hosenball, S. Neil, 299
Hosmer, Rep. Craig, 332–333, 338
Hotz, Robert, 211, 225, 235–236
Hough, Richard R., 387
House of Commons
   Committee of Public Accounts, 230
   Estimates Committee, 237

Housing and Urban Development (HUD), Dept. of, 34, 84, 339
Houston, Tex., 27, 175, 180, 221, 284
Howard, Bailey K., 133
Hsinhua (Chinese press agency), 190
HUD. See Housing and Urban Development, Dept. of.
Hudson Institute, 390–391
Hughes Aircraft Co., 164
   award, 56, 217, 317
   contract, 8, 13, 86, 201
   communications satellite, 249
   Surveyor spacecraft, 113, 265–267, 334
Hughes, Gov. Richard J., 96
Hulbert Center for Space Research, 369
Humphrey, Vice President Hubert H., 53, 125, 291
   astronaut deaths, 25, 29
   MSFC visit, 79, 162
   space program, 76, 79, 153, 224
   technology gap, 70
Humphreys, M/G James W. (USAF), 164, 173
Hundhausen, Dr. A. J., 348
Hungary, 192, 376
Hungate, Rep. William L., 249
Hunt, Dr. Graham R., 227
Hunt, Rep. John E., 18
Huntington Beach, Calif., 260
Huntsville, Ala., 118, 184, 243, 273–274, 330, 388
Husband, Dr. H. C., 312
Huston, M/G Vincent G. (USAF), 87, 200
HYCAT (synthetic compound), 361
Hyde, Rosel H., 96, 387
Hydrofoil, 349, 350
Hydrogen, 61–62
Hydrogen bomb, 190
Hydroskimmer, 96
Hygius Rille (moon), 42
Hynek, Dr. J. Allen, 11, 151–152, 271
Hypersonic aircraft, 8, 17–18, 58, 392

I

IAA. See International Academy of Astronautics.
IAEA. See International Atomic Energy Agency.
IATA. See International Air Transport Assn.
IAU. See International Astronomical Union.
IBM. See International Business Machines Corp.
ICAO. See International Civil Aviation Organization.
Iceland, 196, 204, 224–225
IDA. See Institute for Defense Analyses.
IDC. See Image Dissector Camera.
IDCSP. See Initial Defense Communications Satellite Program.
IEEE. See Institute of Electrical and Electronics Engineers.
Ignatius, Paul R., 234.

IGY. See International Geophysical Year.
Ikeya-Seki (comet), 252
Il-62 (U.S.S.R. airliner), 185, 203, 351
Illinois, Univ. of, 28, 212, 237, 262, 353
  Coordinated Science Laboratory, 306
Image Dissector Camera (IDC), 332
India, 69, 78, 141, 321, 328, 335, 350, 393
Indian Ocean, 126-127, 384
Indiana Dunes National Lakeshore, 181
Indonesia, 184
Industrial College of the Armed Forces, 196, 200, 327
Industrial Research Magazine, 275
Information retrieval, 170
Infrared detector, 60
Ingham, Kenneth T., 217
Initial Defense Communications Satellite Program (IDCSP), 32, 178, 204, 236, 255, 261
  launch, 11, 234
Institute for Defense Analyses (IDA), 154, 157
Institute of Electrical and Electronics Engineers (IEEE), 48, 84
Institute for Strategic Studies, 104, 270
Integrated Medical and Behavioral Laboratory Measurement System, 129
INTELSAT. See International Telecommunications Satellite Consortium.
Intelsat I (Early Bird I) (communications satellite), 7, 41, 83, 100, 134, 157, 193, 262, 282
Intelsat II-A, 7, 83
Intelsat II-B (Pacific I), 21, 41, 83, 133-134, 157, 194, 282
  launch, 7
Intelsat II-C (Atlantic II), 42, 77, 82, 83, 85, 89, 282
Intelsat II-D (Pacific II), 146, 214-215, 281-282
INTELSAT III series, 157
INTELSAT IV, 249, 293, 326
Inter-American Bank, 110-111
Interior, Dept. of, 193, 242, 291-292
International Academy of Astronautics (IAA), 254
International Aeronautical Federation, 261
International Air Transport Assn. (IATA), 6-7, 217-218
International Assn. of Machinists and Aerospace Workers, 175
International Astronautical Congress, 254, 279, 281
International Astronautical Federation, 279-280
International Astronomical Union (IAU), 250, 284
International Atomic Energy Agency (IAEA), 327-328, 354, 364, 371
International Aviation Club, 12
International Buoy Technology Symposium, 274

International Business Machines Corp. (IBM), 84-85, 367, 388
  Federal Systems Div., 312
International Civil Aviation Organization (ICAO), 34
International Commission on Lunar Nomenclature, 56
International Conference on World Crisis in Education, 297
International cooperation, 111
  aircraft, 15-16, 219, 270
  astronomy, 235-236, 284
  communications, 20-21, 113, 121, 133, 161, 184, 192, 193, 203-204, 250
  meteorology, 60, 91, 93
  military, 86, 103, 121, 152, 204
  nuclear power, 216, 299, 354-355, 371
  oceanography, 182, 343
  science and technology, 25-26, 60, 111, 164-165, 216-217, 243
International cooperation, space (see also European Launcher Development Organization; European Space Research Organization; Space law treaty), 227, 387
  conference, proposed, 38-39, 46
  satellite, 98-99, 204-205, 241-242, 393
    Europe, 113, 197
    France-Canada, 161-162
    -Germany, West, 218, 348-349
    NASA-ESRO, 160, 167-168, 356
      -Canada, 197
      -France, 221-222
      -Germany, West, 275
      -Italy, 127-128, 356, 384
    U.S.-U.K., 140, 393
    -U.S.S.R., 60
  sounding rocket, 393
    NASA-Argentina, 393
      -Brazil, 88, 189, 302-303, 326, 393
      -Canada, 25, 27-28, 66, 316, 356
      -Germany, West, 74-75, 189, 356, 393
      -India, 69, 77-78, 393
      -Japan, 93, 393
      -Norway, 258, 393
      -U.S.S.R.-France, 322
  space research, 269, 277, 280, 298, 323, 356, 382, 393
    U.S.-U.S.S.R., 26, 74, 157, 227, 261, 296, 303, 310
    U.S.S.R.-France, 48, 258, 261
  tracking
    U.S.-Australia, 79-80
    U.S.S.R.-Australia, 65
    -North Vietnam, 309
    -U.K., 303-304
International Council of Scientific Unions, 227
International Federation of Automatic Control, 339
International Flat Earth Research Society, 245
International Geophysical Year (IGY), 175-176

International Institute of Space Law, 216–217
International Latex Corp., 246
International Organization for Standardization, 185
International Telecommunications Satellite Consortium (INTELSAT), 205, 241–242, 376, 385, 393
  Interim Communications Satellite Committee (ICSC), 157
International Satellites for Ionospheric Studies program, 197
International Telephone and Telegraph Co. (ITT), 114, 168, 184
International Treaty on Peaceful Uses of Outer Space, 64, 70
International Year of the Quiet Sun (IQSY), 212, 236
Inter-Union Commission of Solar-Terrestrial Physics (IUCSTP), 227, 236
Interrogation, Recording, and Location System (IRLS), 274, 281
Invention, 85
Ionosphere, 25, 39, 85, 89, 91, 127, 191, 197, 212, 227, 324
Iowa, Univ. of, 348, 378
IQSY. See International Year of the Quiet Sun.
IRLS. See Interrogation, Recording, and Location System.
"I–R 100" award, 275
Irving, Dr. G. W., Jr., 167
ISC. See Italian Space Commission.
ISIS–A (Canadian ionosphere satellite), 85, 110, 197
ISIS–B, 197
ISIS–C, 197
"Issues and Answers" (TV program), 200
Istres, France, 384
Itaborai, Brazil, 385
Italian Space Commission (ISC), 128, 384
Italy, 27, 59, 126, 291, 344, 356, 384, 391
Itek Corp., 362, 386
ITT. See International Telephone and Telegraph Corp.
ITT Cable and Radio, Inc.-Puerto Rico, 88, 101
IT&T Federal Electric Corp., 312
ITT World Communications, Inc., 88, 101, 118, 130
IUCSTP. See Inter-Union Commission of Solar-Terrestrial Physics.

J

J–2 (rocket engine), 79, 217, 292
Jackass Flats, Nev., 50–51, 378
Jackson, Sen. Henry M., 299, 331, 360
Jackson, Miss., 52
Jacobs, Kenneth H., 32
Jaffe, Leonard, 67–68, 180–181, 202
Jaffe, Dr. Leonard D., 40, 328–329, 307
Jakimiuk, W. J., 154
James, Jack N., 309, 321
James, Lee B., 89
Jamesburg, Calif., 130, 232

Jane's Fighting Ships, 349
Jane's Surface Skimmer Systems, 349
Japan, 78, 335
  communication, 7, 21, 133–134
  launch
    satellite, 111
    sounding rocket, 93, 393
Japan Air Lines, 177
Jarvis Island, 168
Javelin (sounding rocket), 42, 80, 110, 189, 191, 349, 370
Jeffs, George, 25
Jensen, Paul A., 56
Jet Propulsion Laboratory (JPL) (Cal Tech), 87, 279, 358, 366
  award, 217–218, 309, 317
  contract, 8, 24
  Deep Space Network, 38, 138, 229, 341–342, 353
  Goldstone Tracking Station, 118, 353
  Lunar Orbiter, 38, 59, 63, 92, 229
  Mariner project, 73–74, 78–79, 187, 214, 239, 253, 295, 311, 319
  personnel, 285, 321, 355, 358
  spacecraft research, 261, 303, 372
  Surveyor project, 113, 162, 165, 210, 212, 264–267, 334, 382, 391
  Voyager project, 54, 56, 158, 160
Jodrell Bank Experimental Station (U.K.), 30, 188, 234, 304, 308
Johns Hopkins Univ., 83, 166, 219, 328, 368
Johnsen, Irving A., 19
Johnson, Clarence L., 16
Johnson, Premier Daniel (Quebec), 161
Johnson, Dr. Fred M., 13
Johnson, President Lyndon B., 79, 182, 281, 299
  aeronautics, 380
  appointment, 8, 178, 234, 341, 358
  astronaut, 25
  awards by, 39, 164, 165, 300, 389
  budget, 17, 46, 54, 248, 257, 283, 338
  communications satellite system, 21, 79–80, 236, 241, 276
  cosmonaut, 125
  defense, 221
  disarmament, 9, 48, 60, 253
  education, 297
  international cooperation, 21, 80, 111
  Langley Research Center anniversary, 209
  meteorology, 213
  noise abatement, 34, 35
  nuclear power, 364
  science and technology, 39, 297, 338
  space program, 17, 20, 28, 76, 80, 139, 149, 294, 341, 343, 374–375
  space treaty, 23, 40, 126, 134, 297, 346, 383
  State of the Union Message, 7, 9
  supersonic transport, 32, 134, 157, 393
  transportation, 90
Johnson, Mrs. Lyndon B., 29, 179
Johnson, Dr. Samuel, 326

Johnson, Vincent L., 351
Joint Engineers Council, 50
  Bay Area, 50
Jones, Brendan, 194
Jones, M/G David M. (USAF), 87, 196
Jones, Stacy V., 5
Jones, Dr. Walton L., 76
Journal of the Geochemical Society, 320
JPL. See Jet Propulsion Laboratory.
Jung, Kurt, 37
Junior college, 181
The Junior College and Education in the Sciences (report), 181
Jupiter (planet), 4, 18, 27, 65, 109, 121, 200, 368
Justice Dept., 114, 129

K

Kahn, Herman, 390–391
Kalb, Barry, 241
Kamanin, L/G Nikolay (U.S.S.R.), 147
Kansas, Univ. of, 151
Karachi, Pakistan, 359
Kardashov, Dr. Nikolay, 240
Karth, Rep. Joseph E., 82, 135, 158, 178, 255, 324, 350, 393
Keck, George E., 13
Keegan, Richard J., 269
Kein, Dr. Klaus, 14
Keldysh, Prof. Mstislav V., 124, 233, 303, 308, 310, 322
Kelly, Dr. Albert J., 86, 174, 256
Kelly, Orr, 360
Kemmerer, Walter W., 37
Kemmett, Francis W., 12
Kemp, Klaus, 279
Kennedy Airport, 34
Kennedy, President John F., 226, 255
Kennedy Space Center (KSC) (NASA), 53, 135, 193, 275, 338, 360
  Apollo/Saturn 111, 144, 191, 235, 254, 257
  award, 317
  contract, 170, 171, 188, 189, 220, 226, 292
  fire, 149
    Apollo AS–204, 22, 24, 25, 27, 29, 33, 34, 36, 58, 63, 94, 121, 123, 138
  launch, 18
  personnel, 158, 168, 212
Kentucky, Univ. of, 233
Kenya, Africa, 126, 384, 393
Kepler (moon crater), 1–2
Kettering Grammar School, U.K., 259
Khokhlov, Dr. Rem, 80
Khrushchev, Premier Nikita (U.S.S.R.), 139–140
Kiev Academy of Science, 153
Killian, Dr. James R., Jr., 107, 217, 335
King, Elbert A., Jr., 37
Kirby, Robert H., Jr., 133
Kiruna, Sweden, 74, 94, 344, 354
Kistiakowsky, George B., 389
Kitty Hawk Memorial Award, 372
Kitty Hawk, N.C., 380

Klass, Phillip J., 304
Klein, Milton, 75
Klemin, Dr. Alexander, Award, 172
Klocko, M/G Richard P. (USAF), 341
Klopp, William D., 275
Kluttz, Jerry, 263
Knight, Maj. William J. (USAF), 143, 196, 248, 289–290, 304, 345, 392
Kobin, William, 41
Koelle, Dr. H. H., 280
Koestler, Dr. Alfred C., 39
Komarov, Col. Vladimir M. (U.S.S.R.)
  death, 124–126, 130, 140
  burial, 128
  cause, 134, 143, 158, 239
  editorial comment, 130–131
  honors, 161, 213
  tribute to, 291
Korea, 393
Korean War, 365
Kornberg, Dr. Arthur, 378
Korolev, Sergey, 9
Kosygin, Premier Aleksey (U.S.S.R.), 60, 126, 128
Kotelnikov, V. A., 39
Kourou, French Guiana, 333–334, 382
Kozyrev, Nikolay, 1
Kraft, Christopher C., Jr., 193, 212, 224
Kramer, James S., 317
Kranz, Eugene F., 316
Krasnaya Pakhra, U.S.S.R., 16
Krasnaya Zvezda, 253, 297
Kremlin Wall, 124, 128
Kronauer, B/G Clifford J. (USAF), 87
Krylov, Marshal Nikolai I. (U.S.S.R.), 349
KSC. See Kennedy Space Center.
Kubat, Jerald, 210
Kuiper, Dr. Gerald P., 166
Kupferman, Rep. Theodore R., 281
Kurzweg, Dr. Herman H., 78–79
Kyushu Island, Japan, 39

L

La France, Donald S., 217
Laboratory, 69, 122, 129, 132, 165, 188, 227, 255, 309, 353
Lagoe, Ronnie J., 164
LaHatte, William F., 89
La Jolla, Calif., 37
Lake St. Clair, Mich., 11
Lambda 3H (Japanese sounding rocket), 39
Lambda 4S–3 (Japanese booster), 111
Lamont Geological Observatory, 341
Land, Edwin H., 389
Langley Research Center (LARC) (NASA), 25, 180, 227, 281, 359, 363
  aeronautics research, 13, 152, 218, 348–349
  anniversary, 284, 290
  award, 133, 317
  contract, 160, 168, 182, 221
  facilities, 55, 188
  history, 132
  lunar gravity research, 288

Langley Research Center—Continued
  Lunar Orbiter program, 37, 59, 111, 139, 180
  Planetary Reentry Parachute program, 191, 222
  supersonic transport, 69
  Voyager program, 158, 160
Langley, Samuel Pierpont, 177
Lapp, Ralph, 51
LaRC. See Langley Research Center.
Large Astronomical Satellite (LAS), 302
Large Object Salvage System (LOSS), 369
Lark (missile), 79
LAS. See Large Astronomical Satellite.
Las Vegas, Nev., 111, 153
Laser, 10, 16, 18, 80, 84, 108, 221, 371, 391
LASV. See Low-altitude supersonic vehicle.
Latin American summit conference, 84
Latta, Rep. Delbert L., 192
Latta, M/C William N. (USA), 197
Launch Complex 34, KSC, 23, 26, 81
Launch Complex, 29, 37, 81
Launch Complex, 32, 39, 338, 339
Launch vehicle (see also individual launch vehicles, such as Atlas-Agena, Saturn, etc.), 42, 49, 54, 61, 62, 67, 80, 135, 233, 361
  cost, 135
  procurement, 225
  reusable, 187, 188
  U.S.S.R., 33, 109, 121, 351, 355
Lawrence, J. H., Co., 176
Lawrence Radiation Laboratory (Univ. of Calif.), 15
Lawrence, Maj. Robert H., Jr. (USAF), 197, 371, 373
Leavitt, William, 285
Le Bourget Airfield, France, 162, 174
"Lectures in Aerospace Medicine" conference, 39
Lederberg, Dr. Joshua, 15
Lederer, Jerome F., 173, 369
Lee, Dr. William A., 219
Leeds, Univ. of, 104
Lehigh Univ., 109
Leighton, Dr. Robert B., 250–251
LEM. See Lunar Module.
Le May, Gen. Curtis E. (USAF, Ret.), 344
Lemert, James B., 62
Lemnitzer, Gen. Lyman L. (USA), 204
Leningrad, 327
Lenoir, Dr. William B., 233
Leonard Medal, 320
Leonov, L/C Aleksey (U.S.S.R.), 194, 291
Le Page, Wynn L., 309
LeRC. See Lewis Research Center.
Les V. See Lincoln Experimental Satellite.
Letov, Dr. Alex M., 243
Levine, Arthur L., 315
Lewicki, Dr. George W., 372
Lewis, Anthony, 93
Lewis, Dr. Henry, 10
Lewis Research Center (LeRC) (NASA), 180, 359

Lewis Research Center—Continued
  aircraft research, 11, 261
  Quiet Engine program, 250
  award, 20, 275, 316–317
  booster, 113, 187, 334, 359
  contract, 37, 59, 213
  criticism, 47
  rocket engine, 190
Lewis, Richard, 274
Lewis, W. Deming, 109
Libby, Dr. Willard F., 209, 316
Library of Congress, 74, 139
  Legislative Reference Service, 24, 88, 104
Lick Observatory, 181, 192
Lieblein, Seymour, 19
Lieberman, Henry R., 70
Lifting-body vehicle, 14, 58, 62, 116, 163, 166, 190, 199, 206, 213, 226, 392
Lilly, William E., 75
Lincoln Experimental Satellite (*Les V*), 199, 213
Lincoln Laboratory, MIT, 20
Lindbergh, Charles A., 166, 177, 331
Lindsley, Prof. Donald B., 334
Lindsay, Mayor Robert, 170
Linear accelerator, 267
Ling-Temco-Vought, Inc. (LTV), 221
  Range Systems Div., 205
Link, Edwin A., 316
Literaturnaya Gazeta, 123
Little, Arthur D., Inc., 354
Littleton, Colo., 12
Littlewood, William, 366
Llewellyn, Dr. John A., 233, 234
LM. See Lunar Module.
Local Scientific Survey Module (LSSM), 193
Lockheed California Co., 182
Lockheed-Georgia Co., 268, 301
Lockheed Aircraft Corp., 15, 36, 110, 130, 189, 331
  award, 6
  contract, 61
  SR-71 aircraft, 318
  supersonic transport, 7
Lockheed Missiles and Space Co., 39, 226
  contract, 13, 86, 123, 179, 220, 255, 279
  sounding rocket experiment, 98, 253
Lockheed Propulsion Co., 32, 63
Loftin, Laurence K., Jr., 317
London, U.K., 23, 26, 70, 174, 269, 299, 359
  meeting, 93, 109, 153, 236, 318
  press conference, 219–220, 245
Long, Dr. Frank A., 25
Long Beach, Calif., 64
Lop Nor, Communist China, 386
Loring AFB, Me., 162
Los Alamos Atomic Weapons Laboratory, 322
Los Alamos Scientific Laboratory, 77, 262, 348
Los Angeles, Calif., 48, 199, 312
Los Angeles Chamber of Commerce, 372

LOSS. See Large Object Salvage System.
Lovelace, Dr. W. Randolph, II, 173
Lovelace, W. Randolph, II, Award, 136
Lovell, Sir Bernard, 233, 273–274, 304, 309
Lovell, Capt. James A., Jr. (USN), 178, 187–188, 267, 300, 350
Lovelock, Nev., 318
Low-altitude supersonic vehicle (LASV), 12
Low, Dr. George M., 46, 179, 371
Low-gravity engineering, 15
LSSM. See Local Scientific Survey Module.
LTV. See Ling-Temco-Vought, Inc.
LTV Aerospace Corp., 80, 284
Ludwig, John H., 182
Luedecke, M/G Alvin R. (USAF, Ret.), 285
Lukens, Rep. Donald E., 18
Luna II (U.S.S.R. lunar probe), 56
Luna III, 56
Luna IX, 56, 206
Luna XII, 14
Luna XIII, 1, 2
Lunar eclipse, 309
Lunar Excursion Module (LEM). See Lunar Module.
Lunar Module (LM), 90, 134, 219, 301
  contract, 189, 231, 335–336
  design, 187
  test, 200, 312, 369, 387
Lunar Orbiter (program), 4, 8–9, 18, 24–25, 28–30, 238–239, 350–351
Lunar Orbiter (spacecraft), 9, 23–24
Lunar Orbiter I, 23, 229, 347
  lunar landing, 208–209
  photographs, 8–9, 29, 248, 305
Lunar Orbiter II (Lunar Orbiter B), 347
  lunar landing, 298
  lunar orbit, 108, 112
  photographs, 8, 137, 229, 248, 305
  tracking use, 210
Lunar Orbiter III, 99, 229, 347
  launch, 38
  lunar landing, 298
  lunar orbit, 108, 112
  photographs, 42, 59, 92, 106, 202, 305, 391
  of *Surveyor I*, 72
  tracking use, 210
Lunar Orbiter IV (Lunar Orbiter D)
  launch, 138
  lunar landing, 325
  lunar orbit, 144, 180
  photographs, 173, 229, 248, 305, 347, 391
  tracking use, 210
Lunar Orbiter V (Lunar Orbiter E)
  launch, 229
  lunar orbit, 180, 298
  photographs, 245, 251, 260, 347, 391
Lunar Receiving Laboratory, 37, 77
Luskin, Harold T., 227
Luxembourg, 204
Lyle, Col. V. John (USAF, Ret.), 100
Lynch, Gerald J., 173
Lyon, Dr. Charles, 326
Lyons, Richard D., 166

M

M2–F2 (lifting-body vehicle), 14, 147, 226, 392
McAllister, Joseph W., 85
McCarthy, Dr. John F., 107
McCauley, George D., 212, 376
McClellan, Sen. John L., 289
MacCloskey, B/G Monro (USAF, Ret.), 344
McConnell, James C., 85
McConnell, Gen. John P. (USAF), 195, 207
McCormack, James, 96, 98, 222, 241, 275, 326
McCormick, Robert A., 182
McCoy, M/G John L. (USAF), 199
McDivitt, L/C James A. (USAF), 85, 115, 350
McDonald, Adm. David L. (USN), 207
McDonnell Aircraft Corp. See McDonnell Douglas Corp.
McDonnell Aircraft Astronautics Co. See McDonnell Douglas Corp.
McDonnell Co. See McDonnell Douglas Corp.
McDonnell Douglas Corp., 4, 25, 126, 330
  award, 129
  contract, 48, 126, 149, 159, 161, 190, 199, 201, 268, 292, 370, 383
  DC–8 Super 63, 63, 166
  Delta launch vehicle, 265, 383
  merger, 32, 129
  Saturn V, 13, 41, 387
  Space Systems Center, 260
  STOL airliner, 244
McDonnell, James Smith, 128, 150, 164
Mace A (missile), 3
McGee, Sen. Gale W., 296, 323
McGill Univ., 217, 247
McGraw-Hill, Inc., 150
MacGregor, Rep. Clark, 212
McGuire, Charles W., 248
McIver, L/C Robert C. (USAF), 39
Mack, Dr. Charles, 19, 20
McKay, Richard A., 217
McKee, William F., 104, 204
McKibbin, D. D., 115
McLane, James C., Jr., 36
McLeavy, Roy, 349
McLeod, Ben F., 300
McMurdo Station, Antarctica, 2, 6–7
McNamara, Secretary Robert S., 84
  antimissile defense, 21, 22, 129, 274, 277, 285, 323, 331
  F–111, 123
  Fractional Orbital Bombardment System, 321, 349
  MOL, 140
  nomination to World Bank, 359
  technology gap, 53, 70
  Themis, project, 12
McNaughton, John T., 234
Madras, India, 321
Madrid, Spain, 242, 353
Magazine Publishers Assn., 97
Magliato, Frank J., 299, 363
Magnetic field, 56, 238, 331, 372

Magnetometer, 56, 331
Magnetosphere, 115, 348
Magnuson, Sen. Warren G., 21, 34, 53, 283-284
Magruder, William M., 6
Mahon, Joseph B., 317, 351
Maier, Mayor Henry, 275
Makarov, U., 271
Mali, 376
Malinovsky, Defense Minister Rodion (U.S.S.R.), 21-22, 93
Mallay, George, 25
Mallick, Donald L., 184, 193, 253, 298
MAN Turbo, 270
Man-in-the-Sea, program, 37
Management, 1, 16, 163, 182, 258, 263, 291, 310, 358
Management Services, Inc., 112
Manchester Univ. (U.K.), 30
Manhattan Project, 168
Mankota, Minn., 241
Manned Orbiting Laboratory (MOL), 71, 205, 284, 343, 362
  appropriations, 18, 140, 388
  booster, 275
  contract, 161, 168, 284
  cost, 140
  launch plans, 130, 154, 392
  life support system, 39
  pilots, 197, 344, 371
Manned space flight (see also Apollo program; Astronaut; Cosmonaut; Manned Orbiting Laboratory; and Space Biology), 29, 59, 196, 209, 224, 228, 275, 285, 290, 291, 292, 315, 316, 330-331
  achievements, 48, 235, 293, 316, 337
  cooperation, 243, 261, 280, 296, 356
  criticism, 209, 225, 235, 255-256, 285
  EVA. See Extravehicular activity.
  hazards, 39, 205, 213
  long-duration, 64, 153, 213, 256, 315, 357
  lunar landing, manned. See Moon, landing, manned.
  policy and plans for
    U.S., 40, 48-49, 59-60; 64-65, 80, 132, 178, 225, 235, 243, 256, 257, 275-276, 285, 290-291, 299, 330-331, 356-357
      schedule, 65, 152, 179, 212, 224, 275, 330-331
    U.S.S.R., 225, 243-244, 261, 314, 334
Manned Spacecraft Center (MSC) (NASA), 27, 32, 36, 40, 193, 216, 246, 341, 359, 363, 371
  Apollo Spacecraft Office, 179, 189, 219
  astronauts at, 25, 187, 233, 238
  award, 317
  contract, 88, 142, 189, 191, 205, 220, 231, 284, 335-336
  Lunar Sample Receiving Laboratory, 37, 78
  management, 179, 341, 371
  personnel, 54, 84-85, 108, 133, 136, 205
  spacecraft test. See Apollo (spacecraft).

Manson, Samuel S., 317
Marcotte, Paul G., 317
Mare Tranquillitatis (moon), 266
Marietta, Ga., 268
Marine Resources and Engineering Development Act of 1966, 359
Marine Technology Society, 274
Mariner (program), 8, 17, 27, 30, 59, 68, 82, 90, 338, 393-394
Mariner I (Venus probe), 187
Mariner II (Venus probe), 186-187, 287
Mariner III (Mars probe), 187
Mariner IV (Mars probe), 186-187, 208, 309
  control, 295, 319, 353-354, 385
  micrometeoroid impact, 270-271, 385
  orbit, 239, 265
  photographs, 187, 208, 319, 385
  results, 208, 295, 353-354
  signal, 295, 378
Mariner V (Mariner E) (Venus probe), 183, 185-186, 200, 391
  control, 295, 353-354
  distance traveled, 214, 245, 253, 295
  launch, 185
  results, 229, 239, 287, 310-311, 378-379
Markowitz, William, 271, 282
Marquardt Corp., 12
Marquette Univ., 271
Mars (planet) (see also *Mariner III* and *IV*; Voyager program)
  atmosphere, 37-38
  communication with, 10
  contamination of, 233-234
  craters, 250-251
  exploration, 14, 31, 208, 224, 261, 309
    funding, 246
    manned, 64-65, 81, 256
    nuclear propulsion, 64-65
    plans for, 4, 20, 46, 78-79, 179, 222, 254
    spacecraft, 8, 17, 21, 27, 73-74
    unmanned, 179, 258, 269, 319, 385
  life on, 37-38
  photographs, 319, 385
  surface, 37-38, 250-251, 252
  symposium on, 37-38
  water on, 252
Marshall Space Flight Center (MSFC) (NASA), 239, 260, 280, 359
  Apollo Telescope Mount, 220
  award, 70, 109, 133, 177, 199, 361
  contract, 10, 13, 81-82, 112, 199, 218-219, 239, 274, 368, 386
  launch vehicle. See Saturn.
  management, 6, 55, 69, 92, 102, 180, 187, 194, 239, 273, 274, 294, 341, 373
  personnel, 2, 6, 15, 40-41, 70, 89, 109, 133, 177, 180, 273, 294, 341, 383, 384
  Saturn S-IVB orbital workshop, 199, 220, 260, 292, 373
  Space Sciences Laboratory, 309
  test (see also Saturn), 18, 79, 187, 199, 292, 309
  visitors, 3, 79, 194
Martin Co., 12, 130, 145, 156

Martin, James S., Jr., 317
Martin Marietta Corp., 159–160, 197, 201, 220, 375
Martine (monkey), 65
Maryland, 279
Maryland, Univ. of, 91, 347
Mason-Rust Co., 386
Massachusetts, 251
Massachusetts Institute of Technology (MIT), 84, 126, 133, 183, 202, 217, 335, 358
    Aeronautics and Astronautics, Dept. of, 15, 135
    Instrumentation Laboratory, 317
    Lincoln Laboratory, 20
Masursky, Harold, 235, 305
Material Investigations Reactor (MIR), 367
Materials technology, 52, 54, 88, 107–108, 142, 148–149, 187, 188, 246, 269, 275, 361, 369
Mates, Jack, 247
Mathews, Charles W., 10, 25, 279–280, 357
Mathews, Edward R., 168
Mathias, Rep. Charles McC., Jr., 208
Matthias, Dr. Bernd T., 142
Mattingly, Lt. Thomas K. (USN), 350
Mattoni, Dr. Rudolph H. T., 313
Mauron, France, 231
Maurras, Maj. Donald B. (USAF), 174
Max Planck Institute of Nuclear Physics, 354
Maxwell, M/G Jewell C. (USAF, Ret.), 48, 133, 154, 203, 245
May Day Parade (U.S.S.R.), 133–134
MDA. See Multiple docking adapter.
Medaris, M/G John B. (USA), 118, 184
"Meet the Press" (TV program), 161
Melbourne, Australia, 176
Melbourne, Fla., 281
Memorandum of Understanding, 66–67, 74, 152
Menasco Manufacturing Co., 173–174
Mengel, J. T., 315
Menon, Dr. T. K., 251
Menzel, Dr. Donald H., 284
Merchant, Col. Robert A. (USMC, Ret.), 201
Mercury (planet), 17, 64–65, 354
Mercury (program), 10, 58, 62, 195–196, 290–291, 356–357
Mesa Verde National Park, Colo., 8
Metcalf, Sen. Lee, 48
Meteor, 309
    artificial, 46
"Meteor" (U.S.S.R. meteorological satellite system), 129
Meteorite, 14–15, 92, 240, 320, 341
Meteoritical Society, 320
Meteoroid, 11, 260–261, 328
Meteorological satellite, 75, 324, 391–392
    Ats I, 92, 270
    Ats III, 331–332, 336
    cooperation
        NASA–ESSA, 120, 270, 342, 353, 391–392
        U.S.–U.S.S.R., 60

Meteorological satellite—Continued
    Cosmos CXLIV, 54, 60, 77–78
    Essa I, II, III, and IV, 20, 42, 120
    Essa V, 120, 125, 143, 189, 270
    Essa VI, 342, 353
    Meteor (U.S.S.R.) program, 129
    Nimbus program, 281
        Nimbus II, 155, 224–225, 362
    Tiros program, 120, 190, 342, 353
        Tiros VII, 190
Meteorology, 71, 75, 113–114, 143, 224–225, 343, 350
    balloon use in, 8
    cooperation, 91, 274
        NASA–ESSA, 120, 125, 143, 155, 206, 262, 270, 342
        NASA–Japan, 93
        U.S.–U.S.S.R., 60
    forecasting, 137, 206, 270, 324
    satellite. See Meteorological satellite and individual meteorological satellites.
    sounding rocket experiments (see also individual sounding rockets), 392
        grenade, 28, 375
        vapor cloud, 28, 216, 237, 260, 268, 289
    weather modification. See Weather modification.
    World Meteorological Organization, 92
    World Weather Watch, 18
Mettler, Dr. Ruben F., 136
Meudon Observatory, France, 5, 251
Mexico, 268, 382
Mexico City, 101
Meyer, Karl E., 152
Miami, Fla., 230, 270
Michigan, 11, 158
Michigan, Univ. of, 18, 71, 188, 269, 272–273, 313
Michoud Assembly Facility (MSFC), 6, 239, 260, 373, 382, 386
Microelectronics, 70
Micrometeoroid, 38, 42, 69, 177, 270, 293
Microminiaturization, 55, 121
Middleton, R/A Roderick O. (USN), 212
Midway Island, 260, 321, 356, 378
Mikoyan (U.S.S.R. variable-sweep-wing aircraft), 383–384
Miles, Marvin, 71
Military Sea Transport Service, 381
Milky Way (constellation), 151, 251
Miller, Rep. George P., 33, 72, 74, 178, 238, 310, 334, 351, 378
Mills, Rep. Wilbur D., 297
Millsaps College, 52
Milwaukee, Wisc., 42, 275
Milwitsky, Benjamin, 210, 317, 377
Minnesota, 278, 279
Minnesota, Univ. of, 16, 191, 210, 214–215
Minuteman (ICBM), 3, 62, 169, 294, 338
Minuteman II, 221, 223–224
MIR. See Material Investigations Reactor.
Mirage G (French variable-sweep-wing aircraft), 383–384

Mirage V (French variable-sweep-wing aircraft), 307
MIRV. See Multiple Independently Targetable Reentry Vehicle.
Missile, 3, 60, 248, 329, 385–386
  air-ground, 344
  air-to-surface, 169, 210, 376
  antiaircraft, 336
  antimissile, 7, 21, 27, 129, 169, 204–205, 206, 275, 297, 329–330, 332, 333, 355, 385
  ballistic, intercontinental (ICBM), 3, 9, 63, 109, 204–205, 221, 223–224, 336, 373
  conversion, space-use, 62
  foreign
    Communist China, 8, 204–205, 275, 297, 329–330, 332
    France, 152, 367
    U.K., 152, 185, 367
    U.S.S.R., 7, 9, 21, 27, 204, 206, 267–268, 329, 331, 332, 333, 336, 355–356, 360
  fractional orbital bombardment system (FOBS), 329, 331, 332–333, 348
  launch, 80, 110, 248, 295, 385–386
  medium-range, 8, 336
  nuclear, 60, 152, 185, 206, 267–268, 297, 331
  sea-based antiballistic missile intercept system (SABMIS), 275
  short range attack (SRAM), 63–64
  submarine, missile carrying, 204–205, 206, 338, 367
Missile Sites Labor Commission, 299
Mississippi Test Facility (MTF), 61, 194
Missouri School of Mines, 150–151
MIT. See Massachusetts Institute of Technology.
MIT Club, 151–152
Mitchell, L/Cdr. Edgar D. (USN), 350
Mitchell, Ind., 58
Mitchell, Jesse L., 3–4
Mohole Project, 101
Mojave, Calif., 92, 177, 270, 345
MOL. See Manned Orbiting Laboratory.
Molniya (U.S.S.R. communications satellite), 143, 250
Molniya I (U.S.S.R. communications satellite), 41, 193–194, 392
Molniya 1–2, 80, 165
Molniya 1–5, 165
Molniya 1–6, 290
Molniya 1–7, 313
Molniya II, 143
Molsink (molecular sink of outer space) chamber, 303
Mondale, Sen. Walter F., 121, 145
Mongolia, 8, 376
Monkey experiment, 262
Monochromator, 285
Monsanto Aviation Safety Award, 154
Monsanto Research Corp., 269
Montgomery County, Md., 388
Montreal, 158, 170

Moon, 238
  colony, 1–2, 31
  crater, 1, 227, 260–261, 309, 328–329, 377–378
  distance, 63
  exploration. of, 4, 7, 17, 20, 21, 25–26, 37, 44, 45, 87–88, 172, 206, 214, 377–378, 383
  gravitational field, 38
  landing, 91–92, 147, 212, 261, 278, 288, 291, 299, 377–378
  manned
    U.S., 42, 132, 161, 179, 192–193, 224, 225, 243, 333, 364
    plans for, 38, 40, 59–60, 61–62, 65, 68, 70, 87, 89, 99, 152, 184, 192–193, 209, 339, 341
    U.S.S.R., 87, 89, 147, 243–244, 299
  unmanned, soft
    equipment, 45
    U.S., 40, 45, 113, 118, 119–120, 209, 210, 212, 265, 291, 328–329, 334, 341–342, 350, 391
    U.S.S.R., 291
  landing sites, 8–9, 38, 42, 59, 106
  lunar base, manned, 30–31
  map, 208, 248, 388
  military use, 69
  motion, 5, 63, 391
  nomenclature, 56, 284
  photographs, 4, 141, 258–259
    Luna XII, 14
    Lunar Orbiter I, 9, 229, 305, 347
    Lunar Orbiter II, 9, 137, 229, 305, 347
    Lunar Orbiter III, 38, 42, 72, 92, 106, 202, 229, 305, 347, 391
    Lunar Orbiter IV, 143–144, 173, 229, 305, 347, 391
    Lunar Orbiter V, 229, 235, 238–239, 305, 347, 391
    Ranger VIII, 141
    Ranger IX, 141
    Surveyor I, 40, 305–306
    Surveyor III, 391
    Surveyor V, 265–266, 267, 328–329, 391
    Surveyor VI, 334, 341–342, 391
  probe. See individual probes: Luna XII; Lunar Orbiter I, II, III, IV, and V; Ranger VIII and IX; Surveyor I, III, V, and VI.
  surface, 1–2, 59, 134, 196, 227, 251, 390
    composition, 210, 266–267, 286, 287–288, 328–329, 334, 350–351, 377
    texture, 126, 210, 265–266, 267, 377–378
  topography, 8–9, 38, 99, 126, 210
  water on, 126, 305
Moore, Dr. Henry J., 144
Moorer, Adm. Thomas H. (USN), 330
Morgan, Carroll Z., 217–218
Morgantown, W. Va., 269
Morgenstern, Prof. Oskar, 316
Moritz, Bernard, 158–159
Morozov, P. D., 277
Morris, Thomas D., 234

Morton, Dr. Donald C., 200
Moscow (U.S.S.R.), 9, 16, 26–27, 60, 70, 121, 203, 259, 267–268, 286, 299, 336
  airline service, 170, 185, 351
  comsat transmission to, 133–134, 290
  defense system, 7, 50, 360
  meeting in, 23, 113
  press conference, 124, 240, 309
  television, 133–134, 290
Moscow Aviation Institute, 364–365
Moscow, Univ. of, 80–81
Mosely, Prof. Philip E., 335
Mosher, Rep. Charles A., 12–13, 47–48
Mossinghoff, Gerald J., 368
Motor, rocket (solid-propellant), 189–190
Mt. Palomar Observatory, 16, 153, 181, 327
Mt. Wilson Observatory, 181
MR–12 (U.S.S.R. sounding rocket), 322
MSC. See Manned Spacecraft Center.
MSFC. See Marshall Space Flight Center.
MSFN. See Manned Space Flight Network.
MT–135 (Japanese sounding rocket), 93
MTF. See Mississippi Test Facility.
Mueller, Dr. George E., 27, 39, 100
  Apollo, 36, 40, 53–54, 65, 72–73, 107–108, 110, 116, 144, 145, 147–148, 330–334
  Apollo Applications, 21, 178
  award by, 196
  manned spaceflight, 48–49
  Mars mission, 81
  space program, 7, 288
Mulholland, Dr. J. Derral, 63
Multiple docking adapter (MDA), 60
Multiple Independently Targetable Reentry Vehicle (MIRV), 338, 360
Mump (probe), 18
Murray, Dr. Bruce C., 87, 233–234
Mururoa Atoll, 152, 177, 194
Musgrave, Dr. Franklin S., 233
Musial, Stanley F., 178
MX–774 (missile), 79

N

NAA. See National Aeronautics Assn.
NAA. See North American Rockwell Corp.
NAB. See National Assn. of Broadcasters.
NACA. See National Advisory Committee for Aeronautics.
NAE. See National Academy of Engineering.
Naples, Italy, 204
NAR. See North American Rockwell Corp.
Narwhal (nuclear submarine), 267
NAS. See National Academy of Sciences.
NASA. See National Aeronautics and Space Administration.
NASA–AEC Space Nuclear Propulsion Office (SNPO), 50–51, 61, 155, 281, 359
NASA Communications Center, 179
NASA Engineering Systems Design Summer Faculty Fellowship Program, 180
NASA Lunar and Planetary Missions Advisory Board, 133
NASA Manned Space Flight Network, 92
NASA Network Test and Training Facility (GSFC), 179
NASA Office of Advanced Research and Technology (OART), 56, 73, 76, 78, 190, 204, 215, 366
NASA Office of Manned Space Flight (OMSF), 39, 71, 135, 145, 176, 198, 212, 219, 226
NASA Office of Space Science and Applications (OSSA), 3, 28, 55, 67–68, 98, 113, 134, 180–181, 210, 214, 222, 390
NASC. See National Aeronautics and Space Council.
Nashville, Tenn., 76
Natal, Brazil, 89, 189, 193, 248, 303
Nathan, Dr. Robert, 104
National Academy of Engineering (NAE), 128–129, 200–201
  Aeronautics and Space Engineering Board, 195
  Committee on Science, Engineering, and Regional Development, 353
National Academy of Sciences (NAS), 1, 8, 52, 108–109, 126, 129, 149, 175, 200–201, 254, 310, 313
  Committee for Long-Duration Manned Mission Study, 334
  Committee on Science, Engineering, and Regional Development, 353
  Panel on Applied Science and Technology, 165
  Panel on Astronomical Facilities, 181
National Advisory Committee for Aeronautics (NACA), 315, 366
National Aeronautics Assn. (NAA), 164, 285, 295
  Elder Statesman of Aviation Award, 232
National Aeronautical Establishment (NAE), 348–349
National Aeronautics and Space Act, 366, 374
National Aeronautics and Space Administration (NASA) (see also NASA centers, programs, satellites, and related headings, such as Ames Research Center, Apollo program, *Lunar Orbiter I*, etc.)
  accomplishments, 59, 67, 150, 391
  Ad Hoc Science Advisory Committee, 217
  agreement. See Agreement.
  anniversary, 28, 96, 208, 284, 287, 290
  Antarctic expedition, 2, 6
  Apollo 204 Review Board. See Apollo 204 Review Board.
  astronaut. See Astronaut.
  Astronomy Mission Board, 343

National Aeronautics and Space Administration—Continued
  awards and honors, 19, 41, 45, 54, 56, 70, 85, 86, 109, 129, 133, 136, 137, 142, 175, 177, 181, 184, 190, 193, 196, 212, 215, 217-218, 232, 267, 275, 300-301, 317, 361, 369
  Board of Contract Appeals, 331
  budget, 17-18, 33, 215, 257, 275-276, 279, 293, 318, 332-333, 336-337, 338, 352-353, 356, 378-379, 388
    editorial comment, 20, 87, 204, 211, 224, 226, 227, 235-236, 299, 323, 378-379
    House consideration, 54, 72-73, 120, 150, 156, 177, 192, 213, 225, 230, 237, 243, 246, 248, 249, 250, 296, 314, 320, 352-353
    Senate consideration, 46, 115-116, 145, 179, 194-195, 220, 225, 230, 257, 282-283, 291, 295, 296, 297, 314, 320, 336-337
    Vietnam War, effect of, 13, 255-256, 333
  conference, 239
  contract, 338, 366
    administration, 82
    aircraft, 182
      engine, 254-255
    engine, 8, 37, 149, 189, 220, 221, 231, 235, 239, 292, 360, 383
    facilities, 55, 82, 176, 278, 312, 388
    launch services, 66-67
    noise abatement, 47, 207, 250
    nuclear propulsion, 61, 184
    space equipment, 13, 27, 50, 76-77, 81-82, 159-160, 169-170, 189, 193, 212-213, 218-219, 238, 281, 328, 368
    space station activities, 71
    spacecraft, 23-24, 60-61, 123, 192, 199, 212-213, 216-217, 237, 335-336, 347, 371
    study, 88, 179-180, 255, 386
    support services, 6, 10, 59, 70, 126, 170-171, 188, 189, 196, 205, 226, 284, 312, 370-371
    telemetry, 50
    tracking, 174-175, 179, 195, 211, 242
  cooperation
    AEC, 50-51, 206-207, 281, 378
    ComSatCorp., 215, 391-392
    DOD, 111, 153, 279, 353, 388-389, 392
    ESSA, 20, 143, 206, 262, 270, 287, 342, 343, 353, 391, 392
    FAA, 20, 34, 69
    Interior, Dept. of, 193
    Smithsonian Institution, 72
    USAF, 190, 392
    USN, 287
    Weather Bureau, 262, 270
  cooperation, international. See International cooperation, space and Sounding rocket, international programs.

National Aeronautics and Space Administration—Continued
  criticism, 167, 208, 249, 294, 381, 386
    Apollo AS 204 accident, 40, 42, 49, 53, 54, 71, 101, 102, 103, 106, 114, 117, 118, 121, 122, 128, 130-131, 132, 145, 171, 204, 381
    Apollo spacecraft, 63, 94, 175
    military research, 366
    space program, 45, 82, 87, 139, 147-148, 156, 192-193, 224, 257, 296, 381
  facilities, 37, 99, 188, 358
  Historical Advisory Committee, 175-176, 361
  history, 315
  Incentives Award Fee Board, 347
  information dissemination, 271
  launch
    balloon, 238
    failure, 173
    postponed, 42, 207, 298
    probe
      Lunar Orbiter III, IV, and *V,* 38, 138, 229
      Mariner V, 185, 186-187
      Pioneer VIII, 375
      Surveyor III, IV, V, and *VI,* 113, 209, 210, 265-266, 334
    reentry test, RAM C-1, 311, 392
    satellite, 279
      Apollo 4 (AS-501), 339, 340
      Ariel III, 140
      Ats II and *III,* 98, 331-332
      Biosatellite II, 263
      Esro II, 168
      Essa IV, V, and *VI,* 20, 120, 342-343
      Explorer XXXIV and *XXXV,* 163, 164, 214
      Intelsat II-B, II-C, and *II-D,* 7, 83, 282
      Ogo IV, 222
      Oso III and *IV,* 66, 307, 308
    sounding rocket
      Aerobee 150, 48, 60, 78, 95-96, 125, 140, 142, 146, 173, 177, 191, 201-202, 214-215, 232, 236-238, 240, 253, 268, 285, 289, 293, 311, 320, 344, 350, 368, 369, 372
      Aerobee 150 (Mod I), 98
      Aerobee 150A, 72, 83, 368
      Arcas, 93, 230-231, 248, 270, 272, 276, 301, 311, 313, 316-317, 318
      Arcasonde, 311
      Argo D-4. See Javelin.
      Astrobee 1500, 50, 51
      Boosted Arcas, 316-317
      Javelin, 42, 80, 110, 157, 189, 191, 370
      Nike-Apache, 28, 33, 69-70, 78, 91, 142, 169, 184, 216, 234, 237, 238, 262-263, 272-273, 347, 359, 375
      Nike-Cajun, 28, 33, 91, 97, 103-104, 106, 116, 117, 130, 140, 146, 151, 155, 193, 303, 375

National Aeronautics and Space Administration—Continued
launch—continued
sounding rocket—continued
Nike-Tomahawk, 18, 25, 67, 89, 91, 177, 258, 260, 268, 289, 354, 368
Lunar and Planetary Missions Board, 133, 169
Lunar Orbiter Incentive Evaluation Board, 202
management, 55, 115, 161–162, 163, 167, 183, 211–212, 258, 281, 294
manpower, 167, 257, 272, 279, 359, 360
organization, 55, 102, 138, 299, 383, 388–389
patents, 205
personnel, 5, 38, 45–46, 55–56, 75, 89, 102, 138, 158–159, 168, 174, 176, 210, 211–212, 219, 226, 242, 248, 250, 253, 270, 287, 299, 300, 331, 339, 351, 363, 371, 382, 389
appointment, 12, 14, 18, 39, 59, 100, 155, 173, 174, 198, 201, 204, 211–212, 219, 327, 368, 384, 386
resignation, 39, 59, 86, 158–159, 168, 174, 219, 254, 288, 382
retirement, 388–389
procurement, 366
programs (see also specific programs, such as Apollo, etc., and Space program, national), 13, 28, 29, 54–55, 59, 67–68, 211, 226, 235, 256, 278–279, 282–283, 336
aeronautics, 12, 14, 17–18, 19–20, 69, 73, 84, 94, 153, 218, 315, 392
astronomy, 3–4, 45, 61, 78, 121, 133, 238, 343, 352–353, 391
electronics, 137–138, 242, 271
international, 79, 128, 135, 160, 168, 189, 221–222, 356, 392–393
manned space flight, 6, 10, 17, 21, 30–31, 45, 48, 72–73, 76, 81, 116, 133, 178, 193, 221, 255–256, 273, 275–276, 279–280, 352–353, 356, 391
meteorology, 93, 270, 274, 392
nuclear propulsion, 50–51, 64–65, 73, 120, 378, 392
sounding rocket, 50, 55–56, 93, 392
space medicine, 4, 117, 263, 378, 392
technology utilization, 15, 45, 68, 70, 108–109, 117, 140–141, 163, 242–243, 350, 392
tracking and data acquisition, 79, 117, 207–208, 316, 392
Spacecraft Sterilization Board, 117
supersonic transport, 19, 69, 84, 191–192, 245, 249
test
booster, 13, 61, 79, 187, 232, 246, 311, 319, 382
helicopter, 111
Manned Space Flight Network, 242
motor, solid propellant, 176
nuclear, 378
parachute, 146, 191, 205, 222
spacecraft, 23, 232, 365, 387, 391

National Aeronautics and Space Administration—Continued
universities, 38, 240, 306–307, 343, 346, 363
grants, 51, 132, 151, 288, 306–307, 349
X–15. See X–15.
National Aeronautics and Space Council (NASC), 29, 38, 104, 160, 215, 242, 293, 341
National Aerospace Education Council, 96
National Aerospace Services Association, 136
National Air and Space Museum, 72, 79
Hall of Aerospace Art, 109
National Airspace System, 185
National Association of Air Traffic Specialists, 309
National Association of Broadcasters (NAB), 96, 98–99
National Association of Science Teachers, 81
National Broadcasting Company (NBC), 161
National Center for Air Pollution, 182
National Civil Service League, 85
National Communications System (NCS), 7
National Council on Marine Resources and Engineering Development, 224, 287
National Defense Research Committee, 361
National Educational Television (NET), 41
National Engineering Science Co., 355
National Environmental Satellite Center, 63, 92, 201
National Geographic Society, 174
National Hurricane Center, 270
National Industrial Research Week, 275
National Medal of Science, 39, 389
National Model Rocket Championship, 241
National Newspaper Association (NNA), 193
National Park Service, 8
National Radio Astronomy Observatory (NRAO), 37, 248–249, 251
National Research Corp., 250
National Research Council (NRC), 108–109, 213
Committee on Solar-Terrestrial Research, 32
Space Science Board, 56, 165, 175–176, 233
National Retail Hardware, 216
National Science Teachers Association, 122
National Science Foundation (NSF), 17, 26, 51, 214, 230, 248–249, 259, 263, 361, 389
Office of Economic Manpower Studies, 285
Office of Sea Grant Programs, 354
report on junior colleges, 181
universities and colleges, Federal support of science in, 227–228
National Sea Grant College and Program Act, 354

National security, 53, 160, 221, 228, 286, 359, 367, 373
National Security Industrial Association Symposium, 310
National Sonic Boom Program, 3, 9, 11
National Space Club, 49, 72, 76, 158, 178, 245, 344
National Space Science Data Center (NSSDC), 18
National Storms Laboratory, 137
National Student Conference, 87–88
NATO. See North Atlantic Treaty Organization.
Nature, 52–53
Naugle, Dr. John E., 287, 343
Naval Air Test Center, 146
Naval Oceanographic Office, 287
Naval Ordnance Test Station, 230–231, 248
Naval Research Laboratory (NRL), 78, 146, 151, 169, 191, 201–202, 293, 344
   Atmosphere and Astrophysics Div., 32, 369
Navigation satellite, 30, 134, 169, 207–208, 224, 242, 263, 271–272
NBC. See National Broadcasting Company.
NC–4 (seaplane), 298
NCS. See National Communications System.
Negro astronaut, 371
Neifakr, Alexander, 123
Nelson, Bryce, 123
Nellis AFB, Nev., 213, 239–240, 347, 390
Nelson, Dr. Clifford H., 9, 305, 317
Neon, 259
NERVA. See Nuclear Engine for Rocket Vehicle Application.
Ness, Dr. Norman F., 370
NET. See National Educational Television.
Netherlands, 168
Neutrinos, 35–36
Neutron, 89, 177, 184
New England Assembly on Nuclear Proliferation, 72
New Hampshire, Univ. of, 89, 177
New Jersey, 96
New Mexico State Univ., 252
New Mexico State Univ. Observatory, 296
New Orleans, La., 117
New York Academy of Sciences, 368
New York Airways, 244
New York, N.Y., 12, 16, 19, 27, 37, 38–39, 80–81, 109, 170, 174, 209, 244, 247, 275, 300, 346, 351, 356
New York Port Authority, 351
New York, State Univ. of, 235
New York Univ., 37–38, 61, 142
Newell, Dr. Homer E.
   appointed Associate Administrator, 253, 287
   award to, 317
   sounding rocket, 55–56
   space astronomy, 61

Newell, Dr. Homer E.—Continued
   space programs, scientific, 61, 82, 121, 254–255
   space research, 3, 202
   space results, 9
   spacecraft, 75
   sterilization, 84
News conference. See Press Conference.
Niagara Falls International Airport, 146
Nichols, Donald, 122
Nicks, Oran W., 4, 55
Nieroski, J. S., 280
Nike-Apache (sounding rocket), 74, 347, 349, 354
   electron measurement, 28, 237
   ionospheric experiments, 78, 262–263
   micrometeoroid sampling, 169
   recovery system test, 234
   upper atmosphere data, 28, 33, 69–70, 91, 184, 216, 237, 272–273, 359, 375
Nike-Cajun (sounding rocket), 28, 33, 91, 97, 103, 106, 117, 130, 140, 146, 151, 155, 193, 303, 375
Nike-Iroquois (sounding rocket), 10, 324–325
Nike-Tomahawk (sounding rocket), 18, 25, 67, 89, 91, 177, 258, 260, 268, 289, 354, 368
Nike-X (antimissile missile system), 7, 201, 263, 274, 284–285, 292, 296, 297–298
Nikolayeva-Tereshkova, Valentina, 82
Nimbus II (meteorological satellite), 155, 225, 362
Nimbus-D, 50, 60, 281
Nininger, Dr. Harvey H., 14–15, 320
Niningerite (extraterrestrial mineral), 14–15
Nitrogen, 259
Nitze, Paul H., 111, 182, 234, 332, 338
Nixon, Richard M., 356
NNA. See National Newspaper Assn.
Nobel Peace Prize, 322
Nobel Prize, 322
Noise, 281, 341, 346, 351
Noise, aircraft, 47, 84, 170, 384–385
   contract, 47, 207
   NASA program, 17–18, 34, 153, 207, 315, 392
   regulation, 34, 106–107
   report on, 34
Nomex (nylon fabric), 247
NORAD. See North American Air Defense Command.
Norman, Okla., 137
Norrish, Ronald G. W., 322
North American Air Defense Command (NORAD), 245–246
North American Aviation, Inc. See North American Rockwell Corp.
North American Rockwell Corp. (NAR), 38, 84, 171, 180, 326
   Apollo spacecraft, 23, 25, 36, 63, 72–73, 94, 105, 115–116, 121, 130, 156, 175, 216–217

North American Rockwell Corp.—Con.
  contract 92, 144, 366, 371
  management, 136, 148
  merger, 84, 278
  Phillips Report, 115–116, 117, 138–139, 148, 175, 183
  Rocketdyne Div., 136, 194, 231, 246–247, 294, 312
  test pilot, 3, 9, 19, 174, 267
North Atlantic Assembly, 164–165
North Atlantic Treaty Organization (NATO), 59, 121–122, 204, 236, 257–258, 300
North Carolina, 8, 10
North Dakota State University, 153
Northeast Radio Observatory Corporation (NEROC), 248–249
Northrop Corp., 112, 205, 307, 349
Northrop Systems Laboratories, 90
Northrop Ventura Co., 168
Northwest Cape, Australia, 102
Northwestern University, 151–152, 233
  Astronomy Department, 11
Norton AFB, Calif., 25
Norway, 258, 260, 392–393
Nova Delphini (star), 349
Nowitzky, Albin M., 5
NRAO. See National Radio Astronomy Observatory.
NRC. See National Research Council.
NRDS. See Nuclear Rocket Development Station.
NRL. See Naval Research Laboratory.
NSF. See National Science Foundation.
NSSDC. See National Space Science Data Center.
Nuclear cooling, 191
Nuclear detection, 2–3
Nuclear Engine for Rocket Vehicle Application (NERVA), 17–18, 32, 64–65, 281, 378, 392
  NERVA I, 338, 394
  NERVA II, 54–55, 120–121, 291
Nuclear explosion, 2–3
Nuclear power, 185, 363–364, 371
Nuclear propulsion, 11, 15–16, 17–18, 62, 64–65, 73, 77, 281
Nuclear reactor, 206–207, 363–364, 367, 392
Nuclear Rocket Development Station (NRDS), 206–207
Nuclear submarine, 102, 267, 360
Nuclear test, 177, 194, 372, 386, 394
Nuclear test ban treaty, 60, 129, 134
Nuclear weapons, 104, 146–147, 228, 327–328, 329, 331, 332–333, 343–344, 348, 355–356, 360
Nuttall, Col. James B. (USAF), 29

O

Oahu, Hawaii, 346
Oakland International Airport, 143
OAO. See Orbiting Astronomical Observatory.
OAR. See Office of Aerospace Research.
OART. See NASA Office of Advanced Research and Technology.
Oberth, Hermann, award, 361
Observatory satellite. See Orbiting Astronomical Observatory; Orbiting Geophysical Observatory; Orbiting Solar Observatory.
Ocean data environmental science services acquisition (ODESSA) system, 349
Ocean of Storms (moon), 113, 118, 210
Ocean Systems, Inc., 244
Oceanographer (oceanographic research ship), 182
Oceanography, 53, 122–123, 354
  international law and national security, 359
  NASA, 274, 287, 372
  test, 244, 368–369
  U.S.S.R., 163
O'Connor, B/G Edmund F. (USAF), 273
ODESSA. See Ocean data environmental science services acquisition system.
Office of Aerospace Research (OAR), 35, 179
Office of Naval Research (ONR), 361
Office of Science and Technology (President's) (OST), 34, 84, 88, 227–228, 231, 347
OGO. See Orbiting Geophysical Observatory.
Ogo I (Orbiting Geophysical Observatory), 222, 274, 279
Ogo II, 222, 274, 279
Ogo III, 222, 274, 279
Ogo IV (OGO D)
  contract, 237
  launch, 222, 274, 391–392
  results, 237, 279, 295
  sounding rocket support of, 230–231, 248, 270, 272, 276, 301, 311, 313, 318
OGO-E, 184, 237
OGO-F, 237
Ogonyek, 101
O'Hare International Airport, Ill., 230
Ohio State Univ., 389
Ojai, Calif., 349
O'Keefe, Dr. John A., 377
Oklahoma Center for Continuing Education, 140–141
Oklahoma, Univ. of, 162
Olds, Col. Robert (USAF), 372
O'Leary, Dr. Brian T., 233, 294
Oliver, Covey T., 307
Omega Position Location Experiment (OPLE), 331–332, 388
Omegation Gauge System, 272–273
OMSF. See NASA Office of Manned Space Flight.
Ondrejka, Ronald J., 362
O'Neill, M/G John W. (USAF), 199
ONR. See Office of Naval Research.
Opa Locka Airport, Miami, 230
OPLE. See Omega Position Location Experiment.
Oppenheimer, Dr. J. Robert, 48

"Opportunities for Participation in Space Flight Investigation" (NASA handbook), 196
Optics, 80–81
Orbital Workshop (spacecraft), 178, 199, 220, 260, 291, 373
Orbiting Astronomical Observatory (OAO), 61
Orbiting Geophysical Observatory (OGO) (see also *Ogo I, II, III,* and *IV*), 237, 279
Orbiting Solar Observatory (OSO) (see also *Oso I, II, III,* and *IV*), 308
Ord, L/C John W. (USAF), 39
Order of Lenin, 46
Orenburg Steppe, U.S.S.R., 213
Orionids (meteor shower), 309
Osborne, Stanley de Jongh, 212
OSO. See Orbiting Solar Observatory.
Oso I (Orbiting Solar Observatory), 307–308
Oso II, 308
Oso III (OSO–E), 66, 72, 125, 154–155, 285, 308, 391–392
Oso IV (OSO–D), 307–308, 314–315, 319–320, 355, 377, 391, 392
OSO–G, 344
OSSA. See NASA Office of Space Science and Applications.
Ostapenko, Peter, 295
O'Toole, Thomas, 366
Ottawa, Canada, 348–349
"Our World" (TV program), 193–194
OV I–12, 221
OV I–86, 221
OV III–6 (research satellite), 367
Owens-Corning Fiberglas Corp., 112
Oxygen, 33, 39, 40, 41, 51–52, 105, 107, 118, 122–123, 259
Ozone, 151, 230–231, 248

P

P–1127 Kestrel (U.K. fighter aircraft), 277
Pachette, Arthur, 244
Pacific II (communications satellite), 318
Pacific Ocean
 communications satellite, 7–8, 83
 French atomic test, 152, 177, 194
 Japanese rocket test, 39
 Sealab III experiment, 223
 U.S.S.R. rocket test, 168, 260, 321, 356, 378
 U.S.S.R. seismic test, 253
Packard, Robert F., 269
Page, Dr. Robert M., 136
Pagov, Nikolai M., 321
Paige, Dr. Hilliard V., 368
Pakistan, 359
Palisades, N.Y., 341
Pallman, Dr. Albert, 245
Pan American World Airways, 12, 13, 34, 89–90, 185, 300, 351
Panama, 184
Panofsky, Prof. Wolfgang K. H., 267

Parachute, 63
 Apollo, 149, 205
 steerable, 76, 168
 test, 222
 U.S.S.R., 124, 239, 253
 Voyager, 146, 179, 191, 222
Parawing, 76, 168
Paris, France, 174, 290
Paris International Air and Space Show
 accident, 177
 U.S. exhibit, 116, 162, 166, 174, 218
 U.S.S.R. exhibit, 156, 166, 181–182, 211, 218
Paris, Univ. of, 254
Parker, James, 94
Parker, Dr. Robert A., 233
Parks College of Aeronautical Technology, 224
Parks, Robert J., 309, 317
Parrott, B. F. & Company, 269
Particles, charged, 80, 212, 348, 354
Pastore, Sen. John O., 207, 267
Patagonia, 335
Patuxent, Md., 146
Paumalu, Hawaii, 233, 282
Pay, Rex, 155
Peek, Charles R., 56
Pegasus (meteoroid detection satellite), 274
Peking Aeronautical Institute, 14
Pell, Sen. Claiborne, 343
Pendray, G. Edward, Award, 316
Pennsylvania, Univ. of, 163
Perceived noise in decibels: PNdb
Percy, Sen. Charles H., 32, 194
Performance Technology Corp., 360
Perrine, Calvin H., Jr., 133
Pershing (missile), 268
Peru, 307, 393
Peterborough, U.K., 216
Peterson, A. M., 208
Peterson, Bruce A., 147
Peterson, Maj. Donald H. (USAF), 196
Petrov, Georgi, 323
Pettengill, Dr. Gordon H., 252
Pettis, Rep. Jerry L., 18
Philadelphia, Pa., 190, 210, 351
Philco-Ford Corp., 105, 196, 203, 236, 242
Philippines, 7
Phillips Report, 115–116, 117–118, 128, 131, 139, 145, 148, 175, 183
Phillips, M/G Samuel C. (USAF), 23, 115, 117, 254, 294, 319, 341, 383
Phillips, William H., 317
Phoebus (nuclear reactor), 206–207
 1B, 50–51
Phoenix, Ariz., 179
Phoenix (missile), 111, 161, 307, 330
Piccard, Don, 96
Picciotto, Dr. Edgar E., 389
Pickering, Dr. William H., 74, 321, 358
Pickersgill, John W., 170
Pinson, M/G Ernest A. (USAF), 179, 180
Pioneer VI (interplanetary probe), 115, 206, 376

Pioneer VII, 376
Pioneer VIII (Pioneer C), 375, 392
Pittsburgh, Univ. of, 151
PKS 0237-23 (quasar), 16
Planetary Entry Parachute Program, 191, 222
Planetary Manned Space Flight Network (PMSFN), 315
Planets, life on. See Extraterrestrial life.
Planning Research Corp., 71
Plasma physics, 72
Plateau Station, Antarctica, 389
Plattsburgh, N.Y., 235
Plohr, H. Warren, 317
Plowshare, Project, 15, 72
Plum Brook Station (LeRC), 12-13, 47-48
Plummer, Dr. William, 166
Plymouth, England, 298
PNdb: perceived noise in decibels.
Podgorny, President Nikolay V. (U.S.S.R.), 126, 128
POGO (satellite). See Polar Orbiting Geophysical Observatory.
"Pogo stick" (one-man flying vehicle), 13
Point Barrow, Alaska, 28, 33, 97, 103, 116, 130, 146, 155
Point Mugu, Calif., 230, 248
Poland, 160, 192, 376
Polar Orbiting Geophysical Observatory (POGO), 236
Polaris (missile), 3, 152, 169, 185, 205, 206, 338, 367
A-3, 226
Polaroid Corp., 219, 389
Pollack, Herman, 216
Ponce, Puerto Rico, 77
Ponnamperuma, Dr. Cyril S., 109
Popolo II, 27
Popovich, Pavel R., 286
Porter, George, 322
Porter, Dr. Richard W., 233
Porz-Wahn, W. Germany, 349
Poseidon (missile), 152, 185, 206, 248, 338
Poseidon (NASA barge), 388
Possony, Stefan T., 333
Post, Wiley, 331
Prague, Czechoslovakia, 248, 250, 312
Pratt, Perry W., 66
Pratt & Whitney Div., United Aircraft Corp., 5, 207, 360
Pravda, 6, 9, 213, 288
Precision Recovery Including Maneuvering Entry (Prime), 14, 62, 116
President's Science Advisory Committee (PSAC), 45, 59, 80, 358
Press comment
  aeronautics, 94
  antiballistic missile (ABM) system, 343-344
  Apollo (program), 57, 87, 103, 390
  Apollo AS-204 accident, 26-27, 51-57, 64, 71-72, 131, 132, 171, 204
  Apollo 204 Review Board, 114
  Berkner, Dr. Lloyd V., 176

Press comment—Continued
  communications satellite system, 6, 100, 252
  Lawrence, Maj. Robert H., Jr. (USAF), 373
  Luna XIII, 2
  Lunar Orbiter V, 258
  management, 1
  MOON, 2, 44-45, 89, 137
  NASA, 20, 132, 204, 224, 226-227, 296
  NERVA (program), 64
  nuclear nonproliferation treaty, 327-328
  Paris International Air and Space Show, 181-182
  planetary exploration, 378-380
  proton accelerator (Weston, Ill.), 207
  Soyuz I accident, 130-131, 134
  space law treaty, 2, 25-26, 42, 64, 107, 128, 160
  space rescue treaty, 382, 387
  supersonic transport, 2, 5, 6-7, 31-32, 171-172, 297
  Surveyor III, 123-124
  Surveyor V, 287-288
  technology, 69
  tracking station, 64
Press conference, 187-188, 201-202, 275-276
  Apollo (AS-204), 84-85, 161
  Lunar Module test, 369
  Lunar Orbiter program, 305-306
  Lunar Orbiter III, 99
  Lunar Orbiter V, 235
  space law treaty, 60, 61
  Surveyor III, 162
  U.S. space program, 68, 149, 175, 212, 294, 305-306, 338
  U.S.S.R. space activities, 176, 240
  Venus IV, 322
Prime. See Precision Recovery Including Maneuvering Entry.
Princeton Univ., 17, 27, 49, 72, 200, 249, 316, 326, 369
Pritchard, Wilbur L., 142
Printing Industries of Metropolitan New York, award, 10-11
Probe (see also individual probes, such as *Mariner I, II, III, IV,* and *V*; *Pioneer VI, VII,* and *VIII*; *Venus III, IV*), 14-15, 30, 212, 235, 255-256, 275-276, 347
  contamination by, 5, 74, 87, 117, 118, 233-234
  interplanetary, 30, 66-67, 73-74, 115, 169, 206, 375-376, 378, 392
  Jupiter, 4, 17-18, 30, 31, 74
  lunar. See *Luna II, III, IX, XII,* and *XIII*; *Lunar Orbiter I, II, III, IV,* and *V*; *Surveyor I, II, III, IV, V, VI.*
  Mars, 4, 14, 17, 30-31, 78-79, 81, 87, 233-234, 270-271
  Saturn, 31, 65
  Sunblazer, 3-4, 17, 68
  Venus, 4, 14, 87, 183, 208, 229, 233-234, 287, 304, 308

Propellant. See Fuel.
Proprietary Association, 155
Propulsion (see also Engine), 86, 261, 392
Protective coating, 56
Proton, 42, 348
Proton (U.S.S.R. booster), 120–121
Proton (U.S.S.R. spacecraft), 182, 203
Proxmire, Sen. William F., 46, 179, 235, 294
PSAC. See President's Science Advisory Committee.
Puckett, Dr. Allen E., 164
Puerto Rico, 77, 126, 202, 262
Pulkovo Observatory (U.S.S.R.), 1
Punta del Este, Uruguay, 85, 110, 217
Purdue Univ., 51, 60, 151
Pushchino, U.S.S.R., 16
Pushkov, N. V., 236

Q

Quanah, Texas, 180
Quantum electronics, 84
Quasar (quasi-stellar object), 16, 52, 156, 201
Quebec, Canada, 161
Quincy, Mass., 381

R

Radar, 153, 185, 195, 329, 343
Radiation
  cosmic, 314
  effects, 47, 104, 213, 263, 326, 391
  gamma, 61, 262, 308, 313
  lunar, 38, 138, 214
  measurement, 129, 285
  solar, 35, 72, 129, 174, 206, 229, 262, 279, 285, 287, 288, 308, 358
  space, 38, 174, 213
  Van Allen belt, 20, 39, 129, 314–315
Radiation, Inc., 50, 281
Radio Corporation of America (RCA), 10, 56, 217, 317
  Aerospace Systems Div., 213
  comsat earth station, 101, 118, 130
  contract, 174, 211, 242, 263, 386
  Defense Electronics Products Div., 238
Radioastronomy, 37, 153, 248, 312
Radio Moscow, 26
Radio signal, 251
Radiobiological Factors in Manned Space Flight (Report), 213
Radiotelescope, 30, 53, 188, 263, 312
Raffo, Peter L., 275
Rahmatullah, M., 359
RAM (Radio Attenuation Measurement) Project, 311
RAM C-1 (spacecraft), 311, 370, 392
Ramjet, 8, 12, 17
Ramo, Simon, 16
RAND Corp., 69, 87, 125, 228, 335
Randolph AFB, Tex., 3
Ranger (program), 217
Rathert, George A., Jr., 50
Rauf, Mohammed, Jr., 207–208

Rawicz, Leonard, 250
Raytheon Co., 201, 202–203
RB-207 (aircraft engine), 270
RCA. See Radio Corporation of America.
RCA Communications, Inc., 88–89
RCA Service Co., 112, 170
Read, R/A Albert C. (USN, Ret.), 298
Rechtin, Dr. Eberhardt, 353, 358
Reconnaissance satellite, 64, 76, 205, 210, 218, 303, 361
Record, 3
  altitude, 295, 344, 345
  helicopter, 174
  spacecraft, 136
  speed, 267, 289–290, 295, 300–301, 343, 345, 392
  women's, 267, 301
Le Redoubtable (nuclear-powered submarine), 90
Redstone (booster), 358
Redstone Arsenal, Ala., 79, 118, 162
Reece, L. D., 317
Reed, Col. Charles W. (USAF), 239–240
Reed, Robert D., 317
Reentry, 46, 62–63, 80, 181, 202, 302, 334, 370, 385
Rees, Dr. D., 359
Rees, Eberhard F. M., 361, 393
Regulus (star), 368
Reiffel, Dr. Leonard, 8–9, 99
Reistrup, J. V., 14, 278–279, 386
Relativity, 16–17
Rempel, Robert G., 56
Rendezvous, 21, 136, 228, 334
Renovo, Pa., 380
Report to the Congress on United States Aeronautics and Space Activities, 28–29
Republic Aviation Corp., 253
Research and development (R&D)
  benefits, 256
  expenditures, 151, 284–285, 386
  government role, 151, 254, 310, 386
  national defense, 123, 204–205, 373, 376
  universities and colleges, 227–228, 353
  U.K. and France, 367
Research, military, 123, 204–205
Resolute Bay, Canada, 316–317
Reuters, 50, 86
Reynolds, Dr. Orr E., 4
RF-4C (reconnaissance aircraft), 384
RF-111A (supersonic reconnaissance aircraft), 380
Rhodes, Rep. John J., 213
Rice Univ., 42, 80
Richardson, Keith A., 37
Richland, Wash., 384
Richmond, Va., 193
Ricketts, Henry, 109
Rickover, V/A Hyman G. (USN), 10–11
Rickenbacker, Capt. Edward, 331
Rigg, Sqdn. Ldr. Hugh, 277
Rio de Janeiro, Brazil, 385
Roanoke, Va., 269
Rochester, Pa., 350
Rockaway, N.Y., 298
Rockefeller Public Service Award, 369

Rocket belt, 96
Rocket engine. See Engine, chemical, electric, nuclear, rocket, etc., and individual rocket engines such as F-1, J-2, etc.
Rocket Research Corp., 142
Rockwell, Norman, 109
Rockwell-Standard Corp. See North American Rockwell Corp.
Roderick, M/G Charles R. (USAF, Ret.), 173
Roderick, Robert L., 317
Rogan, J. P., 4
Rogers, Henry H., 317
Rogers, John G., 167–168
Rohini-75 (Indian rocket), 350
Rollins College, 53
Rolls-Royce Co., 270
Romania, 376
Rome, Italy, 27, 218
Rome, Univ. of, 128, 384
Romney, Gov. George, 158
Roseau (French satellite), 333
Rosenbluth, Dr. Marshall L., 72
Rosenfeld, Stephen S., 10
Rosman, N.C., 10, 174–175, 336
Rostow, Eugene V., 241
Rostow, Walt W., 104
Rotary International, 269
Roudebush, Rep. Richard L., 195
Roush, Rep. J. Edward, 181, 185
Rover (program), 206–207, 378
Rover (nuclear rocket engine), 54
Rowan, Dr. Lawrence, 9
Rowlesburg, W. Va., 118
Royal Aeronautical Society, 86, 109, 366
Royal Aircraft Establishment, 222, 230, 291
Royal Radar Establishment, 52–53
Royal Society of Great Britain, 327
Royer, Maj. Harvey J. (USAF), 371
Ruddock, Kenneth A., 56
Rule, Gordon W., 360
Ruleson, Project, 380
Rumsfeld, Rep. Donald, 99–100, 195, 204, 208
Rusk, Secretary of State Dean, 19, 65, 298
Rust Engineering Co., 112
Ruud, Ralph H., 136
Ryall, Dr. Alan, 85–86
Ryan, Rep. William F.
  Apollo (AS-204) accident, 128, 130, 381, 386
  NASA and Congress, 278–279
  NASA appropriations, 250
  NASA management, 105, 108, 139, 148, 168, 183, 211, 294
  SST program, 215
Ryland, Lloyd B., 217

S

SAAB aktiebolag, 197
SABMIS. See Seaborne Anti-Ballistic Missile Intercept System.
SAC. See Strategic Air Command.
Sacramento, Calif., 13, 40–41, 262, 347, 388
Sacramento Peak Observatory, 35
Safety, 40, 51–52, 68–69, 101–102, 139, 154, 315
Sagamore Hill Radio Observatory, 391
Sahara Missile Proving Grounds, 197
Saigon, S. Vietnam, 111
St. Louis, Mo., 150, 209, 347
St. Louis Univ., 224, 245
St. Paul, Minn., 278
Salem, Ore., 262
Salisbury, Dr. John W., 227
Salt Lake City, Utah, 346
San Clemente, Calif., 248
San Clemente Island, 37
San Diego, Calif., 72
San Francisco, Calif., 68, 177, 216, 273
San Marco (Italian satellite project), 344, 384
San Marco I (Italian satellite), 128, 384
San Marco II (San Marco B), 126–128, 384, 392–393
San Marco C, 344, 384
Sandage, Dr. Allan R., 216–217
Sanders Associates, Inc., 10
Sanford, Neal, 293
Santa Barbara, Calif., 280–281
Santa Cruz, Calif., 192
Santa Monica, Calif., 358, 370
Sarnoff, David, Research Center, RCA, 10
Saros (French satellite), 50
SAS. See Small Astronomy Satellite.
Sasso, Mark, 317
Satellite. See individual satellites, satellite programs, and type of satellite such as Aeronomy, Geodetic, Meteorological, Navigation, and Reconnaissance satellite.
Satellite Tracking and Data Acquisition Network (STADAN), 174–175, 179, 211
Satellite, unidentified
  launch vehicle
    Atlas, 221
    Atlas-Agena D, 36, 161, 175
    Atlas F, 321
    Scout, 111, 158, 279
    Thor-Agena D, 9, 50, 91, 146, 169, 188, 219, 235, 271, 327, 371
    Thor-Burner II, 42, 252, 299
    Titan III-B, 50, 191, 244, 274, 318
Sato, Prime Minister Eisaku, 21
Saturn I (booster), 45, 49, 61, 81, 274
Saturn I, Uprated (Saturn IB) (booster), 22, 145
  contract, 112, 196, 239, 292, 368
  engine
    H-1, 196
    J-2, 79, 217, 292
  program, 125, 145, 275, 283, 330, 353, 387
  stage
    S-IB, test, 162, 199, 292, 382
    S-IVB, test, 21, 178

Saturn V (booster), 115, 125
  Apollo/Saturn, 191, 205, 232, 274, 294, 298, 302, 319, 339, 341, 393
  capability, 294, 340, 393
  contract, 10, 36, 112, 180, 220, 235, 368
  cost, 393
  engine
    F-1, 61, 194
    J-2, 79, 217, 292, 340
  plans for, 4, 24, 49, 54, 55, 65, 138, 145, 275, 283, 330–331
  stage
    1st (S-IC), 36, 61, 187, 205, 232, 274, 340
    2nd (S-II), 164, 205, 232, 235, 274, 340, 388
    3rd (S-IVB), 13, 40, 54, 55, 232, 244, 260, 292, 388
Saturn (planet), 5, 31, 65
Saudi Arabia, 141
SCAT Solar Pointing System, 289
Scheer, Julian, 280
Schenectady, N.Y., 92, 160
Scherer, Capt. Lee R., 9, 317, 383
Schirra, Capt. Walter M., Jr. (USN), 23, 85, 144, 149, 161, 244
Schmidt, Dr. Maarten, 16
Schneider, William, 226
Schneiderman, Dan, 393
Schomburg, L/C August (USA, Ret.), 327
Schriever, Gen. Bernard A. (USAF, Ret.), 15, 153, 206, 310, 344
Schuck, O. Hugo, 339
Schutt, John B., 218
Schwartz, Harry, 288
Schwarz, Dr. Francis C., 201
Schweickart, Russell L., 85, 350
Schwinghamer, Robert J., 133
Science, 16, 176, 180
  benefits, 39, 230, 257
  funding, 51, 240
  international cooperation, 110
  national policy, 46, 51, 278, 310
  press coverage, 62
  U.S.S.R., 16, 123
Science Council of Canada, 66
Science Seminar, 180
Scientific and Technical Information Facility (NASA), 284
Scientist-astronaut, 1, 233, 294
Scientists, 26, 39, 123, 153, 231, 259, 278, 285, 313, 348
SCNS. See Self-Contained Navigation System.
Scott, L/C David R. (USAF), 85, 350
Scott, Dr. Ronald F., 120, 162
Scott, Miss Sheila, 267, 301
Scott, William E., 133
Scout (booster), 42, 76, 142, 201, 349, 382
  contract, 221
  launch
    RAM C-1, 311
    satellite
      Ariel III (UK-E), 140
      Esro II, 168

Scout—Continued
  launch—continued
    satellite—continued
      San Marco II, 126
      USAF, 111, 158, 279, 367
  Scramjet, 8
Scoville, Col. Curtis L. (USAF), 163
Scramjet. See Supersonic combustion ramjet.
Sea of Tranquility (moon), 42, 99, 350
Seaborg, Dr. Glenn T., 15, 267, 364, 389
Seaborne Anti-Ballistic Missile Intercept System (SABMIS), 201, 203, 275
Seal Beach, Calif., 164
Sealab II (underwater laboratory), 368
Sealab III, 37, 232, 368
Seamans, Dr. Robert C., Jr., 79, 253
  Apollo (AS-204) accident, 25, 33, 40, 46, 51, 53, 102, 110
  appropriations, 59, 192, 257
  cooperation, 274, 372
  resignation, 288
  space program, 144, 147, 150, 194, 318, 338
SEAR, Project, 309
Sears, G. A., 228
Seattle, Wash., 346
Sedov, Prof. Leonid I., 261, 280, 296, 299, 351
Sehlstedt, Albert, Jr., 13, 333
Seitz, Dr. Frederick, 165, 310
Self-Contained Navigation System (SCNS), 332
Sentinel System (ABM system), 329
Sentinel System Evaluation Agency (SSEA), 329–330
Serpentuator, 219
SERT II (Space Electric Rocket Test) spacecraft, 212
Service Module (SM), 90, 175, 189, 246
Service Propulsion System (SPS), 339
Sevareid, (Arnold) Eric, 29
Shabad, Theodore, 16
Shafer, Edward M., 331
Shai, Charles M., 218
Shea, Dr. Joseph F., 36, 53, 219
Shea, William M., 317
Sheldon, Dr. Charles S., II, 74, 139, 351
Shell Development Corp., 217
Shenton, Samuel, 245
Shepard, Capt. Alan B., Jr. (USN), 27, 175, 177, 207
Shepard, Charles, 328
Shepard, Van, 3, 9, 19, 174
Shinkle, M/C John G. (USA, Ret.), 158–159, 168, 212
Shklovsky, Iosif, 240
Shoemaker, Dr. Eugene, 162, 317, 342
Short-range attack missile (SCRAM), 64
Short takeoff and landing aircraft. See STOL aircraft.
Shurcliff, Dr. William A., 190
Siberia, 250
Sierradyne Corp., 349
Sikorsky Aircraft (United Aircraft Corp.), 262

Sikorsky, Igor I., 262, 285, 380, 389
Silber, Howard, 385
Silva, R. W., 115
Silverstein, Dr. Abe, 13, 48, 316
Sims, Harold, 173
Singer, S. Fred, 291
Sinkiang Province, Communist China, 386
Sinus Medii (moon), 210, 334, 342, 377
Sjogren, William L., 63
Sklarewitz, Norman, 347
Skua (U.K. sounding rocket), 1
Skybolt (missile), 376
Skynet (satellite), 236
SL-1 (solid rocket motor), 190
SL-2, 190
SL-3, 189
Slayton, Maj. Donald K. (USAF), 43, 46, 84
Slomich, Sidney J., 390
Sloop, John L., 135, 350
Sloop, Project, 380
Small Astronomy Satellite (SAS) (program), 328
Small Astronomy Satellite-A (SAS-A), 142
Small Business Administration, 263
Smart, Gen. Jacob E. (USAF, Ret.), 75, 137, 281, 389
Smith, Bradford A., 296
Smith, Francis B., 75
Smith, Dr. Henry, 343
Smith, Sen. Margaret Chase, 75, 145, 175, 257, 296
Smithsonian Astrophysical Observatory (Cambridge, Mass.), 29, 46, 177, 335
Smithsonian Institution, 72, 79, 109, 164, 177, 378
  Rite of Spring celebration, 96
Smoluchowski, Dr. Roman, 27
Smull, Dr. Thomas L. K., 75
SNAP-8 (nuclear reactor), 184
SNAP-10A, 394
Snelling, William, 136
Snetsinger, Dr. Kenneth, 14
Snoddy, William C., 361
Snowden, G., 176
SNPO. See NASA-AEC Space Nuclear Propulsion Office.
Social Security Administration, 382
Société Nationale d'Étude et de Construction des Moteurs d'Aviation (SNECMA), 290
Sohier, Walter D., 54
Solar cell, 56
Solar corona, 17, 355
Solar Distrubance Forecast Center, 206
Solar eclipse. See Eclipse, solar.
Solar flare, 16, 155, 229, 314
Solar Geomagnetic Monitoring Service, 229
Solar Pointing Aerobee Rocket Control System (SARCS), 372
Solar wind, 115, 236, 348, 370
Solid propellant, 63, 361
Soluyanov, F., 298

Sonic boom, 8, 12, 16, 280, 286
  national program, 3, 9, 11, 69, 84, 231
  supersonic transport, 20, 154, 171, 190, 200, 203, 385
Sonmiani, Pakistan, 359
Sounding rocket (see also individual sounding rockets: Aerobee 150, Aerobee 150A, Arcas, Astrobee 1500, Boosted Arcas, Javelin, Nike-Apache, Nike-Cajun, Nike-Iroquois, Nike-Tomahawk), 55, 212, 392
  foreign
    China, Communist, 314
    France, 5, 65
    Japan, 93
    U.S.S.R., 322
  international programs, 113, 227, 368, 382
    ELDO, 368
    ESRO, 38, 94
    NASA-Argentina, 393
      -Brazil, 248, 303, 326, 393
      -Canada, 28, 33, 42, 48, 66, 67, 91, 191
      -Germany, West, 74, 393
      -India, 78, 393
      -Japan, 93, 393
      -Norway, 260, 393
      -Pakistan, 359
    U.S.S.R.-France, 322
South Africa, 335
South Carolina, 279
South Pole, 112
Southeast Asia, 346
Southern Illinois Univ., 62
Southwest Center for Advanced Studies, 16, 110, 142, 191, 370
Southwest Research Institute, 303
Soviet Academy of Sciences, 39, 56, 101, 124, 308, 310, 322, 323
Soviet Armed Forces Day, 50
Soviet Commission for Exploration and Use of Outer Space, 292, 324
Soviet Institute of Oceanology, 163
Soviet Space Programs, 1962-65: Goals and Purposes, Achievements, Plans, and International Implications, 25
Soviet Union. See U.S.S.R.
Soyuz I (U.S.S.R. spacecraft), 124, 134, 147, 158, 161, 175, 213, 291, 334, 392
Space Applications Summer Study, 108
Space biology, 74, 87, 327
  animal experiments, 263, 313, 334
    dog, 104
    mouse, 4, 368
    primate, 5, 30, 39, 65, 262
  atmosphere, artificial, 33, 39, 41, 52, 101, 105, 107, 118, 122, 259
  award, 136
  environment, effects of, 2, 6, 36-37, 39, 47, 76, 129, 136, 263, 326, 334, 368, 378
  laboratory, 129
  life support system, 39, 50, 53, 101, 107, 118, 122, 148, 263, 370

Space biology—Continued
  radiation, effects of, 104, 213, 265, 288, 313, 314, 326, 392
  technology utilization, 104
  weightlessness, effects of
    animals, 2, 5, 65, 104
    humans, 10, 12, 31, 36–37, 56, 71, 76
    organisms, 47, 263–265, 313, 391–392
    plants, 326
Space Biology and Medicine (U.S.S.R. magazine), 259
Space environment, 136
Space law (see also Space rescue treaty), 217, 332
Space law treaty, 2, 54, 64, 71, 293, 343, 361
  editorial comment, 25–26, 42, 64, 107, 128, 134, 160, 299, 322
  effective date, 298
  government comment, 71, 300, 345
  ratification, 23
    France, 239, 392
    U.S., 40, 48, 58, 108, 126, 160
    U.S.S.R., 160
Space, military use of (see also Manned Orbiting Laboratory), 2
  communications, 11–12, 71
  meteorology, 111
  reconnaissance, 64, 76, 160, 205, 210, 296–297
  space station, 63, 69
  U.S., 204–205, 211, 286, 296, 366
  U.S.S.R., 205, 216, 348
  Vietnam War, 111, 366
Space Nuclear Propulsion Office, Nevada (SNPO-N), 200
Space program, national (see also individual programs, such as Apollo program, etc.), 54–55, 201
  accomplishments, 28–29, 59, 67–68, 80, 87–88, 150, 158, 240, 261, 269, 324, 339–341
    international, 80, 135, 150
  management, 3, 90, 258, 263, 285, 293
  manned space flight, 3, 7, 10, 48–49
  budget, 9–10, 17, 17–18, 54, 87, 113, 116, 179, 211, 215, 235, 235–236, 246, 248, 255–256, 257, 336–338, 368
  cost, 73, 140–141, 288–289
  criticism, 51, 80, 94, 114, 114–115, 115–116, 148, 153–154, 158, 178, 209, 364
  education, benefits to, 16, 80, 285, 316, 317, 344
  employment, 138–139, 167, 257, 272, 279, 359
  Humphrey, Vice President Hubert H., 76
  international cooperation. See International cooperation.
  Johnson, President Lyndon B., 17, 17–18, 21, 28, 54, 76, 80, 139, 150, 296, 341, 343, 373–375
  lunar landing. See Moon, landing.
  manned space flight. See Manned space flight.
  military, 19–20, 153–154, 160, 273, 286, 296–297, 361

Space program—Continued
  objectives, 20, 45, 64–65, 76, 88, 143, 235, 235–236, 255–256, 392
  scientific and technological, 15–16, 150, 315–316, 324
  policy, 8, 17, 296, 323, 381
  post-Apollo, 4, 14, 31, 80
    budget decision, 13, 59–60, 132, 192, 192–193, 295–296, 299
    suggested programs, 4, 7, 45, 54–55, 108–109, 184, 381
  significance, 30–31, 134, 135–136, 155, 209, 258, 293, 362, 368
    international, 135–136, 256, 288–289, 356
  U.S. vs. U.S.S.R. See Space race.
  Vietnam War, effect of, 13, 113, 235–236, 255–256, 257, 333
Space race, 116, 130–131, 178, 256, 296, 343, 373–374
  booster, 28, 33, 38, 345, 351, 355
  communications, 70
  cost, 147, 288–289, 293
  criticism, 26–27, 40, 381
  manned space flight, 10, 29
  microminiaturization, 55
  military aspects, 269, 289, 355
  moon, 2, 8, 24–25, 33, 57, 59, 64–65, 89, 130, 351, 364
  planetary, 345, 378
  post-Apollo, 20, 296
Space rescue, 125, 277
Space rescue treaty, 211, 332, 380, 382, 383, 387, 392
Space results (see also Earth; Moon; Mars; Venus; individual probes, satellites, and sounding rockets), 280
  agriculture, 167, 286
  astronomy, 1, 9, 31, 37, 46, 56, 73–74, 81, 115, 201–202, 236, 238, 251, 260, 305, 311, 316, 345, 370, 377
  communications, 21, 32, 80, 98, 150, 209, 224, 242, 251, 345, 356
  earth sciences, 16, 18, 30, 42, 134, 222, 224, 245, 292, 305, 316–317
  engineering, 136
  medicine, 16, 39, 105, 167, 209, 263, 270
  meteoroid detection, 274
  meteorology, 32, 60, 111, 125, 150, 190, 224, 270, 336, 356
  oceanography, 232, 287, 368, 372
  radiation, 136, 279, 287
  social science, 71, 163, 202, 224, 255–256, 280, 285, 291–292, 312, 316, 317, 324, 345
  technology, 30, 46, 69, 70, 105, 140, 202, 209, 224, 255, 263, 271, 272, 280, 302, 368, 390
Space Science Board (NRC), 76
Space simulator, 12
Space station (see also Manned Orbiting Laboratory), 10, 14, 20, 21, 30–31, 33, 88, 274, 280
  cost, 228, 280
  military use, 69, 247, 343

Space suit, 246, 247
Space transportation, 135
Spacecraft (see also individual spacecraft, such as Apollo, Lunar Orbiter, Luna, Mariner, Surveyor, etc.), 391–392
  braking, 124
  design, 36, 51–52, 53–54, 58, 68, 75, 82, 86, 105, 123, 145, 158, 179, 190, 260, 291
  development testing, 68, 339–340
  electrical systems, 40, 92, 101, 108, 149, 269, 339–340, 373
  emergency equipment, 23, 68, 144–145, 339–340
  environment, simulated lunar vehicle, 217, 303
  environmental control system, 303, 319, 339–340, 357, 365, 369, 370, 378
  equipment, 3–4, 60, 68, 81–82, 98, 118, 137, 143, 185, 220, 229, 265–266, 275, 291, 296, 303, 328, 334, 339–340, 353, 354, 365, 375
  escape hatch, 54, 68–69, 82, 102, 148
  extravehicular equipment, 142, 205
  heating, 112
  instrumentation, 75, 129, 137, 274, 279, 295, 303, 306, 307, 308, 310–311, 313, 314, 319, 320, 322, 328–329, 331–332, 334, 336, 342, 345–346, 353, 373, 375, 382–383, 384, 385
  landing system, 124, 162, 178, 191, 205, 245, 260, 321, 339–340, 350, 356, 365, 371
  life support system, 23, 38, 39, 40, 47, 51–52, 73, 118, 125, 148–149, 238, 246
  propulsion. See Engine; Motor, rocket; and individual launch vehicles, such as Saturn, etc.
  reentry control system (see also Reentry), 124, 216
  reusable, 4, 10
  space tool, 219
  sterilization, 74, 84, 86, 117
Spacesuit communications (SSC) systems, 238
Spaco, Inc., 112
Spady, Amos A., Jr., 288
Spain, 1, 242, 293, 335, 353
SPARCS. See Solar Pointing Aerobee Rocket Control System.
Sparta, Project, 121
Spartan (missile), 147, 346
Special Libraries Assn., 170
Spectra-Physics, Inc., 56
Spectrometer, 60, 66, 72, 83, 151, 215, 308, 355, 368
Sperry, Lawrence, Award, 288, 316
Sperry-Rand Corp., 112
  Univac Defense Systems Div., 278
Spica (star), 173
Spilhaus, Dr. Athelstan F., 16, 210
Spin-Scan Cloud Camera (SSCC), 332, 388
Spirit of St. Louis, 166
Spirit of St. Louis medal, 212
Spivak, Jonathan, 31, 299
Spohr, Dr. Arthur, 191

Spradley, L. Harold, 305
Sprint (antimissile missile), 322
SPS. See Service Propulsion System.
Sputnik I (U.S.S.R. satellite), 87, 273, 291, 320
Squier, Carl B., 331
SR–71 (strategic reconnaissance aircraft), 5, 111, 303, 318
SRAM. See Short-Range Attack Missile.
SSCC. See Spin-Scan Cloud Camera.
SST. See Supersonic transport.
STADAN. See Satellite Tracking and Data Acquisition Network.
Stafford, Maj. Thomas P. (USAF), 144, 149, 161
Stalin, Joseph, 143
Stalmach, Charles J., 80
Standardization, 185
Stanford Linear Accelerator Center, 267
Stanford Univ., 37, 130, 151, 188, 267
  Center for Radar Astronomy, 208
  School of Medicine, 15
Star, 35, 60, 89, 95, 151, 192, 216, 326
Star Tracking Rocket Attitude Positioning system. See STRAP III.
Starbird, L/G Alfred D. (USA), 32, 329, 341
State, Dept. of, 29, 185, 216, 228, 307, 339
  Office of Space and Environmental Science Affairs, 269
"Status of the Federal Aircraft Noise Abatement Program" (report), 34–35
Steinbacher, Robert H., 329
Stennis, Sen. John C., 284, 295
Sterling, Project, 3
Sterilization, spacecraft, 74, 84, 86, 117
Sterling, Project, 3
Stevens, Edmund, 134
Stevens Institute of Technology, 142
Stevenson, M/G John D. (USAF, Ret.), 39
Stever, Dr. H. Guyford, 195
Stewart, M/G James T. (USAF), 87
Stine, Howard, 328
Stoddard, Dr. David H., 59
STOL (short takeoff and landing) aircraft, 244
Stolyarov, Gen. Porfiri A. (U.S.S.R.), 361
Stoltenberg, Dr. Gerhard, 219, 276
Stonehouse, Minister of State, John (U.K.), 219, 286
Storms, Harrison A., Jr., 136
Strang, Col. Charles F. (USAF), 25
STRAP III (Star Tracking Rocket Attitude Positioning system), 369
Strategic Air Command (SAC), 358
Stratton, Rep. Samuel S., 49
Strawberry Hill Observatory, 227
Strickler, Dr. Mervin K., Jr., 96
Strong, Dr. John, 166
Strughold, Dr. Hubertus, 39
Sturmthal, L/C Emil (USAF), 184, 265, 327
Stuttgart, West Germany, 70
Submarine, missile-carrying, 90, 201, 203, 292

Subminiature Integrated Antenna (SIA), 121
Sud Aviation, 154, 220
Suitland, Md., 92
Sukhoi (U.S.S.R. variable-sweepwing aircraft), 384
Sullivan, Francis J., 78
Sullivan, Walter, 5, 37, 183, 380
"Summarization of National Aeronautics and Space Administration Management Review of North American Aviation Inc., Activities" (report), 175
Sumter, S.C., 279
Sun (see also Eclipse, solar; Radiation, solar; Solar cell; Solar corona; Solar wind)
  exploration of, 3, 17, 121
  IQSY, 236
  research, 16, 121
  satellite data, 206, 229, 349, 355, 391
Sunblazer (probe), 3, 17, 68
Sunnyvale, Calif., 279
Sunspot, N. Mex., 35
Suomi, Dr. Verner E., 336
Super Guppy (cargo aircraft), 388
Supersonic combustion ramjet (Scramjet), 8
Supersonic transport (SST) (see also Concorde [U.K.-France supersonic transport] and Tu-144)
  benefits, 200, 245, 261–262
  contract, 5, 48, 130, 133, 392
  cost, 2, 33, 48, 89, 176, 220, 224, 230
  criticism, 31–32, 157, 190, 215
  design and development, 5, 6, 357
  editorial comment, 1, 6, 31–32, 171–172
  fares, 224
  foreign, 13, 33, 185, 268
  funds for, 8, 12, 13, 17, 31, 34, 39, 130, 179, 185, 190, 212, 294, 297, 314, 324
  NASA participation in research, 20, 69, 193, 245
  pilots, 2, 237
  reservations, 176
  sonic boom, 190, 200, 203, 231, 280, 385
Supersonic Transport Authority, 8
Surtsey volcano, 224
Surveyor (program), 4, 5, 18, 28, 114, 123, 229, 334, 350
Surveyor (spacecraft), 38, 360
Surveyor I (lunar probe), 120, 210, 234, 342
  communication with, 5, 165–166
  photographs, 29, 40, 72, 106, 119, 266, 342
Surveyor II, 120, 210, 342
Surveyor III (Surveyor C), 210, 234, 342
  communication with, 134, 158, 165, 197
  launch, 113
  lunar landing, 118
  photographs, 118, 134, 162, 266, 391
  results, 123–124
Surveyor IV (Surveyor D), 209, 212, 234, 342

Surveyor V (Surveyor E)
  communication with, 303, 376, 382–283
  launch, 265–266
  lunar landing, 265–266
  photographs, 342, 391
  results, 287, 328, 390
Surveyor VI (Surveyor F), 334, 342, 383, 391
Surveyor-R (spacecraft model), 164
Survival equipment, 125, 261
SV-5 (lifting-body vehicle), 14
SV-5D, 62, 116, 163
SV-5J (manned lifting-body vehicle), 166
SV-5P (manned lifting-body vehicle), 190
Sweden, 38, 74
Swenson, G. W., Jr., 37
Swenson, Loyd S., Jr., 10
Swings, Dr. Pol, 251
Switzerland, 78
Sword Knot (satellite-tracking ship), 46
Sydney, Univ. of, Australia, 7
Symington, Sen. Stuart, 294
Symphonie (Franco-West German communications satellite program), 218, 333
Symposium on Modern Optics, 80
Symposium on Results of the International Years of the Quiet Sun, 212
SYNCOM (communications satellite), 32
Syracuse Univ. Research Corp., 157
Systems engineering, 17, 258, 263, 281, 285, 292, 309

T

T-33 (jet trainer), 348
Taback, Israel, 317
TACV. See Tracked air cushion vehicle.
Taghiev, E. I., 101
Tallahassee, Fla., 293
Tallin Line (U.S.S.R. defense system), 270
Tanyug (Yugoslav press agency), 204
Tanzania, 393
Tarasova, U.S.S.R., 259
Tass, 156, 286
  launch, 124
    probe, 14, 183, 308
    satellite, 104, 290
      Pacific Ocean rocket test, 168, 260, 356
      U.S.S.R. space program, 61, 250, 261, 378, 388
TAT V (underwater cable), 293, 326–327
Tau Puppis (star), 326
Taylor, Frederick, 3
Taylor, Glen, 244
Taylor, Henry J., 257–258
Taylor, Stan, 261
Tchernev, Dr. Dimiter I., 372
TD-1 (ESRO research satellite), 197
TD-2, 197
Teague, Rep. Olin E., 91–92, 95, 192–193, 272, 323
Teague, Olin E., Research Center, 272

Teb. See Triethylborene.
Technica Molodezhi, 271
Technological forecasting, 19, 310
Technology, Economic Growth, and Public Policy, 69
Technology, 230, 242–243, 257, 271
  benefits, 30, 45, 241
  funding, 69, 153–154
  gap, 70, 164–165, 194, 215
  military, 77, 153–154, 299
  policy, 19, 69, 277–278, 390
Technology utilization, space, 30–31, 45, 67–68, 71, 140–141, 202, 209, 255–256, 263–265, 271, 368–369, 390
  air transportation, 262, 271–272, 300
  economic progress, 30–31, 275, 324
  medicine, 104, 209, 263–265
  oceanography, 224, 271–272, 287, 372–373
  science, 30–31, 45, 202, 280, 323
  social progress, 202, 324
  systems engineering, 258, 263–265, 291–292, 309
Tekamah, Neb., 385
Tektite, 320
Telemetry, 4, 50
Telescope, 35, 60, 78, 81–82, 153, 155–156, 181, 182–183, 327, 333, 349
  orbiting, 21, 31
Television (TV), 240–241
  educational, 41, 100, 107, 112, 318
  noncommercial, 6, 96, 96–97, 222–223
  rates, 93
  space probe, use of, 114, 119–120, 123, 334
  via satellite, 21, 41, 83, 85, 98–99, 133–134, 170, 193–194, 236, 250, 262, 277, 282, 290, 326–327, 385
Teller, Dr. Edward, 15, 129, 137
Teltsch, Kathleen, 371
Temperature measurement, 18, 28, 33, 91, 93, 97, 103, 106, 116, 117, 130, 140, 146, 155, 191, 193, 367
Tepper, Dr. Morris, 4
TERLS. See Thumba Equatorial Launching Station.
Terminator Zone, 336
Test ban treaty. See Nuclear test ban treaty.
Teuber, B/G Harold (USAF), 230
Texas, 279, 288
Texas A&M Univ., 272
Texas State Society, 288
Texas, Univ. of, 16
  Astronomy Dept., 294
Texas Symposium on Relativistic Astrophysics, 16
TFX. See F–111.
Thacker School, 349
Thailand, 7, 111
Thant, U, U.N. Secretary General, 293
Themis, Project, 13
Theodolite, 331
Thigpen, David, 311
Thiokol Chemical Corp., 79, 364

This New Ocean: A History of Project Mercury (NASA SP–4201), 10
Thomas, B. K., Jr., 218
Thomas, David D., 154
Thompson, Dr. Floyd L., 25, 85, 102–103, 281, 317
Thompson, Ambassador Llewellyn E., 9, 85
Thompson, Milton O., 184
Thompson, Robert F., 371
Thor-Able (booster), 382
Thor-Agena D (booster), 9, 50, 91, 146, 169, 188, 219, 235, 271, 327, 371
  Thrust-Augmented Thor-Agena, 146, 222
Thor-Burner II (booster), 42, 196, 299
Thor-Delta (booster), 20, 66, 307
  Thrust-Augmented, 120, 342, 375
Thornton, Dr. William E., 233
Thrust-Augmented Delta (booster), 214, 282
Thrust-Augmented Improved Delta, 7, 42, 76, 83, 163, 263
Thrust-Augmented Thor-Agena (booster), 146, 222
Thrust-Augmented Thor-Delta (booster), 120, 342, 375
Thumba Equatorial Launching Station (TERLS), India, 69–70, 78, 350
Thurmond, Sen. Strom, 366
Tier, William, 89
Time zones, 180–181
Tipton, Stuart G., 89, 224, 312
Tiros (program), 217
Tiros VII (meteorological satellite), 190
Tiros Operational Satellite (TOS), 20, 120, 342–343
Tischler, A. O., 215
Titan I (missile), 3
Titan II (missile), 110, 205
Titan III (booster), 268, 338
Titan III–B (booster), 50, 191, 244, 274, 318
Titan III–C (booster), 11–12, 18, 129, 133, 199
Titan III–M, 275, 392
Titan III-Agena D (booster), 367
Titov, L/C Gherman S. (U.S.S.R.), 14
Toftoy, Mrs. Hazel, 330
Toftoy, M/G Holger Nelson (USA, Ret.), 118, 330
Toftoy, General Holger N., award, 361
Tokaty, Prof. G. A., 109
Tokyo, Japan, 21, 133–134, 157, 177
Tolubko, Col. Gen. Vladimir (U.S.S.R.), 348
Tomahawk (nose cone), 375
Topping, Dr. Norman, 315–316
TOS. See Tiros Operational Satellite.
Toulouse, France, 333–334, 373
Townes, Dr. Charles H., 84
TPS–44 (radar system), 185
Tracked air cushion vehicle (TACV), 152
Tracking, 246, 259, 392
  cooperation, 304, 309–310
  deep space, 210, 242, 265, 278
  ship, 242, 380–381

Tracking—Continued
  stations, 207–208, 315
    Australia, 64, 79, 212, 242, 265
    Cuba, 376
    Germany, West, 29
    Mali, 376
    Spain, 242, 353
    United Arab Republic, 376
    U.S., 70, 212, 242, 265, 376, 392
    U.S.S.R., 376
    Vietnam, North, 309–310
Trailblazer (booster), 46
Trailblazer II, 195
Trans-Canada Telephone System, 170
Transit (navigation satellite), 271–272
Transonic compressor, 19
Transponder, 207–208
Transportation, 34, 35, 90–91, 95
Transportation Assn. of America, 35
Transportation, Dept. of (DOT), 34, 90–91, 95, 188, 200, 203–204, 263, 277, 346
  appropriations, 207, 212, 284, 293–294, 314
  cooperation, 20, 339
  program, 95
  Secretary of, appointment, 8
Treasury Dept., 107
Trendex poll, 8, 364
Triethylborene (Teb), 91
Trimble, George S., 135, 300
Troitski, Prof. Vsevolod, 286
Tropical Survival School, 184
Trud, 26–27, 348, 350
TRW, Inc., 192, 265
TRW Systems, Inc., 8, 16, 136, 237, 242, 335–336, 386
Trans World Airlines, 12, 13, 100, 170–171, 267
Tts I (Test and Training Satellite), 375, 392
Tu–144 (U.S.S.R. supersonic transport), 33–34, 66, 116, 185, 268–269
Tu–154 (U.S.S.R. airliner), 156
Tucson, Ariz., 96
Tullahoma, Tenn., 216
Tungsten, 88
Tungsten RHC, 275
Turbotrain, 384
Turin, Italy, 27
Turner, Edwin M., 241
Turner, Richard G., 217
Tycho (lunar crater), 227, 260–261, 309, 377
Tymczyszyn, Joseph J., 2
Tyner, Ralph W., 254
Tyuratam range, 247

U

U–2 (reconnaissance aircraft), 210, 318–319
Uchinoura Range, Japan, 111
Uchinoura Space Center, Japan, 39
Udall, Secretary of the Interior Stewart L., 8, 384–385

UFO. See Unidentified Flying Objects.
Ugolyek (dog), 104
U.K. See United Kingdom.
Ukrainian Academy of Sciences, 89
U.N. See United Nations.
Underground nuclear test, 15
Unidentified flying objects (UFO), 91, 151–152, 271, 282
  Czechoslovakia, 312
  Michigan, 11
  USAF, 10, 11, 151–152, 245–246, 271, 361
  U.S.S.R., 361, 386
Union College, 160
United Air Lines, 13, 347
United Aircraft Corp., 5, 66, 384
  Hamilton Standard Div., 284
  Pratt & Whitney Div., 207, 250
  Sikorsky Aircraft Div., 262, 389
United Arab Republic (U.A.R.), 376
United Kingdom (U.K.), 86, 93, 222, 267, 273, 360
  aircraft, 16, 154, 166, 181–182, 277
    airbus, 219–220, 270
    Concorde, 2, 66, 215, 220, 230, 269
  astronomy, 30, 52–53, 216
  cooperation,
    aircraft, 286
    defense, 16
    space, 168, 304, 308
  launch, 140
  Ministry of Technology, 312
  missile program, 152, 185, 367
  space program, 3, 237
  space treaty, 70, 126, 160, 383, 392
United Nations (U.N.), 23, 65, 77, 128, 239, 343, 345
  Committee on the Peaceful Uses of Outer Space, 46, 380, 382
    Legal Subcommittee, 211, 332, 392
  Conference on Exploration and Peaceful Uses of Outer Space, 38–39, 46
  General Assembly, 70, 247, 293, 332, 380, 383
  Industrial Development Organization, 345
United Nations Assn. of the United States of America (U.N. Assn. of the U.S.A.), 355
United Press International (UPI), 275
United States (U.S.) (see also appropriate agencies)
  Aeronautics and Space Activities, report to Congress on, 28–29
  budget, 17–18, 46, 51, 54, 62, 179, 263
  communications, 6, 21, 63, 80, 100, 101, 107, 241, 242, 252, 293
  criticism, 70, 158, 218, 257
  defense, 7, 9, 104, 129, 169, 221, 231, 273, 275, 324, 329, 346, 360
  disarmament, 41, 60, 85, 252, 327
  education, 10–11, 16, 26, 51, 285, 297, 309, 344
  exhibit, 218
  health, 297, 378

United States—Continued
  international cooperation, 70, 80, 110, 121, 182, 185, 204, 217, 243, 269, 302, 310, 343, 371
  military research, 123, 204
  nuclear weapons, 21, 60, 63, 206, 252, 355, 360
  research and development, 13, 227, 254, 255, 285, 310, 350, 353, 373, 386, 390–391
  science and technology, 39, 62, 70, 176, 188, 201, 202
    Committee on Science, Engineering, and Regional Development, 353
    funds for, 51, 62, 180, 181, 347
    impact of, 6, 9, 10, 15, 19, 53, 77, 214, 230, 242, 278, 323, 350, 353
    information retrieval, 170, 197
    manpower, 26, 153, 231, 234, 259
    planning, 88, 143, 353
  space law treaty. See Space law treaty.
  space program. See Space program, national.
  transportation (see also Supersonic transport), 8, 35, 73
Universities
  graduates, shortage of, 26
  grants, 30, 51, 227–228, 240, 306–307, 347, 354
  NASA program, 306–307, 346, 363
  science in, 165, 227–228, 347, 354
  space exploration, impact of, 317
Themis, Project (DOD), 13
USAF School of Aerospace Medicine (SAM), 29, 39
U.S. Air Force (USAF) (see also individual bases, centers, and commands, such as Air Force Systems Command, Arnold Engineering Development Center, Edwards AFB, etc.), 180, 185, 207, 213, 223, 246, 279, 300, 373
  accident, 29, 83, 177
  agreement, 74, 90–91, 112, 190
  aircraft (see also individual aircraft such as C–5A, F–111A, X–15, XB–70, etc.), 17–18, 213, 347, 390, 392
    accident, 5, 111, 122, 149, 180, 302, 311, 316, 318, 345, 352, 371, 373, 392
    record, 174, 289
  anniversary, 220
  astronaut, 196, 207, 371
  award, 136, 153, 174, 195–196, 200, 207, 213, 300, 372
  booster, 8, 36, 41, 62, 91, 116, 161, 169, 175, 181, 188, 191, 196, 199, 202, 220, 271, 279, 299, 320, 334, 367, 371, 375, 385
  contract, 12, 160, 168, 206, 253, 255, 268, 269, 279, 288
  cooperation, 184, 190, 206, 213, 383

U.S. Air Force—Continued
  launch
    Advanced Ballistic Reentry System, 181, 202, 334, 385
    booster, 11, 129, 195, 199, 279
    comsats, 11, 199
    lifting-body vehicle, 62, 116, 147
    low-altitude supersonic vehicle (LASV), 12
    missile, 3, 12, 46, 80, 110
    postponed, 130
    reentry vehicle, 46, 181, 202, 302, 385
    satellite, 9, 12, 36, 42, 50, 91, 111, 129, 146, 158, 161, 169, 175, 188, 191, 199, 219, 221, 235, 244, 252, 271, 274, 279, 299, 318, 320, 327, 367, 371
    Scramjet, 8
  lifting-body vehicle, 14, 62, 116, 163, 190, 206
  management, 70, 75, 87, 92, 136, 199
  missile program, 64, 109, 195–196, 199, 210, 221, 223, 344
  MOL. See Manned Orbiting Laboratory.
  Office of Aerospace Research, 35, 180
  personnel, 99, 126, 173, 177, 184, 197, 200, 285, 327, 371
  research, 18, 74, 80, 84, 122, 182, 195, 225, 231, 253, 269, 284, 289, 324, 362, 384
  satellite (see also Satellite, unidentified), 9, 36, 41, 91, 161, 169, 175, 188, 191, 199, 202, 255, 271, 299, 320, 367, 371, 390–391
    communications, 11, 32, 178, 182, 204, 213, 236, 250, 255, 261, 392
  sounding rocket, 10
  space program, 62, 118, 161, 169, 177, 196, 199, 210, 220, 253, 255
  spacecraft recovery, 179, 216
  test, 210, 216, 231, 240, 250, 289, 292, 384, 392
  training, 3, 184, 196–197, 233, 240
  UFO, 10, 245–246
  Vietnam War, 111, 239, 269, 347
  Zero Defects Program, 213, 294
U.S. Army (USA), 108, 118, 247, 330
  aircraft, 383
  antimissile system, 263, 268, 329
  cooperation, 213
U.S. Army Missile Command (AMC), 79, 162
U.S. Bureau of Mines, 372
U.S. Chamber of Commerce, 281
U.S. Comptroller General, 338
U.S. Geological Survey (USGS), 9, 17, 99, 144, 162, 224, 235, 305, 317, 342
U.S. Marine Corps (USMC), 201
U.S. Maritime Administration, 272
USN. See U.S. Navy.
U.S. Naval Observatory, 5
U.S. Navy (USN), 153, 178
  aircraft, 111, 146, 289, 294, 352, 383
    F–111B, 111, 122, 149, 174, 307, 311, 380

U.S. Navy—Continued
  award, 105, 207
  communications satellite system, 102, 224, 242, 263, 272
  contract, 201, 263
  cooperation, 213, 289, 352
  Man-in-the-Sea, program, 37, 53, 232
  missile, 201, 203, 248, 307, 330
  nuclear submarine, 267
  personnel, 182, 204, 234, 371
U.S. Navy Air Systems Command, 83, 219
USNS *Huntsville*, 380–381
USNS *Mercury*, 380–381
USNS *Point Barrow*, 388
USNS *Redstone*, 381
USNS *Vanguard*, 365, 381
USNS *Watertown*, 381
U.S. Patent Office, 368
U.S.S. *Bennington*, 339
U.S.S.R. (Union of Soviet Socialist Republics) (see also Soviet Academy of Sciences, etc.), 133, 139, 170, 192, 193, 299
  agreement, 9, 41, 64, 85, 286, 299, 309
  aircraft, 33, 66, 77, 116, 156, 185, 203, 211, 253, 268, 295, 315, 343, 383
  anniversary, 9, 107, 269, 273, 291, 308, 327, 336, 349
  antiaircraft defense, 336
  antimissile missile system, 7, 9, 21, 27, 50, 104, 360
  Armed Forces Day, 50
  astronomy, 89, 182–183, 240, 327
  award, 46
  booster, 33, 109, 351, 355
  communications satellite, 80, 104, 143, 165, 192, 193, 205, 241, 250, 257, 290, 313
  cooperation, 113, 277, 296, 356, 371, 376, 383, 392
    Australia, 64
    Communist-bloc countries, 113, 309, 376
    France, 257, 261, 322
    U.K., 286, 304, 383, 392
    U.S., 25, 60, 236, 241, 243, 261, 296–297, 299, 303, 310, 356, 371, 380, 383, 392
  cosmonaut. See Cosmonaut.
  Cosmonautics Day, 107
  defense system, 270
  disarmament, 60, 210
  exhibit, 182, 203, 211, 218
  hydrofoil, 350
  launch, 247, 391, 393
    probe, 183, 300, 393
    satellite, 304
      Cosmos, 12, 19, 40, 41, 46, 53, 54, 61, 70, 77, 82, 83, 86, 87, 97, 101, 107, 129, 152, 154, 157–158, 161, 173, 176, 180, 183, 188, 189, 201, 210, 218, 225, 237, 238, 239, 252, 258, 268, 272, 275, 278, 281, 299, 304, 309, 316, 320, 321, 327, 352, 355, 365, 380, 382, 387, 388, 392

U.S.S.R.—Continued
  launch—continued
    satellite—continued
      Molniya, 165, 290, 313, 392
      Soyuz, 124
    sounding rocket, 322
  launch site, 259, 348
  lunar exploration (see also *Luna IX*, etc.), 24, 206, 356, 388
    manned, 101, 112, 124, 293
    unmanned, 206
  meteorology, 60, 77, 113–114, 129, 277
  missile and rocket program, 9, 50, 60, 86–87, 104, 168, 206, 267, 270, 273, 336, 358
  nuclear reactor, 367
  photography, 77, 139–140, 309–310
  planetary exploration, 86–87, 183, 233, 309, 345
  probe, 183, 233, 300, 304, 308, 392
  rocket test, 168, 260, 321, 356, 378
  science, 77, 80, 88, 112, 113–114, 123, 153, 163, 168, 233–234, 236, 240, 253, 288, 300, 322, 378
  Soyuz I accident, 124, 130–131, 134, 140, 143, 149, 158, 239
  space biology, 104, 124, 259, 277
  space law treaty. See Space law treaty.
  space program (see also individual probe and satellite programs, such as Cosmos, Luna, Molniya, Venus, etc.), 1–2, 10, 14, 24, 29, 33, 46, 77, 86–87, 101, 112, 116, 121, 134, 143, 147, 158, 168, 206, 247, 253, 259, 260, 280, 288, 293, 304, 314, 316, 321, 322, 324, 334, 345, 346, 348, 351
    manned space flight, 24, 82, 101, 112, 116, 134, 143, 158, 244, 260, 261, 288, 293, 304, 314, 321, 323, 324
  space rescue treaty, 380, 383, 392
  spacecraft, 82, 86, 101, 112, 203, 244, 288, 291
  submarines, 163, 336
  supersonic transport, 33–34, 66, 77, 116, 185, 268
  tracking station, 64, 309, 376
  UFOs, 271 361, 364–365, 386
  U.N. Conference on Exploration and Peaceful Uses of Outer Space, 38–39, 46
  weapons, 77, 104, 107, 203, 206, 211, 247, 257, 304, 313, 323, 329, 331, 332, 333, 336, 343, 348, 355, 358
U.S.S. *Will Rogers*, 226
U.S. Weather Bureau, 137, 262, 270
Utah, 8, 268
UV. See Radiation, ultraviolet.
United Technology Center, 361
Universe, 35, 240
Universities' Research Association, 217
University Club (New York), 277
UPI. See United Press International.
"Urban Government in the Decade Ahead" (colloquium), 163

Urey, Dr. Harold C., 126
Uruguay, 85
USA. See U.S. Army.
USAF. See U.S. Air Force.
USAF Aeronautical Chart and Information Center, 248

V

Van Allen, Dr. James A., 348, 379
Van Allen radiation belt, 20, 39, 129, 314
  simulator, 20
Vance, Cyrus R., 41, 108, 182
Van de Graaff, Dr. Robert J., 11
Vandenberg AFB, Calif. (see also Western Test Range), 12, 279
  facilities, 392
  launch
    missile, 46, 110, 202, 334
    reentry vehicle, 302
    satellite launch vehicle
      Atlas-Agena, 175
      Atlas F, 321
      Scout, 367
      Thor-Agena D, 91, 169, 371
      Thor-Burner II, 42, 299
      Titan III-Agena D, 367
  Scramjet, 8
Vander Jagt, Rep. Guy, 18
Vanguard, Project, 371
Vanguard II (satellite), 335
Vanguard Construction Corp., 276
Variable sweep wing, 13, 16, 203, 213, 218, 239-240, 384
Vasilevskis, Dr. Stanislavs, 192
Vega Baja Airport, Puerto Rico, 262-263, 324-325
Vehicle Assembly Building (VAB) (KSC), 388
Vela (nuclear detection satellite), 129, 133, 153, 348, 392, 394
Velcro (fabric), 247
Velcro Corp., 247
Velcro adhesive pad, 51
Venera IV. See *Venus IV.*
Venezuela, 307
Venus (planet)
  atmosphere, 166, 200, 252, 308-309, 310-311, 322, 368, 392
  contamination of, 234
  exploration, 185-187, 208, 214, 233-234, 253, 354, 391
  funding, 224
  manned, 64-65
  plans for, 4, 14, 45, 74, 90, 121
  unmanned, 308-309, 310-311, 322, 345-346
  life on, 9, 166
  photographs, 119, 200
  radiation, 142
  rotation, 296, 308-309, 310-311
  surface, 166
  temperature, 166, 308-309
  water on, 166, 322
Venus III (U.S.S.R. interplanetary probe), 233-234

Venus IV, 183, 185, 200, 308-309, 310, 322, 348, 392
Vermont, Univ. of, 247
Versatile Information Processor (VIP), 50
Vertical and short takeoff and landing aircraft. See V/STOL aircraft.
Vertical takeoff and landing aircraft. See VTOL aircraft.
Vesta (French sounding rocket), 5, 65
Veterok (dog), 104
Vette, Dr. James I., 18
Vienna, Austria, 46, 328
Vietnam, North, 111, 309
Vietnam war, 3, 26, 32, 63, 199, 225, 362, 366
  aircraft, 239-240, 269, 365
  effects on space budget, 13, 113, 249, 255-256, 257
  troops, 12, 199
Viking (rocket), 79
VIP. See Versatile Information Processor.
"Virginian of the Year" award, 193
Virginia Key, Fla., 132
Virginia Polytechnic Institute (VPI), 243
Virginia Press Association (VPA), 193
Virgo A (galaxy), 201-202
Vitro Corp., 112
Voge, Hervey H., 217
Vogel, Col. Lawrence W. (USA), 75
Vogt, Paul R., 136
Voice Broadcast Satellite, 17, 68
Volcano, 224-225
Volga River Valley, 367
Vonbun, Dr. Friedrich O., 242, 315
Von Braun, Dr. Wernher, 2, 6, 15, 27, 31, 70, 109, 134, 177, 294, 330
Von Diringshofen, Dr. Heinz, 142
Vostok (U.S.S.R. spacecraft), 82, 203
Voyager (program)
  contract, 24, 159-160
  funding, 17, 121, 192, 195, 220, 225, 246, 254, 291, 320
  Planetary Reentry Parachute Program, 146, 179, 222
  plans for, 30, 55, 59, 68, 197, 254, 338
  spacecraft sterilization, 74, 117, 191
Voyager (spacecraft), 82, 393
VPI. See Virginia Polytechnic Institute.
V/STOL (vertical and short takeoff and landing) aircraft, 146, 285-286, 383
VTOL (vertical takeoff and landing) aircraft, 172, 203, 277

W

Wake Island, 89-90
Wake turbulence, 348-349
Waldron, Martin, 137
Wales, 233
Walker, Brian M., 160
Walker, Joseph A., 180, 345
Walker, Mrs. Joseph A., 180
Walker, Richard L., 5

Walker, Robert, 386–387
Wallops Island, Va., 120
Wallops Station (NASA), 128, 348
  award, 85
  contract, 388
  launch
    RAM C-1, 311
    sounding rocket
      Aerobee 150, 285
      Aerobee 150A, 72, 368
      Arcas, 93, 237–238
      Argo D–4 (Javelin), 157, 191
      MT–135, 93
      Nike-Apache, 28, 91, 184, 216, 237–238, 272–273, 347, 375
      Nike-Cajun, 91, 106, 117, 130, 140, 151, 375
      Trailblazer, 46
      Trailblazer II, 195
Walsh, John, 302
Walz, Jay, 247
Warnke, Paul C., 359
Warsaw, Poland, 161
Warschauer, Dr. D. M., 386
Washington, D.C., 95, 209, 247, 329, 351
  awards presented at, 172, 177, 285, 317, 380
  meetings, 13, 16, 33, 36, 48, 49, 93, 115, 146, 151, 172, 175, 178, 180–181, 185, 195, 215, 220, 274, 281, 310, 318, 344
  press conference, 60, 110, 129, 217
  satellite communications with, 12, 21, 199
  space treaty signed at, 23, 25–26, 70, 299
Washington, D.C., Aero Club, 220
Washington, D.C., National Airport, 137
Water, 126
Water injection, 370, 392
Water pollution, 189–190
Waterman, Dr. Alan T., 361
WEAFAX. See Weather facsimile experiment.
Weapon systems, 205, 216, 247, 257–258, 267–268, 329, 331, 343–344, 348
Weather facsimile (WEAFAX) experiment, 270
Weather modification, 113–114, 214, 346–347
Webb Associates, 36–37
Webb, James E., 65
  Apollo (AS–204) accident, 25, 33, 46, 49–50, 53, 103, 114, 130, 139, 171, 279
  appointments by, 14, 173, 212, 253, 299, 327
  astronaut, 232
  award by, 54
  award to, 190
  budget, 18, 20, 220, 221, 224, 243, 272, 275, 283, 284, 338, 352, 353
  cosmonaut, 125
  criticism of, 148, 167, 294, 295
  Seamans, Dr. Robert C., Jr., resignation, 288
  space cooperation, 125, 388

Webb—Continued
  space program, 3, 116, 151, 155, 163, 301, 336, 338, 341, 352, 353
  tribute to, 166, 258
  universities, 150, 306, 307, 363
Webb, Paul, 36
Weightlessness, effects of
  animals, 2, 6, 65, 104
  humans, 10, 12, 31, 36, 56, 71, 76
  organisms, 47, 263, 313, 391
  plants, 326
Weitzel, Frank H., 168
Welles, Benjamin, 85
Welsh, Dr. Edward C., 38, 63, 104, 160, 162, 215, 256, 293, 342
West Palm Beach, Fla., 53
West Point, N.Y., 29
West Point, Neb., 385
West Virginia, 126, 202
Western Electric Co., 70, 263, 284
Western High School Alumni Assn., Washington, D.C., 142
Western Test Range (WTR) (see also Vandenberg AFB, Calif.), 381
  launch
    contract, 126, 371
    lifting-body vehicle, 62
    postponed, 111
    satellite
      launch vehicle
        NASA
          Thor-Delta, 20
          Thrust-Augmented, 120, 143, 342
          Thrust-Augmented Improved Delta, 163
          Thrust-Augmented Thor-Agena, 222
          U.K., Scout, 140
        USAF
          Atlas, 221
          Atlas-Agena D, 36, 161, 175
          Scout, 111, 159, 279
          Thor-Agena D, Thrust-Augmented Thor-Agena, 9, 50, 146, 188, 219, 235, 271, 327
          Thor-Burner II, 196, 236
          Thrust-Augmented Thor-Agena, 146
          Titan III–B, 50, 191, 245, 274, 318
Western Union International, Inc. (WUI), 101, 118, 130
Westinghouse Electric Corp., 213, 244
Weston, Ill., 188, 207
Wheeler, Gen. Earle G. (USA), 108
Whelan, Joseph G., 104
Whipple, Dr. Fred L., 177
White, Alvin S., 12, 267, 300
White, L/C Edward H., II (USAF)
  Apollo (AS–204) accident, 21–22, 24, 25, 26, 26–27, 36, 102
  eulogy, 29
  Gemini IV flight, 387
  honors, 41, 69, 85, 142, 207
  memorial service, 27

White, George C., Jr., 25, 138
White House, 23, 41, 60, 72, 88, 164, 293, 297
White, John R., 106, 108
White, Maurice D., 317
White, Col. Maynard E. (USAF, Ret.), 197–198, 219
White Sands Missile Range (WSMR), N. Mex.
  launch
    Aerobee, 126, 150
      atmospheric data, 60, 191, 214, 236, 368
      failure, 268
      micrometeoroid sampling, 177, 293
      solar astronomy, 125, 146, 289, 293
      stellar data, 60, 140, 344
      ultraviolet astronomy, 60–61, 78, 326, 369
      x-ray astronomy, 202, 350
    Honest John-Nike, 146, 191
    Nike-Apache, 169, 234
    Sprint, 323
  Sentinel System Evaluation Agency, 329
  test
    Lunar Module, 201–202
    SCAT Solar Pointing System, 289
    SPARCS (Solar Pointing Aerobee Rocket Control System), 372
    STRAP III stellar pointing control, 368
    Voyager parachute, 146, 191, 222
White, Gen. Thomas D., Space Trophy, 174
Whitford, Albert E., 181
Wiener, Anthony J., 390–391
Wiesner, Dr. Jerome B., 358
Wilford, John N., 14, 20, 36, 138
Wilkinson, R. L., 79
Williams, Maj. Clifton C., Jr. (USMC), 293
Williams, Sen. Harrison A., Jr., 58
Williams, John, 25
Williams, Walter C., 62
Williamsburg, Va., 55, 166–167, 297
Williamson, David, 75
Wilson, George C., 130, 210, 365
Wilson, Prime Minister Harold (U.K.), 152, 185
Wilson, Richard, 156
Windsor Locks, Conn., 284
Wing, aircraft, 203, 218
  delta, 60–61, 373
  variable sweep, 13, 16, 203, 213, 218, 239, 383–384
Wings Club, 12
Winn, Rep. Larry, Jr., 18
Winters, Bruce, 336
Wisconsin, Univ. of, 143, 233, 336
Wisner, Neb., 385
Witkin, Richard, 330
Withington, H. H., 357
Witze, Claude, 390
Witzke, Walter R., 375
WMO. See World Meteorological Organization.
Woeller, Fritz, 109
Wolfe, Dr. John H., 115
Wolfle, Dael, 302
Wolper Productions (TV documentaries), 112
Women's Forum on National Security, 16
Woods, George, 359
Woods Hole, Mass., 56, 108
Woodward, William H., 102
Wooley, Bennie C., 37
Woomera Rocket Range, Australia, 3, 121, 224, 233, 358, 368, 369
Worden, Capt. Alfred M. (USAF), 212, 350
World Bank, 359
World Meteorological Organization (WMO), 92
World Petroleum Congress, 101
World Weather, Inc., 346
World Weather Watch, 18
"The Worth of the Space Program" (AIAA Forum), 316
Wresat I (Australian Weapons Research Establishment Satellite), 358, 393
Wright Brothers Medal, 366
Wright Brothers Memorial Trophy, 285
Wright, Orville, 126, 142, 177, 380
Wright-Patterson AFB, Ohio, 18–19, 96, 241, 253, 358, 384
Wright, Wilbur, 142, 177, 380
Wrightsville Marine Bio-Medical Laboratory, 122
WSMR. See White Sands Missile Range.
WTR. See Western Test Range.
Wuenscher, Hans, 219
WUI. See Western Union International, Inc.
Wyatt, DeMarquis D., 75, 253
Wydler, Rep. John W., 346
Wyld Propulsion Award, 215

X

X–3 (missile), 344
X–15 (rocket research aircraft)
  flight, 392
    No. 1, 83, 129, 158, 188, 196
    No. 2, 143, 248, 290
    No. 3, 193, 214, 254, 292, 305
      accident, 344, 352
  funding, 17
  pilots, 238, 305
  record, 289, 344, 352
  test, 58
    accelerometer, 157, 215, 249, 254
    camera, 249, 292, 305
    electrical loads, 83, 130
    guidance, 157, 193, 214, 254, 292, 304
    heat, 157, 193, 214
    horizon scanner, 129, 188
    landing skid, 83, 130
    load, 158, 215, 254
    micrometeoroid, 292, 305
    pilot checkout, 83, 130, 188
    pressure attitude indicator, 84
    ramjet, 143, 250, 290

X-15—Continued
  test—continued
    sonic boom, 83, 158, 193
    stability, 143, 249, 290
    stabilizer, 83, 129, 188
    thermocouple, 143
X-15 Accident Investigating Board, 344
X-22A (v/STOL aircraft), 146, 382
X-24A (SV-5P) (lifting-body vehicle), 206
XB-70 (supersonic aircraft)
  accident, 180, 344
  contract, 91, 99
  flight, 3, 9, 11, 19, 28, 126, 252, 253, 265, 298, 327
  pilots, 267, 300
  research program, 12, 17, 75, 184, 392
X-ray, 61, 104
  antimissile use, 28, 147, 358
  source, 17, 88, 98, 202, 253, 288, 308, 329, 350

Y

Yaffee, Michael L., 250
Yak-40 (U.S.S.R. airliner), 156
Yakovlev, Alexander, 218
Yardley, John, 25
Yarnema, U.S.S.R., 259
Yates, Harold W., 201
Yegorov, Col. Gen. Nikolai V. (U.S.S.R.), 349
Yeshiva Univ., 16, 38
Yilmaz, Dr. Huseyin, 354
"Yolo Wonder" (pepper plant), 326
Young, John D., 263
Young, Cdr. John W. (USN), 144
Young, Sen. Stephen M., 156
Youth Science Congress, 122
"Youth Wants To Know" (TV program), 46
Yugoslavia, 204, 376

Z

Zabelin, Igor M., 6
Zemlya i Vselennaya, 147
Zero Defects Achievement Award, 295
Zeta Ophiuchi (star), 140, 173
Zeta Perseid/Anetid (meteor showed), 169
Zeta Puppis (star), 369
Zhukov, Gennadi, 61
Zigel, Prof. Feliks, 364
Zimmer, Harro, 29
Le Zinc (aerotrain), 367
Zisch, William E., 14
Zheludev, Leonid, 269
Zoeckler, M/G John L. (USAF), 358
Zond II (U.S.S.R. space probe), 234
Zond III, 56

NASA HISTORICAL PUBLICATIONS

Histories:
  Management History Series:
    Robert L. Rosholt, *An Administrative History of NASA, 1958–1963*, NASA SP–4101, 1966, $4.00.*
  Program History Series:
    Loyd S. Swenson, James M. Grimwood, and Charles C. Alexander, *This New Ocean: A History of Project Mercury*, NASA SP–4201, 1966, $5.50.
    Constance McL. Green and Milton Lomask, *Vanguard: A History* (1969).
  Center History Series:
    Alfred Rosenthal, *Venture Into Space: Early Years of Goddard Space Flight Center*, NASA SP–4301, 1968, $2.50.
Historical Studies:
  Eugene M. Emme (ed.), *History of Rocket Technology*, special issue of *Technology and Culture* (Fall 1963); augmented and published by Society for the History of Technology (Detroit: Wayne State University, 1964).
  Mae Mills Link, *Space Medicine in Project Mercury*, NASA SP–4003, 1965, $1.00.
  Historical Sketch of NASA, NASA EP–29, 1965 (Out of Print).
  Katherine M. Dickson, *History of Aeronautics and Astronautics: A Bibliography*, in conjunction with the Library of Congress, NASA SP–4010 (1969).
Chronologies:
  Aeronautics and Astronautics: An American Chronology of Science and Technology in the Exploration of Space, 1915–1960, compiled by Eugene M. Emme, Washington: NASA, 1961 (Out of Print).
  Aeronautical and Astronautical Events of 1961, published by the House Committee on Science and Astronautics, 1962 (Out of Print).
  Astronautical and Aeronautical Events of 1962, published by the House Committee on Science and Astronautics, 1963, $1.00.
  Astronautics and Aeronautics, 1963, NASA SP–4004, 1964, $1.75.
  Astronautics and Aeronautics, 1964, NASA SP–4005, 1965, $1.75.
  Astronautics and Aeronautics, 1965, NASA SP–4006, 1966, $2.25.
  Astronautics and Aeronautics, 1966, NASA SP–4007, 1967, $1.50.
  Project Mercury: A Chronology, by James M. Grimwood, NASA SP–4001, 1963, $1.50.
  Project Gemini Technology and Operations: A Chronology, by James M. Grimwood and Barton C. Hacker, with Peter J. Vorzimmer, NASA SP–4002 (1969).
  The Apollo Spacecraft: A Chronology, Vol. I, *through November 7, 1962*, by Ivan D. Ertel and Mary L. Morse, NASA SP–4009 (1969).

*All titles with prices can be ordered from the Superintendent of Documents, Government Printing Office, Washington, D.C. 20402.

www.ingramcontent.com/pod-product-compliance
Lightning Source LLC
Chambersburg PA
CBHW081714170526
45167CB00009B/3577